Lecture Notes in Mathematics 1691

Springer
Berlin
Heidelberg
New York
Barcelona
Budapest
Hong Kong
London
Milan
Paris
Singapore
Tokyo

Roman Bezrukavnikov
Michael Finkelberg
Vadim Schechtman

Factorizable Sheaves and Quantum Groups

Springer

Authors

Roman Bezrukavnikov
School of Mathematical Sciences
Tel-Aviv University
Ramat-Aviv, 69978 Israel
e-mail: roman@math.tau.ac.il

Vadim Schechtman
Max-Planck-Institut für Mathematik
Gottfried-Claren-Strasse 26
D-53225 Bonn, Germany
e-mail: vadik@mpim-bonn.mpg.de

Michael Finkelberg
Independent University of Moskow
Bolshoj Vlasjevskij
Pereulok, dom 11
Moscow 121002, Russia
e-mail: fnklberg@mccme.ru

Cataloging-in-Publication Data applied for

Die Deutsche Bibliothek - CIP-Einheitsaufnahme

Bezrukavnikov, Roman:
Factorizable sheaves and quantum groups / Roman Bezrukavnikov ; Michael Finkelberg ; Vadim Schechtman. - Berlin ; Heidelberg ; New York ; Barcelona ; Budapest ; Hong Kong ; London ; Milan ; Paris ; Santa Clara ; Singapore ; Tokyo : Springer, 1998
(Lecture notes in mathematics ; 1691)
ISBN 3-540-64619-1

Mathematics Subject Classification (1991):
81R50, 17B37, 52B30, 55N33, 14H10, 17B35

ISSN 0075-8434
ISBN 3-540-64619-1 Springer-Verlag Berlin Heidelberg New York

Typesetting: Camera-ready TEX output by the authors
SPIN: 10649911 46/3143-543210 - Printed on acid-free paper

To Alexandre Grothendieck on his 70th birthday, with admiration

Table of contents

Part I. INTERSECTION COHOMOLOGY OF REAL ARRANGEMENTS

Part II. CONFIGURATION SPACES AND QUANTUM GROUPS

Part III. TENSOR CATEGORIES ARISING FROM CONFIGURATION SPACES

Part IV. LOCALIZATION OVER $\mathbb{P}^1$

Introduction

The aim of this work is to present a geometric construction of modules over Lusztig's small quantum groups. To give a better idea of a problem, we first make some historical remarks.

I.1. The story begins with **Knizhnik-Zamolodchikov** system of differential equations

$$\frac{\partial\phi(z)}{\partial z_i} = \frac{1}{\kappa}\sum_{j\neq i}\frac{\Omega_{ij}\phi(z)}{z_i - z_j}, \qquad (I.1)$$

$i = 1, \ldots, n$. Here $\phi(z) = \phi(z_1, \ldots, z_n)$ is a function of complex variables with values in a tensor product $M = M_1 \otimes M_2 \otimes \ldots M_n$ of $\mathfrak{g}$-modules, $\mathfrak{g}$ being a semisimple complex Lie algebra. One fixes an invariant symmetric inner product on $\mathfrak{g}$, given by a tensor $\Omega \in \mathfrak{g} \otimes \mathfrak{g}$. For $i \neq j$, Ω_{ij} is the endomorphism of M induced by the multiplication by Ω, acting on $M_i \otimes M_j$, and κ is a non-zero complex parameter. The system (I.1) is of primary importance. It was discovered by physicists, [KZ], as a system of differential equations on correlation functions in Wess-Zumino-Witten models of two dimensional conformal field theory.

As was discovered in the other physical paper [DF] (based on the fundamental work by Feigin and Fuchs, [FF]), the correlation functions in conformal field theories may be expressed in terms of the following integrals of hypergeometric type

$$I(z_1, \ldots, z_n) = \int r(z_1, \ldots, z_n; t_1, \ldots, t_N) \prod (z_i - t_j)^{a_{ij}} \prod (t_p - t_q)^{b_{pq}} dt \qquad (I.2)$$

where $r(z;t)$ is a rational function, a_{ij}, b_{pq} are complex numbers. So, we are dealing with integrals over homology cycles of some one-dimensional local systems over a *configuration space*

$$X^N(z_1, \ldots, z_n) = \mathbb{C}^N \setminus \bigcup_{1\leq i\leq n;\ 1\leq j\leq N} \{z_i = t_j\} \setminus \bigcup_{1\leq p<q\leq N} \{t_p = t_q\} \qquad (I.3)$$

(the coordinates on $\mathbb{C}^N$ being t_i), these integrals depending on $z_1, \ldots, z_n$ as on parameters. The integrals $I(z_1, \ldots, z_n)$, as functions of z, satisfy the *Gauss-Manin* differential equations. These results have been obtained by the powerful technique of *bosonization* in the representation theory of Virasoro and Kac-Moody Lie algebras.

Inspired by these (and other) physical works, one wrote down in [SV1] the complete (for generic κ) set of solutions of KZ equations in terms of the integrals (I.2). In this paper, an interesting phenomenon has been found. It turned out that the the contragradient Verma modules over $\mathfrak{g}$ and the irreducible ones could be realized in some spaces of logarithmic differential forms over the spaces $X^N(z)$ (in the same manner, their tensor products appeared over the spaces $X^N(z_1, \ldots, z_N)$). The picture resembled very much the classical Beilinson-Bernstein theory of localization of $\mathfrak{g}$-modules, [B1]. The role of the flag space G/B is played by the above configuration spaces; in order to restore the whole module, one has to consider the spaces $X^N(z)$ for all N. The Verma modules are connected with the !-extensions of some standard local systems, their contragradient duals — with $*$-extensions.

The other part of the story is the important **Kohno-Drinfeld** theorem, [K], [D], which says that the monodromy of the solutions of (I.1) is expressible in terms of the R-matrix of the quantized enveloping algebra $U_q\mathfrak{g}$. The quantization parameter q is connected with the parameter κ in (I.1) through the identity

$$q = \exp(2\pi i/\kappa) \qquad (I.4)$$

Another proof of this theorem follows from the results of [SV1], since the monodromy of the integrals (I.2) may be explicitely calculated. Moreover, it turned out that the results of [SV1] may be transferred into the topological world, with cycles replacing differential forms, and this way one got the realization of standard representations of the quantum group $U_q\mathfrak{g}$, cf. [SV2]. As B.Feigin put it at that time, "differential forms are responsible for the Lie algebra $\mathfrak{g}$, and cycles — for the quantum group".

The idea to realize standard representations of quantum groups in certain cohomology of configuration spaces appeared also in [FW]. The important role of the adjoint representation and certain higher-dimensional local systems over the configuration spaces of curves of higher genus was first revealed in [CFW].

One noted (cf. [S]) that in the topological world, the standard representations of a quantum group are realized in certain spaces of *vanishing cycles*, and the operators of the quantum group as the Lefschetz-Deligne *canonical* and *variation* maps between them. This implied the idea that some categories of representations of quantum groups could be realized geometrically as categories of compatible systems of perverse sheaves (smooth along some natural stratifications) over various configuration spaces. Here the main point of inspiration was Beilinson's *Gluing theorem*, allowing one to glue perverse sheaves from various spaces of vanishing cycles, [B2]. In the present work, we suggest a construction of such geometric category.

I.2. Here is a brief guide through the text. It consists of six parts. Part 0 contains a detailed overview of Parts I — V, which constitute the main body of the work. We recommend the reader to look through this part. (Practically) all the results of the work are precisely formulated, and the exposition may be in some respects more clear. We also recommend to look through the introductions to all Parts, where their contents is described.

The plan of Parts I and II resembles that of [SV1]. In part I, we give some general results concerning perverse sheaves on the affine plane smooth along the stratification associated by an arrangement of hyperplanes. This is a nice subject in itself, maybe of independent interest. To each perverse sheaf as above, we assign a linear algebra datum, by means of which one can compute the cohomology of this sheaf.

Part II contains some important results concerning the perverse sheaves on configuration spaces. Technically, it is a base of our main results. In particular, it contains the proof of all results announced in [SV2] (so, it may be instructive to look through *op. cit.* before reading Part II).

Part III contains our main definition — of the tensor category $\mathcal{FS}$ of **factorizable sheaves**. A factorizable sheaf is certain compatible collection of perverse sheaves over configuration spaces, cf. Definition 4.2. We recommend to read Part 0, Introduction, 1.2, where this definition is explained in a slightly different language. This is the main hero of the present work.

The main theorem of this work claims that this category is equivalent to the category $\mathcal{C}$ connected with Lusztig's small quantum group, cf. Theorem 17.1. See also Theorem 18.4 for a similar result, but with the generic quantization parameter ζ.

In Parts IV and V, we want to globalize our constructions. The category $\mathcal{FS}$ is a local object — factorizable sheaves leave on a (formal) disk. As it is usual in conformal field theory, we can place these sheaves into points of an arbitrary algebraic curve C. As a result, we get some interesting perverse sheaves on symmetric powers of C.

In Part IV, we study the case of a projective line. It turns out that the cohomology of the above "glued" perverse sheaves are expressible in terms of the Arkhipov's "semiinfinite" cohomology of the quantum group, cf. Theorem 8.4. In Part V, we study the case of an arbitrary curve C, and families of curves. As an application, we get the semisimplicity of the local systems of conformal blocks in Wess-Zumino-Witten models, cf. Corollary 19.9.

I.3. In this work, we live in the topological world. To go back to the world of differential equations, one should pass from perverse sheaves to $\mathcal{D}$-modules. If one translates our constructions to the language of $\mathcal{D}$-modules, one gets a geometric description of $\mathfrak{g}$-modules (at least for a generic κ), cf. [KS1], [KS2]. If one could say that the present work puts [SV2] in a natural framework, then [KS1], [KS2] play the similar role for [SV1]. Kohno-Drinfeld theorem then translates into Riemann-Hilbert correspondence.

I.4. We believe that the geometry studied in this work is an instance of the *semiinfinite geometry of the configuration space of a curve.* We learnt of this concept from A. Beilinson in 1991. It is intimately, though mysteriously, related to the geometry of *semiinfinite flag space* we learnt from B. Feigin since 1989. For example, the definition of the category $\mathcal{PS}$ of factorizable sheaves on the semiinfinite flag space just copies the definition of factorizable sheaves in the present work, cf. [FM]. Also, the semiinfinite cohomology of the small quantum group arise persistently in the computation of IC-hypercohomology of both configuration spaces and Quasimaps' spaces (finite-dimensional approximations of the semiinfinite flag space), cf. [FK] and [FFKM]. The more striking parallels between the two theories are accounted for in [FKM].

I.5. The ideas of A. Beilinson and B. Feigin obviously penetrate this whole work. During the preparation of the text, the authors benefited from the ideas and suggestions of P. Deligne, G. Felder and D. Kazhdan. We are very obliged to G. Lusztig for the permission to use his unpublished results. The second author is happy to thank L. Positselski whose friendly help came at the crucial moment. Our special gratitude goes to D. Kazhdan and Yu.I. Manin for the important incouragement at the initial (resp. final) stages of the work.

References

[B1] A. Beilinson, Localization of representations of reductive Lie algebras, in: Proceedings of the ICM Warsaw 1983, 699-710, PWN, Warsaw, 1984.

[B2] A. Beilinson, How to glue perverse sheaves, in: K-theory, arithmetic and geometry (Moscow, 1984-1986), 42-51, *Lect. Notes in Math.* **1289**, Springer-Verlag, Berlin-Heidelberg-New York, 1987.

[CFW] M. Crivelli, G. Felder, C. Wieczerkowski, Generalized hypergeometric functions on the torus and the adjoint representation of $U_q sl_2$, *Comm. Math. Phys.* **154** (1993), 1-23.

[DF] Vl.S. Dotsenko, V.A. Fateev, Conformal algebra and multipoint correlation functions in 2D statistical models, *Nucl. Phys.* **B240** (1984), 312-348.

[D] V.G. Drinfeld, Quasi-Hopf algebras, Algebra i Analiz **1** (1989), 114-148 (russian).

[FF] B.L. Feigin, D.B. Fuchs, Representations of the Virasoro algebra, in: Representations of Lie groups and related topics, *Adv. Stud. Contemp. Math.* **7**, Gordon and Breach, New York, 1990.

[FFKM] B.Feigin, M.Finkelberg, A.Kuznetsov, I.Mirković, Semiinfinite Flags. II. Local and global Intersection Cohomology of Quasimaps spaces, Preprint alg-geom/9711009.

[FW] G. Felder, C. Wieczerkowski, Topological representations of the quantum group $U_q sl_2$, *Comm. Math. Phys.* **138** (1991), 683-605.

[FK] M.Finkelberg, A.Kuznetsov, Global Intersection Cohomology of Quasimaps' spaces, *Int. Math. Res. Notices* **1997**, no. 7, 301-328.

[FKM] M.Finkelberg, A.Kuznetsov, I.Mirković, Frobenius realization of $U(\check{\mathfrak{n}})$, in preparation.

[FM] M.Finkelberg, I.Mirković, Semiinfinite Flags. I. Case of global curve $\mathbb{P}^1$, Preprint alg-geom/9707010.

[KS1] S. Khoroshkin, V. Schechtman, Factorizable $\mathcal{D}$-modules, *Math. Res. Letters* **4** (1997), 239-257.

[KS2] S. Khoroshkin, V. Schechtman, Non-resonance $\mathcal{D}$-modules over arrangements of hyperplanes, Preprint MPI 98-14; math.AG/9801134.

[KZ] V.G. Knizhnik, A.B. Zamolodchikov, Current algebra and Wess-Zumino model in two dimensions, *Nucl Phys.* **B247** (1984), 83-103.

[K] T. Kohno, Monodromy representations of braid groups and Yang-Baxter equations, *Ann. Inst. Fourier* (Grenoble) **37** (1987), 139-160.

[S] V. Schechtman, Vanishing cycles and quantum groups. I, *Int. Math. Res. Notices* **1992**, no. 3, 39-49; II, *ibid.* **1992**, no. 10, 207-215.

[SV1] V. Schechtman, A. Varchenko, Arrangements of hyperplanes and Lie algebra homology, *Inv. Math.* **106** (1991), 139-194.

[SV2] V. Schechtman, A. Varchenko, Quantum groups and homology of local systems, in: Algebraic geometry and analytic geometry (Tokyo, 1990), 182-197, ICM-90 Satell. Conf. Proc., Springer, Tokyo, 1991.

Acknowledgement

During the work on this project, the first author visited Independent Moscow University, the second author visited Harvard University, Stony Brook University and Tel-Aviv University, and the third author visited Harvard University and Max-Planck-Institut für Mathematik. We thank these institutions for the hospitality.

The second author was partially supported by the grants from AMS, INTAS94-4720, CRDF RM1-265. The third author was partially supported by NSF.

Part 0. OVERVIEW

1. Introduction

1.1. Let $(I, \cdot)$ be a Cartan datum of finite type and $(Y, X, \ldots)$ the simply connected root datum of type $(I, \cdot)$, cf. [L1].

Let $l > 1$ be an integer. Set $\ell = l/(l, 2)$; to simplify the exposition, we will suppose in this Introduction that $d_i := i \cdot i/2$ divides ℓ for all $i \in I$; we set $\ell_i := \ell/d_i$. We suppose that all $\ell_i > 3$. Let $\rho \in X$ be the half-sum of positive roots; let $\rho_\ell \in X$ be defined by $\langle i, \rho_\ell \rangle = \ell_i - 1$ $(i \in I)$. We define a lattice $Y_\ell \subset Y$ by $Y_\ell = \{\mu \in X |$ for all $\mu' \in X, \mu \cdot \mu' \in \ell\mathbb{Z}\}$, and set $\mathrm{d}_\ell = \mathrm{card}(X/Y_\ell)$.

We fix a base field k of characteristic not dividing l. We suppose that k contains a primitive l-th root of unity ζ, and fix it. Starting from these data, one defines certain category $\mathcal{C}$. Its objects are finite dimensional X-graded k-vector spaces equipped with an action of Lusztig's "small" quantum group $\mathfrak{u}$ (cf. [L2]) such that the action of its Cartan subalgebra is compatible with the X-grading. Variant: one defines certain algebra $\dot{\mathfrak{u}}$ which is an "X-graded" version of $\mathfrak{u}$ (see 12.2), and an object of $\mathcal{C}$ is a finite dimensional $\dot{\mathfrak{u}}$-module. For the precise definition of $\mathcal{C}$, see 2.11, 2.13. For l prime and $\mathrm{char}\,\mathsf{k} = 0$, the category $\mathcal{C}$ was studied in [AJS], for $\mathrm{char}\,\mathsf{k} > 0$ and arbitrary l, $\mathcal{C}$ was studied in [AW]. The category $\mathcal{C}$ admits a remarkable structure of a *ribbon*[1] category (Lusztig).

1.2. The main aim of this work is to introduce certain tensor category $\mathcal{FS}$ of geometric origin, which is equivalent to $\mathcal{C}$. Objects of $\mathcal{FS}$ are called **(finite) factorizable sheaves**.

A notion of a factorizable sheaf is the main new concept of this work. Let us give an informal idea, what kind of an object is it. Let D be the unit disk in the complex plane[2]. Let $\mathcal{D}$ denote the space of positive Y-valued divisors on D. Its points are formal linear combinations $\sum \nu \cdot x$ $(\nu \in Y^+ := \mathbb{N}[I], x \in D)$. This space is a disjoint union

$$\mathcal{D} = \coprod_{\nu \in Y^+} D^\nu$$

where D^ν is the subspace of divisors of degree ν. Variant: D^ν is the configuration space of ν points running on D; its points are (isomorphism clases of) pairs of maps

$$(J \longrightarrow D,\ j \mapsto x_j;\ \pi : J \longrightarrow I,\ j \mapsto i_j),$$

J being a finite set. We say that we have a finite set $\{x_j\}$ of (possibly equal) points of D, a point x_j being "coloured" by the colour i_j. The sum (in Y) of colours of all points should be equal to ν. The space D^ν is a smooth affine analytic variety; it carries a canonical stratification defined by various intersections of hypersurfaces $\{x_j = 0\}$ and $\{x_{j'} = x_{j''}\}$. The open stratum (the complement of all the hypersurfaces above) is denoted by $A^{\nu\circ}$.

[1] in other terminology, braided balanced rigid tensor

[2] One could also take a complex affine line or a formal disk; after a suitable modification of the definitions, the resulting categories of factorizable sheaves are canonically equivalent.

One can imagine $\mathcal{D}$ as a Y^+-graded smooth stratified variety. Let $\mathcal{A} = \coprod A^\nu \subset \mathcal{D}$ be the open Y^+-graded subvariety of positive Y-valued divisors on $D - \{0\}$. We have an open Y^+-graded subvariety $\mathcal{A}^\circ = \coprod A^{\nu\circ} \subset \mathcal{A}$.

Let us consider the $(Y^+)^2$-graded variety

$$\mathcal{A} \times \mathcal{A} = \coprod A^{\nu_1} \times A^{\nu_2};$$

we define another $(Y^+)^2$-graded variety $\widetilde{\mathcal{A} \times \mathcal{A}}$, together with two maps

(a) $\mathcal{A} \times \mathcal{A} \xleftarrow{p} \widetilde{\mathcal{A} \times \mathcal{A}} \xrightarrow{m} \mathcal{A}$

respecting the Y^+-gradings[3]; the map p is a homotopy equivalence. One can imagine the diagram (a) above as a "homotopy multiplication"

$$m_{\mathcal{A}} : \mathcal{A} \times \mathcal{A} \longrightarrow \mathcal{A};$$

this "homotopy map" is "homotopy associative"; the meaning of this is explained in the main text. We say that $\mathcal{A}$ is a (Y^+-graded) "homotopy monoid"; $\mathcal{A}^\circ \subset \mathcal{A}$ is a "homotopy submonoid".

The space $\mathcal{D}$ is a "homotopy $\mathcal{A}$-space": there is a "homotopy map"

$$m_{\mathcal{D}} : \mathcal{D} \times \mathcal{A} \longrightarrow \mathcal{D}$$

which is, as above, a diagram of usual maps between Y^+-graded varieties

(b) $\mathcal{D} \times \mathcal{A} \xleftarrow{p} \widetilde{\mathcal{D} \times \mathcal{A}} \xrightarrow{m} \mathcal{D}$,

p being a homotopy equivalence.

For each $\mu \in X$, $\nu \in Y^+$, one defines a one-dimensional k-local system $\mathcal{I}^\nu_\mu$ over $A^{\nu\circ}$. Its monodromies are defined by variuos scalar products of "colours" of running points and the colour μ of the origin $0 \in D$. These local systems have the following compatibility. For each $\mu \in X, \nu_1, \nu_2 \in Y^+$, a *factorization isomorphism*

$$\phi_\mu(\nu_1, \nu_2) : m^* \mathcal{I}^{\nu_1+\nu_2}_\mu \xrightarrow{\sim} p^*(\mathcal{I}^{\nu_1}_\mu \boxtimes \mathcal{I}^{\nu_2}_{\mu-\nu_1})$$

is given (where m, p are the maps in the diagram (a) above), these isomorphisms satisfying certain *(co)associativity property*. The collection of local systems $\{\mathcal{I}^\nu_\mu\}$ is an X-graded local system $\mathcal{I} = \oplus\ \mathcal{I}_\mu$ over $\mathcal{A}^\circ$. One could imagine the collection of factorization isomorphisms $\{\phi_\mu(\nu_1, \nu_2)\}$ as an isomorphism

$$\phi :\ m^*_{\mathcal{A}^\circ} \mathcal{I} \xrightarrow{\sim} \mathcal{I} \boxtimes \mathcal{I}$$

We call $\mathcal{I}$ the *braiding local system.*

Let $\mathcal{I}^\bullet$ be the perverse sheaf on $\mathcal{A}$ which is the Goresky-MacPherson extension of the local system $\mathcal{I}$. The isomorphism ϕ above induces a similar isomorphism

$$\phi^\bullet :\ m^*_{\mathcal{A}} \mathcal{I}^\bullet \xrightarrow{\sim} \mathcal{I}^\bullet \boxtimes \mathcal{I}^\bullet$$

This sheaf, together with $\phi^\bullet$, looks like a "coalgebra"; it is an incarnation of the quantum group $\overset{\bullet}{\mathfrak{u}}$.

A factorizable sheaf is a couple

($\mu \in X$ ("the highest weight"); a perverse sheaf $\mathcal{X}$ over $\mathcal{D}$, smooth along the canonical stratification).

[3]a $(Y^+)^2$-graded space is considered as a Y^+-graded by means of the addition $(Y^+)^2 \longrightarrow Y^+$.

Thus, $\mathcal{X}$ is a collection of sheaves $\mathcal{X}^\nu$ over $\mathcal{D}^\nu$. These sheaves should be connected by *factorization isomorphisms*

$$\psi(\nu_1,\nu_2): \ m^*\mathcal{X}^{\nu_1+\nu_2} \xrightarrow{\sim} p^*(\mathcal{X}^{\nu_1}\boxtimes\mathcal{I}^{\nu_2\bullet}_{\mu-\nu_1})$$

satisfying an associativity property. Here m,p are as in the diagram (b) above. One could imagine the whole collection $\{\psi(\nu_1,\nu_2)\}$ as an isomorphism

$$\psi: m^*_{\mathcal{D}}\mathcal{X}\xrightarrow{\sim}\mathcal{I}^\bullet\boxtimes\mathcal{X}$$

satisfying a (co)associativity property. We impose also certain finiteness (of singularities) condition on $\mathcal{X}$. So, this object looks like a "comodule" over $\mathcal{I}^\bullet$. It is an incarnation of an $\overset{\bullet}{\mathfrak{u}}$-module.

We should mention one more important part of the structure of the space $\mathcal{D}$. It comes from natural closed embeddings $\iota_\nu: \mathcal{D}\hookrightarrow\mathcal{D}[\nu]$ (were $[\nu]$ denotes the shift of the grading); these mappings define certain inductive system, and a factorizable sheaf is a sheaf on its inductive limit.

The latter inductive limit is an example of a **"semiinfinite space"**.

For the precise definitions, see Sections 3, 4 below.

The category $\mathcal{FS}$ has a structure of a braided balanced tensor category coming from geometry. The tensor structure on $\mathcal{FS}$ is defined using the functors of nearby cycles. The tensor equivalence

$$\Phi:\mathcal{FS}\xrightarrow{\sim}\mathcal{C}$$

is defined using vanishing cycles functors. It respects the braidings and balances. This is the contents of Parts I–III.

1.3. Factorizable sheaves are local objects. It turns out that one can "glue" them along complex curves. More precisely, given a finite family $\{\mathcal{X}_a\}_{a\in A}$ of factorizable sheaves and a smooth proper curve C together with a family $\{x_a,\tau_a\}_A$ of distinct points $x_a\in C$ with non-zero tangent vectors τ_a at them (or, more generally, a family of such objects over a smooth base S), one can define a perverse sheaf, denoted by

$$\boxed{\times}_A^{(C)}\ \mathcal{X}_a$$

on the (relative) configuration space C^ν. Here $\nu\in Y^+$ is defined by

(a) $\nu=\sum_{a\in A}\ \mu_a+(2g-2)\rho_\ell$

where g is the genus of C, μ_a is the highest weight of $\mathcal{X}_a$. We *assume* that the right hand side of the equality belongs to Y^+.

One can imagine this sheaf as an "exterior tensor product" of the family $\{\mathcal{X}_a\}$ along C. It is obtained by "planting" the sheaves $\mathcal{X}_a$ into the points x_a. To glue them together, one needs a "glue". This glue is called the **Heisenberg local system** $\mathcal{H}$; it is a sister of the braiding local system $\mathcal{I}$. Let us describe what it is.

For a finite set A, let $\mathcal{M}_A$ denote the moduli stack of punctured curves $(C,\{x_a,\tau_a\}_A)$ as above; let $\eta: C_A\longrightarrow\mathcal{M}_A$ be the universal curve. Let $\delta_A=\det(R\eta_*\mathcal{O}_{C_A})$ be the determinant line bundle on $\mathcal{M}_A$, and $\mathcal{M}_{A;\delta}\longrightarrow\mathcal{M}_A$ be the total space of δ_A with the zero section removed. For $\nu\in Y^+$, let $\eta^\nu: C^\nu_A\longrightarrow\mathcal{M}_A$ be the relative configuration space of ν points running on C_A; let $C^{\nu\circ}_A\subset C^\nu_A$ be the open stratum (where the running points are distinct and distinct from the punctures x_a). The complementary subscript $()_\delta$ will denote the base change under $\mathcal{M}_{A;\delta}\longrightarrow\mathcal{M}_A$. The complementary subscript

$()_g$ will denote the base change under $\mathcal{M}_{A,g} \hookrightarrow \mathcal{M}_A$, $\mathcal{M}_{A,g}$ being the substack of curves of genus g.

The Heisenberg local system is a collection of local systems $\mathcal{H}_{\vec{\mu};A,g}$ over the stacks $C^{\nu o}_{A,g;\delta}$. Here $\vec{\mu}$ is an A-tuple $\{\mu_a\} \in X^a$; $\nu = \nu(\vec{\mu}; g)$ is defined by the equality (a) above. We assume that $\vec{\mu}$ is such that the right hand side of this equality really belongs to Y^+. The dimension of $\mathcal{H}_{\vec{\mu};A,g}$ is equal to d^g_ℓ; the monodromy around the zero section of the determinant line bundle is equal to

(b) $c = (-1)^{\mathrm{card}(I)} \zeta^{-12\rho\cdot\rho}$.

These local systems have a remarkable compatibility ("fusion") property which we do not specify here, see Section 15.

1.4. In Part IV we study the sheaves $\boxed{\times}^{(C)}_A\ \mathcal{X}_a$ for $C = \mathbb{P}^1$. Their cohomology, when expressed algebraically, turn out to coincide with certain "semiinfinite" Tor spaces in the category $\mathcal{C}$ introduced by S.M.Arkhipov. Due to the results of Arkhipov, this enables one to prove that "spaces of conformal blocks" in WZW models are the natural subquotients of such cohomology spaces.

1.5. In Part V we study the sheaves $\boxed{\times}^{(C)}_A\ \mathcal{X}_a$ for arbitrary smooth families of punctured curves. Let $\boxed{\times}_{A,g}\ \mathcal{X}_a$ denotes the "universal exterior product" living on $C^\nu_{A,g;\delta}$ where $\nu = \nu(\vec{\mu}; g)$ is as in 1.3 (a) above, $\vec{\mu} = \{\mu_a\}$, μ_a being the highest weight of $\mathcal{X}_a$. Let us integrate it: consider

$$\int_{C^\nu} \boxed{\times}_{A,g}\ \mathcal{X}_a := R\eta^\nu_*(\boxed{\times}_{A,g}\ \mathcal{X}_a);$$

it is a complex of sheaves over $\mathcal{M}_{A,g;\delta}$ with smooth cohomology; let us denote it $\langle \otimes_A\ \mathcal{X}_a \rangle_g$. The cohomology sheaves of such complexes define a *fusion structure*[4] on $\mathcal{FS}$ (and hence on $\mathcal{C}$).

The classical WZW model fusion category is, in a certain sense, a subquotient of one of them. The number c, 1.3 (b), coincides with the "multiplicative central charge" of the model (in the sense of [BFM]).

As a consequence of this geometric description and of the Purity Theorem, [BBD], the local systems of conformal blocks (in arbitrary genus) are semisimple. The Verdier duality induces a canonical non-degenerate Hermitian form on them (if $\mathsf{k} = \mathbb{C}$).

1.6. We should mention a very interesting related work of G.Felder and collaborators. The idea of realizing quantum groups' modules in the cohomology of configuration spaces appeared independently in [FW]. The Part V of the present work arose from our attempts to understand [CFW].

1.7. This work consists of 6 parts. Part 0 is an overview of the Parts I–V. More precisely, Chapter 0.1 surveys the Parts I–III; Chapter 0.2 is an exposition of Part IV, and Chapter 0.3 is an exposition of Part V. The proofs are given in the Parts I–V. Part 0 contains no proofs, but we hope that it clarifies the general picture to some extent. Each part starts with a separate Introduction. The references within one Part look like 1.1.1, and the references to other parts look like I.1.1.1. The first author would like to stress that his contribution is limited to chapters 1,2 of Part V.

[4] actually a family of such structures (depending on the degree of cohomology)

The remaining authors would like to note that these chapters consitute the core of Part V.

Chapter 1. Local.

2. The category $\mathcal{C}$

2.1. We will follow Lusztig's terminology and notations concerning root systems, cf. [L1].

We fix an irreducible Cartan datum $(I, \cdot)$ of finite type. Thus, I is a finite set together with a symmetric $\mathbb{Z}$-valued bilinear form $\nu_1, \nu_2 \mapsto \nu_1 \cdot \nu_2$ on the free abelian group $\mathbb{Z}[I]$, such that for any $i \in I$, $i \cdot i$ is even and positive, for any $i \neq j$ in I, $2i \cdot j / i \cdot i \leq 0$ and the $I \times I$-matrix $(i \cdot j)$ is positive definite. We set $d_i = i \cdot i/2$.

Let $d = \max_{i \in I} d_i$. This number is equal to the least common multiple of the numbers d_i, and belongs to the set $\{1, 2, 3\}$. For a simply laced root datum $d = 1$.

We set $Y = \mathbb{Z}[I], X = \mathrm{Hom}(Y, \mathbb{Z})$; $\langle , \rangle : Y \times X \longrightarrow \mathbb{Z}$ will denote the obvious pairing. The obvious embedding $I \hookrightarrow Y$ will be denoted by $i \mapsto i$. We will denote by $i \mapsto i'$ the embedding $I \hookrightarrow X$ given by $\langle i, j' \rangle = 2i \cdot j / i \cdot i$. (Thus, $(Y, X, \dots)$ is the simply connected root datum of type $(I, \cdot)$, in the terminology of Lusztig.)

The above embedding $I \subset X$ extends by additivity to the embedding $Y \subset X$. We will regard Y as the sublattice of X by means of this embedding. For $\nu \in Y$, we will denote by the same letter ν its image in X. We set $Y^+ = \mathbb{N}[I] \subset Y$.

We will use the following partial order on X. For $\mu_1, \mu_2 \in X$, we write $\mu_1 \leq \mu_2$ if $\mu_2 - \mu_1$ belongs to Y^+.

2.2. We fix a base field k, an integer $l > 1$ and a primitive root of unity $\zeta \in \mathsf{k}$ as in the Introduction.

We set $\ell = l$ if l is odd and $\ell = l/2$ if l is even. For $i \in I$, we set $\ell_i = \ell/(\ell, d_i)$. Here (a, b) stands for the greatest common divisor of a, b. We set $\zeta_i = \zeta^{d_i}$. We will assume that $\ell_i > 1$ for any $i \in I$ and $\ell_i > -\langle i, j' \rangle + 1$ for any $i \neq j$ in I.

We denote by ρ (resp. ρ_ℓ) the element of X such that $\langle i, \rho \rangle = 1$ (resp. $\langle i, \rho_\ell \rangle = \ell_i - 1$) for all $i \in I$.

For a coroot $\beta \in Y$, there exists an element w of the Weyl group W of our Cartan datum and $i \in I$ such that $w(i) = \beta$. We set $\ell_\beta := \frac{\ell}{(\ell, d_i)}$; this number does not depend on the choice of w and i.

We have $\rho = \frac{1}{2} \sum \alpha$; $\rho_\ell = \frac{1}{2} \sum (\ell_\alpha - 1)\alpha$, the sums over all positive roots $\alpha \in X$.

For $a \in \mathbb{Z}, i \in I$, we set $[a]_i = 1 - \zeta_i^{-2a}$.

2.3. We use the same notation $\mu_1, \mu_2 \mapsto \mu_1 \cdot \mu_2$ for the unique extension of the bilinear form on Y to a $\mathbb{Q}$-valued bilinear form on $Y \otimes_{\mathbb{Z}} \mathbb{Q} = X \otimes_{\mathbb{Z}} \mathbb{Q}$.

We define a lattice $Y_\ell = \{\lambda \in X |$ for all $\mu \in X, \lambda \cdot \mu \in \ell\mathbb{Z}\}$.

2.4. Unless specified otherwise, a "vector space" will mean a vector space over k; $\otimes$ will denote the tensor product over k. A "sheaf" (or a "local system") will mean a sheaf (resp. local system) of k-vector spaces.

If $(T, \mathcal{S})$ is an open subspace of the space of complex points of a separate scheme of finite type over $\mathbb{C}$, with the usual topology, together with an algebraic stratification $\mathcal{S}$

satisfying the properties [BBD] 2.1.13 b), c), we will denote by $\mathcal{M}(T;\mathcal{S})$ the category of perverse sheaves over T lisse along $\mathcal{S}$, with respect to the middle perversity, cf. [BBD] 2.1.13, 2.1.16.

2.5. Let $'\mathfrak{f}$ be the free associative k-algebra with 1 with generators θ_i $(i \in I)$. For $\nu = \sum \nu_i i \in \mathbb{N}[I]$, let $'\mathfrak{f}_\nu$ be the subspace of $'\mathfrak{f}$ spanned by the monomials $\theta_{i_1} \cdot \ldots \cdot \theta_{i_a}$ such that $\sum_j i_j = \nu$ in $\mathbb{N}[I]$.

Let us regard $'\mathfrak{f} \otimes {}'\mathfrak{f}$ as a k-algebra with the product $(x_1 \otimes x_2)(y_1 \otimes y_2) = \zeta^{\nu\cdot\mu} x_1 y_1 \otimes x_2 y_2$ $(x_2 \in {}'\mathfrak{f}_\nu, y_1 \in {}'\mathfrak{f}_\mu)$. Let r denote a unique homomorphism of k-algebras $'\mathfrak{f} \longrightarrow {}'\mathfrak{f} \otimes {}'\mathfrak{f}$ carrying θ_i to $1 \otimes \theta_i + \theta_i \otimes 1$ $(i \in I)$.

2.6. **Lemma-definition.** *There exists a unique* k*-valued bilinear form* $(\cdot,\cdot)$ *on* $'\mathfrak{f}$ *such that*

(i) $(1,1) = 1$; $(\theta_i, \theta_j) = \delta_{ij}$ $(i, j \in I)$; (ii) $(x, yy') = (r(x), y \otimes y')$ *for all* $x, y, y' \in {}'\mathfrak{f}$.

This bilinear form is symmetric. □

In the right hand side of the equality (ii) we use the same notation $(\cdot,\cdot)$ for the bilinear form on $'\mathfrak{f} \otimes {}'\mathfrak{f}$ defined by $(x_1 \otimes x_2, y_1 \otimes y_2) = (x_1, y_1)(x_2, y_2)$.

The radical of the form $'\mathfrak{f}$ is a two-sided ideal of $'\mathfrak{f}$.

2.7. Let us consider the associative k-algebra $\mathfrak{u}$ (with 1) defined by the generators ϵ_i, θ_i $(i \in I)$, K_ν $(\nu \in Y)$ and the relations (a) — (e) below.

(a) $K_0 = 1$, $K_\nu \cdot K_\mu = K_{\nu+\mu}$ $(\nu, \mu \in Y)$;

(b) $K_\nu \epsilon_i = \zeta^{\langle \nu, i' \rangle} \epsilon_i K_\nu$ $(i \in I, \nu \in Y)$;

(c) $K_\nu \theta_i = \zeta^{-\langle \nu, i' \rangle} \theta_i K_\nu$ $(i \in I, \nu \in Y)$;

(d) $\epsilon_i \theta_j - \zeta^{i\cdot j} \theta_j \epsilon_i = \delta_{ij}(1 - \tilde{K}_i^{-2})$ $(i, j \in I)$.

Here we use the notation $\tilde{K}_\nu = \prod_i K_{d_i \nu_i i}$ $(\nu = \sum \nu_i i)$.

(e) If $f(\theta_i) \in {}'\mathfrak{f}$ belongs to the radical of the form $(\cdot,\cdot)$ then $f(\theta_i) = f(\epsilon_i) = 0$ in $\mathfrak{u}$.

2.8. There is a unique k-algebra homomorphism $\Delta : \mathfrak{u} \longrightarrow \mathfrak{u} \otimes \mathfrak{u}$ such that

$$\Delta(\epsilon_i) = \epsilon_i \otimes 1 + \tilde{K}_i^{-1} \otimes \epsilon_i;\ \Delta(\theta_i) = \theta_i \otimes 1 + \tilde{K}_i^{-1} \otimes \theta_i;\ \Delta(K_\nu) = K_\nu \otimes K_\nu$$

for any $i \in I, \nu \in Y$. Here $\mathfrak{u} \otimes \mathfrak{u}$ is regarded as an algebra in the standard way.

There is a unique k-algebra homomorphism $e : \mathfrak{u} \longrightarrow$ k such that $e(\epsilon_i) = e(\theta_i) = 0, e(K_\nu) = 1$ $(i \in I, \nu \in Y)$.

2.9. There is a unique k-algebra homomorphism $s : \mathfrak{u} \longrightarrow \mathfrak{u}^{\mathrm{opp}}$ such that

$$s(\epsilon_i) = -\epsilon_i \tilde{K}_i;\ s(\theta_i) = -\theta_i \tilde{K}_i;\ s(K_\nu) = K_{-\nu}\ (i \in I, \nu \in Y).$$

There is a unique k-algebra homomorphism $s' : \mathfrak{u} \longrightarrow \mathfrak{u}^{\mathrm{opp}}$ such that

$$s'(\epsilon_i) = -\tilde{K}_i \epsilon_i;\ s'(\theta_i) = -\tilde{K}_i \theta_i;\ s'(K_\nu) = K_{-\nu}\ (i \in I, \nu \in Y).$$

2.10. The algebra $\mathfrak{u}$ together with the additional structure given by the comultiplication Δ, the counit e, the antipode s and the skew-antipode s', is a Hopf algebra.

2.11. Let us define a category $\mathcal{C}$ as follows. An object of $\mathcal{C}$ is a u-module M which is finite dimensional over k, with a given direct sum decomposition $M = \oplus_{\lambda \in X} M_\lambda$ (as a vector space) such that $K_\nu x = \zeta^{\langle \nu, \lambda \rangle} x$ for any $\nu \in Y, \lambda \in X, x \in M_\lambda$. A morphism in $\mathcal{C}$ is a u-linear map respecting the X-gradings.

Alternatively, an object of $\mathcal{C}$ may be defined as an X-graded finite dimensional vector space $M = \oplus M_\lambda$ equipped with linear operators

$$\theta_i : M_\lambda \longrightarrow M_{\lambda - i'}, \ \epsilon_i : M_\lambda \longrightarrow M_{\lambda + i'} \ (i \in I, \lambda \in X)$$

such that

(a) for any $i, j \in I, \lambda \in X$, the operator $\epsilon_i \theta_j - \zeta^{i \cdot j} \theta_j \epsilon_i$ acts as the multiplication by $\delta_{ij} [\langle i, \lambda \rangle]_i$ on M_λ.

Note that $[\langle i, \lambda \rangle]_i = [\langle d_i i, \lambda \rangle] = [i' \cdot \lambda]$.

(b) If $f(\theta_i) \in {}'\mathfrak{f}$ belongs to the radical of the form $(\cdot, \cdot)$ then the operators $f(\theta_i)$ and $f(\epsilon_i)$ act as zero on M.

2.12. In [L2] Lusztig defines an algebra $\mathbf{u}_\mathcal{B}$ over the ring $\mathcal{B}$ which is a quotient of $\mathbb{Z}[v, v^{-1}]$ (v being an indeterminate) by the l-th cyclotomic polynomial. Let us consider the k-algebra $\mathbf{u}_\mathsf{k}$ obtained from $\mathbf{u}_\mathcal{B}$ by the base change $\mathcal{B} \longrightarrow \mathsf{k}$ sending v to ζ. The algebra $\mathbf{u}_\mathsf{k}$ is generated by certain elements E_i, F_i, K_i $(i \in I)$. Here $E_i = E^{(1)}_{\alpha_i}, F_i = F^{(1)}_{\alpha_i}$ in the notations of *loc.cit.*

Given an object $M \in \mathcal{C}$, let us introduce the operators E_i, F_i, K_i on it by

$$E_i = \frac{\zeta_i}{1 - \zeta_i^{-2}} \epsilon_i \tilde{K}_i, \ \ F_i = \theta_i, K_i = K_i.$$

2.13. **Theorem.** *The above formulas define the action of the Lusztig's algebra* $\mathbf{u}_\mathsf{k}$ *on an object* M.

This rule defines an equivalence of $\mathcal{C}$ *with the category whose objects are* X*-graded finite dimensional* $\mathbf{u}_\mathsf{k}$*-modules* $M = \oplus M_\lambda$ *such that* $K_i x = \zeta^{\langle i, \lambda \rangle} x$ *for any* $i \in I, \lambda \in X, x \in M_\lambda$. □

2.14. The structure of a Hopf algebra on u defines canonically a *rigid tensor* structure on $\mathcal{C}$ (cf. [KL]IV, Appendix).

The Lusztig's algebra u_k also has an additional structure of a Hopf algebra. It induces the same rigid tensor structure on $\mathcal{C}$.

We will denote the duality in $\mathcal{C}$ by $M \mapsto M^*$. The unit object will be denoted by $\mathbf{1}$.

2.15. Let u^- (resp. u^+, u^0) denote the k-subalgebra generated by the elements θ_i $(i \in I)$ (resp. ϵ_i $(i \in I)$, K_ν $(\nu \in Y)$). We have the triangular decomposition $\mathrm{u} = \mathrm{u}^- \mathrm{u}^0 \mathrm{u}^+ = \mathrm{u}^+ \mathrm{u}^0 \mathrm{u}^-$.

We define the "Borel" subalgebras $\mathrm{u}^{\leq 0} = \mathrm{u}^- \mathrm{u}^0$, $\mathrm{u}^{\geq 0} = \mathrm{u}^+ \mathrm{u}^0$; they are the Hopf subalgebras of u.

Let us introduce the X-grading $\mathrm{u} = \oplus \mathrm{u}_\lambda$ as a unique grading compatible with the structure of an algebra such that $\theta_i \in \mathrm{u}_{-i'}$, $\epsilon_i \in \mathrm{u}_{i'}$, $K_\nu \in \mathrm{u}_0$. We will use the induced gradings on the subalgebras of u.

2.16. Let $\mathcal{C}^{\leq 0}$ (resp. $\mathcal{C}^{\geq 0}$) be the category whose objects are X-graded finite dimensional $\mathfrak{u}^{\leq 0}$- (resp. $\mathfrak{u}^{\geq 0}$-) modules $M = \oplus M_\lambda$ such that $K_\nu x = \zeta^{\langle \nu, \lambda \rangle} x$ for any $\nu \in Y, \lambda \in X, x \in M_\lambda$. Morphisms are $\mathfrak{u}^{\leq 0}$- (resp. $\mathfrak{u}^{\geq 0}$-) linear maps compatible with the X-gradings.

We have the obvious functors $\mathcal{C} \longrightarrow \mathcal{C}^{\leq 0}$ (resp. $\mathcal{C} \longrightarrow \mathcal{C}^{\geq 0}$). These functors admit the exact left adjoints $\mathrm{ind}^{\mathfrak{u}}_{\mathfrak{u}^{\leq 0}} : \mathcal{C}^{\leq 0} \longrightarrow \mathcal{C}$ (resp. $\mathrm{ind}^{\mathfrak{u}}_{\mathfrak{u}^{\geq 0}} : \mathcal{C}^{\geq 0} \longrightarrow \mathcal{C}$).

For example, $\mathrm{ind}^{\mathfrak{u}}_{\mathfrak{u}^{\geq 0}}(M) = \mathfrak{u} \otimes_{\mathfrak{u}^{\geq 0}} M$. The triangular decomposition induces an isomorphism of graded vector spaces $\mathrm{ind}^{\mathfrak{u}}_{\mathfrak{u}^{\geq 0}}(M) \cong \mathfrak{u}^- \otimes M$.

2.17. For $\lambda \in X$, let us consider on object $\mathsf{k}^\lambda \in \mathcal{C}^{\geq 0}$ defined as follows. As a graded vector space, $\mathsf{k}^\lambda = \mathsf{k}^\lambda_\lambda = \mathsf{k}$. The algebra $\mathfrak{u}^{\geq 0}$ acts on k^λ as follows: $\epsilon_i x = 0, K_\nu x = \zeta^{\langle \nu, \lambda \rangle} x$ $(i \in I, \nu \in Y, x \in \mathsf{k}^\lambda)$.

The object $M(\lambda) = \mathrm{ind}^{\mathfrak{u}}_{\mathfrak{u}^{\geq 0}}(\mathsf{k}^\lambda)$ is called a *(baby) Verma module.* Each $M(\lambda)$ has a unique irreducible quotient object, to be denoted by $L(\lambda)$. The objects $L(\lambda)$ $(\lambda \in X)$ are mutually non-isomorphic and every irreducible object in $\mathcal{C}$ is isomorphic to one of them. Note that the category $\mathcal{C}$ is *artinian*, i.e. each object of $\mathcal{C}$ has a finite filtration with irreducible quotients.

For example, $L(0) = \mathbf{1}$.

2.18. Recall that a *braiding* on $\mathcal{C}$ is a collection of isomorphisms

$$R_{M,M'} : M \otimes M' \xrightarrow{\sim} M' \otimes M \ (M, M' \in \mathcal{C})$$

satisfying certain compatibility with the tensor structure (see [KL]IV, A.11).

A *balance* on $\mathcal{C}$ is an automorphism of the identity functor $b = \{b_M : M \xrightarrow{\sim} M \ (M \in \mathcal{C})\}$ such that for every $M, N \in \mathcal{C}$, $b_{M \otimes N} \circ (b_M \otimes b_N)^{-1} = R_{N,M} \circ R_{M,N}$ (see *loc. cit.*).

2.19. Let ϖ denote the determinant of the $I \times I$-matrix $(\langle i, j' \rangle)$. From now on we assume that char(k) does not divide 2ϖ, and k contains an element ζ' such that $(\zeta')^{2\varpi} = \zeta$; we fix such an element ζ'. For a number $q \in \frac{1}{2\varpi}\mathbb{Z}$, ζ^q will denote $(\zeta')^{2\varpi q}$.

2.20. **Theorem.** (G. Lusztig) *There exists a unique braided structure* $\{R_{M,N}\}$ *on the tensor category* $\mathcal{C}$ *such that for any* $\lambda \in X$ *and* $M \in \mathcal{C}$, *if* $\mu \in X$ *is such that* $M_{\mu'} \neq 0$ *implies* $\mu' \leq \mu$, *then*

$$R_{L(\lambda),M}(x \otimes y) = \zeta^{\lambda \cdot \mu} y \otimes x$$

for any $x \in L(\lambda), y \in M_\mu$. □

2.21. Let $n : X \longrightarrow \frac{1}{2\varpi}\mathbb{Z}$ be the function defined by

$$n(\lambda) = \frac{1}{2}\lambda \cdot \lambda - \lambda \cdot \rho_\ell.$$

2.22. **Theorem.** *There exists a unique balance* b *on* $\mathcal{C}$ *such that for any* $\lambda \in X$, $b_{L(\lambda)} = \zeta^{n(\lambda)}$. □

2.23. The rigid tensor category $\mathcal{C}$, together with the additional structure given by the above braiding and balance, is a *ribbon category* in the sense of Turaev, cf. [K] and references therein.

3. Braiding local systems

3.1. For a topological space T and a finite set J, T^J will denote the space of all maps $J \longrightarrow T$ (with the topology of the cartesian product). Its points are J-tuples (t_j) of points of T. We denote by T^{Jo} the subspace consisting of all (t_j) such that for any $j' \neq j''$ in J, $t_{j'} \neq t_{j''}$.

Let $\nu = \sum \nu_i i \in Y^+$. Let us call an *unfolding* of ν a map of finite sets $\pi : J \longrightarrow I$ such that $\mathrm{card}(\pi^{-1}(i)) = \nu_i$ for all i. Let Σ_π denote the group of all automorphisms $\sigma : J \xrightarrow{\sim} J$ such that $\pi \circ \sigma = \pi$.

The group Σ_π acts on the space T^J in the obvious way. We denote by T^ν the quotient space T^J/Σ_ν. This space does not depend, up to a unique isomorphism, on the choice of an unfolding π. The points of T^ν are collections (t_j) of I-colored points of T, such that for any I, there are ν_i points of color i. We have the canonical projection $T^J \longrightarrow T^\nu$, also to be denoted by π. We set $T^{\nu o} = \pi(T^{Jo})$. The map π restricted to T^{Jo} is an unramified Galois covering with the Galois group Σ_π.

3.2. For a real $r > 0$, let $D(r)$ denote the open disk on the complex plane $\{t \in \mathbb{C} |\ |t| < r\}$ and $\bar{D}(r)$ its closure. For $r_1 < r_2$, denote by $A(r_1, r_2)$ the open annulus $D(r_2) - \bar{D}(r_1)$.

Set $D = D(1)$. Let $\dot{D}$ denote the punctured disk $D - \{0\}$.

3.3. For an integer $n \geq 1$, consider the space

$$E_n = \{(r_0, \dots, r_n) \in \mathbb{R}^{n+1} |\ 0 = r_0 < r_1 < \dots < r_n = 1\}.$$

Obviously, the space E_n is contractible.

Let $J_1, \dots, J_n$ be finite sets. Set $J = \coprod_{a=1}^n J_a$. Note that $D^J = D^{J_1} \times \dots \times D^{J_n}$. Let us untroduce the subspace $D^{J_1,\dots,J_n} \subset E_n \times D^J$. By definition it consists of points $((r_a); (t_j^a) \in D^{J_a},\ a = 1, \dots, n)$ such that $t_j^a \in A(r_{a-1}, r_a)$ for $a = 1, \dots, n$. The canonical projection $E_n \times D^J \longrightarrow D^J$ induces the map

$$m(J_1, \dots, J_n) : D^{J_1,\dots,J_n} \longrightarrow D^J.$$

The image of the above projection lands in the subspace $D^{J_1} \times \dot{D}^{J_2} \times \dots \times \dot{D}^{J_n} \subset D^{J_1} \times \dots \times D^{J_n} = D^J$. The induced map

$$p(J_1, \dots, J_n) : D^{J_1,\dots,J_n} \longrightarrow D^{J_1} \times \dot{D}^{J_2} \times \dots \times \dot{D}^{J_n}$$

is homotopy equivalence.

Now assume that we have maps $\pi_a : J_a \longrightarrow I$ which are unfoldings of the elements ν_a. Then their sum $\pi : J \longrightarrow I$ is an unfolding of $\nu = \nu_1 + \dots + \nu_n$. We define the space $D^{\nu_1,\dots,\nu_n} \subset E_n \times D^\nu$ as the image of $D^{J_1,\dots,J_n}$ under the projection $\mathrm{Id} \times \pi : E_n \times D^J \longrightarrow E_n \times D^\nu$. The maps $m(J_1, \dots, J_n)$ and $p(J_1, \dots, J_n)$ induce the maps

$$m(\nu_1, \dots, \nu_n) : D^{\nu_1,\dots,\nu_n} \longrightarrow D^{\nu_1 + \dots + \nu_n}$$

and

$$p(\nu_1, \dots, \nu_n) : D^{\nu_1,\dots,\nu_n} \longrightarrow D^{\nu_1} \times \dot{D}^{\nu_2} \times \dots \times \dot{D}^{\nu_n}$$

respectively, the last map being homotopy equivalence.

3.4. We define the open subspaces

$$\dot{D}^{\nu_1,\dots,\nu_n} = D^{\nu_1,\dots,\nu_n} \cap (E_n \times \dot{D}^{\nu})$$

and

$$\dot{D}^{\nu_1,\dots,\nu_n o} = D^{\nu_1,\dots,\nu_n} \cap (E_n \times \dot{D}^{\nu o}).$$

We have the maps

$$m_a(\nu_1,\dots,\nu_n) : D^{\nu_1,\dots,\nu_n} \longrightarrow D^{\nu_1,\dots,\nu_{a-1},\nu_a+\nu_{a+1},\nu_{a+2},\dots,\nu_n}$$

and

$$p_a(\nu_1,\dots,\nu_n) : D^{\nu_1,\dots,\nu_n} \longrightarrow D^{\nu_1,\dots,\nu_a} \times \dot{D}^{\nu_{a+1},\dots,\nu_n}$$

$(a = 1,\dots,n-1)$ defined in an obvious manner. They induce the maps

$$m_a(\nu_1,\dots,\nu_n) : \dot{D}^{\nu_1,\dots,\nu_n} \longrightarrow \dot{D}^{\nu_1,\dots,\nu_{a-1},\nu_a+\nu_{a+1},\nu_{a+2},\dots,\nu_n}$$

and

$$p_a(\nu_1,\dots,\nu_n) : \dot{D}^{\nu_1,\dots,\nu_n} \longrightarrow \dot{D}^{\nu_1,\dots,\nu_a} \times \dot{D}^{\nu_{a+1},\dots,\nu_n}$$

and similar maps between "o"-ed spaces. All the maps p are homotopy equivalences.

All these maps satisfy some obvious compatibilities. We will need the following particular case.

3.5. The *rhomb* diagram below commutes.

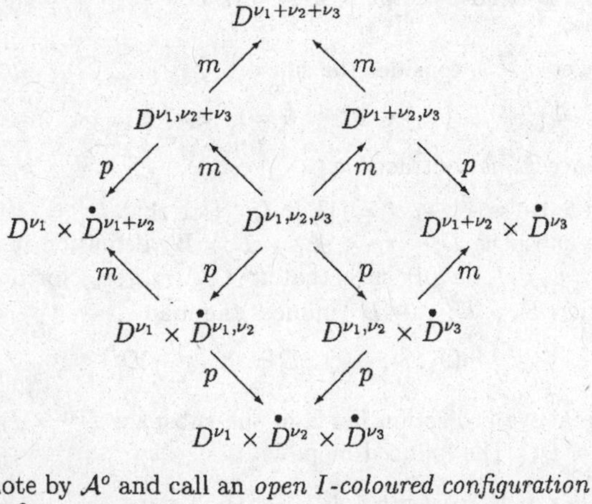

3.6. We will denote by $\mathcal{A}^o$ and call an *open I-coloured configuration space* the collection of all spaces $\{\dot{D}^{\nu_1,\dots,\nu_n o}\}$ together with the maps $\{m_a(\nu_1,\dots,\nu_n), p_a(\nu_1,\dots,\nu_n)\}$ between their various products.

We will call a *local system* over $\mathcal{A}^o$, or a *braiding local system* a collection of data (a), (b) below satisfying the property (c) below.

(a) A local system $\mathcal{I}^{\nu}_{\mu}$ over $\dot{D}^{\nu o}$ given for any $\nu \in Y^+, \mu \in X$.

(b) An isomorphism $\phi_\mu(\nu_1,\nu_2) : m^*\mathcal{I}^{\nu_1+\nu_2}_{\mu} \xrightarrow{\sim} p^*(\mathcal{I}^{\nu_1}_{\mu} \boxtimes \mathcal{I}^{\nu_2}_{\mu-\nu_1})$ given for any $\nu_1,\nu_2 \in Y^+, \mu \in X$.

Here $p = p(\nu_1,\nu_2)$ and $m = m(\nu_1,\nu_2)$ are the arrows in the diagram $\dot{D}^{\nu_1 o} \times \dot{D}^{\nu_2 o} \xleftarrow{p} \dot{D}^{\nu_1,\nu_2 o} \xrightarrow{m} \dot{D}^{\nu_1+\nu_2 o}$.

The isomorphisms $\phi_\mu(\nu_1,\nu_2)$ are called the *factorization isomorphisms.*

(c) (The *associativity* of factorization isomorphisms.) For any $\nu_1,\nu_2,\nu_3\in Y^+,\mu\in X$, the octagon below commutes. Here the maps m,p are the maps in the rhombic diagram above, with D, $\overset{\bullet}{D}$ replaced by $\overset{\bullet}{D}{}^o$.

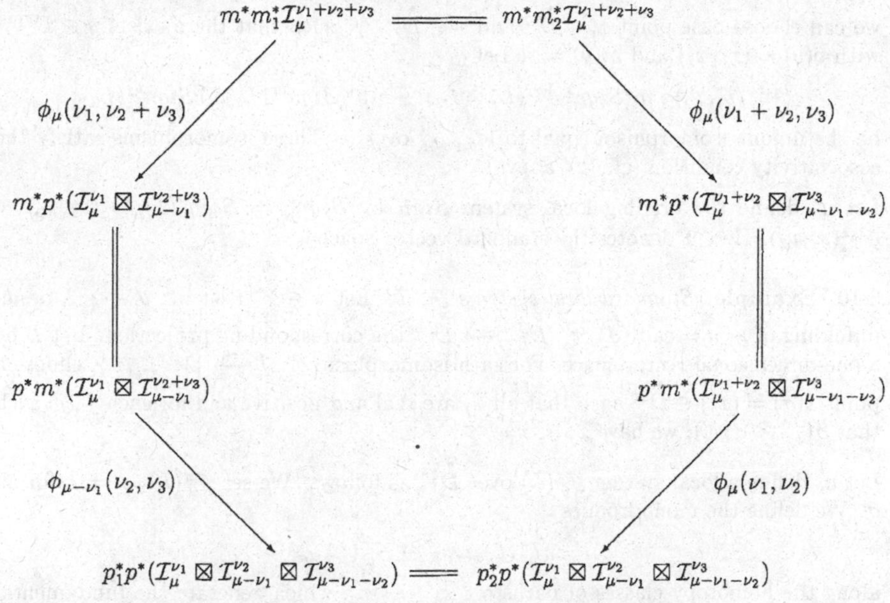

Written more concisely, the axiom (c) reads as a "cocycle" condition

(c)′ $\phi_\mu(\nu_1,\nu_2)\circ\phi_\mu(\nu_1+\nu_2,\nu_3)=\phi_{\mu-\nu_1}(\nu_2,\nu_3)\circ\phi_\mu(\nu_1,\nu_2+\nu_3)$.

3.7. **Correctional lemma.** *Assume we are given the data* (a), (b) *as above, with one-dimensional local systems* $\mathcal{I}_\mu^\nu$. *Then there exist a collection of constants* $c_\mu(\nu_1,\nu_2)\in\mathsf{k}^*$ $(\mu\in X,\nu_1,\nu_2\in Y^+)$ *such that the corrected isomorphisms* $\phi'_\mu(\nu_1,\nu_2)=c_\mu(\nu_1,\nu_2)\phi_\mu(\nu_1,\nu_2)$ *satisfy the associativity axiom* (c). □

3.8. The notion of a morphism between two braiding local systems is defined in an obvious way. This defines the category of braiding local systems, $\mathcal{B}ls$.

Suppose that $\mathcal{I}=\{\mathcal{I}_\mu^\nu;\phi_\mu(\nu_1,\nu_2)\}$ and $\mathcal{J}=\{\mathcal{J}_\mu^\nu;\psi_\mu(\nu_1,\nu_2)\}$ are two braiding local systems. Their *tensor product* $\mathcal{I}\otimes\mathcal{J}$ is defined by $(\mathcal{I}\otimes\mathcal{J})_\mu^\nu=\mathcal{I}_\mu^\nu\otimes\mathcal{J}_\mu^\nu$, the factorization isomorphisms being $\phi_\mu(\nu_1,\nu_2)\otimes\psi_\mu(\nu_1,\nu_2)$. This makes $\mathcal{B}ls$ a tensor category. The subcategory of one-dimensional braiding local systems is a Picard category.

3.9. **Example.** *Sign local system.* Let L be a one-dimensional vector space. Let $\nu\in Y^+$. Let $\pi:J\longrightarrow I$ be an unfolding of ν, whence the canonical projection $\pi:\overset{\bullet}{D}{}^{Jo}\longrightarrow\overset{\bullet}{D}{}^{\nu o}$. Pick a point $\tilde{x}=(x_j)\in\overset{\bullet}{D}{}^{Jo}$ with all x_j being *real*; let $x=\pi(\tilde{x})$. The choice of base points defines the homomorphism $\pi_1(\overset{\bullet}{D}{}^{\nu o};x)\longrightarrow\Sigma_\pi$. Consider its composition with the sign map $\Sigma_\pi\longrightarrow\mu_2\hookrightarrow\mathsf{k}^*$. Here $\mu_2=\{\pm1\}$ is the group of square roots of 1 in k^*. We get a map $s:\pi_1(\overset{\bullet}{D}{}^{\nu o};x)\longrightarrow\mathsf{k}^*$. It defines a local system

$\mathcal{S}ign^\nu(L)$ over $\overset{\bullet}{D}^{\nu\circ}$ whose stalk at x is equal to L. This local system does not depend (up to the unique isomorphism) on the choices made.

Given a diagram

$$\overset{\bullet}{D}^{\nu_1\circ} \times \overset{\bullet}{D}^{\nu_2\circ} \overset{p}{\longleftarrow} \overset{\bullet}{D}^{\nu_1,\nu_2\circ} \overset{m}{\longrightarrow} \overset{\bullet}{D}^{\nu_2+\nu_2\circ}$$

we can choose base point $x_i \in \overset{\bullet}{D}^{\nu_i}$ and $x \in \overset{\bullet}{D}^{\nu_1+\nu_2\circ}$ such that there exists $y \in \overset{\bullet}{D}^{\nu_1,\nu_2}$ with $p(y) = (x_1, x_2)$ and $m(y) = x$. Let

$$\phi^{L_1,L_2}(\nu_1,\nu_2) : m^*\mathcal{S}ign^{\nu_1+\nu_2}(L_1 \otimes L_2) \overset{\sim}{\longrightarrow} p^*(\mathcal{S}ign^{\nu_1}(L_1) \boxtimes \mathcal{S}ign^{\nu_2}(L_2))$$

be the unique isomorphism equal to $\mathrm{Id}_{L_1\otimes L_2}$ over y. These isomorphisms satisfy the associativity condition (cf. (c)$'$ above).

Let us define the brading local system $\mathcal{S}ign$ by $\mathcal{S}ign^\nu_\mu = \mathcal{S}ign^\nu(\mathbf{1})$, $\phi_\mu(\nu_1,\nu_2) = \phi^{\mathbf{1},\mathbf{1}}(\nu_1,\nu_2)$. Here $\mathbf{1}$ denotes the standard vector space k.

3.10. **Example.** *Standard local systems* $\mathcal{J}$, $\mathcal{I}$. Let $\nu \in Y^+$, let $\pi : J \longrightarrow I$ be an unfolding of ν, $n = \mathrm{card}(J)$, $\pi : \overset{\bullet}{D}^{J\circ} \longrightarrow \overset{\bullet}{D}^{\nu\circ}$ the corresponding projection. Let L be a one-dimensional vector space. For each isomorphism $\sigma : J \overset{\sim}{\longrightarrow} \{1, \dots, n\}$, choose a point $x(\sigma) = (x_j) \in \overset{\bullet}{D}^{J\circ}$ such that all x_j are real and positive and for each j', j'' such that $\sigma(j') < \sigma(j'')$, we have $x_{j'} < x_{j''}$.

Let us define a local system $\mathcal{J}^\pi_\mu(L)$ over $\overset{\bullet}{D}^{J\circ}$ as follows. We set $\mathcal{J}^\pi_\mu(L)_{x(\sigma)} = L$ for all σ. We define the monodromies

$$T_\gamma : \mathcal{J}^\pi_\mu(L)_{x(\sigma)} \longrightarrow \mathcal{J}^\pi_\mu(L)_{x(\sigma')}$$

along the homotopy classes of paths $\gamma : x_\sigma \longrightarrow x_{\sigma'}$ which generate the fundamental groupoid. Namely, let $x_{j'}$ and $x_{j''}$ be some neighbour points in $x(\sigma)$, with $x_{j'} < x_{j''}$. Let $\gamma(j',j'')^+$ (resp. $\gamma(j',j'')^-$) be the paths corresponding to the movement of $x_{j''}$ in the upper (resp. lower) hyperplane to the left from $x_{j'}$ position. We set

$$T_{\gamma(j',j'')^\pm} = \zeta^{\mp\pi(j')\cdot\pi(j'')}.$$

Let x_j be a point in $x(\sigma)$ closest to 0. Let $\gamma(j)$ be the path corresponding to the counterclockwise travel of x_j around 0. We set

$$T_{\gamma(j)} = \zeta^{2\mu\cdot\pi(j)}.$$

The point is that the above fromulas give a well defined morphism from the fundamental groupoid to the groupoid of one-dimensional vector spaces.

The local system $\mathcal{J}^\pi_\mu(L)$ admits an obvious Σ_π-equivariant structure. This defines the local system $\mathcal{J}^\nu_\mu(L)$ over $\overset{\bullet}{D}^{\nu\circ}$. Given a diagram

$$\overset{\bullet}{D}^{\nu_1\circ} \times \overset{\bullet}{D}^{\nu_2\circ} \overset{p}{\longleftarrow} \overset{\bullet}{D}^{\nu_1+\nu_2\circ} \overset{m}{\longrightarrow} \overset{\bullet}{D}^{\nu_1+\nu_2\circ},$$

let

$$\phi^{L_1,L_2}_\mu(\nu_1,\nu_2) : m^*\mathcal{J}^{\nu_1+\nu_2}_\mu(L_1 \otimes L_2) \overset{\sim}{\longrightarrow} p^*(\mathcal{J}^{\nu_1}_\mu(L_1) \boxtimes \mathcal{J}^{\nu_2}_{\mu-\nu_1}(L_2))$$

be the unique isomorphism equal to $Id_{L_1\otimes L_2}$ over compatible base points. Here the compatibility is understood in the same sense as in the previous subsection. These isomorphisms satisfy the associativity condition.

We define the braiding local system $\mathcal{J}$ by $\mathcal{J}^\nu_\mu = \mathcal{J}^\nu_\mu(\mathbf{1})$, $\phi_\mu(\nu_1,\nu_2) = \phi^{\mathbf{1},\mathbf{1}}_\mu(\nu_1,\nu_2)$.

We define the braiding local system $\mathcal{I}$ by $\mathcal{I} = \mathcal{J} \otimes \mathcal{S}ign$. The local systems $\mathcal{J}, \mathcal{I}$ are called the standard local systems. In the sequel we will mostly need the local system $\mathcal{I}$.

4. Factorizable sheaves

4.1. In the sequel for each $\nu \in Y^+$, we will denote by $\mathcal{S}$ the stratification on the space D^ν the closures of whose strata are various intersections of hypersurfaces given by the equations $t_j = 0$, $t_{j'} = t_{j''}$. The same letter will denote the induced stratifications on its subspaces.

4.2. Set $\mathcal{I}_\mu^{\nu\bullet} = j_{!*}\mathcal{I}_\mu^\nu[\dim \overset{\bullet}{D}{}^{\nu o}] \in \mathcal{M}(\overset{\bullet}{D}{}^\nu; \mathcal{S})$ where $j : \overset{\bullet}{D}{}^{\nu o} \hookrightarrow \overset{\bullet}{D}{}^\nu$ is the embedding. Given a diagram

$$\overset{\bullet}{D}{}^{\nu_1} \times \overset{\bullet}{D}{}^{\nu_2} \overset{p}{\longleftarrow} \overset{\bullet}{D}{}^{\nu_1,\nu_2} \overset{m}{\longrightarrow} \overset{\bullet}{D}{}^{\nu_1+\nu_2},$$

the structure isomorphisms $\phi_\mu(\nu_1, \nu_2)$ of the local system $\mathcal{I}$ induce the isomorphisms, to be denoted by the same letters,

$$\phi_\mu(\nu_1, \nu_2) : m^*\mathcal{I}^{\nu_1+\nu_2\bullet} \xrightarrow{\sim} p^*(\mathcal{I}^{\nu_1\bullet} \boxtimes \mathcal{I}^{\nu_2\bullet}).$$

Obviously, these isomorphisms satisfy the associativity axiom.

4.3. Let us fix a coset $c \in X/Y$. We will regard c as a subset of X. We will call a *factorizable sheaf $\mathcal{M}$ supported at* c a collection of data (w), (a), (b) below satisfying the axiom (c) below.

(w) An element $\lambda = \lambda(\mathcal{M}) \in c$.

(a) A perverse sheaf $\mathcal{M}^\nu \in \mathcal{M}(D^\nu; \mathcal{S})$ given for each $\nu \in Y^+$.

(b) An isomorphism $\psi(\nu_1, \nu_2) : m^*\mathcal{M}^{\nu_1+\nu_2} \xrightarrow{\sim} p^*(\mathcal{M}^{\nu_1} \boxtimes \mathcal{I}_{\lambda-\nu_1}^{\nu_2\bullet})$ given for any $\nu_1, \nu_2 \in Y^+$.

Here p, m denote the arrows in the diagram $D^{\nu_1} \times \overset{\bullet}{D}{}^{\nu_2} \overset{p}{\longleftarrow} D^{\nu_1,\nu_2} \overset{m}{\longrightarrow} D^{\nu_1+\nu_2}$.

The isomorphisms $\psi(\nu_1, \nu_2)$ are called the *factorization isomorphisms.*

(c) For any $\nu_1, \nu_2, \nu_3 \in Y^+$, the following *associativity condition* is fulfilled:

$$\psi(\nu_1, \nu_2) \circ \psi(\nu_1 + \nu_2, \nu_3) = \phi_{\lambda-\nu_1}(\nu_2, \nu_3) \circ \psi(\nu_1, \nu_2 + \nu_3).$$

We leave to the reader to draw the whole octagon expressing this axiom.

4.4. Let $\mathcal{M} = \{\mathcal{M}^\nu;\ \psi(\nu_1, \nu_2)\}$ be a factorizable sheaf supported at a coset $c \in X/Y$, $\lambda = \lambda(\mathcal{M})$. For each $\lambda' \geq \lambda, \nu \in Y^+$, define a sheaf $\mathcal{M}_{\lambda'}^\nu \in \mathcal{M}(D^\nu; \mathcal{S})$ by

$$\mathcal{M}_{\lambda'}^\nu = \begin{cases} \iota(\lambda' - \lambda)_*\mathcal{M}^{\nu-\lambda'+\lambda} & \text{if } \nu - \lambda' + \lambda \in Y^+ \\ 0 & \text{otherwise.} \end{cases}$$

Here

$$\iota(\nu') : D^\nu \longrightarrow D^{\nu+\nu'}$$

denotes the closed embedding adding ν' points sitting at the origin. The factorization isomorphisms $\psi(\nu_1, \nu_2)$ induce similar isomorphisms

$$\psi_{\lambda'}(\nu_1, \nu_2) : m^*\mathcal{M}_{\lambda'}(\nu_1 + \nu_2) \xrightarrow{\sim} p^*(\mathcal{M}_{\lambda'}^{\nu_1} \boxtimes \mathcal{I}_{\lambda'-\nu_1}^{\nu_2})\ (\lambda' \geq \lambda)$$

4.5. Let $\mathcal{M}, \mathcal{N}$ be two factorizable sheaves supported at c. Let $\lambda \in X$ be such that $\lambda \geq \lambda(\mathcal{M})$ and $\lambda \geq \lambda(\mathcal{N})$. For $\nu \geq \nu'$ in Y^+, consider the following composition

$$\tau_\lambda(\nu, \nu') : \mathrm{Hom}(\mathcal{M}^\nu_\lambda, \mathcal{N}^\nu_\lambda) \xrightarrow{m_*} \mathrm{Hom}(m^*\mathcal{M}^\nu_\lambda, m^*\mathcal{N}^\nu_\lambda) \xrightarrow[\sim]{\psi(\nu',\nu-\nu')_*}$$

$$\xrightarrow{\sim} \mathrm{Hom}(p^*(\mathcal{M}^{\nu'}_\lambda \boxtimes \mathcal{I}^{\nu-\nu'\bullet}_{\lambda-\nu'}), p^*(\mathcal{N}^{\nu'}_\lambda \boxtimes \mathcal{I}^{\nu-\nu'\bullet}_{\lambda-\nu'})) = \mathrm{Hom}(\mathcal{M}^{\nu'}_\lambda, \mathcal{N}^{\nu'}_\lambda).$$

Let us define the space of homomorphisms $\mathrm{Hom}(\mathcal{M}, \mathcal{N})$ by

$$\mathrm{Hom}(\mathcal{M}, \mathcal{N}) = \lim_{\to \lambda} \lim_{\leftarrow \nu} \mathrm{Hom}(\mathcal{M}^\nu_\lambda, \mathcal{N}^\nu_\lambda)$$

Here the inverse limit is taken over $\nu \in Y^+$, the transition maps being $\tau_\lambda(\nu, \nu')$ and the direct limit is taken over $\lambda \in X$ such that $\lambda \geq \lambda(\mathcal{M}), \lambda \geq \lambda(\mathcal{N})$, the transition maps being induced by the obvious isomorphisms

$$\mathrm{Hom}(\mathcal{M}^\nu_\lambda, \mathcal{N}^\nu_\lambda) = \mathrm{Hom}(\mathcal{M}^{\nu+\nu'}_{\lambda+\nu'}, \mathcal{N}^{\nu+\nu'}_{\lambda+\nu'}) \ (\nu' \in Y^+).$$

With these spaces of homomorphisms and the obvious compositions, the factorizable sheaves supported at c form the category, to be denoted by $\widetilde{\mathcal{FS}}_c$. By definition, the category $\widetilde{\mathcal{FS}}$ is the direct product $\prod_{c \in X/Y} \widetilde{\mathcal{FS}}_c$.

4.6. **Finite sheaves.** Let us call a factorizable sheaf $\mathcal{M} = \{\mathcal{M}^\nu\} \in \widetilde{\mathcal{FS}}_c$ *finite* if there exists only a finite number of $\nu \in Y^+$ such that the conormal bundle of the origin $O \in \mathcal{A}^\nu$ is contained in the singular support of $\mathcal{M}^\nu$. Let $\mathcal{FS}_c \subset \widetilde{\mathcal{FS}}_c$ be the full subcategory of finite factorizable sheaves. We define the category $\mathcal{FS}$ by $\mathcal{FS} = \prod_{c \in X/Y} \mathcal{FS}_c$. One proves (using the lemma below) that $\mathcal{FS}$ is an abelian category.

4.7. **Stabilization Lemma.** *Let $\mathcal{M}, \mathcal{N} \in \mathcal{FS}_c$, $\mu \in X_c, \mu \geq \lambda(\mathcal{M}), \mu \geq \lambda(\mathcal{N})$. There exists $\nu_0 \in Y^+$ such that for all $\nu \geq \nu_0$ the canonical maps*

$$\mathrm{Hom}(\mathcal{M}, \mathcal{N}) \longrightarrow \mathrm{Hom}(\mathcal{M}^\nu_\mu, \mathcal{N}^\nu_\mu)$$

are isomorphisms. □

4.8. **Standard sheaves.** Given $\mu \in X$, let us define the "standard sheaves" $\mathcal{M}(\mu), \mathcal{DM}(\mu)$ and $\mathcal{L}(\mu)$ supported at the coset $\mu + Y$, by $\lambda(\mathcal{M}(\mu)) = \lambda(\mathcal{DM}(\mu)) = \lambda(\mathcal{L}(\mu)) = \mu$;

$$\mathcal{M}(\mu)^\nu = j_! \mathcal{I}^{\nu\bullet}_\mu; \ \mathcal{DM}(\mu)^\nu = j_* \mathcal{I}^{\nu\bullet}_\mu; \ \mathcal{L}(\mu)^\nu = j_{!*} \mathcal{I}^{\nu\bullet}_\mu,$$

j being the embedding $\overset{\bullet}{D}{}^\nu \hookrightarrow D^\nu$. The factorization maps are defined by functoriality from the similar maps for $\mathcal{I}^\bullet$.

One proves that all these sheaves are finite.

5. Tensor product

In this section we will give (a sketch of) the construction of the tensor structure on the category $\widetilde{\mathcal{FS}}$. We will make the assumption of 2.19[5].

5.1. For $z \in \mathbb{C}$ and a real positive r, let $D(z; r)$ denote the open disk $\{t \in \mathbb{C} | \ |t - z| < r\}$ and $\bar{D}(z; r)$ its closure.

[5]Note that these assumptions are not necessary for the construction of the tensor structure. They are essential, however, for the construction of braiding.

5.2. For $\nu \in Y^+$, let us define the space $D^\nu(2)$ as the product $\overset{\bullet}{D} \times D^\nu$. Its points will be denoted $(z;(t_j))$ where $z \in \overset{\bullet}{D}, (t_j) \in D^\nu$. Let us define the open subspaces

$$\overset{\bullet}{D}{}^\nu(2) = \{(z;(t_j)) \in D^\nu(2) |\ t_j \neq 0, z \text{ for all } j\};\ \overset{\bullet}{D}{}^\nu(2)^\circ = \overset{\bullet}{D}{}^\nu(2) \cap (\overset{\bullet}{D} \times \overset{\bullet}{D}{}^{\nu\circ}).$$

For $\nu, \nu' \in Y^+$, let us define the space $D^{\nu,\nu'}(2)$ as the subspace of $\mathbb{R}_{>0} \times D^{\nu+\nu'}(2)$ consisting of all elements $(r;z;(t_j))$ such that $|z| < r < 1$ and ν of the points t_j live inside the disk $D(r)$ and ν' of them inside the annulus $A(r,1)$.

We have a diagram

(a) $D^\nu(2) \times \overset{\bullet}{D}{}^{\nu'} \overset{p}{\longleftarrow} D^{\nu,\nu'}(2) \overset{m}{\longrightarrow} D^{\nu+\nu'}(2)$.

Here $p((r;z;(t_j))) = ((z;(t_{j'})),(t_{j''}))$ where $t_{j'}$ (resp. $t_{j''}$) being the points from the collection (t_j) lying in $D(r)$ (resp. in $A(r,1)$); $m((r;z;(t_j))) = (z;(t_j))$. The map p is a homotopy equivalence.

For $\nu_1, \nu_2, \nu \in Y^+$, let $D^{\nu_1;\nu_2;\nu}(2)$ be the subspace of $\mathbb{R}_{>0} \times \mathbb{R}_{>0} \times D^{\nu_1+\nu_2+\nu}(2)$ consisting of all elements $(r_1;r_2;z;(t_j))$ such that $\bar{D}(r_1) \cup \bar{D}(z;r_2) \subset D$; $\bar{D}(r_1) \cap \bar{D}(z;r_2) = \emptyset$; ν_1 of the points (t_j) lie inside $D(r_1)$, ν_2 of them lie inside $D(z;r_2)$ and ν of them lie inside $D - (\bar{D}(r_1) \cup \bar{D}(z;r_2))$.

We have a diagram

(b) $D^{\nu_1} \times D^{\nu_2} \times \overset{\bullet}{D}{}^\nu(2) \overset{p}{\longleftarrow} D^{\nu_1;\nu_2;\nu}(2) \overset{m}{\longrightarrow} D^{\nu_1+\nu_2+\nu}(2)$.

Here $p((r_1;r_2;z;(t_j))) = ((t_{j'});(t_{j''}-z);(t_{j'''}))$ where $t_{j'}$ (resp. $t_{j''}, t_{j'''}$) are the points lying inside $D(r_1)$ (resp. $D(z;r_2), D-(\bar{D}(r_1)\cup\bar{D}(z;r_2))$); $m((r_1;r_2;z;(t_j))) = (z;(t_j))$. The map p is a homotopy equivalence.

5.3. We set $\overset{\bullet}{D}{}^{\nu,\nu'}(2) = D^{\nu,\nu'}(2) \cap (\mathbb{R}_{>0} \times \overset{\bullet}{D}{}^{\nu+\nu'}(2))$; $\overset{\bullet}{D}{}^{\nu,\nu'}(2)^\circ = D^{\nu,\nu'}(2) \cap (\mathbb{R}_{>0} \times \overset{\bullet}{D}{}^{\nu+\nu'}(2)^\circ)$; $\overset{\bullet}{D}{}^{\nu_1;\nu_2;\nu}(2) = D^{\nu_1;\nu_2;\nu}(2) \cap (\mathbb{R}_{>0} \times \mathbb{R}_{>0} \times \overset{\bullet}{D}{}^{\mu_1+\nu_2+\nu}(2))$; $\overset{\bullet}{D}{}^{\nu_1;\nu_2;\nu}(2)^\circ = D^{\nu_1;\nu_2;\nu}(2) \cap (\mathbb{R}_{>0} \times \mathbb{R}_{>0} \times \overset{\bullet}{D}{}^{\nu_1+\nu_2+\nu}(2)^\circ)$.

5.4. Given $\mu_1, \mu_2 \in X, \nu \in Y^+$, choose an unfolding $\pi : J \longrightarrow I$ of the element ν. In the same manner as in 3.10, we define the one-dimensional local system $\mathcal{J}^\nu_{\mu_1,\mu_2}$ over $\overset{\bullet}{D}{}^\nu(2)^\circ$ with the following monodromies: the monodromy around a loop corresponding to the counterclockwise travel of the point z around 0 (resp. t_j around 0, t_j around z, $t_{j'}$ around $t_{j''}$) is equal to the multiplication by $\zeta^{-2\mu_1\cdot\mu_2}$ (resp. $\zeta^{2\mu_1\cdot\pi(j)}$, $\zeta^{2\mu_2\cdot\pi(j)}$, $\zeta^{-2\pi(j')\cdot\pi(j'')}$).

As in *loc. cit.*, one defines isomorphisms

(a) $\phi_{\mu_1,\mu_2}(\nu,\nu') : m^*\mathcal{J}^{\nu+\nu'}_{\mu_1,\mu_2} \overset{\sim}{\longrightarrow} p^*(\mathcal{J}^\nu_{\mu_1,\mu_2} \boxtimes \mathcal{J}^{\nu'}_{\mu_1+\mu_2-\nu})$

where p, m are the morphisms in the diagram 5.2 (a) (restricted to the $\overset{\bullet}{D}{}^\circ$-spaces) and

(b) $\phi_{\mu_1,\mu_2}(\nu_1;\nu_2;\nu) : m^*\mathcal{J}^{\nu_1+\nu_2+\nu}_{\mu_1,\mu_2} \overset{\sim}{\longrightarrow} p^*(\mathcal{J}^{\nu_1}_{\mu_1} \boxtimes \mathcal{J}^{\nu_2}_{\mu_2} \boxtimes \mathcal{J}^\nu_{\mu_1-\nu_1,\mu_2-\nu_2})$

where p, m are the morphisms in the diagram 5.2 (b) (restricted to the $\overset{\bullet}{D}{}^\circ$-spaces), which satisfy the cocycle conditions

(c) $\phi_{\mu_1,\mu_2}(\nu,\nu') \circ \phi_{\mu_1,\mu_2}(\nu+\nu',\nu'') = \phi_{\mu_1+\mu_2-\nu}(\nu',\nu'') \circ \phi_{\mu_1,\mu_2}(\nu,\nu'+\nu'')$

(d) $(\phi_{\mu_1}(\nu_1,\nu_1') \boxtimes \phi_{\mu_2}(\nu_2,\nu_2')) \circ \phi_{\mu_1,\mu_2}(\nu_1+\nu_1';\nu_2+\nu_2';\nu) =$
$= \phi_{\mu_1-\nu_1,\mu_2-\nu_2}(\nu_1';\nu_2';\nu) \circ \phi_{\mu_1,\mu_2}(\nu_1;\nu_2;\nu+\nu_1'+\nu_2')$
(we leave to the reader the definition of the corresponding spaces).

5.5. Let us consider the sign local systems introduced in 3.9. We will keep the same notation $\mathcal{S}ign^\nu$ for the inverse image of the local system $\mathcal{S}ign^\nu$ under the forgetting of z map $\overset{\bullet}{D}{}^\nu(2)^\circ \longrightarrow \overset{\bullet}{D}{}^{\nu\circ}$. We have the factorization isomorphisms
(a) $\phi^{\mathcal{S}ign}(\nu,\nu') : m^*\mathcal{S}ign^{\nu+\nu'} \xrightarrow{\sim} p^*(\mathcal{S}ign^\nu \boxtimes \mathcal{S}ign^{\nu'})$;
(b) $\phi^{\mathcal{S}ign}(\nu_1;\nu_2;\nu) : m^*\mathcal{S}ign^{\nu_1+\nu_2+\nu} \xrightarrow{\sim} p^*(\mathcal{S}ign^{\nu_1} \boxtimes \mathcal{S}ign^{\nu_2} \boxtimes \mathcal{S}ign^\nu)$
which satisfy the cocycle conditions similar to (c), (d) above.

5.6. We define the local systems $\mathcal{I}^\nu_{\mu_1,\mu_2}$ over the spaces $\overset{\bullet}{D}{}^\nu(2)^\circ$ by
$\mathcal{I}^\nu_{\mu_1,\mu_2} = \mathcal{J}^\nu_{\mu_1,\mu_2} \otimes \mathcal{S}ign^\nu$.
The collection of local systems $\{\mathcal{I}^\nu_{\mu_1,\mu_2}\}$ together with the maps $\phi^{\mathcal{I}}_{\mu_1,\mu_2}(\nu,\nu') = \phi^{\mathcal{J}}_{\mu_1,\mu_2}(\nu,\nu') \otimes \phi^{\mathcal{S}ign}(\nu,\nu')$ and $\phi^{\mathcal{I}}_{\mu_1,\mu_2}(\nu_1;\nu_2;\nu) = \phi^{\mathcal{J}}_{\mu_1,\mu_2}(\nu_1;\nu_2;\nu) \otimes \phi^{\mathcal{S}ign}(\nu_1;\nu_2;\nu)$, forms an object $\mathcal{I}(2)$ which we call a *standard braiding local system over the configuration space* $\mathcal{A}(2)^\circ = \{\overset{\bullet}{D}{}^\nu(2)^\circ\}$. It is unique up to a (non unique) isomorphism. We fix such a local system.

5.7. We set $\mathcal{I}^{\nu\bullet}_{\mu_1,\mu_2} = j_{!*}\mathcal{I}^\nu_{\mu_1,\mu_2}[\dim \overset{\bullet}{D}{}^\nu(2)^\circ]$ where $j : \overset{\bullet}{D}{}^\nu(2)^\circ \hookrightarrow \overset{\bullet}{D}{}^\nu(2)$ is the open embedding. It is an object of the category $\mathcal{M}(\overset{\bullet}{D}{}^\nu(2);\mathcal{S})$ where $\mathcal{S}$ is the evident stratification. The factorization isomorphisms for the local system $\mathcal{I}$ induce the analogous isomorphisms between these sheaves, to be denoted by the same letter. The collection of these sheaves and factorization isomorphisms will be denoted $\mathcal{I}(2)^\bullet$.

5.8. Suppose we are given two factorizable sheaves $\mathcal{M},\mathcal{N}$. Let us call their *gluing*, and denote by $\mathcal{M}\boxtimes\mathcal{N}$, the collection of perverse sheaves $(\mathcal{M}\boxtimes\mathcal{N})^\nu$ over the spaces $D^\nu(2)$ $(\nu \in Y^+)$ together with isomorphisms

$$\psi(\nu_1;\nu_2;\nu) : m^*(\mathcal{M}\boxtimes\mathcal{N})^\nu \xrightarrow{\sim} p^*(\mathcal{M}^{\nu_1} \boxtimes \mathcal{N}^{\nu_2} \boxtimes \mathcal{I}^{\nu\bullet}_{\lambda(\mathcal{M})-\nu_1,\lambda(\mathcal{N})-\nu_2}),$$

p, m being the maps in the diagram 5.2 (b), which satisfy the cocycle condition
$(\psi^{\mathcal{M}}(\nu_1,\nu_1') \boxtimes \psi^{\mathcal{N}}(\nu_2,\nu_2')) \circ \psi(\nu_1+\nu_1';\nu_2+\nu_2';\nu) =$
$= \phi_{\lambda(\mathcal{M})-\nu_1,\lambda(\mathcal{N})-\nu_2}(\nu_1';\nu_2';\nu) \circ \psi(\nu_1;\nu_2;\nu+\nu_1'+\nu_2')$
for all $\nu_1,\nu_1',\nu_2,\nu_2',\nu \in Y^+$.

Such a gluing exists and is unique, up to a unique isomorphism. The factorization isomorphisms $\phi_{\mu_1,\mu_2}(\nu_1;\nu_2;\nu)$ for $\mathcal{I}(2)^\bullet$ and the ones for $\mathcal{M},\mathcal{N}$, induce the isomorphisms

$$\psi^{\mathcal{M}\boxtimes\mathcal{N}}(\nu,\nu') : m^*(\mathcal{M}\boxtimes\mathcal{N})^{\nu+\nu'} \xrightarrow{\sim} p^*((\mathcal{M}\boxtimes\mathcal{N})^\nu \boxtimes \mathcal{I}^{\nu'\bullet}_{\lambda(\mathcal{M})+\lambda(\mathcal{N})-\nu}).$$

satisfying the obvious cocycle condition.

5.9. Now we can define the tensor product $\mathcal{M}\otimes\mathcal{N}\in\widetilde{\mathcal{FS}}$. Namely, set $\lambda(\mathcal{M}\otimes\mathcal{N}) = \lambda(\mathcal{M})+\lambda(\mathcal{N})$. For each $\nu\in Y^+$, set

$$(\mathcal{M}\otimes\mathcal{N})^\nu = \Psi_{z\to 0}((\mathcal{M}\boxtimes\mathcal{N})^\nu).$$

Here $\Psi_{z\to 0}:\mathcal{M}(D^\nu(2))\longrightarrow\mathcal{M}(D^\nu)$ denotes the functor of nearby cycles for the function $D^\nu(2)\longrightarrow D$ sending $(z;(t_j))$ to z. Note that

$$\Psi_{z\to 0}(\mathcal{I}^\nu_{\mu_1,\mu_2}) = \mathcal{I}^\nu_{\mu_1+\mu_2}.$$

The factorization isomorphisms $\psi^{\mathcal{M}\boxtimes\mathcal{N}}$ induce the factorization isomorphisms between the sheaves $(\mathcal{M}\otimes\mathcal{N})^\nu$. This defines a factorizable sheaf $\mathcal{M}\otimes\mathcal{N}$.

One sees at once that this construction is functorial; thus it defines a functor of tensor product $\otimes:\widetilde{\mathcal{FS}}\times\widetilde{\mathcal{FS}}\longrightarrow\widetilde{\mathcal{FS}}$.

The subcategory $\mathcal{FS}\subset\widetilde{\mathcal{FS}}$ is stable under the tensor product. The functor $\otimes$: $\mathcal{FS}\times\mathcal{FS}\longrightarrow\mathcal{FS}$ extends uniquely to a functor $\otimes:\ \mathcal{FS}\otimes\mathcal{FS}\longrightarrow\mathcal{FS}$ (for the discussion of the tensor product of abelian categories, see [D2] 5).

5.10. The half-circle travel of the point z around 0 from 1 to -1 in the upper halfplane defines the braiding isomorphisms

$$R_{\mathcal{M},\mathcal{N}}:\mathcal{M}\otimes\mathcal{N}\overset{\sim}{\longrightarrow}\mathcal{N}\otimes\mathcal{M}.$$

We will not describe here the precise definition of the associativity isomorphisms for the tensor product $\otimes$. We just mention that to define them one should introduce into the game certain configuration spaces $D^\nu(3)$ whose (more or less obvious) definition we leave to the reader.

The unit of this tensor structure is the sheaf $\mathbf{1}=\mathcal{L}(0)$ (cf. 4.8).

Equipped with these complementary structures, the category $\widetilde{\mathcal{FS}}$ becomes a *braided tensor category.*

6. Vanishing cycles

GENERAL GEOMETRY

6.1. Let us fix a finite set J, and consider the space D^J. Inside this space, let us consider the subspaces $D^J_{\mathbb{R}} = D^J\cap\mathbb{R}^J$ and $D^{J+}=D^J\cap\mathbb{R}^J_{\geq 0}$.

Let $\mathcal{H}$ be the set (*arrangement*) of all real hyperplanes in $D^J_{\mathbb{R}}$ of the form $H_j:\ t_j=0$ or $H_{j'j''}:\ t_{j'}=t_{j''}$. An *edge* L of the arrangement $\mathcal{H}$ is a subspace of $D^J_{\mathbb{R}}$ which is a non-empty intersection $\bigcap H$ of some hyperplanes from $\mathcal{H}$. We denote by L° the complement $L-\bigcup L'$, the union over all edges $L'\subset L$ of smaller dimension. A *facet* of $\mathcal{H}$ is a connected component F of some L°. We call a facet *positive* if it lies entirely inside D^{J+}.

For example, we have a unique smallest facet O — the origin. For each $j\in J$, we have a positive one-dimensional facet F_j given by the equations $t_{j'}=0$ $(j'\neq j)$; $t_j\geq 0$.

Let us choose a point w_F on each positive facet F. We call a *flag* a sequence of embedded positive facets $\mathbf{F}:\ F_0\subset\ldots F_p$; we say that $\mathbf{F}$ *starts* from F_0. To such a flag we assign the simplex $\Delta_{\mathbf{F}}$ — the convex hull of the points $w_{F_0},\ldots,w_{F_p}$.

To each positive facet F we assign the following two spaces: $D_F = \bigcup \Delta_{\mathbf{F}}$, the union over all flags $\mathbf{F}$ starting from F, and $S_F = \bigcup \Delta_{\mathbf{F}'}$, the union over all flags $\mathbf{F}'$ starting from a facet which properly contains F. Obviously, $S_F \subset D_F$.

6.2. Given a complex $\mathcal{K} \in \mathcal{D}(D^J; \mathcal{S})$ and a positive facet F, we introduce a complex of vector spaces $\Phi_F^+(\mathcal{K})$ by

$$\Phi_F^+(\mathcal{K}) = R\Gamma(D_F, S_F; \mathcal{K})[-\dim F].$$

This is a well defined object of the bounded derived category $\mathcal{D}(*)$ of finite dimensional vector spaces, not depending on the choice of points w_F. It is called the *complex of vanishing cycles of $\mathcal{K}$ across F.*

6.3. **Theorem.** *We have canonically $\Phi_F^+(D\mathcal{K}) = D\Phi_F^+(\mathcal{K})$ where D denotes the Verdier duality in the corresponding derived categories.* □

6.4. **Theorem.** *If $\mathcal{M} \in \mathcal{M}(D^J; \mathcal{S})$ then $H^i(\Phi_F^+(\mathcal{M})) = 0$ for $i \neq 0$. Thus, Φ_F^+ induces an exact functor*

$$\Phi_F^+ : \mathcal{M}(D^J; \mathcal{S}) \longrightarrow \mathcal{V}ect$$

to the category $\mathcal{V}ect$ of finite dimensional vector spaces. □

6.5. Given a positive facet E and $\mathcal{K} \in \mathcal{D}(D^J, \mathcal{S})$, we have $S_E = \bigcup_{F \in \mathcal{F}^1(E)} D_F$, the union over the set $\mathcal{F}^1(E)$ of all positive facets $F \supset E$ with $\dim F = \dim F + 1$, and

$$R\Gamma(S_E, \bigcup_{F \in \mathcal{F}^1(E)} S_F; \mathcal{K}) = \oplus_{F \in \mathcal{F}^1(E)} R\Gamma(D_F, S_F; \mathcal{K}).$$

6.6. For two positive facets E and $F \in \mathcal{F}^1(E)$, and $\mathcal{K}(D^J; \mathcal{S})$, let us define the natural map

$$u = u_E^F(\mathcal{K}) : \Phi_F^+(\mathcal{K}) \longrightarrow \Phi_E^+(\mathcal{K})$$

called *canonical*, as the composition

$$R\Gamma(D_F, S_F; \mathcal{K})[-p] \longrightarrow R\Gamma(S_E, \bigcup_{F' \in \mathcal{F}^1(E)} S_{F'}; \mathcal{K})[-p] \longrightarrow R\Gamma(S_E; \mathcal{K})[-p] \longrightarrow$$

$$\longrightarrow R\Gamma(D_E, S_E)[-p+1]$$

where $p = \dim F$, the first arrow being induced by the equality in 6.5, the last one being the coboundary map.

Define the natural map

$$v = v_F^E(\mathcal{K}) : \Phi_E^+(\mathcal{K}) \longrightarrow \Phi_F^+(\mathcal{K})$$

called *variation*, as the map dual to the composition

$$D\Phi_F^+(\mathcal{K}) = \Phi_F^+(D\mathcal{K}) \overset{u(D\mathcal{K})}{\longrightarrow} \Phi_E^+(D\mathcal{K}) = D\Phi_E^+(\mathcal{K}).$$

BACK TO FACTORIZABLE SHEAVES

6.7. Let $\nu \in Y$. We are going to give two equivalent definitions of an exact functor, called *vanishing cycles at the origin*

$$\Phi : \mathcal{M}(D^\nu; \mathcal{S}) \longrightarrow \mathcal{V}ect.$$

First definition. Let $f : D^\nu \longrightarrow D$ be the function $f((t_j)) = \sum t_j$. For an object $\mathcal{K} \in \mathcal{D}(D^\nu; \mathcal{S})$, the Deligne's complex of vanishing cycles $\Phi_f(\mathcal{K})$ (cf. [D3]) is concentrated at the origin of the hypersurface $f^{-1}(0)$. It is t-exact with respect to the middle t-structure. We set by definition, $\Phi(\mathcal{M}) = H^0(\Phi_f(\mathcal{M}))$ $(\mathcal{M} \in \mathcal{M}(D^\nu; \mathcal{S}))$.

Second definition. Choose an unfolding of ν, $\pi : J \longrightarrow I$. Let us consider the canonical projection $\pi : D^J \longrightarrow D^\nu$. For $\mathcal{K} \in \mathcal{D}(D^\nu; \mathcal{S})$, the complex $\pi^*\mathcal{K}$ is well defined as an element of the Σ_π-equivariant derived category, hence $\Phi^+_O(\pi^*(\mathcal{K}))$ is a well defined object of the Σ_π-equivariant derived category of vector spaces (O being the origin facet in D^J). Therefore, the complex of Σ_π-invariants $\Phi^+_O(\pi^*\mathcal{K})^{\Sigma_\pi}$ is a well defined object of $\mathcal{D}(*)$. If $\mathcal{K} \in \mathcal{M}(D^\nu; \mathcal{S})$ then all the cohomology of $\Phi^+_O(\pi^*\mathcal{K})^{\Sigma_\pi}$ in non-zero degree vanishes. We set

$$\Phi(\mathcal{M}) = H^0(\Phi^+_O(\pi^*\mathcal{M})^{\Sigma_\pi})$$

$(\mathcal{M} \in \mathcal{M}(D^\nu; \mathcal{S}))$. The equivalence of the two definitions follows without difficulty from the proper base change theorem. In computations the second definition is used.[6]

6.8. Let $\mathcal{M}$ be a factorizable sheaf supported at $c \in X/Y$, $\lambda = \lambda(\mathcal{M})$. For $\nu \in Y^+$, define a vector space $\Phi(\mathcal{M})_{\lambda-\nu}$ by

$$\Phi(\mathcal{M})_{\lambda-\nu} = \Phi(\mathcal{M}^\nu).$$

If $\mu \in X$, $\mu \not\leq \lambda$, set $\Phi(\mathcal{M})_\mu = 0$. One sees easily that this way we get an exact functor Φ from $\widetilde{\mathcal{FS}}_c$ to the category of X-graded vector spaces with finite dimensional components. We extend it to the whole category $\widetilde{\mathcal{FS}}$ by additivity.

6.9. Let $\mathcal{M} \in \widetilde{\mathcal{FS}}_c$, $\lambda = \lambda(\mathcal{M})$, $\nu = \sum \nu_i i \in Y^+$. Let $i \in I$ be such that $\nu_i > 0$. Pick an unfolding of ν, $\pi : J \longrightarrow I$. For each $j \in \pi^{-1}(i)$, the restriction of π, $\pi_j : J - \{j\} \longrightarrow I$, is an unfolding of $\nu - i$.

For each $j \in \pi^{-1}(i)$, we have canonical and variation morphisms

$$u_j : \Phi^+_{F_j}(\pi^*\mathcal{M}^\nu) \rightleftharpoons \Phi^+_O(\pi^*\mathcal{M}^\nu) : v_j$$

(the facet F_j has been defined in 6.1). Taking their sum over $\pi^{-1}(i)$, we get the maps

$$\sum u_j : \oplus_{j\in\pi^{-1}(i)} \Phi^+_{F_j}(\pi^*\mathcal{M}^\nu) \rightleftharpoons \Phi^+_O(\pi^*\mathcal{M}) : \sum v_j$$

Note that the group Σ_π acts on both sides and the maps respect this action. After passing to Σ_π-invariants, we get the maps

(a) $\Phi^+_{F_j}(\pi^*\mathcal{M}^\nu)^{\Sigma_{\pi_j}} = (\oplus_{j'\in\pi^{-1}(i)} \Phi^+_{F_{j'}}(\pi^*\mathcal{M}^\nu))^{\Sigma_\pi} \rightleftharpoons \Phi^+_O(\pi^*\mathcal{M}^\nu)^{\Sigma_\pi} = \Phi(\mathcal{M})_\nu$.

Here $j \in \pi^{-1}(i)$ is an arbitrary element. Let us consider the space

$$F_j^\perp = \{(t_{j'}) \in D^J | t_j = r, t_{j'} \in D(r') \text{ for all } j' \neq j\}$$

where r, r' are some fixed real numbers such that $0 < r' < r < 1$. The space $F_j^\perp$ is transversal to F_j and may be identified with $D^{J-\{j\}}$. The factorization isomorphism induces the isomorphism

$$\pi^*\mathcal{M}^\nu|_{F_j^\perp} \cong \pi_j^*\mathcal{M}^{\nu-i} \otimes (\mathcal{I}^i_{\lambda-\nu+i})_{\{r\}} = \pi^*\mathcal{M}^{\nu-i}$$

[6]its independence of the choice of an unfolding follows from its equivalence to the first definition.

which in turn induces the isomorphism

$$\Phi^+_{F_j}(\pi^*\mathcal{M}^\nu) \cong \Phi^+_O(\pi_j^*\mathcal{M}^{\nu-i}).$$

which is Σ_{π_j}-equivariant. Taking Σ_{π_j}-invariants and composing with the maps (a), we get the maps

(b) $\epsilon_i : \Phi(\mathcal{M})_{\nu-i} = \Phi^+_O(\pi_j^*\mathcal{M}^{\nu-i})^{\Sigma_j} \rightleftharpoons \Phi(\mathcal{M})_\nu : \theta_i$

which do not depend on the choice of $j \in \pi^{-1}(i)$.

6.10. For an arbitrary $\mathcal{M} \in \widetilde{\mathcal{FS}}$, let us define the X-graded vector space $\Phi(\mathcal{M})$ as $\Phi(\mathcal{M}) = \oplus_{\lambda\in X}\Phi(\mathcal{M})_\lambda$.

6.11. **Theorem.** *The operators ϵ_i, θ_i ($i \in I$) acting on the X-graded vector space $\Phi(\mathcal{M})$ satisfy the relations 2.11* (a), (b). □

6.12. A factorizable sheaf $\mathcal{M}$ is finite iff the space $\Phi(\mathcal{M})$ is finite dimensional. The previous theorem says that Φ defines an exact functor

$$\Phi : \mathcal{FS} \longrightarrow \mathcal{C}.$$

One proves that Φ is a tensor functor.

6.13. **Example.** For every $\lambda \in X$, the factorizable sheaf $\mathcal{L}(\lambda)$ (cf. 4.8) is finite. It is an irreducible object of $\mathcal{FS}$, and every irreducible object in $\mathcal{FS}$ is isomorphic to some $\mathcal{L}(\lambda)$. We have $\Phi(\mathcal{L}(\lambda)) = L(\lambda)$.

The next theorem is the main result of the present work.

6.14. **Theorem.** *The functor Φ is an equivalence of braided tensor categories.* □

6.15. **Remark.** As a consequence, the category $\mathcal{FS}$ is rigid. We do not know a geometric construction of the rigidity; it would be very interesting to find one.

Chapter 2. Global (genus 0)

7. Cohesive local systems

7.1. From now on until the end of part 0 we make the assumptions of 2.19. In the operadic notations below we partially follow [BD].

7.2. **Operad $\mathcal{D}$.** For a nonempty finite set J, let $D(J)$ denote the space whose points are J-tuples $\{x_j, \tau_j\}$ $(j \in J)$ where $x_j \in D$ and τ_j is a non-zero tangent vector at x_j, such that all points x_j are distinct.

Let $\tilde{D}(J)$ be the space whose points are J-tuples $\{\phi_j\}$ of holomorphic maps $\phi_j : D \longrightarrow D$ $(j \in J)$, each ϕ_j having the form $\phi_j(z) = x_j + \tau_j z$ $(x_j \in D, \tau_j \in \mathbb{C}^*)$, such that $\phi_j(D) \cap \phi_{j'}(D) = \emptyset$ for $j \neq j'$. We shall identify the j-tuple $\{\phi_j\}$ with the J-tuple $\{x_j, \tau_j\}$, and consider τ_j as a non-zero tangent vector from $T_{x_j}D$, thus identifying $T_{x_j}D$ with $\mathbb{C}$ using the local coordinate $z - x_j$. So, τ_j is the image under ϕ_j of the unit tangent vector at 0. We have an obvious map $p(J) : \tilde{D}(J) \longrightarrow D(J)$ which is a homotopy equivalence.

If $\rho : K \longrightarrow J$ is an epimorphic map of finite sets, the composition defines a holomorphic map

$$m(\rho) : \prod_J \tilde{D}(K_j) \times \tilde{D}(J) \longrightarrow \tilde{D}(K)$$

where $K_j := \rho^{-1}(j)$. If $L \xrightarrow{\sigma} K \xrightarrow{\rho} J$ are two epimorphisms of finite sets, the square

$$\begin{array}{ccc}
\prod_K \tilde{D}(L_k) \times \prod_J \tilde{D}(K_j) \times \tilde{D}(J) & \xrightarrow{m(\sigma)} & \prod_K \tilde{D}(L_k) \times \tilde{D}(K) \\
\prod m(\sigma_j) \downarrow & & \downarrow m(\sigma) \\
\prod_J \tilde{D}(L_j) \times \tilde{D}(J) & \xrightarrow{m(\rho\sigma)} & \tilde{D}(L)
\end{array}$$

commutes. Here $\sigma_j : L_j \longrightarrow K_j$ are induced by σ.

Let $*$ denote the one element set. The space $\tilde{D}(*)$ has a marked point, also to be denoted by $*$, corresponding to the identity map $\phi : D \longrightarrow D$.

If $\rho : J' \xrightarrow{\sim} J$ is an isomorphism, it induces in the obvious way an isomorphism $\rho^* : \tilde{D}(J) \xrightarrow{\sim} \tilde{D}(J')$ (resp. $D(J) \xrightarrow{\sim} D(J')$). The first map coincides with $m(\rho)$ restricted to $(\prod_J *) \times \tilde{D}(J)$. In particular, for each J, the group Σ_J of automorphisms of the set J, acts on the spaces $\tilde{D}(J), D(J)$.

The map $m(J \longrightarrow *)$ restricted to $* \times \tilde{D}(J)$ is the identity of $\tilde{D}(J)$.

We will denote the collection of the spaces and maps $\{\tilde{D}(J), m(\rho)\}$ by $\mathcal{D}$, and call it the *operad of disks with tangent vectors.*

7.3. **Coloured local systems over $\mathcal{D}$.** If $\rho : K \longrightarrow J$ is an epimorphism of finite sets and $\pi : K \longrightarrow X$ is a map of sets, we define the map $\rho_*\pi : J \longrightarrow X$ by $\rho_*\pi(j) = \sum_{K_j} \pi(k)$. For $j \in J$, we denote $K_j := \rho^{-1}(j)$ as above, and $\pi_j : K_j \longrightarrow X$ will denote the restriction of π.

Let us call an *X-coloured local system $\mathcal{J}$* over $\mathcal{D}$ a collection of local systems $\mathcal{J}(\pi)$ over the spaces $\tilde{D}(J)$ given for every map $\pi : J \longrightarrow X$, J being a non-empty finite set, together with *factorization isomorphisms*

$$\phi(\rho) : m(\rho)^*\mathcal{J}(\pi) \xrightarrow{\sim} \boxed{\times}_J \mathcal{J}(\pi_j) \boxtimes \mathcal{J}(\rho_*\pi)$$

given for every epimorphism $\rho : K \longrightarrow J$ and $\pi : K \longrightarrow X$, which satisfy the properties (a), (b) below.

(a) *Associativity.* Given a map $\pi : L \longrightarrow X$ and a pair of epimorphisms $L \xrightarrow{\sigma} K \xrightarrow{\rho} J$, the square below commutes.

$$\begin{array}{ccc} m(\rho)^* m(\sigma)^* \mathcal{J}(\pi) & \xrightarrow{\phi(\sigma)} & \boxed{\times}_K \mathcal{J}(\pi_k) \boxtimes m(\rho)^* \mathcal{J}(\sigma_* \pi) \\ \phi(\rho\sigma) \downarrow & & \downarrow \phi(\rho) \\ \boxed{\times}_J m(\sigma_j)^* \mathcal{J}(\pi_j) \boxtimes \mathcal{J}(\rho_* \sigma_* \pi) & \xrightarrow{\boxtimes \phi(\sigma_j)} & \boxed{\times}_K \mathcal{J}(\pi_k) \boxtimes \boxed{\times}_J \mathcal{J}((\sigma_* \pi)_j) \boxtimes \mathcal{J}(\rho_* \sigma_* \pi) \end{array}$$

Note that $\pi_{j*}\sigma_j = (\sigma_*\pi)_j$.

For $\mu \in X$, let $\pi_\mu : * \longrightarrow X$ be defined by $\pi_\mu(*) = \mu$. The isomorphisms $\phi(\mathrm{id}_*)$ restricted to the marked points in $D(*)$, give the isomorphisms $\mathcal{J}(\pi_\mu)_* \xrightarrow{\sim} \mathsf{k}$ (and imply that the local systems $\mathcal{J}(\pi_\mu)$ are one-dimensional).

(b) For any $\pi : J \longrightarrow X$, the map $\phi(\mathrm{id}_J)$ restricted to $(\prod_J *) \times \tilde{D}(J)$, equals $\mathrm{id}_{\mathcal{J}(\pi)}$.

The map $\phi(J \longrightarrow *)$ restricted to $* \times \tilde{D}(J)$, equals $\mathrm{id}_{\mathcal{J}(\pi)}$.

7.4. The definition above implies that the local systems $\mathcal{J}(\pi)$ are functorial with respect to isomorphisms. In particular, the action of the group Σ_π on $\tilde{D}(J)$ lifts to $\mathcal{J}(\pi)$.

7.5. **Standard local system over $\mathcal{D}$.** Let us define the "standard" local systems $\mathcal{J}(\pi)$ by a version of the construction 3.10.

We will use the notations of *loc. cit.* for the marked points and paths. We can identify $D(J) = D^{J\circ} \times (\mathbb{C}^*)^J$ where we have identified all tangent spaces T_xD $(x \in D)$ with $\mathbb{C}$ using the local coordinate $z - x$.

We set $\mathcal{J}(\pi)_{x(\sigma)} = \mathsf{k}$, the monodromies being $T_{\gamma(j',j'')\pm} = \zeta^{\mp\pi(j')\cdot\pi(j'')}$, and the monodromy T_j corresponding to the counterclockwise circle of a tangent vector τ_j is $\zeta^{-2n\pi(j)}$. The factorization isomorphisms are defined by the same condition as in *loc. cit.*

This defines the *standard local system* $\mathcal{J}$ over $\mathcal{D}$. Below, the notation $\mathcal{J}$ will be reserved for this local system.

7.6. Set $P = \mathbb{P}^1(\mathbb{C})$. We pick a point $\infty \in P$ and choose a global coordinate $z : \mathbb{A}^1(\mathbb{C}) = P - \{\infty\} \xrightarrow{\sim} \mathbb{C}$. This gives local coordinates: $z - x$ at $x \in \mathbb{A}^1(\mathbb{C})$ and $1/z$ at ∞.

For a non-empty finite set J, let $P(J)$ denote the space of J-tuples $\{x_j, \tau_j\}$ where x_j are distinct points on P, and τ_j is a non-zero tangent vector at x_j. Let $\tilde{P}(J)$ denote the space whose points are J-tuples of holomorphic embeddings $\phi_j : D \longrightarrow P$ with non-intersecting images such that each ϕ_j is a restriction of an algebraic morphism $P \longrightarrow P$. We have the 1-jet projections $\tilde{P}(J) \longrightarrow P(J)$. We will use the notation $P(n)$ for $P(\{1, \ldots, n\})$, etc.

An epimorphism $\rho : K \longrightarrow J$ induces the maps

$$m_P(\rho) : \prod_J \tilde{D}(K_j) \times \tilde{P}(J) \longrightarrow \tilde{P}(K)$$

and

$$\bar{m}_P(\rho): \prod_J D(K_j) \times \tilde{P}(J) \longrightarrow P(K).$$

For a pair of epimorphisms $L \xrightarrow{\sigma} K \xrightarrow{\rho} J$, the square

$$\begin{array}{ccc} \prod_K \tilde{D}(L_k) \times \prod_J \tilde{D}(K_j) \times \tilde{P}(J) & \xrightarrow{m_P(\sigma)} & \prod_K \tilde{D}(L_k) \times \tilde{P}(K) \\ \prod m(\sigma_j) \downarrow & & \downarrow m_P(\sigma) \\ \prod_J \tilde{D}(L_j) \times \tilde{P}(J) & \xrightarrow{m_P(\rho\sigma)} & \tilde{P}(L) \end{array}$$

commutes.

If $\rho : J' \xrightarrow{\sim} J$ is an isomorphism, it induces in the obvious way an isomorphism $\rho^* : \tilde{P}(J) \xrightarrow{\sim} \tilde{P}(J')$ (resp. $P(J) \xrightarrow{\sim} P(J')$). This last map coincides with $m_P(\rho)$ (resp. $\bar{m}_P(\rho)$) restricted to $(\prod_J *) \times \tilde{P}(J)$ (resp. $(\prod_J *) \times P(J)$).

The collection of the spaces and maps $\{\tilde{P}(J), m_P(\rho)\}$ form an object $\tilde{P}$ called a *right module* over the operad $\mathcal{D}$.

7.7. Fix an element $\mu \in X$. Let us say that a map $\pi : J \longrightarrow X$ has *level* μ if $\sum_J \pi(j) = \mu$.

A *cohesive local system* $\mathcal{H}$ *of level* μ over P is a collection of local systems $\mathcal{H}(\pi)$ over $\tilde{P}(J)$ given for every $\pi : J \longrightarrow X$ of level μ, together with *factorization isomorphisms*

$$\phi_P(\rho) : m_P(\rho)^*\mathcal{H}(\pi) \xrightarrow{\sim} \boxed{\times}_J \mathcal{J}(\pi_j) \boxtimes \mathcal{H}(\rho_*\pi)$$

given for every epimorphism $\rho : K \longrightarrow J$ and $\pi : K \longrightarrow X$ of level μ, which satisfy the properties (a), (b) below.

(a) *Associativity.* Given a map $\pi : L \longrightarrow X$ and a pair of epimorphisms $L \xrightarrow{\sigma} K \xrightarrow{\rho} J$, the square below commutes.

$$\begin{array}{ccc} m_P(\rho)^* m_P(\sigma)^* \mathcal{H}(\pi) & \xrightarrow{\phi_P(\sigma)} & \boxed{\times}_K \mathcal{J}(\pi_k) \boxtimes m_P(\rho)^*\mathcal{H}(\sigma_*\pi) \\ \phi_P(\rho\sigma) \downarrow & & \downarrow \phi_P(\rho) \\ \boxed{\times}_J m(\sigma_j)^* \mathcal{J}(\pi_j) \boxtimes \mathcal{H}(\rho_*\sigma_*\pi) & \xrightarrow{\boxtimes\phi(\sigma_j)} & \boxed{\times}_K \mathcal{J}(\pi_k) \boxtimes \boxed{\times}_J \mathcal{J}((\sigma_*\pi)_j) \boxtimes \mathcal{H}(\rho_*\sigma_*\pi) \end{array}$$

(b) For any $\pi : J \longrightarrow X$, the map $\phi_P(\mathrm{id}_J)$ restricted to $(\prod_J *) \times \tilde{P}(J)$, equals $\mathrm{id}_{\mathcal{H}(\pi)}$.

7.8. The definition above implies that the local systems $\mathcal{H}(\pi)$ are functorial with respect to isomorphisms. In particular, the action of the group Σ_π on $\tilde{P}(J)$, lifts to $\mathcal{H}(\pi)$.

7.9. **Theorem.** *For each $\mu \in X$ such that $\mu \equiv 2\rho_\ell \bmod Y_\ell$, there exists a unique up to an isomorphism one-dimensional cohesive local system $\mathcal{H} = \mathcal{H}^{(\mu)}$ of level μ over P.* □

The element $\rho_\ell \in X$ is defined in 2.2 and the lattice Y_ℓ in 2.3.

¿From now on, let us fix such a local system $\mathcal{H}^{(\mu)}$ for each μ as in the theorem.

7.10. Note that the obvious maps $p(J) : \tilde{P}(J) \longrightarrow P(J)$ are homotopy equivalences. Therefore the local systems $\mathcal{H}(\pi)$ $(\pi : J \longrightarrow X)$ descend to the unique local systems over $P(J)$, to be denoted by the same letter.

8. Gluing

8.1. Let us fix a finite set K. For $\nu \in Y^+$, pick an unfolding of ν, $\pi : J \longrightarrow I$. Consider the space P^ν; its points are formal linear combinations $\sum_J \pi(j)x_j$ ($x_j \in P$). We define the space $P^\nu(K) = P(K) \times P^\nu$; its points are tuples $\{y_k, \tau_k; \sum \pi(j)x_j\}$ ($k \in K, y_k \in P, \tau_k \neq 0$ in $T_{y_k}P$), all y_k being distinct. Let $P^\nu(K)^\circ$ (resp. $P^\nu(K)^\bullet$) be the subspace whose points are tuples as above with all x_j distinct from y_k and pairwise distinct (resp. all x_j distinct from y_k). We will use the notation $P^\nu(n)$ for $P^\nu(\{1, \dots, n\})$, etc.

Let $\nu' \in Y^+$ and $\vec{\nu} = \{\nu_k\} \in (Y^+)^K$ be such that $\sum_K \nu_k + \nu' = \nu$. Define the space $P^{\vec{\nu};\nu'} \subset \tilde{P}(K) \times P^\nu$ consisting of tuples $\{\phi_k, \sum \pi(j)x_j\}$ such that for each k, ν_k of the points x_j lie inside $\phi_k(D)$ and ν' of them lie outside all closures of these disks.

Let $\vec{0} \in (Y^+)^K$ be the zero K-tuple. Define the space $\tilde{P}^\nu(K) := P^{\vec{0};\nu}$. We have obvious maps

(a) $\prod_K D^{\nu_k} \times \tilde{P}^{\nu'}(K) \overset{p(\vec{\nu};\nu')}{\longleftarrow} P^{\vec{\nu};\nu'} \overset{m(\vec{\nu};\nu')}{\longrightarrow} P^\nu(K)$.

Let $\vec{\nu}^1, \vec{\nu}^2 \in (Y^+)^K$ be such that $\vec{\nu}^1 + \vec{\nu}^2 = \vec{\nu}$. Define the space

$P^{\vec{\nu}^1;\vec{\nu}^2;\nu} \subset \tilde{P}(K) \times \mathbb{R}^K_{>0} \times \tilde{P}(K) \times P^\nu$ consisting of all tuples $\{\phi_k; r_k; \phi'_k; \sum \pi(j)x_j\}$ such that $r_k < 1$; $\phi_k(z) = \phi'_k(r_k z)$, and ν^1_k (resp. ν^2_k, ν') from the points x_j lie inside $\phi_k(D)$ (resp. inside the annulus $\phi'_k(D) - \overline{\phi_k(D)}$, inside $P - \bigcup \overline{\phi'_k(D)}$). We set $\tilde{P}^{\vec{\nu};\nu'} := P^{\vec{0};\vec{\nu};\nu'}$, cf. 5.2.

We have a commutative romb (cf. 3.5):

(b)

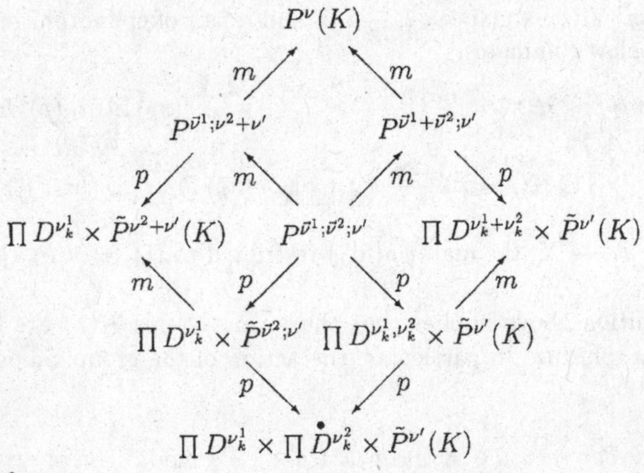

Here $\nu^2 := \sum \nu^2_k$.

8.2. Let $P(K; J)$ be the space consisting of tuples $\{y_k; \tau_k; x_j\}$ ($k \in K, j \in J, y_k, x_j \in P, \tau_k \neq 0$ in $T_{y_k}P$), where all points x_k, y_j are distinct. We have $P^\nu(K)^\circ = P(K; J)/\Sigma_\pi$. We have an obvious projection $P(K \coprod J) \longrightarrow P(K; J)$.

8.3. Let $\{\mathcal{M}_k\}$ be a K-tuple of factorizable sheaves supported at some cosets in X/Y; let $\mu_k = \lambda(\mathcal{M}_k)$.

Let $\tilde{\pi}: K \coprod J \longrightarrow X$ be a map defined by $\tilde{\pi}(k) = \mu_k$, $\tilde{\pi}(j) = -\pi(j) \in I \hookrightarrow X$. The local system $\mathcal{H}(\tilde{\pi})$ over $P(K \coprod J)$ descends to $P(K; J)$ since $\zeta^{2n(-i)} = 1$ for all $i \in I$; this one in turn descends to the unique local system $\tilde{\mathcal{H}}^{\nu}_{\vec{\mu}}$ over $P^{\nu}(K)^{\circ}$, due to Σ_{π}-equivariance. Let us define the local system $\mathcal{H}^{\nu}_{\vec{\mu}} := \tilde{\mathcal{H}}^{\nu}_{\vec{\mu}} \otimes \mathcal{S}ign^{\nu}$. Here $\mathcal{S}ign^{\nu}$ denotes the inverse image of the sign local system on $P^{\nu o}$ (defined in the same manner as for the disk, cf. 3.9) under the forgetful map $P^{\nu}(K)^{\circ} \longrightarrow P^{\nu o}$.

Let $\mathcal{H}^{\nu\bullet}_{\vec{\mu}}$ be the perverse sheaf over $P^{\nu}(K)^{\bullet}$ which is the middle extension of $\mathcal{H}^{\nu}_{\vec{\mu}}[\dim P^{\nu}(K)]$. Let us denote by the same letter the inverse image of this perverse sheaf on the space $\tilde{P}^{\nu}(K)$ with respect to the evident projection $\tilde{P}^{\nu}(K) \longrightarrow P^{\nu}(K)^{\bullet}$.

8.4. Let us call an element $\nu \in Y^+$ *admissible* (for a K-tuple $\{\mu_k\}$) if $\sum \mu_k - \nu \equiv 2\rho_{\ell} \bmod Y_{\ell}$, see 2.3.

8.5. **Theorem - definition.** *For each admissible ν, there exists a unique, up to a unique isomorphism, perverse sheaf, denoted by $\boxed{\times}^{(\nu)}_K \mathcal{M}_k$, over $P^{\nu}(K)$, equipped with isomorphisms*

$$\psi(\vec{\nu};\nu') : m^* \boxed{\times}^{(\nu)}_K \mathcal{M}_k \xrightarrow{\sim} p^*(\boxed{\times}_K \mathcal{M}^{\nu_k}_k \boxtimes \mathcal{H}^{\nu'\bullet}_{\vec{\mu}-\vec{\nu}})$$

given for every diagram 8.1 (a) such that for each rhomb 8.1 (b) the cocycle condition

$$\phi(\vec{\nu}^2;\nu') \circ \psi(\vec{\nu}^1;\nu^2+\nu') = (\boxed{\times}_K \psi^{\mathcal{M}_k}(\nu^1_k,\nu^2_k)) \circ \psi(\vec{\nu}^1+\vec{\nu}^2;\nu')$$

holds. □

8.6. The sheaf $\boxed{\times}^{(\nu)}_K \mathcal{M}_k$ defines for each K-tuple of $\vec{y} = \{y_k, \tau_k\}$ of points of P with non-zero tangent vectors, the sheaf $\boxed{\times}^{(\nu)}_{\vec{y}} \mathcal{M}_k$ over P^{ν}, to be called the *gluing of the factorizable sheaves $\mathcal{M}_k$ into the points (y_k, τ_k)*.

8.7. **Example.** The sheaf $\boxed{\times}^{(\nu)}_K \mathcal{L}(\mu_k)$ is equal to the middle extension of the sheaf $\mathcal{H}^{\nu\bullet}_{\vec{\mu}}$.

9. Semiinfinite cohomology

In this section we review the theory of semiinfinite cohomology in the category $\mathcal{C}$, due to S. M. Arkhipov, cf. [Ark].

9.1. Let $\mathcal{C}_r$ be a category whose objects are *right* u-modules N, finite dimensional over k, with a given X-grading $N = \oplus_{\lambda \in X} N_{\lambda}$ such that $xK_{\nu} = \zeta^{-\langle\nu,\lambda\rangle}x$ for any $\nu \in Y, \lambda \in X, x \in N_{\lambda}$. All definitions and results concerning the category $\mathcal{C}$ given above and below, have the obvious versions for the category $\mathcal{C}_r$.

For $M \in \mathcal{C}$, define $M^{\vee} \in \mathcal{C}_r$ as follows: $(M^{\vee})_{\lambda} = (M_{-\lambda})^*$ (the dual vector space); the action of the operators θ_i, ϵ_i being the transpose of their action on M. This way we get an equivalence $^{\vee} : \mathcal{C}^{opp} \xrightarrow{\sim} \mathcal{C}_r$.

9.2. Let us call an object $M \in \mathcal{C}$ $\mathfrak{u}^-$- (resp. $\mathfrak{u}^+$-) *good* if it admits a filtration whose successive quotients have the form $\mathrm{ind}^{\mathfrak{u}}_{\mathfrak{u}_{\geq 0}}(M')$ (resp. $\mathrm{ind}^{\mathfrak{u}}_{\mathfrak{u}_{\leq 0}}(M''))$ for some $M' \in \mathcal{C}^{\geq 0}$ (resp. $M'' \in \mathcal{C}^{\leq 0}$) (cf. 2.16). These classes of objects are stable with respect to the tensor multiplication by an arbitrary object of $\mathcal{C}$.

If M is $\mathfrak{u}^-$- (resp. $\mathfrak{u}^+$-) good then M^* is $\mathfrak{u}^-$- (resp. $\mathfrak{u}^+$-) good.

If M is $\mathfrak{u}^-$-good and M' is $\mathfrak{u}^+$-good then $M \otimes M'$ is a projective object in $\mathcal{C}$.

9.3. Let us say that a complex $M^\bullet$ in $\mathcal{C}$ is *concave* (resp. *convex*) if

(a) there exists $\mu \in X$ such that all nonzero components $M^\bullet_\lambda$ have the weight $\lambda \geq \mu$ (resp. $\lambda \leq \mu$);

(b) for any $\lambda \in X$, the complex $M^\bullet_\lambda$ is finite.

9.4. For an object $M \in \mathcal{C}$, we will call a *left* (resp. *right*) $\mathfrak{u}^\pm$*-good resolution* of M an exact complex

$$\ldots \longrightarrow P^{-1} \longrightarrow P^0 \longrightarrow M \longrightarrow 0$$

(resp.

$$0 \longrightarrow M \longrightarrow R^0 \longrightarrow R^1 \longrightarrow \ldots)$$

such that all P^i (resp. R^i) are $\mathfrak{u}^\pm$-good.

9.5. **Lemma.** *Each object $M \in \mathcal{C}$ admits a convex $\mathfrak{u}^-$-good left resolution, a concave $\mathfrak{u}^+$-good left resolution, a concave $\mathfrak{u}^-$-good right resolution and a convex $\mathfrak{u}^+$-good right resolution.* □

9.6. For $N \in \mathcal{C}_r$, $M \in \mathcal{C}$, define a vector space $N \otimes_{\mathcal{C}} M$ as the zero weight component of the tensor product $N \otimes_{\mathfrak{u}} M$ (which has an obvious X-grading).

For $M, M' \in \mathcal{C}$, we have an obvious perfect pairing

(a) $\mathrm{Hom}_{\mathcal{C}}(M, M') \otimes (M'^{\vee} \otimes_{\mathcal{C}} M) \longrightarrow \mathsf{k}$.

9.7. $M, M' \in \mathcal{C}$, $N \in \mathcal{C}_r$ and $n \in \mathbb{Z}$, define the *semiinifinite* Ext *and* Tor spaces

$$\mathrm{Ext}_{\mathcal{C}}^{\frac{\infty}{2}+n}(M, M') = H^n(\mathrm{Hom}_{\mathcal{C}}(R^\bullet_{\searrow}(M), R^\bullet_{\nearrow}(M')))$$

where $R^\bullet_{\searrow}(M)$ (resp. $R^\bullet_{\nearrow}(M')$) is an arbitrary $\mathfrak{u}^+$-good convex right resolution of M (resp. $\mathfrak{u}^-$-good concave right resolution of M'),

$$\mathrm{Tor}^{\mathcal{C}}_{\frac{\infty}{2}+n}(N, M) = H^{-n}(P^\bullet_{\swarrow}(N) \otimes_{\mathcal{C}} R^\bullet_{\searrow}(M))$$

where $P^\bullet_{\swarrow}(N)$ is an arbitrary $\mathfrak{u}^-$-good convex left resolution of N.

This definition does not depend, up to a unique isomorphism, upon the choice of resolutions, and is functorial.

These spaces are finite dimensional and are non-zero only for finite number of degrees n.

The pairing 9.6 (a) induces perfect pairings

$$\mathrm{Ext}_{\mathcal{C}}^{\frac{\infty}{2}+n}(M, M') \otimes \mathrm{Tor}^{\mathcal{C}}_{\frac{\infty}{2}+n}(M'^{\vee}, M) \longrightarrow \mathsf{k}\ (n \in \mathbb{Z}).$$

10. Conformal blocks (genus 0)

In this section we suppose that $\mathsf{k} = \mathbb{C}$.

10.1. Let $M \in \mathcal{C}$. We have a canonical embedding of vector spaces

$$\mathrm{Hom}_{\mathcal{C}}(\mathbf{1}, M) \hookrightarrow M$$

which identifies $\mathrm{Hom}_{\mathcal{C}}(\mathbf{1}, M)$ with the maximal trivial subobject of M. Here "trivial" means "isomorphic to a sum of a few copies of the object $\mathbf{1}$". Dually, we have a canonical epimorhism

$$M \longrightarrow \mathrm{Hom}_{\mathcal{C}}(M, \mathbf{1})^*$$

which identifies $\mathrm{Hom}_{\mathcal{C}}(M, \mathbf{1})^*$ with the maximal trivial quotient of M. Let us denote by $\langle M \rangle$ the image of the composition

$$\mathrm{Hom}_{\mathcal{C}}(\mathbf{1}, M) \longrightarrow M \longrightarrow \mathrm{Hom}_{\mathcal{C}}(M, \mathbf{1})^*$$

Thus, $\langle M \rangle$ is canonically a subquotient of M.

One sees easily that if $N \subset M$ is a trivial direct summand of M which is maximal, i.e. not contained in greater direct summand, then we have a canonical isomorphism $\langle M \rangle \xrightarrow{\sim} N$. For this reason, we will call $\langle M \rangle$ *the maximal trivial direct summand* of M.

10.2. Let $\gamma_0 \in Y$ denote the highest coroot and $\beta_0 \in Y$ denote the coroot dual to the highest root ($\gamma_0 = \beta_0$ for a simply laced root datum).

Let us define the *first alcove* $\Delta_\ell \subset X$ by

$$\Delta_\ell = \{\lambda \in X |\ \langle i, \lambda + \rho \rangle > 0 \text{ for all } i \in I;\ \langle \gamma_0, \lambda + \rho \rangle < \ell\}$$

if d does not divide ℓ, i.e. if $\ell_i = \ell$ for all $i \in I$, and by

$$\Delta_\ell = \{\lambda \in X |\ \langle i, \lambda + \rho \rangle > 0 \text{ for all } i \in I;\ \langle \beta_0, \lambda + \rho \rangle < \ell_{\beta_0}\}$$

otherwise, cf. [AP] 3.19 (d is defined in 2.1, and ℓ_{β_0} in 2.2). Note that $\ell_{\beta_0} = \ell/d$.

10.3. For $\lambda_1, \dots, \lambda_n \in \Delta_\ell$, define the *space of conformal blocks* by

$$\langle L(\lambda_1), \dots, L(\lambda_n) \rangle := \langle L(\lambda_1) \otimes \dots \otimes L(\lambda_n) \rangle.$$

In fact, due to the ribbon structure on $\mathcal{C}$, the right hand side is a *local system* over the space $P(n) := P(\{1, \dots, n\})$ (cf. [D1]). It is more appropriate to consider the previous equality as the definition of the *local system of conformal blocks* over $P(n)$.

10.4. **Theorem.** (Arkhipov) *For each $\lambda_1, \dots, \lambda_n \in \Delta_\ell$, the space of conformal blocks $\langle L(\lambda_1), \dots, L(\lambda_n) \rangle$ is naturally a subquotient of the space*

$$\mathrm{Tor}^{\mathcal{C}}_{\frac{\infty}{2}+0}(\mathbf{1}_r, L(\lambda_1) \otimes \dots \otimes L(\lambda_n) \otimes L(2\rho_\ell)).$$

More precisely, due to the ribbon structure on $\mathcal{C}$, the latter space is a stalk of a local system over $P(n+1)$, and inverse image of the local system $\langle L(\lambda_1), \dots, L(\lambda_n) \rangle$ under the projection onto the first coordinates $P(n+1) \longrightarrow P(n)$, is a natural subquotient of this local system.

Here $\mathbf{1}_r$ is the unit object in $\mathcal{C}_r$.

Examples, also due to Arkhipov, show that the local systems of conformal blocks are in general *proper* subquotients of the corresponding Tor local systems.

This theorem is an immediate consequence of the next lemma, which in turn follows from the geometric theorem 11.2 below, cf. 11.3.

10.5. **Lemma.** *We have* $\mathrm{Tor}^{\mathcal{C}}_{\frac{\infty}{2}+n}(\mathbf{1}_r, L(2\rho_\ell)) = \mathsf{k}$ *if* $n = 0$, *and* 0 *otherwise.* □

10.6. Let $\mathfrak{g}$ be the simple Lie algebra (over k) associated with our Cartan datum; let $\hat{\mathfrak{g}}$ be the corresponding affine Lie algebra.

Let $\mathcal{MS}$ denote the category of integrable $\hat{\mathfrak{g}}$-modules of central charge $\kappa - h$. Here κ is a fixed positive integer, h is the dual Coxeter number of our Cartan datum. $\mathcal{MS}$ is a semisimple abelian category whose irreducible objects are $\mathfrak{L}(\lambda)$, $\lambda \in \Delta_\ell$ where $l = 2d\kappa$, i.e. $\ell = d\kappa$ (we are grateful to Shurik Kirillov who pointed out the necessity of the factor d here) and $\mathfrak{L}(\lambda)$ is the highest weight module with a highest vector v whose "top" part $\mathfrak{g} \cdot v$ is the irreducible $\mathfrak{g}$-module of the highest weight λ.

According to Conformal field theory, $\mathcal{MS}$ has a natural structure of a ribbon category, cf. [MS], [K].

The usual local systems of conformal blocks in the WZW model may be defined as

$$\langle \mathfrak{L}(\lambda_1), \dots, \mathfrak{L}(\lambda_n)\rangle = \mathrm{Hom}_{\mathcal{MS}}(\mathbf{1}, \mathfrak{L}(\lambda_1) \otimes \dots \otimes \mathfrak{L}(\lambda_n))$$

the structure of a local system on the right hand side is due to the ribbon structure on $\mathcal{MS}$.

10.7. Let $\zeta = \exp(\pi\sqrt{-1}/d\kappa)$. We have an exact functor

$$\phi:\ \mathcal{MS} \longrightarrow \mathcal{C}$$

sending $\mathfrak{L}(\lambda)$ to $L(\lambda)$. This functor identifies $\mathcal{MS}$ with a full subcategory of $\mathcal{C}$.

The functor ϕ does not respect the tensor structures. It admits the left and right adjoints, $\phi^\flat, \phi^\sharp$. For $M \in \mathcal{C}$, let $\langle M\rangle_{\mathcal{MS}}$ denotes the image of the composition

$$\phi \circ \phi^\flat(M) \longrightarrow M \longrightarrow \phi \circ \phi^\sharp(M).$$

We have the following comparison theorem.

10.8. **Theorem.** *We have naturally*

$$\phi(\mathfrak{M} \otimes \mathfrak{M}') = \langle \phi(\mathfrak{M}) \otimes \phi(\mathfrak{M}')\rangle_{\mathcal{MS}}.$$

This follows from the combination of the results of [AP], [KL], [L3] and [F2].

10.9. **Corollary.** *For any* $\lambda_1, \dots, \lambda_n \in \Delta_\ell$, *the functor* ϕ *induces an isomorphism of local systems*

$$\langle \mathfrak{L}(\lambda_1), \dots, \mathfrak{L}(\lambda_n)\rangle = \langle \mathfrak{L}(\lambda_1), \dots, \mathfrak{L}(\lambda_n)\rangle. \ \square$$

11. Integration

We keep the notations of the previous section.

11.1. Let K be a finite set, $m = \mathrm{card}(K)$, $\{\mathcal{M}_k\}$ a K-tuple of finite factorizable sheaves, $\mathcal{M}_k \in \mathcal{FS}_{c_k}$, $\mu_k := \lambda(\mathcal{M}_k)$. Assume that $\nu := \sum_K \mu_k - 2\rho_\ell$ belongs to Y^+.

Let $\eta : P^\nu(K) \longrightarrow P(K)$ be the projection.

11.2. **Theorem.** *We have canonical isomorphisms of local systems over $P(K)$*

$$R^{a-2m}\eta_*(\boxed{\times}_K^{(\nu)} \mathcal{M}_k) = \mathrm{Tor}^{\mathcal{C}}_{\frac{\infty}{2}-a}(\mathbf{1}_r, \otimes_K \Phi(\mathcal{M}_k))\ (a \in \mathbb{Z}),$$

the structure of a local system on the right hand side being induced by the ribbon structure on $\mathcal{C}$. □

11.3. **Proof of Lemma 10.5.** Apply the previous theorem to the case when the K-tuple consists of one sheaf $\mathcal{L}(2\rho_\ell)$ and $\nu = 0$. □

11.4. From now until the end of the section, $\mathbf{k} = \mathbb{C}$ and $\zeta = \exp(\pi\sqrt{-1}/d\kappa)$. Let $\lambda_1, \dots, \lambda_n \in \Delta_\ell$. Let $\nu = \sum_{m=1}^n \lambda_m$; assume that $\nu \in Y^+$. Set $\vec{\mu} = \{\lambda_1, \dots, \lambda_n, 2\rho_\ell\}$.

Let η be the projection $P^\nu(n+1) := P^\nu(\{1, \dots, n+1\}) \longrightarrow P(n+1)$ and $p : P(n+1) \longrightarrow P(n)$ be the projection on the first coordinates.

Let $\mathcal{H}^{\nu\sharp}_{\vec{\mu}}$ denote the middle extension of the sheaf $\mathcal{H}^{\nu\bullet}_{\vec{\mu}}$. By the Example 8.7,

$$\mathcal{H}^{\nu\sharp}_{\vec{\mu}} = \boxed{\times}^{(\nu)}_{1\leq a\leq n} \mathcal{L}(\lambda_a) \boxtimes \mathcal{L}(2\rho_\ell).$$

11.5. **Theorem.** *The local system $p^*\langle L(\lambda_1), \dots, L(\lambda_n)\rangle$ is canonically a subquotient of the local system*

$$R^{-2n-2}\eta_*\mathcal{H}^{\nu\sharp}_{\vec{\mu}}.$$

This theorem is an immediate corollary of the previous one and of the Theorem 10.4.

11.6. **Corollary.** *In the notations of the previous theorem, the local system $\langle L(\lambda_1), \dots, L(\lambda_n)\rangle$ is semisimple.*

Proof. The local system $p^*\langle L(\lambda_1), \dots, L(\lambda_n)\rangle$ is a subquotient of the geometric local system $R^{-2n-2}\eta_*\mathcal{H}^{\nu\sharp}_{\vec{\mu}}$, and hence is semisimple by the Beilinson-Bernstein-Deligne-Gabber Decomposition theorem, [BBD], Théorème 6.2.5. Therefore, the local system $\langle L(\lambda_1), \dots, L(\lambda_n)\rangle$ is also semisimple, since the map p induces the surjection on the fundamental groups. □

11.7. For a sheaf $\mathcal{F}$, let $\bar{\mathcal{F}}$ denote the sheaf obtained from $\mathcal{F}$ by the complex conjugation on the coefficients.

If a perverse sheaf $\mathcal{F}$ on P^ν is obtained by gluing some irreducible factorizable sheaves into some points of P then its Verdier dual $D\mathcal{F}$ is canonically isomorphic to $\bar{\mathcal{F}}$. Therefore, the Poincaré-Verdier duality induces a perfect hermitian pairing on $R\Gamma(P^\nu; \mathcal{F})$.

Therefore, in notations of theorem 11.5, The Poincaré-Verdier duality induces a non-degenerate hermitian form on the local system $R^{-2n-2}\eta_*\mathcal{H}^{\nu\sharp}_{\vec{\mu}}$.

By a little more elaborated argument using fusion, one can introduce a canonical hermitian form on the systems of conformal blocks.

Compare [K], where a certain hermitian form on the spaces of conformal blocks (defined up to a positive constant) has been introduced.

11.8. By the similar reasons, the Verdier duality defines a hermitian form on all irreducible objects of $\mathcal{C}$ (since the Verdier duality commutes with Φ, cf. Theorem 6.3).

12. Regular representation

12.1. From now on we are going to modify slightly the definition of the categories $\mathcal{C}$ and $\mathcal{FS}$. Let X_ℓ be the lattice

$$X_\ell = \{\mu \in X \otimes \mathbb{Q} |\ \mu \cdot \mathbb{Y}_\ell \in \ell\mathbb{Z}\}$$

We have obviously $X \subset X_\ell$, and $X = X_\ell$ if $d|\ell$.

In this Chapter we will denote by $\mathcal{C}$ a category of X_ℓ*-graded* (instead of X-graded) finite dimensional vector spaces $M = \oplus_{\lambda \in X_\ell} M_\lambda$ equipped with linear operators $\theta_i : M_\lambda \longrightarrow M_{\lambda - i'}$, $\epsilon_i : M_\lambda \longrightarrow M_{\lambda + i'}$ which satisfy the relations 2.11 (a), (b). This makes sense since $\langle d_i i, \lambda \rangle = i' \cdot \lambda \in \mathbb{Z}$ for each $i \in I, \lambda \in X_\ell$.

Also, in the definition of $\mathcal{FS}$ we replace X by X_ℓ. All the results of the previous Chapters hold true *verbatim* with this modification.

We set $\mathrm{d}_\ell = \mathrm{card}(X_\ell/Y_\ell)$; this number is equal to the determinant of the form $\mu_1, \mu_2 \mapsto \frac{1}{\ell}\mu_1 \cdot \mu_2$ on Y_ℓ.

12.2. Let $\tilde{\mathrm{u}} \subset \mathrm{u}$ be the k-subalgebra generated by $\tilde{K}_i, \epsilon_i, \theta_i$ $(i \in I)$. Following the method of [L1] 23.1, define a new algebra $\dot{\mathrm{u}}$ (without unit) as follows.

If $\mu', \mu'' \in X_\ell$, we set

$$_{\mu'}\tilde{\mathrm{u}}_{\mu''} = \tilde{\mathrm{u}}/(\sum_{i\in I}(\tilde{K}_i - \zeta^{i\cdot\mu'})\tilde{\mathrm{u}} + \sum_{i\in I}\tilde{\mathrm{u}}(\tilde{K}_i - \zeta^{i\cdot\mu''}));\ \dot{\mathrm{u}} = \oplus_{\mu',\mu''\in X_\ell}(_{\mu'}\tilde{\mathrm{u}}_{\mu''}).$$

Let $\pi_{\mu',\mu''} : \tilde{\mathrm{u}} \longrightarrow {}_{\mu'}\tilde{\mathrm{u}}_{\mu''}$ be the canonical projection. We set $1_\mu = \pi_{\mu,\mu}(1) \in \dot{\mathrm{u}}$. The structure of an algebra on $\dot{\mathrm{u}}$ is defined as in *loc. cit.*

As in *loc. cit.*, the category $\mathcal{C}$ may be identified with the category of finite dimensional (over k) (left) $\dot{\mathrm{u}}$-modules M which are *unital*, i.e.

(a) for every $x \in M$, $\sum_{\mu \in X_\ell} 1_\mu x = x$.

If M is such a module, the X_ℓ-grading on M is defined by $M_\mu = 1_\mu M$.

Let u' denote the quotient algebra of the algebra $\tilde{\mathrm{u}}$ by the relations $\tilde{K}_i^{l_i} = 1$ $(i \in I)$. Here $l_i := \frac{l}{(l,d_i)}$. We have an isomorphism of vector spaces $\dot{\mathrm{u}} = \mathrm{u}' \otimes \mathsf{k}[Y_\ell]$, cf. 12.5 below.

12.3. Let $a : \mathcal{C} \xrightarrow{\sim} \mathcal{C}_r$ be an equivalence defined by $aM = M$ $(M \in \mathcal{C})$ as an X_ℓ-vector space, $mx = s(x)m$ $(x \in \mathfrak{u}, m \in M)$. Here $s : \mathfrak{u} \longrightarrow \mathfrak{u}^{\mathrm{opp}}$ is the antipode. We will use the same notation a for a similar equivalence $\mathcal{C}_r \xrightarrow{\sim} \mathcal{C}$.

Let us consider the category $\mathcal{C} \otimes \mathcal{C}$ (resp. $\mathcal{C} \otimes \mathcal{C}_r$) which may be identified with the category of finite dimensional $\dot{\mathfrak{u}}\otimes\dot{\mathfrak{u}}$- (resp. $\dot{\mathfrak{u}}\otimes(\dot{\mathfrak{u}})^{\mathrm{opp}}$-) modules satisfying a "unitality" condition similar to (a) above. Let us consider the algebra $\dot{\mathfrak{u}}$ itself as a regular $\dot{\mathfrak{u}} \otimes (\dot{\mathfrak{u}})^{\mathrm{opp}}$-module. It is infinite dimensional, but is a union of finite dimensional modules, hence it may be considered as an object of the category $\mathrm{Ind}(\mathcal{C} \otimes \mathcal{C}_r) = \mathrm{Ind}\mathcal{C} \otimes \mathrm{Ind}\mathcal{C}_r$ where Ind denotes the category of Ind-objects, cf. [D4] §4. Let us denote by $\mathbf{R}$ the image of this object under the equivalence $\mathrm{Id} \otimes a : \mathrm{Ind}\mathcal{C} \otimes \mathrm{Ind}\mathcal{C}_r \xrightarrow{\sim} \mathrm{Ind}\mathcal{C} \otimes \mathrm{Ind}\mathcal{C}$.

Every object $\mathcal{O} \in \mathcal{C} \otimes \mathcal{C}$ induces a functor $F_{\mathcal{O}} : \mathcal{C} \longrightarrow \mathcal{C}$ defined by

$$F_{\mathcal{O}}(M) = a(aM \otimes_{\mathcal{C}} \mathcal{O}).$$

The same formula defines a functor $F_{\mathcal{O}} : \mathrm{Ind}\mathcal{C} \longrightarrow \mathrm{Ind}\mathcal{C}$ for $\mathcal{O} \in \mathrm{Ind}(\mathcal{C} \otimes \mathcal{C})$.

We have $F_{\mathbf{R}} = \mathrm{Id}_{\mathrm{Ind}\mathcal{C}}$.

We can consider a version of the above formalism using semiinfinite Tor's. An object $\mathcal{O} \in \mathrm{Ind}(\mathcal{C} \otimes \mathcal{C})$ defines functors $F_{\mathcal{O};\frac{\infty}{2}+n} : \mathrm{Ind}\mathcal{C} \longrightarrow \mathrm{Ind}\mathcal{C}$ $(n \in \mathbb{Z})$ defined by

$$F_{\mathcal{O};\frac{\infty}{2}+n}(M) = a\mathrm{Tor}^{\mathcal{C}}_{\frac{\infty}{2}+n}(aM, \mathcal{O}).$$

12.4. **Theorem.** (i) *We have* $F_{\mathbf{R};\frac{\infty}{2}+n} = \mathrm{Id}_{\mathrm{Ind}\mathcal{C}}$ *if* $n = 0$, *and* 0 *otherwise.*

(ii) *Conversely, suppose we have an object* $Q \in \mathrm{Ind}(\mathcal{C} \otimes \mathcal{C})$ *together with an isomorphism of functors* $\phi : F_{\mathbf{R};\frac{\infty}{2}+\bullet} \xrightarrow{\sim} F_{Q;\frac{\infty}{2}+\bullet}$. *Then* ϕ *is induced by the unique isomorphism* $\mathbf{R} \xrightarrow{\sim} Q$. □

12.5. **Adjoint representation.** For $\mu \in Y_\ell$, let $T(\mu)$ be a one-dimensional $\dot{\mathfrak{u}}\otimes(\dot{\mathfrak{u}})^{\mathrm{opp}}$-module equal to $L(\mu)$ (resp. to $aL(-\mu)$) as a $\dot{\mathfrak{u}}$- (resp. $(\dot{\mathfrak{u}})^{\mathrm{opp}}$-) module. Let us consider the module $T_\mu\mathbf{R} = \mathbf{R} \otimes T(\mu) \in \mathrm{Ind}(\mathcal{C} \otimes \mathcal{C})$. This object represents the same functor $\mathrm{Id}_{\mathrm{Ind}\mathcal{C}}$, hence we have a canonical isomorphism $t_\mu : \mathbf{R} \xrightarrow{\sim} T_\mu\mathbf{R}$.

Let us denote by $\hat{\mathbf{ad}} \in \mathrm{Ind}\mathcal{C}$ the image of $\mathbf{R}$ under the tensor product $\otimes : \mathrm{Ind}(\mathcal{C} \otimes \mathcal{C}) \longrightarrow \mathrm{Ind}\mathcal{C}$. The isomorphisms t_μ above induce an action of the lattice Y_ℓ on $\hat{\mathbf{ad}}$. Set $\mathbf{ad} = \hat{\mathbf{ad}}/Y_\ell$. This is an object of $\mathcal{C} \subset \mathrm{Ind}\mathcal{C}$ which is equal to the algebra $\mathfrak{u}'$ considered as a $\dot{\mathfrak{u}}$-module by means of the adjoint action.

In the notations of 10.7, let us consider an object

$$\mathbf{ad}_{\mathcal{MS}} := \oplus_{\mu\in\Delta_\ell}\langle L(\mu) \otimes L(\mu)^*\rangle_{\mathcal{MS}} \in \mathcal{MS},$$

cf. [BFM] 4.5.3.

12.6. **Theorem.** *We have a canonical isomorphism* $\langle\mathbf{ad}\rangle_{\mathcal{MS}} = \mathbf{ad}_{\mathcal{MS}}$. □

13. Regular sheaf

13.1. **Degeneration of quadrics.** The construction below is taken from [KL]II 15.2. Let us consider the quadric $Q \subset \mathbb{P}^1 \times \mathbb{P}^1 \times \mathbb{A}^1$ given by the equation $uv = t$ where (u, v, t) are coordinates in the triple product. Let $f : Q \longrightarrow \mathbb{A}^1$ be the projection to the third coordinate; for $t \in \mathbb{A}^1$ denote $Q_t := f^{-1}(t)$. For $t \neq 0$, Q_t is isomorphic to $\mathbb{P}^1$; the fiber Q_0 is a union of two projective lines clutched at a point: $Q_0 = Q_u \cup Q_v$ where Q_u (resp. Q_v) is an irreducible component given (in Q_0) by the equation $v = 0$ (resp. $u = 0$) and is isomorphic to $\mathbb{P}^1$; their intersection being a point. We set $'Q = f^{-1}(\mathbb{A}^1 - \{0\})$.

We have two sections $x_1, x_2 : \mathbb{A}^1 \longrightarrow Q$ given by $x_1(t) = (\infty, 0, t)$, $x_2(t) = (0, \infty, t)$. Consider two "coordinate charts" at these points: the maps $\phi_1, \phi_2 : \mathbb{P}^1 \times \mathbb{A}^1 \longrightarrow Q$ given by

$$\phi_1(z, t) = (\frac{tz}{z-1}, \frac{z-1}{z}, t);\ \phi_2(z, t) = (\frac{z-1}{z}, \frac{tz}{z-1}, t).$$

This defines a map

(a) $\phi : \mathbb{A}^1 - \{0\} \longrightarrow \tilde{P}(2)$,

in the notations of 7.6.

13.2. For $\nu \in Y^+$, let us consider the corresponding (relative over $\mathbb{A}^1$) configuration scheme $f^\nu : Q^\nu_{/\mathbb{A}^1} \longrightarrow \mathbb{A}^1$. For the brevity we will omit the subscript $_{/\mathbb{A}^1}$ indicating that we are dealing with the relative version of configuration spaces. We denote by $Q^{\nu\bullet}$ (resp. $Q^{\nu\circ}$) the subspace of configurations with the points distinct from x_1, x_2 (resp. also pairwise distinct). We set $'Q^{\nu\circ} = Q^{\nu\circ}|_{\mathbb{A}^1 - \{0\}}$, etc.

The map ϕ above, composed with the canonical projection $\tilde{P}(2) \longrightarrow P(2)$, induces the maps

$$'Q^{\nu\circ} \longrightarrow P^\nu(2)\ (\nu \in Y^+)$$

(in the notations of 8.1). For $\nu \in Y^+$ and $\mu_1, \mu_2 \in X_\ell$ such that $\mu_1 + \mu_2 - \nu = 2\rho_\ell$, let $\mathcal{J}^\nu_{\mu_1,\mu_2}$ denote the local system over $'Q^{\nu\circ}$ which is the inverse image of the local system $\mathcal{H}^\nu_{\mu_1,\mu_2}$ over $P^\nu(2)^\circ$. Let $\mathcal{J}^{\nu\bullet}_{\mu_1,\mu_2}$ denote the perverse sheaf over $'Q^{\nu\bullet}(\mathbb{C})$ which is the middle extension of $\mathcal{J}^\nu_{\mu_1,\mu_2}[\dim Q^\nu]$.

13.3. Let us take the nearby cycles and get a perverse sheaf $\Psi_{f^\nu}(\mathcal{J}^{\nu\bullet}_{\mu_1,\mu_2})$ over $Q^{\nu\bullet}_0(\mathbb{C})$. Let us consider the space $Q^{\nu\bullet}_0$ more attentively. This is a reducible scheme which is a union

$$Q^{\nu\bullet}_0 = \bigcup_{\nu_1+\nu_2=\nu} \mathbb{A}^{\nu_1} \times \mathbb{A}^{\nu_2},$$

the component $\mathbb{A}^{\nu_1} \times \mathbb{A}^{\nu_2}$ corresponding to configurations where ν_1 (resp. ν_2) points are running on the affine line $Q_u - x_1(0)$ (resp. $Q_v - x_2(0)$). Here we identify these affine lines with a "standard" one using the coordinates u and v respectively. Using this decomposition we can define a closed embedding

$$i_\nu : Q^{\nu\bullet}_0 \hookrightarrow \mathbb{A}^\nu \times \mathbb{A}^\nu$$

whose restriction to a component $\mathbb{A}^{\nu_1} \times \mathbb{A}^{\nu_2}$ sends a configuration as above, to the configuration where all remaining points are equal to zero. Let us define a perverse sheaf

$$\mathcal{R}^{\nu,\nu}_{\mu_1,\mu_2} = i_{\nu*}\Psi_{f^\nu}(\mathcal{J}^{\nu\bullet}_{\mu_2,\mu_1}) \in \mathcal{M}(\mathbb{A}^\nu(\mathbb{C}) \times \mathbb{A}^\nu(\mathbb{C}); \mathcal{S})$$

Let us consider the collection of sheaves $\{\mathcal{R}^{\nu,\nu}_{\mu_1,\mu_2} | \mu_1, \mu_2 \in X_\ell, \nu \in Y^+, \mu_1 + \mu_2 - \nu = 2\rho_\ell\}$. One can complete this collection to an object $\mathcal{R}$ of the category $\mathrm{Ind}(\mathcal{FS} \otimes \mathcal{FS})$ where $\mathcal{FS}\otimes\mathcal{FS}$ is understood as a category of finite factorizable sheaves corresponding to the *square* of our initial Cartan datum, i.e. $I \coprod I$, etc. For a precise construction, see Part V.

13.4. **Theorem.** *We have* $\Phi(\mathcal{R}) = \mathbf{R}$. $\square$

Chapter 3. Modular

14. Heisenberg local systems

In this section we sketch a construction of certain remarkable cohesive local systems on arbitrary smooth families of compact smooth curves, to be called the *Heisenberg local systems.*

In the definition and construction of local sustems below we will have to assume that our base field k contains roots of unity of sufficiently high degree; the characteristic of k is assumed to be prime to this degree.

14.1. From now on until 14.11 we fix a smooth proper morphism $f : C \longrightarrow S$ of relative dimension 1, S being a smooth connected scheme over $\mathbb{C}$. For $s \in S$, we denote $C_s := f^{-1}(s)$. Let g be the genus of fibres of f.

Let S_δ denote the total space of the determinant line bundle $\delta_{C/S} = \det Rf_*\Omega^1_{C/S}$ without the zero section. For any object (?) over S (e.g., a scheme over S, a sheaf over a scheme over S, etc.), we will denote by $(?)_\delta$ its base change under $S_\delta \longrightarrow S$.

Below, if we speak about a scheme as a topological (analytic) space, we mean its set of $\mathbb{C}$-points with the usual topology (resp. analytic structure).

14.2. We will use the relative versions of configuration spaces; to indicate this, we will use the subscript $_{/S}$. Thus, if J is a finite set, $C^J_{/S}$ will denote the J-fold fibered product of C with itself over S, etc.

Let $C(J)_{/S}$ denote the subscheme of the J-fold cartesian power of the relative tangent bundle $T_{C/S}$ consisting of J-tuples $\{x_j, \tau_j\}$ where $x_j \in C$ and $\tau_j \neq 0$ in $T_{C/S,x}$, the points x_j being pairwise distinct. Let $\tilde{C}(J)_{/S}$ denote the space of J-tuples of holomorphic embeddings $\phi_j : D \times S \longrightarrow C$ over S with disjoint images; we have the 1-jet maps $\tilde{C}(J)_{/S} \longrightarrow C(J)_{/S}$.

An epimorphism $\rho : K \longrightarrow J$ induces the maps

$$m_{C/S}(\rho) : \prod_J \tilde{D}(K_j) \times \tilde{C}(J)_{/S} \longrightarrow \tilde{C}(J)_{/S}$$

and

$$\bar{m}_{C/S}(\rho) : \prod_J D(K_j) \times \tilde{C}(J)_{/S} \longrightarrow C(K)_{/S}$$

which satisfy the compatibilities as in 7.6.

14.3. We extend the function n to X_ℓ (see 12.1) by $n(\mu) = \frac{1}{2}\mu \cdot \mu - \mu \cdot \rho_\ell$ ($\mu \in X_\ell$). We will denote by $\mathcal{J}$ the X_ℓ-coloured local system over the operad of disks $\mathcal{D}$ which is defined exactly as in 7.5, with X replaced by X_ℓ.

14.4. A *cohesive local system* $\mathcal{H}$ *of level* $\mu \in X_\ell$ *over* C/S is a collection of local systems $\mathcal{H}(\pi)$ over the spaces $C(J)_{/S;\delta}$ given for every map $\pi : J \longrightarrow X_\ell$ of level μ (note the base change to S_δ!), together with the factorization isomorphisms

$$\mathring{\phi}_C(\rho) : m_{C/S}(\rho)^*\mathcal{H}(\pi) \xrightarrow{\sim} \boxed{\times}_J \mathcal{J}(\pi_j) \boxtimes \mathcal{H}(\rho_*\pi).$$

Here we have denoted by the same letter $\mathcal{H}(\pi)$ the lifting of $\mathcal{H}(\pi)$ to $\tilde{C}_{/S;\delta}$. The factorization isomorphisms must satisfy the obvious analogs of properties 7.7 (a), (b).

14.5. Now we will sketch a construction of certain cohesive local system over C/S of level $(2-2g)\rho_\ell$. For alternative beautiful constructions of $\mathcal{H}$, see [BP].

To simplify the exposition we will assume below that $g \geq 2$ (the construction for $g \leq 1$ needs some modification, and we omit it here, see Part V). Let us consider the group scheme $\mathrm{Pic}(C/S) \otimes X_\ell$ over S. Here $\mathrm{Pic}(C/S)$ is the relative Picard scheme. The group of connected components $\pi_0(\mathrm{Pic}(C/S) \otimes X_\ell)$ is equal to X_ℓ. Let us denote by $\mathcal{J}ac$ the connected component corresponding to the element $(2-2g)\rho_\ell$; this is an abelian scheme over S, due to the existence of the section $S \longrightarrow \mathcal{J}ac$ defined by $\Omega^1_{C/S} \otimes (-\rho_\ell)$.

For a scheme S' over S, let $H_1(S'/S)$ denote the local system of the first relative integral homology groups over S. We have $H_1(\mathcal{J}ac/S) = H_1(C/S) \otimes X_\ell$. We will denote by ω the polarization of $\mathcal{J}ac$ (i.e. the skew symmetric form on the latter local system) equal to the tensor product of the standard form on $H_1(C/S)$ and the form $(\mu_1, \mu_2) \mapsto \frac{\mathrm{d}^g_\ell}{\ell}\mu_1 \cdot \mu_2$ on X_ℓ. Note that the assumption $g \geq 2$ implies that $\frac{\mathrm{d}^g_\ell}{\ell}\mu_1 \cdot \mu_2 \in \mathbb{Z}$ for any $\mu_1, \mu_2 \in X_\ell$. Since the latter form is positive definite, ω is relatively ample (i.e. defines a relatively ample invertible sheaf on $\mathcal{J}ac$).

14.6. Let $\alpha = \sum n_\mu \cdot \mu \in \mathbb{N}[X_\ell]$; set $\mathrm{Supp}(\alpha) = \{\mu |\ n_\mu \neq 0\}$. Let us say that α is *admissible* if $\sum n_\mu \mu = (2-2g)\rho_\ell$. Let us denote by

$$\mathrm{aj}_\alpha : C^\alpha_{/S} \longrightarrow \mathrm{Pic}(C/S) \otimes X_\ell$$

the Abel-Jacobi map sending $\sum \mu \cdot x_\mu$ to $\sum x_\mu \otimes \mu$. If α is admissible then the map aj_α lands in $\mathcal{J}ac$.

Let D^α denote the following relative divisor on $C^\alpha_{/S}$

$$D^\alpha = \frac{\mathrm{d}^g_\ell}{\ell}\Big(\sum_{\mu \neq \nu} \mu \cdot \nu \Delta_{\mu\nu} + \frac{1}{2}\sum_\mu \mu \cdot \mu \Delta_{\mu\mu}\Big).$$

Here $\Delta_{\mu\nu}$ ($\mu, \nu \in \mathrm{Supp}(\alpha)$) denotes the corresponding diagonal in $C^\alpha_{/S}$. Note that all the multiplicities are integers.

Let $\pi : J \longrightarrow X_\ell$ be an unfolding of α. We will denote by D^π the pull-back of D^α to $C^J_{/S}$. Let us introduce the following line bundles

$$\mathcal{L}(\pi) = \otimes_{j \in J} \mathcal{T}_j^{\otimes \frac{\mathrm{d}^g_\ell}{\ell} n(\pi(j))} \otimes \mathcal{O}(D^\pi)$$

on $C^J_{/S}$, and

$$\mathcal{L}_\alpha = \mathcal{L}(\pi)/\Sigma_\pi$$

on $C^\alpha_{/S}$ (the action of Σ_π is an obvious one). Here $\mathcal{T}_j$ denotes the relative tangent line bundle on $C^J_{/S}$ in the direction j. Note that the numbers $\frac{\mathrm{d}^g_\ell}{\ell} n(\mu)$ ($\mu \in X_\ell$) are integers.

14.7. **Proposition.** *There exists a unique line bundle $\mathcal{L}$ on $\mathcal{J}ac$ such that for each admissible α, we have $\mathcal{L}_\alpha = \mathrm{aj}^*_\alpha(\mathcal{L})$. The first Chern class $c_1(\mathcal{L}) = -[\omega]$.* □

14.8. In the sequel if $\mathcal{L}_0$ is a line bundle, let $\overset{\bullet}{\mathcal{L}}_0$ denote the total space its with the zero section removed.

The next step is the construction of a certain local system $\mathfrak{H}$ over $\overset{\bullet}{\mathcal{L}}_\delta$. Its dimension is equal to d_ℓ^g and the monodromy around the zero section of $\mathcal{L}$ (resp. of the determinant bundle) is equal to $\zeta^{-2\ell/\mathrm{d}_\ell^g}$ (resp. $(-1)^{\mathrm{rk}(X)}\zeta^{-12\rho_\ell\cdot\rho_\ell}$). The construction of $\mathfrak{H}$ is outlined below.

The previous construction assigns to a triple

(a lattice Λ, a symmetric bilinear form $(\ ,\ ):\ \Lambda\times\Lambda\longrightarrow\mathbb{Z}$, $\nu\in\Lambda$)

an abelian scheme $\mathcal{J}ac_\Lambda := (\mathrm{Pic}(C/S)\otimes\Lambda)_{(2g-2)\nu}$ over S, together with a line bundle $\mathcal{L}_\Lambda$ on it (in the definition of $\mathcal{L}_\Lambda$ one should use the function $n_\nu(\mu)=\frac{1}{2}\mu\cdot\mu+\mu\cdot\nu$). We considered the case $\Lambda=X_\ell,\ (\mu_1,\mu_2)=\frac{\mathrm{d}_\ell^g}{\ell}\mu_1\cdot\mu_2,\ \nu=-\rho_\ell$.

Now let us apply this construction to the lattice $\Lambda=X_\ell\oplus Y_\ell$, the bilinear form $((\mu_1,\nu_1),(\mu_2,\nu_2))=-\frac{1}{\ell}(\nu_1\cdot\mu_2+\nu_2\cdot\mu_1+\nu_1\cdot\nu_2)$ and $\nu=(-\rho_\ell,0)$. The first projection $\Lambda\longrightarrow X_\ell$ induces the morphism

$$p:\ \mathcal{J}ac_\Lambda\longrightarrow\mathcal{J}ac$$

the fibers of p are abelian varieties $\mathcal{J}ac(C_s)\otimes Y_\ell$ $(s\in S)$.

14.9. **Theorem.** (i) *The line bundle $\mathcal{L}_\Lambda$ is relatively ample with respect to p. The direct image $\mathcal{E}:=p_*\mathcal{L}_\Lambda$ is a locally free sheaf of rank d_ℓ^g.*

(ii) *We have an isomorphism*

$$\det(\mathcal{E})=\mathcal{L}\otimes\delta^{\mathrm{d}_\ell^g(-\frac{1}{2}\mathrm{rk}(X_\ell)+6\frac{\rho_\ell\cdot\rho_\ell}{\ell})}.$$

□

Here δ denotes the pull-back of the determinant bundle $\delta_{C/S}$ to $\mathcal{J}ac$.

14.10. Let us assume for a moment that $\mathsf{k}=\mathbb{C}$ and $\zeta=\exp(-\frac{\pi\sqrt{-1}}{\ell})$. By the result of Beilinson-Kazhdan, [BK] 4.2, the vector bundle $\mathcal{E}$ carries a canonical flat projective connection. By *loc. cit.* 2.5, its lifting to $\det(\mathcal{E})^\bullet$ carries a flat connection with the scalar monodromy around the zero section equal to $\exp(\frac{2\pi\sqrt{-1}}{\mathrm{d}_\ell^g})$. We have an obvious map

$$m:\overset{\bullet}{\mathcal{L}}_\delta\longrightarrow\mathcal{L}\otimes\delta^{\mathrm{d}_\ell^g(-\frac{1}{2}\mathrm{rk}(X_\ell)+6\frac{\rho_\ell\cdot\rho_\ell}{\ell})}.$$

By definition, $\mathfrak{H}$ is the local system of horizontal sections of the pull-back of $\mathcal{E}$ to $\overset{\bullet}{\mathcal{L}}_\delta$. The claim about its monodromies follows from part (ii) of the previous theorem.

This completes the construction of $\mathfrak{H}$ for $\mathsf{k}=\mathbb{C}$ and $\zeta=\exp(-\frac{\pi\sqrt{-1}}{\ell})$. The case of arbitrary k (of sufficiently large characteristic) and ζ follows from this one.

14.11. Let us consider an obvious map $q:C(J)_{/S;\delta}\longrightarrow C^J_{/S;\delta}$. The pull-back $q^*\mathcal{L}(\pi)$ has a canonical non-zero section s. Let $\tilde{\mathcal{H}}(\pi)$ be the pull-back of the local system $\mathfrak{H}$ to $q^*\mathcal{L}(\pi)$. By definition, we set $\mathcal{H}(\pi)=s^*\tilde{\mathcal{H}}(\pi)$. For the construction of the factorization isomorphisms, see Part V.

14.12. Let $\mathfrak{g}$ be the simple Lie algebra connected with our Cartan datum. Assume that $\zeta = \exp(\frac{\pi\sqrt{-1}}{d\kappa})$ for some positive integer κ, cf. 10.6 (d is defined in 2.1).

We have $12\rho_\ell \cdot \rho_\ell \equiv 12\rho \cdot \rho \mod l$, and $\mathrm{rk}(X) \equiv \dim \mathfrak{g} \mod 2$. By the strange formula of Freudenthal-de Vries, we have $12\rho \cdot \rho = dh \dim \mathfrak{g}$ where h is the dual Coxeter number of our Cartan datum. It follows that the monodromy of $\mathcal{H}$ around the zero section of the determinant line bundle is equal to $\exp(\pi\sqrt{-1}\frac{(\kappa-h)\dim \mathfrak{g}}{\kappa})$. This number coincides with the multiplicative central charge of the conformal field theory associated with the affine Lie algebra $\hat{\mathfrak{g}}$ at level κ (see [BFM] 4.4.1, 6.1.1, 2.1.3, [TUY] 1.2.2), cf 16.2 below.

UNIVERSAL HEISENBERG SYSTEMS

14.13. Let us define a category *Sew* as follows (cf. [BFM] 4.3.2). Its object A is a finite set $\overline{A}$ together with a collection $N_A = \{n\}$ of non-intersecting two-element subsets $n \subset \overline{A}$. Given such an object, we set $A^1 = \bigcup_{N_A} n$, $A^0 = \overline{A} - A^1$. A morphism $f : A \longrightarrow B$ is an embedding $i_f : \overline{B} \hookrightarrow \overline{A}$ and a collection N_f of non-intersecting two-element subsets of $\overline{B} - \overline{A}$ such that $N_A = N_B \coprod N_f$. The composition of morphisms is obvious. (*Sew* coincides with the category $Sets^\sharp/\emptyset$, in the notations of [BFM] 4.3.2.)

For $A \in Sew$, let us call an *A-curve* a data $(C, \{x_a, \tau_a\}_{A^0})$ where C is a smooth proper (possibly disconnected) complex curve, $\{x_a, \tau_a\}_{A^0}$ is an A^0-tuple of distinct points $x_a \in C$ together with non-zero tangent vectors τ_a at them. For such a curve, let $\overline{C}_A$ denote the curve obtained from C by clutching pairwise the points $x_{a'}$ with $x_{a''}$ ($n = \{a', a''\}$) for all sets $n \in N_A$. Thus, the set N_A is in the bijection with the set of nodes of the curve $\overline{C}_A$ ($\overline{C}_A = C$ if $N_A = \emptyset$).

Let us call an *enhanced graph* a pair $\Gamma = (\overline{\Gamma}, \mathbf{g})$ here $\overline{\Gamma}$ is a non-oriented graph and $\mathbf{g} = \{g_v\}_{v \in \mathrm{Vert}(\overline{\Gamma})}$ is a $\mathbb{N}$- valued 0-chain of $\overline{\Gamma}$. Here $\mathrm{Vert}(\overline{\Gamma})$ denotes the set of vertices of $\overline{\Gamma}$. Let us assign to a curve $\overline{C}_A$ an enhanced graph $\Gamma(\overline{C}_A) = (\overline{\Gamma}(\overline{C}_A), \mathbf{g}(\overline{C}_A))$. By definition, $\overline{\Gamma}(\overline{C}_A)$ is a graph with $\mathrm{Vert}(\overline{\Gamma}(\overline{C}_A)) = \pi_0(C) = \{$ the set of irreducible components of $\overline{C}_A\}$ and the set of edges $\mathrm{Edge}(\overline{\Gamma}(\overline{C}_A)) = N_A$, an edge $n = \{a', a''\}$ connecting the vertices corresponding to the components of the points $x_{a'}, x_{a''}$. For $v \in \pi_0(C)$, $g(\overline{C}_A)_v$ is equal to the genus of the corresponding component $C_v \subset C$.

14.14. Let $\mathcal{M}_A$ denote the moduli stack of A-curves $(C, \ldots)$ such that the curve $\overline{C}_A$ is stable in the sense of [DM] (in particular connected). The stack $\mathcal{M}_A$ is smooth; we have $\mathcal{M}_A = \coprod_{g \geq 0} \mathcal{M}_{A,g}$ where $\mathcal{M}_{A,g}$ is a substack of A-curves C with $\overline{C}_A$ having genus (i.e. $\dim H^1(\overline{C}_A, \mathcal{O}_{\overline{C}_A})$) equal to g. In turn, we have the decomposition into connected components

$$\mathcal{M}_{A,g} = \coprod_{\Gamma, \vec{A^0}} \mathcal{M}_{\vec{A^0},g,\Gamma}$$

where $\mathcal{M}_{\vec{A^0},g,\Gamma}$ is the stack of A-curves $(C, \{x_a, \tau_a\})$ as above, with $\Gamma(\overline{C}_A) = \Gamma$, $\vec{A^0} = \{A^0_v\}_{v \in \mathrm{Vert}(\overline{\Gamma})}$, $A^0 = \coprod A^0_v$, such that x_a lives on the connected component C_v for $a \in A^0_v$.

We denote by $\eta : C_{A,g} \longrightarrow \mathcal{M}_{A,g}$ (resp. $\bar{\eta} : \overline{C}_{A,g} \longrightarrow \mathcal{M}_{A,g}$) the universal smooth curve (resp. stable surve). For $\nu \in Y^+$, we have the corresponding *relative* configuration spaces $C^\nu_{A,g}, C^{\nu\circ}_{A,g}, \overline{C}^\nu_{A,g}$. For brevity we omit the relativeness subscript $/\mathcal{M}_{A,g}$

from these notations. The notation $C^{\nu}_{\vec{A}^0,g,\Gamma}$ etc., will mean the restriction of these configuration spaces to the component $\mathcal{M}_{\vec{A}^0,g,\Gamma}$.

Let $\mathcal{M}_g$ be the moduli stack of smooth connected curves of genus g, and $\overline{\mathcal{M}}_g$ be its Grothendieck-Deligne-Mumford-Knudsen compactification, i.e. the moduli stack of stable curves of genus g. Let $\bar{\eta}:\overline{C}_g\longrightarrow\overline{\mathcal{M}}_g$ be the universal stable curve; let $\overline{\delta}_g=\det R\bar{\eta}_*(\omega_{\overline{C}_g/\overline{\mathcal{M}}_g})$ be the determinant line bundle; let $\overline{\mathcal{M}}_{g;\delta}\longrightarrow\overline{\mathcal{M}}_g$ be its total space with the zero section removed.

We have obvious maps $\mathcal{M}_{A,g}\longrightarrow\overline{\mathcal{M}}_g$. Let the complementary subscript $(\cdot)_\delta$ denote the base change of all the above objects under $\overline{\mathcal{M}}_{g;\delta}\longrightarrow\overline{\mathcal{M}}_g$.

14.15. Let us consider the configuration space $C^{\nu\circ}_{A,g}$; it is the moduli stack of ν distinct points running on A-curves $(C,\{x_a,\tau_a\})$ and not equal to the marked points x_a. This stack decomposes into connected components as follows:

$$C^{\nu\circ}_{A,g}=\coprod_{\Gamma,\vec{A}^0,\vec{\nu}} C^{\vec{\nu}}_{\vec{A}^0,g,\Gamma}$$

where $\vec{\nu}=\{\nu_v\}_{v\in\mathrm{Vert}(\Gamma)}$ and $C^{\vec{\nu}}_{\vec{A}^0,g,\Gamma}$ being the moduli stack of objects as above, with $\Gamma(\overline{C}_A)=\Gamma$ and ν_v points running on the component C_v. The decomposition is taken over appropriate graphs Γ, decompositions $A^0=\coprod A^0_v$ and the tuples $\vec{\nu}$ with $\sum\nu_v=\nu$.

Let us call an A_0-tuple $\vec{\mu}=\{\mu_a\}\in X_\ell^{A^0}$ (g,ν)-*good* if

(a) $\sum_{a\in A^0}\mu_a-\nu\equiv(2-2g)\rho_\ell \mod Y_\ell$.

Given such a tuple, we are going to define certain local system $\mathcal{H}^{\nu}_{\vec{\mu};A,g}$ over $C^{\nu\circ}_{A,g;\delta}$. Let us describe its restriction $\mathcal{H}^{\vec{\nu}}_{\vec{\mu};\vec{A}^0,g,\Gamma}$ to a connected component $C^{\vec{\nu}}_{\vec{A}^0,g,\Gamma;\delta}$.

Let Γ' be the first subdivision of $\overline{\Gamma}$. We have $\mathrm{Vert}(\Gamma')=\mathrm{Vert}(\overline{\Gamma})\coprod\mathrm{Edge}(\overline{\Gamma})=\pi_0(C)\coprod N_A$. The edges of Γ' are indexed by the pairs (n,a) where $n\in N_A, a\in n$, the corresponding edge $e_{n,a}$ having the ends a and n. Let us define an orientation of Γ' by the requierement that a is the beginning of $e_{n,a}$. Consider the chain complex

$$C_1(\Gamma';X_\ell/Y_\ell)\xrightarrow{d}C_0(\Gamma';X_\ell/Y_\ell).$$

Let us define a 0-chain $c=c^{\vec{\nu}}_{\vec{\mu}}\in C_0(\Gamma';X_\ell/Y_\ell)$ by

$$c(v)=\sum_{a\in A^0_v}\mu_a+(2g_v-2)\rho_\ell-\nu_v\ (v\in\pi_0(C));\ c(n)=2\rho_\ell\ (n\in N_A).$$

The goodness assumption (a) ensures that c is a boundary. By definition,

$$\mathcal{H}^{\vec{\nu}}_{\vec{\mu};\vec{A}^0,g,\Gamma}=\oplus_{\chi:\ d\chi=c}\mathcal{H}_\chi$$

Note that the set $\{\chi|\ d\chi=c\}$ is a torsor over the group $H_1(\Gamma';X_\ell/Y_\ell)=H_1(\Gamma;X_\ell/Y_\ell)$.

The local system $\mathcal{H}_\chi$ is defined below, in 14.18, after a little notational preparation.

14.16. Given two finite sets J,K, let $C(J;K)_g$ denote the moduli stack of objects $(C,\{x_j\},\{y_k,\tau_k\})$. Here C is a smooth proper connected curve of genus g, $\{x_j\}$ is a J-tuple of distinct points $x_j\in C$ and $\{y_k,\tau_k\}$ is a K-tuple of distinct points $y_k\in C$ together with non-zero tangent vectors $\tau_k\in T_{y_k}C$. We suppose that $y_k\neq x_j$ for all k,j.

We set $C(J)_g:=C(\emptyset;J)_g$. We have the forgetful maps $C(J\coprod K)_g\longrightarrow C(J;K)_g$.

The construction of 14.5 — 14.11 defines the Heisenberg system $\mathcal{H}(\pi)$ over the smooth stack $C(J)_{g;\delta}$ for each $\pi : J \longrightarrow X_\ell$.

Given $\nu \in Y^+$, choose an unfolding of ν, $\pi : J \longrightarrow I$, and set $C^\nu(K)^\circ_g := C(J;K)/\Sigma_\pi$. Given a K-tuple $\vec{\mu} = \{\mu_k\} \in X_\ell^K$, define a map $\tilde{\pi} : J \coprod K \longrightarrow X_\ell$ by $\tilde{\pi}(j) = -\pi(j) \in -I \subset X_\ell$ $(j \in J)$, $\tilde{\pi}(k) = \mu_k$ $(k \in K)$. The local system $\mathcal{H}(\tilde{\pi})$ over $C(J \coprod K)_{g;\delta}$ descends to $C(J;K)_{g;\delta}$ since $\zeta^{2n(-i)} = 1$, and then to $C^\nu(K)^\circ_{g;\delta}$, by Σ_π-equivariance; denote the latter local system by $\tilde{\mathcal{H}}^\nu_{\vec{\mu}}$, and set $\mathcal{H}^\nu_{\vec{\mu}} = \tilde{\mathcal{H}}^\nu_{\vec{\mu}} \otimes Sign^\nu$, cf. 8.3.

14.17. **Lemma.** *If $\vec{\mu} \equiv \vec{\mu}'$ mod Y_ℓ^K then we have canonical isomorphisms $\mathcal{H}^\nu_{\vec{\mu}} = \mathcal{H}^\nu_{\vec{\mu}'}$.* $\square$

Therefore, it makes sense to speak about $\mathcal{H}^\nu_{\vec{\mu}}$ for $\vec{\mu} \in (X_\ell/Y_\ell)^K$.

14.18. Let us return to the situation at the end of 14.15. We have $\Gamma = (\overline{\Gamma}, \{g_v\}_{v\in \mathrm{Vert}(\overline{\Gamma})})$. Recall that $A^0 = \coprod A^0_v$; we have also $A^1 = \coprod A^1_v$ where $A^1_v := \{a \in A^1 | \ x_a \in C_v\}$ $(v \in \mathrm{Vert}(\overline{\Gamma})$. Set $\overline{A}_v = A^0_v \coprod A^1_v$, so that $\overline{A} = \coprod \overline{A}_v$. We have an obvious map

(a) $\mathcal{M}_{\vec{A}^0,g,\Gamma} \longrightarrow \prod_{\mathrm{Vert}(\overline{\Gamma})} C(\overline{A}_v)_{g_v}$,

and a map

(b) $C^{\vec{\nu}}_{\vec{A}^0,g,\Gamma} \longrightarrow \prod_{\mathrm{Vert}(\overline{\Gamma})} C^{\nu_v}(\overline{A}_v)^\circ_{g_v}$

over (a). For each v, define an $\overline{A}_v$-tuple $\vec{\mu}(\chi; v)$ equal to μ_a at $a \in A^0_v$ and $\chi(e_{a,n})$ at $a \in n \subset A^1$. By definition, the local system $\mathcal{H}_\chi$ over $C^{\vec{\nu}}_{\vec{A}^0,g,\Gamma;\delta}$ is the inverse image of the product

$$\boxtimes_{\mathrm{Vert}(\overline{\Gamma})} \mathcal{H}^{\nu_v}_{\vec{\mu}(\chi;v)}$$

under the map (b) (pulled back to the determinant bundle).

This completes the definition of the local systems $\mathcal{H}^\nu_{\vec{\mu};A,g}$. They have a remarkable compatibility property (when the object A varies) which we are going to describe below, see theorem 14.23.

14.19. Let $\tilde{\mathcal{T}}^\nu_{A,g}$ denote the fundamental groupoid $\pi(C^{\nu\circ}_{A,g;\delta})$. We are going to show that these groupoids form a cofibered category over $\mathcal{S}ew$.

14.20. A morphism $f : A \longrightarrow B$ in $\mathcal{S}ew$ is called a *sewing* (resp. *deleting*) if $A^0 = B^0$ (resp. $N_f = \emptyset$). A sewing f with card$(N_f) = 1$ is called *simple*. Each morphism is a composition of a sewing and a deleting; each sewing is a composition of simple ones.

(a) Let $f : A \longrightarrow B$ be a simple sewing. We have canonical morphisms

$$\wp_f : \mathcal{M}_{A,g;\delta} \longrightarrow \dot{T}_{\partial\overline{\mathcal{M}}_{B,g;\delta}} \overline{\mathcal{M}}_{B,g;\delta}$$

and

$$\wp_f^{(\nu\circ)} : C^{\nu\circ}_{A,g;\delta} \longrightarrow \dot{T}_{\partial\overline{C^{\nu\circ}_{B,g;\delta}}} \overline{C^{\nu\circ}_{B,g;\delta}},$$

over $\wp_f$, cf. [BFM] 4.3.1. Here $\overline{\mathcal{M}}_{B,g}$ (resp. $\overline{C^{\nu\circ}_{B,g}}$) denotes the Grothendieck-Deligne-Mumford-Knudsen compactification of $\mathcal{M}_{B,g}$ (resp. of $C^{\nu\circ}_{B,g}$) and $\partial\overline{\mathcal{M}}_{B,g}$ (resp. $\partial\overline{C^{\nu\circ}_{B,g}}$) denotes the smooth locus of the boundary $\overline{\mathcal{M}}_{B,g} - \mathcal{M}_{B,g}$ (resp. $\overline{C^{\nu\circ}_{B,g}} - \overline{C}^\nu_{B,g}$). The subscript $_\delta$ indicates the base change to the determinant bundle, as before.

Composing the specialization with the inverse image under $\wp_f^{(\nu o)}$, we get the canonical map $f_* : \tilde{T}^\nu_{A,g} \longrightarrow \tilde{T}^\nu_{B,g}$.

(b) Let $f : A \longrightarrow B$ be a deleting. It induces the obvious morphisms (denoted by the same letter)

$$f : \mathcal{M}_{A,g;\delta} \longrightarrow \mathcal{M}_{B,g;\delta}$$

and

$$f : C^{\nu o}_{A,g;\delta} \longrightarrow C^{\nu o}_{B,g;\delta}.$$

The last map induces $f_* : \tilde{T}^\nu_{A,g} \longrightarrow \tilde{T}^\nu_{B,g}$.

Combining the constructions (a) and (b) above, we get a category $\tilde{T}^\nu_g$ cofibered in groupoids over $\mathcal{S}ew$, with fibers $\tilde{T}^\nu_{A,g}$.

14.21. Let $\mathcal{R}ep^\nu_{c;A,g}$ be the category of finite dimensional representations of $\tilde{T}^\nu_{A,g}$ (over k) with the monodromy $c \in \mathsf{k}^*$ around the zero section of the determinant bundle. The previous construction shows that these categories form a fibered category $\mathcal{R}ep^\nu_{c;g}$ over $\mathcal{S}ew$.

14.22. For $A \in \mathcal{S}ew$, let us call an A^0-tuple $\vec{\mu} = \{\mu_a\} \in X_\ell^{A^0}$ *good* if $\sum_{A^0} \mu_a \in Y$. If $f : B \longrightarrow A$ is a morphism, define a B^0-tuple $f^*\vec{\mu} = \{\mu'_b\}$ by $\mu'_b = \mu_{i_f^{-1}(b)}$ if $b \in i_f(A^0)$, and 0 otherwise. Obviously, $\sum_{B^0} \mu'_b = \sum_{A^0} \mu_a$.

Given a good $\vec{\mu}$, let us pick an element $\nu \in Y^+$ such that $\nu \equiv \sum \mu_a + (2g-2)\rho_\ell \mod Y_\ell$. We can consider the local system $\mathcal{H}^\nu_{\vec{\mu};A,g}$ as an object of $\mathcal{R}ep^\nu_{c;A,g}$ where $c = (-1)^{\mathrm{card}(I)}\zeta^{-12\rho\cdot\rho}$.

14.23. **Theorem.** *For any morphism $f : B \longrightarrow A$ in $\mathcal{S}ew$ and a g-good $\vec{\mu} \in X_\ell^{A^0}$, we have the canonical isomorphism*

$$f^*\mathcal{H}^\nu_{\vec{\mu};A,g} = \mathcal{H}^\nu_{f^*\vec{\mu};B,g}.$$

In other words, the local systems $\mathcal{H}^\nu_{\vec{\mu};B,g}$ define a **cartesian section** *of the fibered category $\mathcal{R}ep^\nu_{c;g}$ over $\mathcal{S}ew/A$. Here $c = (-1)^{\mathrm{card}(I)}\zeta^{-12\rho\cdot\rho}$.* □

15. Fusion structures on $\mathcal{FS}$

15.1. Below we will construct a family of "fusion structures" on the category $\mathcal{FS}$ (and hence, due to the equivalence Φ, on the category $\mathcal{C}$) indexed by $m \in \mathbb{Z}$. We should explain what a fusion structure is. This is done in 15.8 below. We will use a modification of the formalism from [BFM] 4.5.4.

15.2. Recall that we have a regular object $\mathcal{R} \in \mathrm{Ind}(\mathcal{FS}^{\otimes 2})$, cf. Section 13. We have the canonical isomorphism $t(\mathcal{R}) = \mathcal{R}$ where $t : \mathrm{Ind}(\mathcal{FS}^{\otimes 2}) \xrightarrow{\sim} \mathrm{Ind}(\mathcal{FS}^{\otimes 2})$ is the permutation, hence an object $\mathcal{R}_n \in \mathrm{Ind}(\mathcal{FS}^{\otimes 2})$ is well defined for any two-element set n.

For an object $A \in \mathcal{S}ew$, we set $\tilde{A} = A^0 \coprod N_A$. Let us call an *A-collection* of factorizable sheaves an $\tilde{A}$-tuple $\{\mathcal{X}_{\tilde{a}}\}_{\tilde{a}\in\tilde{A}}$ where $\mathcal{X}_{\tilde{a}} \in \mathcal{FS}_{c_{\tilde{a}}}$ if $\tilde{a} \in A^0$ and $\mathcal{X}_{\tilde{a}} = \mathcal{R}_{\tilde{a}}$ if $\tilde{a} \in N_A$.

We impose the condition that $\sum_{a\in A^0} c_a = 0 \in X_\ell/Y$. We will denote such an object $\{\mathcal{X}_a; \mathcal{R}_n\}_A$. It defines an object

$$\otimes_A \{\mathcal{X}_a; \mathcal{R}_n\} := (\otimes_{a\in A^0} \mathcal{X}_a) \otimes (\otimes_{n\in N_A} \mathcal{R}_n) \in \mathrm{Ind}(\mathcal{FS}^{\otimes \bar{A}}).$$

If $f : B \longrightarrow A$ is a morphism in $\mathcal{S}ew$, we define a B-collection $f^*\{\mathcal{X}_a; \mathcal{R}_n\}_B = \{\mathcal{Y}_{\tilde{b}}\}_{\tilde{B}}$ by $\mathcal{Y}_{\tilde{b}} = \mathcal{X}_{i_f^{-1}(\tilde{b})}$ for $\tilde{b} \in i_f(A^0)$, $\mathbf{1}$ if $\tilde{b} \in B^0 - i_f(A^0)$ and $\mathcal{R}_{\tilde{b}}$ if $\tilde{b} \in N_B = N_A \cup N_f$.

15.3. Given an A-collection $\{\mathcal{X}_a, \mathcal{R}_n\}_A$ with $\lambda(\mathcal{X}_a) = \mu_a$ such that

(a) $\nu := \sum_{A^0} \mu_a + (2g-2)\rho_\ell \in Y^+$,

one constructs (following the pattern of 8.5) a perverse sheaf $\boxed{\times}^{(\nu)}_{A,g} \{\mathcal{X}_a; \mathcal{R}_n\}$ over $\overline{C}^\nu_{A,g;\delta}$. It is obtained by planting factorizable sheaves $\mathcal{X}_a$ into the universal sections x_a of the stable curve $\overline{C}_{A,g}$, the regular sheaves $\mathcal{R}_n$ into the nodes n of this curve and pasting them together into one sheaf by the Heisenberg system $\mathcal{H}^\nu_{\vec{\mu};A,g}$.

15.4. Given $A \in \mathcal{S}ew$ and an A-collection $\{\mathcal{X}_a; \mathcal{R}_n\}_A$, choose elements $\mu_a \geq \lambda(\mathcal{X}_a)$ in X_ℓ such that (a) above holds (note that $2\rho_\ell \in Y$). Below ν will denote the element as in (a) above.

Let $\mathcal{X}'_a$ denote the factorizable sheaf isomorphic to $\mathcal{X}_a$ obtained from it by the change of the highest weight from $\lambda(\mathcal{X}_a)$ to μ_a. For each $m \in \mathbb{Z}$, define a local system $\langle \otimes_A \{\mathcal{X}_a; \mathcal{R}_n\}\rangle^{(m)}_g$ over $\mathcal{M}_{A,g;\delta}$ as follows.

Let $\langle \otimes_A \{\mathcal{X}_a; \mathcal{R}_n\}\rangle^{(m)}_{\vec{A}^0,g,\Gamma}$ denote the restriction of the local system to be defined to the connected component $\mathcal{M}_{\vec{A}^0,g,\Gamma;\delta}$. By definition,

$$\langle \otimes_A \{\mathcal{X}_a; \mathcal{R}_n\}\rangle^{(m)}_{\vec{A}^0,g,\Gamma} := R^{m-\dim \mathcal{M}_{\vec{A}^0,g,\Gamma;\delta}} \bar{\eta}^\nu_* (\boxed{\times}^{(\nu)}_{\vec{A}^0,g,\Gamma} \{\mathcal{X}'_a; \mathcal{R}_n\}).$$

Here $\boxed{\times}^{(\nu)}_{\vec{A}^0,g,\Gamma} \{\mathcal{X}'_a; \mathcal{R}_n\}$ denotes the perverse sheaf $\boxed{\times}^{(\nu)}_{A,g} \{\mathcal{X}'_a; \mathcal{R}_n\}$ restricted to the subspace $\overline{C}^\nu_{\vec{A}^0,g,\Gamma;\delta}$. This definition does not depend on the choice of the elements μ_a.

15.5. Given a morphism $f : B \longrightarrow A$, acting as in 14.19, 14.21, a perverse sheaf $f^*(\boxed{\times}^{(\nu)}_{A,g} \{\mathcal{X}_a; \mathcal{R}_n\})$ over $\overline{C}^\nu_{B,g;\delta}$ and local systems $f^*\langle \otimes_A \{\mathcal{X}_a; \mathcal{R}_n\}\rangle^{(m)}_g$ over $\mathcal{M}_{B,g;\delta}$.

15.6. **Theorem.** *In the above notations, we have canonical isomorphisms*

$$f^*(\boxed{\times}^{(\nu)}_{A,g} \{\mathcal{X}_a; \mathcal{R}_n\}) = \boxed{\times}^{(\nu)}_{B,g} f^*\{\mathcal{X}_a; \mathcal{R}_n\}.$$

This is a consequence of Theorem 14.23 above and the definition of the regular sheaf $\mathcal{R}$ as a sheaf of nearby cycles of the braiding local system, 13.3.

15.7. **Corollary.** *We have canonical isomorphisms of local systems*

$$f^*\langle \otimes_A \{\mathcal{X}_a; \mathcal{R}_n\}\rangle^{(m)}_g = \langle \otimes_B f^*\{\mathcal{X}_a; \mathcal{R}_n\}\rangle^{(m)}_g \ (m \in \mathbb{Z}). \ \square$$

15.8. The previous corollary may be expressed as follows. The various A-collections of factorizable sheaves (resp. categories $\widetilde{\mathcal{R}ep}_{c;A}$ of finite dimensional representations of "Teichmüller groupoids" $\widetilde{\mathcal{T}eich}_A = \pi(\mathcal{M}_{A;\delta})$ having monodromy c around the zero section of the determinant bundle) define a fibered category $\mathcal{FS}^\sharp$ (resp. $\widetilde{\mathcal{R}ep}_c$) over $\mathcal{S}ew$.

For any $m \in \mathbb{Z}$, the collection of local systems $\langle \otimes_A \{\mathcal{X}_a; \mathcal{R}_n\} \rangle_g^{(m)}$ and the canonical isomorphisms of the previous theorem define a **cartesian functor**

$$\langle\ \rangle^{(m)} : \mathcal{FS}^\sharp \longrightarrow \widetilde{\mathcal{R}ep}_c$$

where $c = (-1)^{\mathrm{card}(I)} \zeta^{-12\rho\cdot\rho}$. We call such a functor a *fusion structure of multiplicative central charge* c on the category $\mathcal{FS}$. The category $\mathcal{FS}$ with this fusion structure will be denoted by $\mathcal{FS}^{(m)}$.

The difference from the definition of a fusion structure given in [BFM] 4.5.4 is that our fibered categories live over $\mathcal{S}ew = \mathcal{S}ets^\sharp/\emptyset$ and not over $\mathcal{S}ets^\sharp$, as in *op. cit.*

15.9. **Example.** For $A \in \mathcal{S}ew$, let us consider an A-curve $P = (\mathbb{P}^1, \{x_a, \tau_a\})$; it defines a geometric point Q of the stack $\mathcal{M}_{A,g}$ where $g = \mathrm{card}(N_A)$ and hence a geometric point $P = (Q, 1)$ of the stack $\mathcal{M}_{A,g;\delta}$ since the determinant bundle is canonically trivialized at Q.

15.9.1. **Theorem.** *For an A-collection $\{\mathcal{X}_a; \mathcal{R}_n\}$, the stalk of the local system $\langle \otimes_A \mathcal{X}_a \rangle_g^{(m)}$ at a point P is isomorphic to*

$$\mathrm{Tor}^{\mathcal{C}}_{\frac{\infty}{2}-m}(\mathbf{1}_r, (\otimes_{A^0} \Phi(\mathcal{X}_a)) \otimes \mathbf{ad}^{\otimes g}).$$

To prove this, one should apply theorem 11.2 and the following remark. Degenerating all nodes into cusps, one can include the nodal curve $\overline{P}_A$ into a one-parameter family whose special fiber is $\mathbb{P}^1$, with $\mathrm{card}(A^0) + g$ marked points. The nearby cycles of the sheaf $\boxed{\times}_A \{\mathcal{X}_a; \mathcal{R}_n\}$ will be the sheaf obtained by the gluing of the sheaves $\mathcal{X}_a$ and g copies of the sheaf $\Phi^{-1}(\mathbf{ad})$ into these marked points.

15.10. Note that an arbitrary A-curve may be degenerated into a curve considered in the previous example. Due to 15.7, this determines the stalks of all our local systems (up to a non canonical isomorphism).

16. Conformal blocks (higher genus)

In this section we assume that $\mathsf{k} = \mathbb{C}$.

16.1. Let us make the assumptions of 10.6, 14.12. Consider the full subcategory $\mathcal{MS} \subset \mathcal{C}$. Let us define the *regular object* $\mathbf{R}_{\mathcal{MS}}$ by

$$\mathbf{R}_{\mathcal{MS}} = \oplus_{\mu \in \Delta_\ell} (\mathfrak{L}(\mu) \otimes \mathfrak{L}(\mu)^*) \in \mathcal{MS}^{\otimes 2}.$$

As in the previous section, we have a notion of A-collection $\{L_a; \mathbf{R}_{\mathcal{MS}n}\}_A$ ($A \in \mathcal{S}ew$, $\{L_a\} \in \mathcal{MS}^{A^0}$). The classical fusion structure on $\mathcal{MS}$, [TUY], defines for each A-collection as above, a local system

$$\langle \otimes_A \{L_a; \mathbf{R}_{\mathcal{MS}n}\} \rangle_{\mathcal{MS}}$$

on the moduli stack $\mathcal{M}_{A;\delta}$.

We have $A = (\overline{A}, N_A)$; let us define another object $A' = (\overline{A} \cup \{*\}, N_A)$. We have an obvious deleting $f_A : A' \longrightarrow A$. Given an A-collection as above, define an A'-collection in $\{L_a; L(2\rho_\ell); \mathbf{R}_n\}_{A'}$ in the category $\mathcal{C}$. Using the equivalence Φ, we transfer to $\mathcal{C}$ the fusion structures defined in the previous section on $\mathcal{FS}$; we denote them $\langle\ \rangle_{\mathcal{C}}^{(m)}$.

The following theorem generalizes Theorem 11.5 to higher genus.

16.2. **Theorem.** *For each $A \in \mathcal{S}ew$, the local system $f_A^*\langle \otimes_A \{L_a; \mathbf{R}_{\mathcal{MS}n}\}\rangle_{\mathcal{MS}}$ on $\mathcal{M}_{A';\delta}$ is a canonical subquotient of the local system $\langle \otimes_{A'} \{L_a; L(2\rho_\ell); \mathbf{R}_n\}\rangle_{\mathcal{C}}^{(0)}$.* □

16.3. Let us consider the special case $A = (A^0, \emptyset)$; for an A-collection $\{L_a\}_{A^0}$ in $\mathcal{MS}$, we have the classical local systems of conformal blocks $\langle \otimes_{A^0} \{L_a\}\rangle_{\mathcal{MS}}$ on $\mathcal{M}_{A;\delta}$.

16.4. **Corollary.** *The local systems $\langle \otimes_{A^0} \{L_a\}\rangle_{\mathcal{MS}}$ are semisimple. They carry a canonical non-degenerate Hermitian form.*

In fact, the local system $f_A^*\langle \otimes_{A^0} \{L_a\}\rangle_{\mathcal{MS}}$ is semisimple by the previous theorem and by Beilinson-Bernstein-Deligne-Gabber, [BBD] 6.2.5. The map of fundamental groupoids $f_{A*} : \widetilde{\mathcal{T}eich_{A'}} \longrightarrow \widetilde{\mathcal{T}eich_A}$ is surjective; therefore the initial local system is semisimple.

The Hermitian form is defined in the same manner as in genus zero, cf. 11.7.

Part I. INTERSECTION COHOMOLOGY OF REAL ARRANGEMENTS

1. Introduction

1.1. Let $\mathbb{A}$ be a complex affine space, $\mathcal{H}$ a finite collection of hyperplanes in $\mathbb{A}$. We suppose that all $H \in \mathcal{H}$ are given by real equations, and denote by $H_{\mathbb{R}} \subset H$ the subspace of real points. An arrangement $\mathcal{H}$ induces naturally a stratification of $\mathbb{A}$ denoted by $\mathcal{S}_{\mathcal{H}}$ (for precise definitions see section 2 below). The main goal of this part is the study of the category $\mathcal{M}(\mathbb{A}; \mathcal{S}_{\mathcal{H}})$ of perverse sheaves (of vector spaces over a fixed ground field) over $\mathbb{A}$ which are smooth along $\mathcal{S}_{\mathcal{H}}$.

In *section 1* we collect topological notations and known facts we will need in the sequel.

Section 2. To each object $\mathcal{M} \in \mathcal{M}(\mathbb{A}; \mathcal{S}_{\mathcal{H}})$ we assign "a linear algebra data". Namely, for each *facet* of $\mathcal{H}$ — i.e. a connected component of an intersection of some $H_{\mathbb{R}}$'s — we define a vector space $\Phi_F(\mathcal{M})$; if F is contained and has codimension 1 in the closure of another facet E, we define two linear operators between $\Phi_F(\mathcal{M})$ and $\Phi_E(\mathcal{M})$ acting in opposite directions. These data contain all information about $\mathcal{M}$ we need.

In fact, it is natural to expect that the sheaf $\mathcal{M}$ may be *uniquely reconstructed* from the linear algebra data above (one can check this in the simplest cases).

The spaces $\Phi_F(\mathcal{M})$ are defined using certain relative cohomology. They are similar to a construction by R. MacPherson (see [B2]). The spaces $\Phi_F(\mathcal{M})$ are analogous to functors of vanishing cycles.

The main technical properties of functors $\mathcal{M} \mapsto \Phi_F(\mathcal{M})$ are (a) *commutation with Verdier duality*, and (b) *exactness.* In fact, (b) follows from (a) without difficulty; the proof of (a) is the principal geometrical result of this part (see Theorem 3.5). Actually, we prove a more general statement concerning all complexes smooth along $\mathcal{S}_{\mathcal{H}}$. This is a result of primary importance for us.

Using these linear algebra data, we define an exact functor from $\mathcal{M}(\mathbb{A}; \mathcal{S}_{\mathcal{H}})$ to complexes of finite dimensional vector spaces computing cohomology $R\Gamma(\mathbb{A}; \bullet)$, see Theorem 3.14. This is the main result we will need below. The idea of constructing such functorial complexes on categories of perverse sheaves (quite analogous to the usual computation of singular cohomology using cell decompositions) is due to A. Beilinson, [B1]. It was an important source of inspiration for us.

A similar problem was considered in [C].

In *section 3* we present explicit computations for standard extensions of an arbitrary one-dimensional local system over an open stratum. The main result is the computation of *intersection cohomology*, see Theorem 4.17. In our computations we use some simple geometric ideas from M. Salvetti's work, [Sa]. However, we do not use a more difficult main theorem of this paper. On the contrary, maybe our considerations shed some new light on it.

1.2. The idea of this work appeared several years ago, in some discussions with Hélène Esnault and Eckart Viehweg. We use this occasion to express to them our sincere gratitude. We are also very grateful to Paul Bressler for a stimulating discussion. We thank Kari Vilonen for pointing out an error in an earlier version of the manuscript.

2. Topological preliminaries

In this section we introduce our notations and recall some basic facts from topology. Main references are [KS], [BBD].

2.1. Throughout this work, all our topological spaces will be *locally compact*, in particular *Hausdorff*. $\{pt\}$ will denote a one-point space. For a topological space X, $a_X : X \longrightarrow \{pt\}$ will be the unique map. If $Y \subset X$ is a subspace, $\bar{Y}$ will denote the closure of Y.

Throughout this chapter we fix an arbitrary ground field k. *A vector space* will mean a vector space over k. For a finite dimensional vector space V, V^* will denote a dual space. $\mathcal{V}ect$ will denote a category of vector spaces.

A sheaf will mean a sheaf of vector spaces. For a topological space X, $\mathcal{S}h(X)$ will denote a category of sheaves over X, $\mathcal{D}^*(X)$ will denote the derived category of $\mathcal{S}h(X)$, $* = +, -, b$ or $\emptyset$ will have the usual meaning.

For $p \in \mathbb{Z}$, $\mathcal{D}^{\leq p}(X) \subset \mathcal{D}(X)$ will denote a full subcategory of complexes $\mathcal{K}$ with $H^i(\mathcal{K}) = 0$ for all $i > p$.

If $\mathcal{A}, \mathcal{B}$ are abelian categories, we will say that a left exact functor $F : \mathcal{D}^*(\mathcal{A}) \longrightarrow \mathcal{D}(\mathcal{B})$ ($* = +, -, b$ or $\emptyset$) has *cohomological dimension* $\leq r$, and write $\mathrm{cd}(F) \leq r$, if $H^i(F(A)) = 0$ for any $A \in \mathcal{A}$ and $i > r$. (Left exactness here means that $H^i(F(A)) = 0$ for $i < 0$).

2.2. Let $f : X \longrightarrow Y$ be a continuous map. We will use the following notations for standard functors.

$f^* : \mathcal{D}(Y) \longrightarrow \mathcal{D}(X)$ — the inverse image; $f_* : \mathcal{D}^+(X) \longrightarrow \mathcal{D}^+(Y)$ — (the right derived of) the direct image; $f_! : \mathcal{D}^+(X) \longrightarrow \mathcal{D}^+(Y)$ — (the right derived of) the direct image with proper supports; $f^! : \mathcal{D}^+(Y) \longrightarrow \mathcal{D}^+(X)$ — the right adjoint to $f_!$ (defined when $f_!$ has finite cohomological dimension), see [KS], 3.1.

We will denote the corresponding functors on sheaves as follows: $f^* : \mathcal{S}h(Y) \longrightarrow \mathcal{S}h(X)$; $R^0f_* : \mathcal{S}h(X) \longrightarrow \mathcal{S}h(Y)$; $R^0f_! : \mathcal{S}h(X) \longrightarrow \mathcal{S}h(Y)$.

We will denote $R\Gamma(X;\cdot) := p_{X*}$; $R\Gamma_c(X;\cdot) := p_{X!}$. For $\mathcal{K} \in \mathcal{D}^+(X)$, $i \in \mathbb{Z}$, we set $H^i(X;\mathcal{K}) := H^i(R\Gamma(X;\mathcal{K}))$; $H^i_c(X;\mathcal{K}) := H^i(R\Gamma_c(X;\mathcal{K}))$.

For $V \in \mathcal{V}ect = \mathcal{S}h(\{pt\})$ we denote by V_X the constant sheaf p_X^*V.

If Y is a subspace of X, we will use a notation $i_{Y,X}$, or simply i_Y if X is understood, for the embedding $Y \hookrightarrow X$. If Y is open in X, we will also write j_Y instead of i_Y.

For $\mathcal{K} \in \mathcal{D}(X)$, we will use notations $\mathcal{K}|_Y := i^*_{Y,X}\mathcal{K}$; $R\Gamma(Y;\mathcal{K}) := R\Gamma(Y;\mathcal{K}|_Y)$, $H^i(Y;\mathcal{K}) := H^i(Y;\mathcal{K}_Y)$, etc.

2.3. We have functors

$$R\mathcal{H}om : \mathcal{D}^-(X)^{opp} \times \mathcal{D}^+(X) \longrightarrow \mathcal{D}^+(X), \tag{1}$$

$$\otimes : \mathcal{D}^-(X) \times \mathcal{D}^*(X) \longrightarrow \mathcal{D}^*(X) \tag{2}$$

where $* =, -, b$ or $\emptyset$;

$$R\mathrm{Hom} = R\Gamma \circ R\mathcal{H}om : \mathcal{D}^-(X)^{opp} \times \mathcal{D}^+(X) \longrightarrow \mathcal{D}^+(\mathcal{V}ect) \tag{3}$$

For $\mathcal{K}, \mathcal{L} \in \mathcal{D}^b(X)$ we have

$$\mathrm{Hom}_{\mathcal{D}(X)}(\mathcal{K}, \mathcal{L}) = H^0(R\mathrm{Hom}(\mathcal{K}, \mathcal{L})) \tag{4}$$

We denote $\mathcal{K}^* := R\mathcal{H}om(\mathcal{K}, \mathsf{k}_X)$.

2.4. Let X be a topological space, $j : U \longrightarrow X$ an embedding of an open subspace, $i : Y \longrightarrow X$ an embedding of the complement. In this case $i_* = i_!$ and $j^! = j^*$. $R^0 i_*$ and $R^0 j_!$ are exact, so we omit R^0 from their notation. $j_!$ is the functor of extension by zero. $i^!$ is the right derived of the functor $R^0 i^! : \mathcal{S}h(X) \longrightarrow \mathcal{S}h(Y)$ of the subsheaf of sections with supports in Y (in notations of [KS], $R^0 i^!(F) = \Gamma_Y(F)$).

Let $\mathcal{K} \in \mathcal{D}^+(X)$. We will use the following notations for relative cohomology: $R\Gamma(X, Y; \mathcal{K}) := R\Gamma(X; j_!\mathcal{K}|_U)$; $H^i(X, Y; \mathcal{K}) := H^i(R\Gamma(X, Y; \mathcal{K}))$.

If $Z \hookrightarrow Y$ is a closed subspace, we have a canonical exact triangle

$$i_{X-Y!}\mathcal{K}|_{X-Y} \longrightarrow i_{X-Z!}\mathcal{K}|_{X-Z} \longrightarrow i_{Y-Z!}\mathcal{K}|_{Y-Z} \longrightarrow i_{X-Y!}\mathcal{K}|_{X-Y}[1] \tag{5}$$

(of course, $i_{X-Y} = j$) inducing a long cohomology sequence

$$\ldots \longrightarrow H^i(X, Y; \mathcal{K}) \longrightarrow H^i(X, Z; \mathcal{K}) \longrightarrow H^i(Y, Z; \mathcal{K}) \xrightarrow{\partial} H^{i+1}(X, Y; \mathcal{K}) \longrightarrow \ldots \tag{6}$$

2.5. Let X be a topological space such that the functor $R\Gamma_c(X; \cdot)$ has a finite cohomological dimension. We define *a dualizing complex*:

$$\omega_X := a_X^! \mathsf{k} \in \mathcal{D}^b(X) \tag{7}$$

For $\mathcal{K} \in \mathcal{D}^b(X)$ we set $D_X\mathcal{K} = R\mathcal{H}om(\mathcal{K}, \omega_X) \in \mathcal{D}^b(X)$. If there is no risk of confusion we denote $D\mathcal{K}$ instead of $D_X\mathcal{K}$. We get a functor

$$D : \mathcal{D}^b(X)^{opp} \longrightarrow \mathcal{D}^b(X) \tag{8}$$

which comes together with a natural transformation

$$\mathrm{Id}_{\mathcal{D}^b(X)} \longrightarrow D \circ D \tag{9}$$

2.6. **Orientations.** (Cf. [KS], 3.3.) Let X be an n-dimensional C^0-manifold. We define an *orientation sheaf* $\mathcal{O}r_X$ as the sheaf associated to a presheaf $U \mapsto \mathrm{Hom}_{\mathbb{Z}}(H^n_c(U; \mathbb{Z}), \mathbb{Z})$. It is a locally constant sheaf of abelian groups of rank 1. Its tensor square is constant. By definition, *orientation of X* is an isomorphism $\mathcal{O}r_X \xrightarrow{\sim} \mathbb{Z}_X$.

We have a canonical isomorphism

$$\omega_X \xrightarrow{\sim} \mathcal{O}r_X \otimes_{\mathbb{Z}} \mathsf{k}[n] \tag{10}$$

2.7. **Homology.** Sometimes it is quite convenient to use homological notations. Let $Y \subset X$ be a closed subspace of a topological space X (Y may be empty), $\mathcal{K} \in \mathcal{D}^b(X)$. We define homology groups as

$$H_i(X,Y;\mathcal{F}) := H^i(X,Y;\mathcal{F}^*)^*.$$

These groups behave covariantly with respect to continuous mappings.

Let σ be *a relative singular n-cell*, i.e. a continuous mapping

$$\sigma : (D^n, \partial D^n) \longrightarrow (X, Y)$$

where, D^n denotes a standard closed unit ball in $\mathbb{R}^n$. We supply D^n with the standard orientation. Let $\mathring{D}^n$ denote the interior of D^n.

Suppose that $\sigma^*\mathcal{K}|_{\mathring{D}^n}$ is constant. Then by Poincaré duality we have isomorphisms

$$H_n(D^n, \partial D^n; \sigma^*\mathcal{K}) = H_c^n(\mathring{D}^n; \sigma^*\mathcal{K}^*)^* \xrightarrow{\sim} H^0(\mathring{D}^n; \sigma^*\mathcal{K}).$$

(recall that $\mathring{D}^n$ is oriented). Thus, given an element $s \in H^0(\mathring{D}^n; \mathcal{K})$, we can define a homology class

$$cl(\sigma; s) := \sigma_*(s) \in H_n(X, Y; \mathcal{K}).$$

We will call the couple $(\sigma; s)$ *a relative singular n-cell for* $\mathcal{K}$.

These classes enjoy the following properties.

2.7.1. *Homotopy.* Let us call to cells $(\sigma_0; s_0)$ and (σ_1, s_1) *homotopic* if there exists a map

$$\sigma : (D^n \times I, \partial D^n \times I) \longrightarrow (X, Y)$$

(where I denotes a unit interval) such that $\sigma^*\mathcal{K}|_{\mathring{D}^n \times I}$ is constant, and a section

$$s \in H^0(\mathring{D}^n \times I; \sigma^*\mathcal{K})$$

such that (σ, s) restricted to $D^n \times \{i\}$ is equal to $(\sigma_i; s_i)$, $i = 0, 1$.

We have

(H) *if* $(\sigma_0; s_0)$ *is homotopic to* $(\sigma_1; s_1)$ *then* $cl(\sigma_0; s_0) = cl(\sigma_1; s_1)$.

2.7.2. *Additivity.* Suppose D^n is represnted as a union of its upper and lower half-balls $D^n = D_1^n \cup D_2^n$ where $D_1^n = \{(x_1, \dots, x_n) \in D^n |\ x_n \geq 0\}$ and $D_2^n = \{(x_1, \dots, x_n) \in D^n |\ x_n \leq 0\}$. Let us supply D_i^n with the induced orientation.

Suppose we are given a relative n-cell $(\sigma; s)$ such that $\sigma(D_1^n \cap D_2^n) \subset Y$. Then its restriction to D_i^n gives us two relative n-cells $(\sigma_i; s_i)$, $i = 1, 2$. We have

(A) $cl(\sigma; s) = cl(\sigma_1; s_1) + cl(\sigma_2; s_2)$.

2.8. We will call *a local system* a locally constant sheaf of finite rank.

Let X be a topological space, $i : Y \hookrightarrow X$ a subspace. We will say that $\mathcal{K} \in \mathcal{D}(X)$ is *smooth* along Y if all cohomology sheaves $H^p(i^*\mathcal{K})$, $p \in \mathbb{Z}$, are local systems.

We will call a *stratification* of X a partition $X = \bigcup_{S \in \mathcal{S}} S$ into a finite disjoint of locally closed subspaces — *strata* — such that the closure of each stratum is the union of strata.

We say that $\mathcal{K} \in \mathcal{D}(X)$ is *smooth along* $\mathcal{S}$ if it is smooth along each stratum. We will denote by $\mathcal{D}^*(X;\mathcal{S})$ the full subcategory of $\mathcal{D}^*(X)$ ($* = +, -, b$ or $\emptyset$) consisting of complexes smooth along $\mathcal{S}$.

2.9. Let us call a stratification $\mathcal{S}$ of a topological space X *good* if the following conditions from [BBD], 2.1.13 hold.

(b) All strata are equidimensional topological varieties. If a stratum S lies in the closure of a stratum T, $\dim S < \dim T$.

(c) If $i : S \hookrightarrow X$ is a stratum, the functor $i_* : \mathcal{D}^+(S) \longrightarrow \mathcal{D}^+(X)$ has a finite cohomological dimension. If $\mathcal{F} \in \mathcal{S}h(S)$ is a local system, $i_*(\mathcal{F})$ is smooth along $\mathcal{S}$.

Let $\mathcal{S}$ be a good stratification such that all strata have even dimension. We will denote by $\mathcal{M}(X;\mathcal{S}) \subset \mathcal{D}^b(X;\mathcal{S})$ the category of smooth along $\mathcal{S}$ perverse sheaves corresponding to the middle perversity $p(S) = -\dim\ S/2$, cf. *loc. cit.*

2.10. Let X be a complex algebraic variety. Let us call its stratification $\mathcal{S}$ *algebraic* if all strata are algebraic subvarieties. Following [BBD], 2.2.1, let us call a sheaf $\mathcal{F} \in \mathcal{S}h(X)$ *constructible* if it is smooth along some algebraic stratification. (According to Verdier, every algebraic stratification admits a good refinement.) We denote by $\mathcal{D}^b_c(X)$ a full subcategory of $\mathcal{D}^b(X)$ consisting of complexes with constructible cohomology.

We denote by $\mathcal{M}(X) \subset \mathcal{D}^b_c(X)$ the category of perverse sheaves corresponding to the middle perversity. It is a filtered union of categories $\mathcal{M}(X;\mathcal{S})$ over all good algebraic stratifications $\mathcal{S}$, cf. *loc.cit.*

2.11. Let X be a complex manifold, $f : X \longrightarrow \mathbb{C}$ a holomorphic map. Set $Y = f^{-1}(0)$, $U = X - Y$, let $i : Y \hookrightarrow X$, $j : U \hookrightarrow X$ denote the embeddings.

We define a functor of *nearby cycles*

$$\Psi_f : \mathcal{D}^b(U) \longrightarrow \mathcal{D}^b(Y) \tag{11}$$

as $\Psi_f(\mathcal{K}) = \psi_f j_*(\mathcal{K})[-1]$ where ψ_f is defined in [KS], 8.6.1.

We define a functor of *vanishing cycles*

$$\Phi_f : \mathcal{D}^b(X) \longrightarrow \mathcal{D}^b(Y) \tag{12}$$

as ϕ_f from *loc. cit.*

2.12. **Lemma.** *Let X be topological space, $Y \hookrightarrow X$ a closed subspace, $\mathcal{F} \in \mathcal{S}h(X)$. Then natural maps*

$$\lim_{\rightarrow} H^i(U;\mathcal{F}) \longrightarrow H^i(Y;\mathcal{F}),\ i \in \mathbb{Z},$$

where U ranges through all the open neighbourhoods of U, are isomorphisms.

Proof. See [KS], 2.5.1, 2.6.9. □

2.13. **Conic sheaves.** (Cf. [KS], 3.7.) Let $\mathbb{R}^{*+}$ denote the multiplicative group of positive real numbers. Let X be a topological space endowed with an $\mathbb{R}^{*+}$-action.

Following *loc.cit.*, we will call a sheaf $\mathcal{F}$ over X *conic* (with respect to the given $\mathbb{R}^{*+}$-action) if its restriction to every $\mathbb{R}^{*+}$-orbit is constant. We will call a complex $\mathcal{K} \in \mathcal{D}(X)$ *conic* if all its cohomology sheaves are conic.

We will denote by $Sh_{\mathbb{R}^{*+}}(X) \subset Sh(X)$, $\mathcal{D}^{*}_{\mathbb{R}^{*+}}(X) \subset \mathcal{D}^{*}(X)$, $* = b, +, -$ or $\emptyset$, the full subcategories of conic objects.

2.13.1. **Lemma.** *Let $U \hookrightarrow X$ be an open subset. Suppose that for every $\mathbb{R}^{*+}$-orbit $O \subset X$, $O \cap U$ is contractible (hence, non-empty). Then for every conic $\mathcal{K} \in \mathcal{D}^{+}(X)$ the restriction morphism*

$$R\Gamma(X; \mathcal{K}) \longrightarrow R\Gamma(U; \mathcal{K})$$

is an isomorphism.

Proof. See *loc. cit.*, 3.7.3. □

3. Vanishing cycles functors

3.1. **Arrangements.** Below we use some terminology from [Sa]. Let $\mathbb{A}_{\mathbb{R}}$ be a real affine space $\mathbb{A}_{\mathbb{R}}$ of dimension N, and $\mathcal{H}_{\mathbb{R}} = \{H_{\mathbb{R}}\}$ a finite set of distinct real affine hyperplanes in $\mathbb{A}_{\mathbb{R}}$. Such a set is called a *real arrangement.* We pick once and for all a square root $i = \sqrt{-1} \in \mathbb{C}$.

Let $\mathbb{A} = \mathbb{A}_{\mathbb{R}} \otimes_{\mathbb{R}} \mathbb{C}$ denote the complexification of $\mathbb{A}_{\mathbb{R}}$, and $\mathcal{H} = \{H\}$ where $H := H_{\mathbb{R}} \otimes_{\mathbb{R}} \mathbb{C}$. (A finite set of complex hyperplanes in a complex affine space will be called a *complex arrangement*).

We will say that $\mathcal{H}$ is *central* if $\bigcap_{H \in \mathcal{H}} H$ consists of one point.

For a subset $\mathcal{K} \subset \mathcal{H}$ denote

$$H_{\mathcal{K}} = \bigcap_{H \in \mathcal{K}} H; \ {}_{\mathcal{K}}H = \bigcup_{H \in \mathcal{K}} H,$$

and

$$H_{\mathbb{R},\mathcal{K}} = \bigcap_{H \in \mathcal{K}} H_{\mathbb{R}}, \ {}_{\mathcal{K}}H_{\mathbb{R}} = \bigcup_{H \in \mathcal{K}} H_{\mathbb{R}}.$$

The nonempty subspaces $H_{\mathcal{K}}$ and $H_{\mathcal{K},\mathbb{R}}$ are called complex and real *edges* respectively. Set

$$\overset{\circ}{H}_{\mathcal{K}} = H_{\mathcal{K}} - \cup L,$$

the union over all the complex edges $L \subset H_{\mathcal{K}}$, $L \neq H_{\mathcal{K}}$. We set $\overset{\circ}{H}_{\mathbb{R},\mathcal{K}} = H_{\mathbb{R},\mathcal{K}} \cap \mathbb{A}_{\mathbb{R}}$. Connected components of $\overset{\circ}{H}_{\mathbb{R},\mathcal{K}}$ are called *facets* of $\mathcal{H}_{\mathbb{R}}$. Facets of codimension 0 (resp., 1) are called *chambers* (resp., *faces*).

Let us denote by $\mathcal{S}_{\mathcal{H}}$ a stratification of $\mathbb{A}$ whose strata are all non-empty $\overset{\circ}{H}_{\mathcal{K}}$. We will denote by $\overset{\circ}{\mathbb{A}}_{\mathcal{H}}$ a unique open stratum

$$\overset{\circ}{\mathbb{A}}_{\mathcal{H}} = \overset{\circ}{H}_{\emptyset} = \mathbb{A} - {}_{\mathcal{H}}H.$$

In this Section we will study categories of sheaves $\mathcal{D}(\mathbb{A}; \mathcal{S}_{\mathcal{H}})$ and $\mathcal{M}(\mathbb{A}; \mathcal{S}_{\mathcal{H}})$.

We will denote by $\mathcal{S}_{\mathcal{H}_{\mathbb{R}}}$ a stratification of $\mathbb{A}_{\mathbb{R}}$ whose strata are all facets. We set

$$\overset{\circ}{\mathbb{A}}_{\mathcal{H},\mathbb{R}} = \mathbb{A}_{\mathbb{R}} - {}_{\mathcal{H}}H_{\mathbb{R}}.$$

It is a union of all chambers.

3.2. **Dual cells.** (cf. [Sa]). Let us fix a point ${}^F w$ on each facet F. We will call this set of points $\mathbf{w} = \{{}^F w\}$ *marking* of our arrangement.

For two facets F, E let us write $F < E$ if $F \subset \overline{E}$ and dim $F <$ dim E. We will say that E is *adjacent* to F. We will denote by $\mathrm{Ch}(F)$ the set of all chambers adjacent to F.

Let us call a *flag* a sequence of $q - p + 1$ facets $\mathbf{F} = (F_p < F_{p+1} < \ldots < F_q)$ with dim $F_i = i$. We say that F_p is *the beginning* and F_q *the end* of $\mathbf{F}$.

Let us denote by ${}^{\mathbf{F}}\Delta$ a closed $(q-p)$-symplex with vertices ${}^{F_p}w, \ldots, {}^{F_q}w$. Evidently, ${}^{\mathbf{F}}\Delta \subset \overline{F}_q$.

Suppose we are given two facets $F_p < F_q$, $\dim F_i = i$. We will denote

$$D_{F_p < F_q} = \bigcup {}^{\mathbf{F}}\Delta,$$

the sum over all flags beginning at F_p and ending at F_q. This is a $(q-p)$-dimensional cell contained in $\overline{F}_q$.

For a facet F let us denote

$$D_F = \bigcup_{C \in \mathrm{Ch}(F)} D_{F<C}$$

We set

$$S_F := \bigcup_{E,C: F<E<C;} D_{E<C},$$

the union over all facets E and chambers C. The space S_F is contained in D_F (in fact, in the defintion of S_F it is enough to take the union over all facets E such that dim $E =$ dim $F + 1$). If $q =$ codim F, then D_F is homeomorphic to a q-dimensional disc and S_F — to a $(q-1)$- dimensional sphere, cf. [Sa], Lemma 6. We denote $\overset{\circ}{D}_F := D_F - S_F$. We will call D_F *a dual cell corresponding to* F.

We set $\overset{\circ}{D}_{F<C} := D_{F<C} \cap \overset{\circ}{D}_F$.

3.3. **Generalized vanishing cycles.** Let $\mathcal{K} \in \mathcal{D}^b(\mathbb{A}; \mathcal{S}_{\mathcal{H}})$, F a p-dimensional facet. Let us introduce a complex

$$\Phi_F(\mathcal{K}) := R\Gamma(D_F, S_F; \mathcal{K})[-p] \in \mathcal{D}^b(pt) \tag{13}$$

This complex will be called *a complex of generalized vanishing cycles of* $\mathcal{K}$ *at a facet* F.

Formally, the definition of functor Φ_F depends upon the choice of a marking $\mathbf{w}$. However, functors defined using two different markings are canonically isomorphic. This is evident. Because of this, we omit markings from the notations.

3.4. **Transversal slices.** Let F be a facet of dimension p which is a connected component of a real edge $M_{\mathbb{R}}$ with the complexification M. Let us choose a real affine subspace $L_{\mathbb{R}}$ of codimension p transversal to F and passing through ${}^F w$. Let L be its complexification.

Let us consider a small disk $L_\epsilon \subset L$ with the centrum at ${}^F w = L \cap M$. We identify L_ϵ with an affine space by dilatation. Our arrangement induces a central arrangement $\mathcal{H}_L$ in L_ϵ. Given $\mathcal{K} \in \mathcal{D}^b(\mathbb{A}; \mathcal{S}_{\mathcal{H}})$, define $\mathcal{K}_L := i^*_{L_\epsilon}\mathcal{K}[-p]$.

3.4.1. **Lemma.** *We have a natural isomorphism*

$$D(\mathcal{K}_L) \xrightarrow{\sim} (D\mathcal{K})_L \tag{14}$$

Proof. Consider an embedding of smooth complex manifolds $i_L : L \longrightarrow \mathbb{A}$. Let us consider the following complexes: $\omega_{L/\mathbb{A}} := i_L^! \mathsf{k}_\mathbb{A}$ (cf. [KS], 3.1.16 (i)) and $\mathcal{O}r_{L/\mathbb{A}} := \mathcal{O}r_L \otimes_\mathbb{Z} \mathcal{O}r_\mathbb{A}$. We have canonical isomorphism

$$\omega_{L/\mathbb{A}} \xrightarrow{\sim} \mathcal{O}r_{L/\mathbb{A}}[-2p] \otimes \mathsf{k}.$$

The chosen orientation of $\mathbb{C}$ enables us to identify $\mathcal{O}r_\mathbb{A}$ and $\mathcal{O}r_L$, and hence $\mathcal{O}r_{L/\mathbb{A}}$ with constant sheaves; consequently, we get an isomorphism

$$\omega_{L/\mathbb{A}} \xrightarrow{\sim} \mathsf{k}[-2p].$$

The canonical map

$$i_L^! \mathcal{K} \longrightarrow i^* \mathcal{K} \otimes \omega_{L/\mathbb{A}} \tag{15}$$

(cf. [KS], (3.1.6)) is an isomorphism since singularities of $\mathcal{K}$ are transversal to L (at least in the neighbourhood of ${}^F w$). Consequently we get an isomorphism $i_L^! \mathcal{K} \xrightarrow{\sim} i^* \mathcal{K}_L[-2p]$.

Now we can compute:

$$D(\mathcal{K}_L) = D(i_L^* \mathcal{K}[-p]) \xrightarrow{\sim} D(i_L^* \mathcal{K})[2p] \xrightarrow{\sim} i_L^! D\mathcal{K}[p] \xrightarrow{\sim} i_L^* D\mathcal{K}[-p] = (D\mathcal{K})_L,$$

QED. □

3.4.2. **Lemma.** *We have a natural isomorphism*

$$\Phi_F(\mathcal{K}) \cong \Phi_{\{{}^F w\}}(\mathcal{K}_L) \tag{16}$$

This follows directly from the definition of functors Φ.

This remark is often useful for reducing the study of functors Φ_F to the case of a central arrangement.

3.4.3. **Lemma.** If $\mathcal{M} \in \mathcal{M}(\mathbb{A}; \mathcal{S}_\mathcal{H})$ then $\mathcal{M}_L \in \mathcal{M}(L_\epsilon; \mathcal{S}_{\mathcal{H}_{L_\epsilon}})$.

Proof follows from transversality of L_ϵ to singularities of $\mathcal{M}$. □

3.5. **Duality. Theorem.** *Functor Φ_F commutes with Verdier duality. More precisely, for every $\mathcal{K} \in \mathcal{D}^b(\mathbb{A}, \mathcal{S}_\mathcal{H})$ there exists a natural isomorphism*

$$\Phi_F(D\mathcal{K}) \xrightarrow{\sim} D\Phi_F(\mathcal{K}) \tag{17}$$

This is the basic property of our functors.

3.6. **1-dimensional case.** To start with the proof, let us treat first the simplest case. Consider an arrangement consisting of one point — origin — in a one-dimensional space $\mathbb{A}$. It has one 0-face F and two 1-faces $E_\pm$ — real rays $\mathbb{R}_{>0}$ and $\mathbb{R}_{<0}$. A marking consists of two points $w_\pm \in E_\pm$ and F. We set $w_\pm = \pm 1$ (we pick a coordinate on $\mathbb{A}$).

For a positive r denote $\mathbb{A}_{\leq r} := \{z \in \mathbb{A} |\ |z| \leq r\}$, $S_r := \partial \mathbb{A}_{\leq r}$; $\mathbb{A}_{<r}$, $\mathbb{A}_{\geq r}$, etc. have an evident meaning. We also set $\mathbb{A}_{(r',r'')} := \mathbb{A}_{>r'} \cap \mathbb{A}_{<r''}$. A subscript $\mathbb{R}$ will denote an intersection of these subsets with $\mathbb{A}_\mathbb{R}$. Evidently, $D_F = \mathbb{A}_{\leq 1,\mathbb{R}}$, $S_F = S_{1,\mathbb{R}}$. Define $D_F^{opp} := \mathbb{A}_{\geq 1,\mathbb{R}}$, $Y := i \cdot D_F^{opp}$.

One sees easily (cf. *infra*, Lemma 3.7) that one has isomorphisms

$$\Phi_F(\mathcal{K}) \cong R\Gamma(\mathbb{A}, S_F; \mathcal{K}) \xrightarrow{\sim} R\Gamma(\mathbb{A}, D_F^{opp}; \mathcal{K}). \tag{18}$$

Let us choose real numbers ϵ, r', r'' such that $0 < \epsilon < r' < 1 < r''$. Set $Y := \epsilon i \cdot D_F^{opp}$. Denote $j := j_{\mathbb{A}-S_F}$. We have natural isomorphisms

$$D\Phi_F(\mathcal{K}) \cong DR\Gamma(\mathbb{A}, S_F; \mathcal{K}) \cong R\Gamma_c(\mathbb{A}; j_*j^*D\mathcal{K})$$

(Poincaré duality)

$$\cong R\Gamma(\mathbb{A}, \mathbb{A}_{\geq r''}; j_*j^*D\mathcal{K}) \cong R\Gamma(\mathbb{A}, Y \cup \mathbb{A}_{\geq r''}; j_*j^*D\mathcal{K})$$

(homotopy). Consider the restriction map

$$res : R\Gamma(\mathbb{A}, Y \cup \mathbb{A}_{\geq r''}; j_*j^*D\mathcal{K}) \longrightarrow R\Gamma(\mathbb{A}_{\leq r'}, Y \cap \mathbb{A}_{\leq r'}; D\mathcal{K}) \tag{19}$$

We claim that res is an isomorphism. In fact, Cone(res) is isomorphic to

$$R\Gamma(\mathbb{A}, \mathbb{A}_{\leq r'} \cup \mathbb{A}_{\geq r''} \cup Y; j_*j^*D\mathcal{K}) = R\Gamma_c(\mathbb{A}_{<r''}, \mathbb{A}_{\leq r'} \cup Y; j_*j^*D\mathcal{K}) \cong$$
$$\cong DR\Gamma(\mathbb{A}_{<r''} - (\mathbb{A}_{\leq r'} \cup Y); j_!j^*\mathcal{K})$$

We have by definition

$$R\Gamma(\mathbb{A}_{<r''} - (\mathbb{A}_{\leq r'} \cup Y); j_!j^*\mathcal{K}) = R\Gamma(\mathbb{A}_{(r',r'')} - Y, S_F; \mathcal{K})$$

On the other hand, evidently $\mathcal{K}$ is smooth over $\mathbb{A}_{(r',r'')}$, and we have an evident retraction of $\mathbb{A}_{(r',r'')}$ on S_F (see Fig. 1). Therefore, $R\Gamma(\mathbb{A}_{(r',r'')} - Y, S_F; \mathcal{K}) = 0$ which proves the claim.

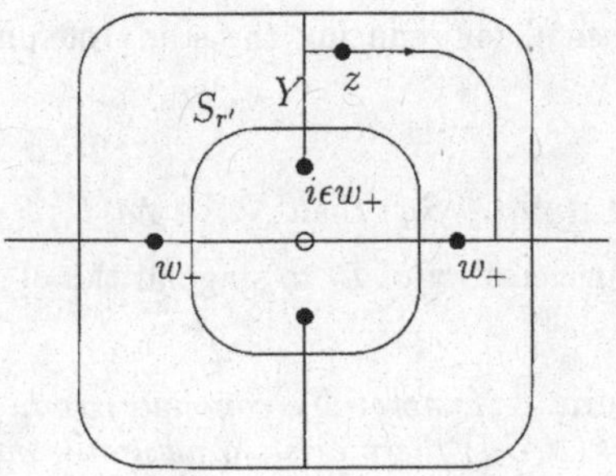

Fig. 1

A clockwise rotation by $\pi/2$ induces an isomorphism

$$R\Gamma(\mathbb{A}_{\leq r'}, Y \cap \mathbb{A}_{\leq r'}; D\mathcal{K}) \cong R\Gamma(\mathbb{A}_{\leq r'}, \epsilon \cdot D_F^{opp}; D\mathcal{K}),$$

and the last complex is isomorphic to $\Phi_F(D\mathcal{K})$ by dilatation and (18). This proves the theorem for Φ_F. The statement for functors $\Phi_{E_\pm}$ is evident.

Let us return to the case of an arbitrary arrangement.

3.7. **Lemma.** *Suppose that $\mathcal{H}$ is central. Let F be the unique* 0- *dimensional facet. The evident restriction maps induce canonical isomorphisms*

$$\Phi_F(\mathcal{K}) \overset{(1)}{\cong} R\Gamma(\mathbb{A}_{\mathbb{R}}, S_F; \mathcal{K}) \overset{(2)}{\cong} R\Gamma(\mathbb{A}, S_F; \mathcal{K}) \overset{(3)}{\cong} R\Gamma(\mathbb{A}, D_F^{opp}; \mathcal{K}) \overset{(4)}{\cong} R\Gamma(\mathbb{A}_{\mathbb{R}}, D_F^{opp}; \mathcal{K})$$

where $D_F^{opp} = \mathbb{A}_{\mathbb{R}} - \overset{\circ}{D}_F$.

Proof. Let us fix a coordinate system with the origin at F, and hence a metric on $\mathbb{A}$. For $\epsilon > 0$ let $U_\epsilon \subset \mathbb{A}_{\mathbb{R}}$ denote the set of points $x \in \mathbb{A}_{\mathbb{R}}$ having distance $< \epsilon$ from D_F.

It follows from 2.12 that

$$\Phi_F(\mathcal{K}) = R\Gamma(D_F, S_F; \mathcal{K}) \cong \lim_{\rightarrow \epsilon} R\Gamma(U_\epsilon, S_F; \mathcal{K}).$$

On the other hand, from 2.13.1 it follows that restriction maps

$$R\Gamma(\mathbb{A}_{\mathbb{R}}, S_F; \mathcal{K}) \longrightarrow R\Gamma(U_\epsilon, S_F; \mathcal{K})$$

are isomorphisms. This establishes an isomorphism (1). To prove (3), one remarks that its cone is acyclic.

The other isomorphisms are proven by the similar arguments. We leave the proof to the reader. □

3.8. **Proof of 3.5.** First let us suppose that $\mathcal{H}$ is central and F is its 0-dimensional facet. We fix a coordinate system in $\mathbb{A}$ as in the proof of the lemma above. We have a decomposition $\mathbb{A} = \mathbb{A}_{\mathbb{R}} \oplus i \cdot \mathbb{A}_{\mathbb{R}}$. We will denote by $\Re, \Im : \mathbb{A} \longrightarrow \mathbb{A}_{\mathbb{R}}$ the evident projections.

Let $\mathbb{A}_{\leq r}$, etc. have the meaning similar to the one-dimensional case above. Let us denote $j := j_{\mathbb{A}-S_F}$. We proceed as in one-dimensional case.

Let us choose positive numbers r', r'', ϵ, such that

$$\epsilon D_F \subset \mathbb{A}_{<r'} \subset \overset{\circ}{D}_F \subset D_F \subset \mathbb{A}_{<r''}$$

Let us introduce a subspace

$$Y = \epsilon i \cdot D_F^{opp}$$

We have isomorphisms

$$D\Phi_F(\mathcal{K}) \cong DR\Gamma(\mathbb{A}, S_F; \mathcal{K})$$

(by Lemma 3.7)

$$\cong R\Gamma_c(\mathbb{A}; j_*j^*D\mathcal{K}) \cong R\Gamma(\mathbb{A}, \mathbb{A}_{\geq r''}; j_*j^*D\mathcal{K}) \cong R\Gamma(\mathbb{A}, Y \cup \mathbb{A}_{\geq r''}; j_*j^*D\mathcal{K})$$

(homotopy). Consider the restriction map

$$res : R\Gamma(\mathbb{A}, Y \cup \mathbb{A}_{\geq r''}; j_*j^*D\mathcal{K}) \longrightarrow R\Gamma(\mathbb{A}_{\leq r'}, Y \cap \mathbb{A}_{\leq r'}; D\mathcal{K}) \qquad (20)$$

Cone(res) is isomorphic to

$$R\Gamma(\mathbb{A}, \mathbb{A}_{\leq r'} \cup \mathbb{A}_{\geq r''} \cup Y; j_*j^*D\mathcal{K}) = R\Gamma_c(\mathbb{A}_{<r''}, \mathbb{A}_{\leq r'} \cup Y; j_*j^*D\mathcal{K}) \cong$$
$$\cong DR\Gamma(\mathbb{A}_{<r''} - (\mathbb{A}_{\leq r'} \cup Y); j_!j^*\mathcal{K})$$

We have by definition

$$R\Gamma(\mathbb{A}_{<r''} - (\mathbb{A}_{\leq r'} \cup Y); j_!j^*\mathcal{K}) = R\Gamma(\mathbb{A}_{(r',r'')} - Y, S_F; \mathcal{K})$$

3.8.1. **Lemma.** *Set* $\mathbb{A}'_{\mathbb{R}} := \mathbb{A}_{\mathbb{R}} - \{0\}$. *We have* $R\Gamma(\mathbb{A} - i \cdot \mathbb{A}_{\mathbb{R}}, \mathbb{A}'_{\mathbb{R}}; \mathcal{K}) = 0$.

Proof. We have to prove that the restriction map

$$R\Gamma(\mathbb{A} - i \cdot \mathbb{A}_{\mathbb{R}}; \mathcal{K}) \longrightarrow R\Gamma(\mathbb{A}'_{\mathbb{R}}; \mathcal{K}) \tag{21}$$

is an isomorphism. Consider projection to the real part

$$\pi : \mathbb{A} - i \cdot \mathbb{A}_{\mathbb{R}} \longrightarrow \mathbb{A}'_{\mathbb{R}}.$$

Evidently,

$$R\Gamma(\mathbb{A} - i \cdot \mathbb{A}_{\mathbb{R}}; \mathcal{K}) = R\Gamma(\mathbb{A}'_{\mathbb{R}}, \pi_*\mathcal{K}).$$

On the other hand, since $\mathcal{K}$ is conic along fibers of π, we have

$$\pi_*\mathcal{K} \xrightarrow{\sim} e^*\mathcal{K}$$

where $e : \mathbb{A}'_{\mathbb{R}} \longrightarrow \mathbb{A} - i \cdot \mathbb{A}_{\mathbb{R}}$ denotes the embedding (cf. 2.13). This implies our lemma. □

It follows easily that Cone(*res*) is acyclic, therefore the map *res* is an isomorphism. In other words, we have constructed an isomorphism

$$D\Phi_F(\mathcal{K}) \xrightarrow{\sim} R\Gamma(\mathbb{A}_{\leq r'}, Y \cap \mathbb{A}_{\leq r'}; D\mathcal{K}).$$

A clockwise rotation by $\pi/2$ induces an isomorphism

$$R\Gamma(\mathbb{A}_{\leq r'}, Y \cap \mathbb{A}_{\leq r'}; D\mathcal{K}) \cong R\Gamma(\mathbb{A}_{\leq r'}, \epsilon \cdot D_F^{opp}; D\mathcal{K}),$$

and the last complex is isomorphic to $\Phi_F(D\mathcal{K})$ by dilatation and 3.7 (3).

This proves the theorem for Φ_F. Note that we have constructed an explicit isomorphism.

3.8.2. If F is an arbitrary facet, we consider a transversal slice L as in 3.4. By the results of *loc.cit.*, and the above proven case, we have natural isomorphisms

$$D\Phi_F(\mathcal{K}) \xrightarrow{\sim} D\Phi_{\{F_w\}}(\mathcal{K}_L) \xrightarrow{\sim} \Phi_{\{F_w\}}(D\mathcal{K}_L) \xrightarrow{\sim} \Phi_{\{F_w\}}((D\mathcal{K})_L) \xrightarrow{\sim} \Phi_F(D\mathcal{K}).$$

This proves the theorem. □

3.9. **Theorem.** *For every* $\mathcal{M} \in \mathcal{M}(\mathbb{A}; \mathcal{S}_{\mathcal{H}})$ *and every facet* F *we have* $H^i(\Phi_F(\mathcal{M})) = 0$ *for all* $i \neq 0$.

In other words, functors Φ_F are t-exact with respect to the middle perversity.

Proof. First let us suppose that $\mathcal{H}$ is central and F is its 0-dimensional facet. Let us prove that Φ_F is right exact, that is, $H^i(\Phi_F(\mathcal{M})) = 0$ for $i > 0$ and every $\mathcal{M}$ as above.

In fact, we know that

3.9.1. *if S is any stratum of $\mathcal{S}_{\mathcal{H}}$ of dimension p then $\mathcal{M}|_S \in \mathcal{D}^{\leq -p}(S)$*

by the condition of perversity. In particular, $R\Gamma(\mathbb{A};\mathcal{M}) = i_F^*\mathcal{M} \in \mathcal{D}^{\leq 0}(\{pt\})$.

On the other hand, one deduces from 3.9.1 that $R\Gamma(S_F;\mathcal{M}) \in \mathcal{D}^{\leq -1}(\{pt\})$. In fact, by definition S_F is a union of certain simplices Δ, all of whose edges lie in strata of positive dimension. This implies that $R\Gamma(\Delta;\mathcal{M}|_\Delta) \in \mathcal{D}^{\leq -1}(\{pt\})$, and one concludes by Mayer-Vietoris argument, using similar estimates for intersections of simplices.

Consequently we have

$$\Phi_F(\mathcal{M}) \cong R\Gamma(\mathbb{A}, S_F;\mathcal{M}) \in \mathcal{D}^{\leq 0}(\{pt\}),$$

as was claimed.

On the other hand by Duality theorem 3.5 we have an opposite inequality, which proves that Φ_F is exact in our case.

The case of an arbitrary facet is reduced immediately to the central one by noting that an operation of the restriction to a transversal slice composed with a shift by its codimension is t-exact and using (16). The theorem is proved. □

3.10. By the above theorem, the restriction of functors Φ_F to the abelian subcategory $\mathcal{M}(\mathbb{A},\mathcal{S}_{\mathcal{H}})$ lands in subcategory $\mathcal{V}ect \subset \mathcal{D}(\{pt\})$.

In other words, we get exact functors

$$\Phi_F : \mathcal{M}(\mathbb{A},\mathcal{S}_{\mathcal{H}}) \longrightarrow \mathcal{V}ect \tag{22}$$

These functors commute with Verdier duality.

We will also use the notation $\mathcal{M}_F$ for $\Phi_F(\mathcal{M})$.

3.11. **Canonical and variation maps.** Suppose we have a facet E. Let us denote by $\mathcal{F}ac^1(E)$ the set of all facets F such that $E < F$, dim $F =$ dim $E + 1$. We have

$$S_E = \bigcup_{F\in\mathcal{F}ac^1(E)} D_F \tag{23}$$

Suppose we have $\mathcal{K} \in \mathcal{D}(\mathbb{A},\mathcal{S}_{\mathcal{H}})$.

3.11.1. **Lemma.** *We have a natural isomorphism*

$$R\Gamma(S_E, \bigcup_{F\in\mathcal{F}ac^1(E)} S_F;\mathcal{K}) \cong \oplus_{F\in\mathcal{F}ac^1 E} R\Gamma(D_F, S_F;\mathcal{K})$$

Proof. Note that $S_E - \bigcup_{F\in\mathcal{F}ac^1(E)} S_F = \bigcup_{F\in\mathcal{F}ac^1(E)} \overset{\circ}{D}_F$ (disjoint union). The claim follows now from the Poincaré duality. □

Therefore, for any $F \in \mathcal{F}ac^1(E)$ we get a natural inclusion map

$$i_E^F : R\Gamma(D_F, S_F;\mathcal{K}) \hookrightarrow R\Gamma(S_E, \bigcup_{F'\in\mathcal{F}ac^1(E)} S_{F'};\mathcal{K}) \tag{24}$$

Let us define a map

$$u_E^F(\mathcal{K}) : \Phi_F(\mathcal{K}) \longrightarrow \Phi_E(\mathcal{K})$$

as a composition

$$R\Gamma(D_F, S_F; \mathcal{K})[-p] \xrightarrow{i_E^F} R\Gamma(S_E, \bigcup_{F' \in \mathcal{F}ac^1(E)} S_{F'}; \mathcal{K})[-p] \longrightarrow R\Gamma(S_E; \mathcal{K})[-p] \longrightarrow R\Gamma(D_E, S_E)[-p+1]$$

where the last arrow is the coboundary map for the couple (S_E, D_E), and the second one is evident.

This way we get a natural transormation

$$u_E^F : \Phi_F \longrightarrow \Phi_E \tag{25}$$

which will be called a *canonical map*.

We define a *variation map*

$$v_F^E : \Phi_E \longrightarrow \Phi_F \tag{26}$$

as follows. By definition, $v_F^E(\mathcal{K})$ is the map dual to the composition

$$D\Phi_F(\mathcal{K}) \xrightarrow{\sim} \Phi_F(D\mathcal{K}) \xrightarrow{u_E^F(D\mathcal{K})} \Phi_E(D\mathcal{K}) \xrightarrow{\sim} D\Phi_E(\mathcal{K}).$$

3.12. **Lemma.** *Suppose we have 4 facets A, B_1, B_2, C such that $A < B_1 < C$, $A < B_2 < C$ and* dim $A =$ dim $B_i - 1 =$ dim $C - 2$ *(see Fig. 2). Then*

$$u_A^{B_1} \circ u_{B_1}^C = -u_A^{B_2} \circ u_{B_2}^C$$

and

$$v_C^{B_1} \circ v_{B_1}^A = -v_C^{B_2} \circ v_{B_2}^A.$$

For a proof, see below, 3.13.2.

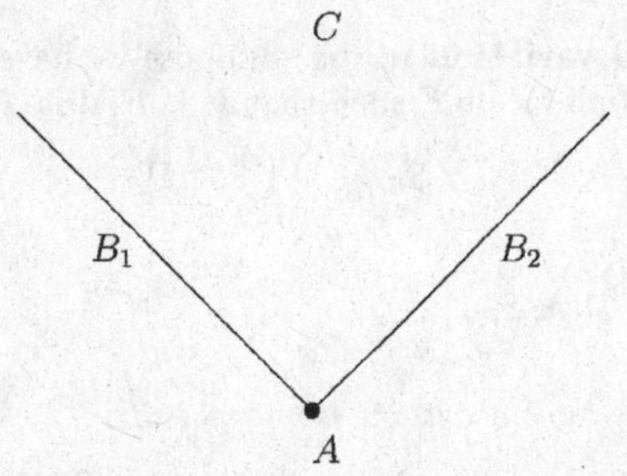

Fig. 2

3.13. **Cochain complexes.** For each integer p, $0 \leq p \leq N$, and $\mathcal{M} \in \mathcal{M}(\mathbb{A}, \mathcal{S}_{\mathcal{H}})$ introduce vector spaces

$$C_{\mathcal{H}}^{-p}(\mathbb{A}; \mathcal{M}) = \oplus_{F:\dim F = p} \mathcal{M}_F \tag{27}$$

For $i > 0$ or $i < -N$ set $C_{\mathcal{H}}^i(\mathbb{A}; \mathcal{M}) = 0$.

Define operators

$$d : C_{\mathcal{H}}^{-p}(\mathbb{A}; \mathcal{M}) \longrightarrow C_{\mathcal{H}}^{-p+1}(\mathbb{A}; \mathcal{M})$$

having components u_E^F.

3.13.1. **Lemma.** $d^2 = 0$.

Proof. Let us denote $X_p := \bigcup_{F:\dim F=p} D_F$. We have $X_p \supset X_{p+1}$. Evident embeddings of couples $(D_F, S_F) \hookrightarrow (X_p, X_{p+1})$ induce maps

$$R\Gamma(X_p, X_{p+1}; \mathcal{M}) \longrightarrow \oplus_{F:\dim F=p} R\Gamma(D_F, S_F; \mathcal{M})$$

which are easily seen to be isomorphisms. Thus, we can identify $C^{-p}(\mathbb{A}; \mathcal{M})$ with $R\Gamma(X_p, X_{p+1}; \mathcal{M})$. In these terms, d is a boundary homomorphism for the triple $X_p \supset X_{p+1} \supset X_{p+2}$. After this description, the equality $d^2 = 0$ is a general fact from homological algebra. □

3.13.2. **Proof of 3.12.** The above lemma is equivalent to the statement of 3.12 about maps u, which is thus proven. The claim for variation maps follows by duality. □

This way we get a complex $C^\bullet_{\mathcal{H}}(\mathbb{A}; \mathcal{M})$ lying in degrees from $-N$ to 0. It will be called the *cochain complex* of our arrangement $\mathcal{H}_{\mathbb{R}}$ with coefficients in $\mathcal{M}$.

3.14. **Theorem.** *(i) A functor*

$$\mathcal{M} \mapsto C^\bullet_{\mathcal{H}}(\mathbb{A}; \mathcal{M})$$

is an exact functor from $\mathcal{M}(\mathbb{A}; \mathcal{S}_{\mathcal{H}})$ to the category of complexes of vector spaces.
(ii) We have a canonical natural isomorphism in $\mathcal{D}(\{pt\})$

$$C^\bullet_{\mathcal{H}}(\mathbb{A}; \mathcal{M}) \xrightarrow{\sim} R\Gamma(\mathbb{A}; \mathcal{M})$$

Proof. (i) is obvious from the exactness of functors Φ_F, cf. Thm. 3.9. To prove (ii), let us consider the filtration

$$\mathbb{A} \supset X_0 \supset X_1 \supset \ldots X_N \supset 0.$$

It follows easily from homotopy argument (cf. 2.12, 2.13) that the restriction

$$R\Gamma(\mathbb{A}; \mathcal{M}) \longrightarrow R\Gamma(X_0; \mathcal{M})$$

is an isomorphism. On the other hand, a "Cousin" interpretation of $C^\bullet_{\mathcal{H}}(\mathbb{A}; \mathcal{M})$ given in the proof of Lemma 3.13.1, shows that one has a canonical isomorphism $R\Gamma(X_0; \mathcal{M}) \xrightarrow{\sim} C^\bullet_{\mathcal{H}}(\mathbb{A}; \mathcal{M})$. □

4. Computations for standard sheaves

4.1. Suppose we have a connected locally simply connected topological space X and a subspace $Y \subset X$ such that each connected component of Y is simply connected. Recall that a *groupoid* is a category all of whose morphisms are isomorphisms. Let us define a *Poincaré groupoid* $\pi_1(X; Y)$ as follows.

We set $\mathrm{Ob}\pi_1(X; Y) = \pi_0(Y)$. To define morphisms, let us choose a point y_i on each connected component $Y_i \subset Y$. By definition, for two connected components Y_i and Y_j, the set of homomorphisms $\mathrm{Hom}_{\pi_1(X,Y)}(Y_i, Y_j)$ is the set of all homotopy classes of paths in X starting at y_i and ending at y_j.

A different choice of points y_i gives a canonically isomorphic groupoid. If Y is reduced to one point we come back to a usual definition of the fundamental group.

Given a local system $\mathcal{L}$ on X, we may assign to it a "fiber" functor

$$F_{\mathcal{L}} : \pi_1(X;Y) \longrightarrow \mathcal{V}ect,$$

carrying Y_i to the fiber $\mathcal{L}_{y_i}$. This way we get an equivalence of the category of local systems on X and the category of functors $\pi_1(X,Y) \longrightarrow \mathcal{V}ect$.

4.2. Return to the situation of the previous section. It is known (cf. [Br]) that the homology group $H_1(\overset{\circ}{\mathbb{A}}_{\mathcal{H}};\mathbb{Z})$ is a free abelian group with a basis consisting of classes of small loops around hyperplanes $H \in \mathcal{H}$. Consequently, for each map

$$\mathbf{q} : \mathcal{H} \longrightarrow \mathbb{C}^* \tag{28}$$

there exists a one-dimensional local system $\mathcal{L}(\mathbf{q})$ whose monodromy around $H \in \mathcal{H}$ is equal to $\mathbf{q}(H)$. Such a local system is unique up to a non-unique isomorphism.

Let us construct such a local system explicitely, using a language of the previous subsection.

4.3. From now on we fix a real equation for each $H \in \mathcal{H}$, i.e. a linear function $f_H : \mathbb{A}_{\mathbb{R}} \longrightarrow \mathbb{R}$ such that $H_{\mathbb{R}} = f^{-1}(0)$. We will denote also by f_H the induced function $\mathbb{A} \longrightarrow \mathbb{C}$.

The hyperplane $H_{\mathbb{R}}$ divides $\mathbb{A}_{\mathbb{R}}$ into two halfspaces: $\mathbb{A}^+_{\mathbb{R},H} = \{x \in \mathbb{A}_{\mathbb{R}} | f_H(x) > 0\}$ and $\mathbb{A}^-_{\mathbb{R},H} = \{x \in \mathbb{A}_{\mathbb{R}} | f_H(x) < 0\}$.

Let $F \subset H_{\mathbb{R}}$ be a facet of dimension $N-1$. We have two chambers $F_\pm$ adjacent to F, where $F_\pm \subset \mathbb{A}^\pm_{\mathbb{R},H}$. Pick a point $w \in F$. Let us choose a real affine line $l_{\mathbb{R}} \subset \mathbb{A}_{\mathbb{R}}$ transversal to $H_{\mathbb{R}}$ and passing through w. Let l denote its complexification.

The function f_H induces isomorphism $l \xrightarrow{\sim} \mathbb{C}$, and $f_H^{-1}(\mathbb{R}) \cap l = l_{\mathbb{R}}$. Let us pick a real $\epsilon > 0$ such that two points $f_H^{-1}(\pm\epsilon) \cap l_{\mathbb{R}}$ lie in $F_\pm$ respectively. Denote these points by $w_\pm$.

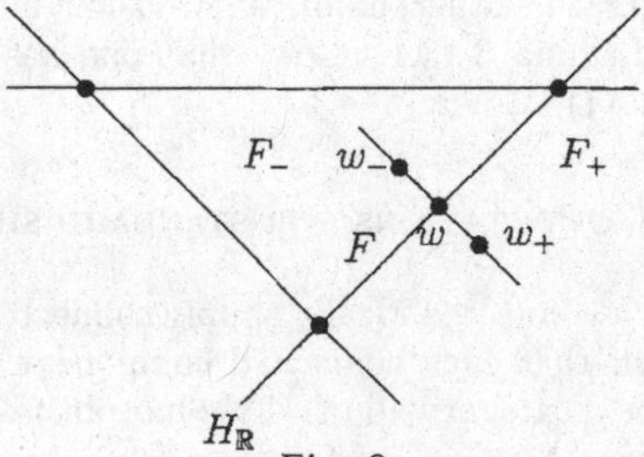

Fig. 3

Let us denote by τ^+ (resp., τ^-) a counterclockwise (resp., clockwise) path in the upper (resp., lower) halfplane connecting ϵ with $-\epsilon$. Let us denote

$$\tau_F^\pm = f_H^{-1}(\tau^\pm)$$

This way we get two well-defined homotopy classes of paths connecting chambers F_+ and F_-. The argument arg f_H increases by $\mp\frac{\pi}{2}$ along $\tau_F^\pm$.

Note that if H' is any other hyperplane of our arrangement then $\arg f_{H'}$ gets no increase along $\tau_F^{\pm}$.

4.4. Now suppose we have $\mathbf{q}$ as in (28). Note that all connected components of $\mathring{\mathbb{A}}_{\mathcal{H},\mathbb{R}}$ — chambers of our arrangement — are contractible. Let us define a functor

$$F(\mathbf{q}^2) : \pi_1(\mathring{\mathbb{A}}_{\mathcal{H}}, \mathring{\mathbb{A}}_{\mathcal{H},\mathbb{R}}) \longrightarrow \mathcal{V}ect_{\mathbb{C}}$$

as follows. For each chamber C we set $F(\mathbf{q}^2)(C) = \mathbb{C}$. For each facet F of codimension 1 which lies in a hyperplane H we set

$$F(\mathbf{q}^2)(\tau_F^{\pm}) = \mathbf{q}(H)^{\pm 1}$$

It follows from the above remark on the structure of $H_1(\mathring{\mathbb{A}}_{\mathcal{H}})$ that we get a correctly defined functor.

The corresponding abelian local system over $\mathbb{A}_{\mathcal{H}}$ will be denoted $\mathcal{L}(\mathbf{q}^2)$; it has a monodromy $\mathbf{q}(H)^2$ around $H \in \mathcal{H}$. If all numbers $\mathbf{q}(H)$ belong to some subfield $\mathsf{k} \subset \mathbb{C}$ then the same construction gives a local system of k-vector spaces.

4.5. From now on until the end of this section we fix a map

$$\mathbf{q} : \mathcal{H} \longrightarrow \mathsf{k}^*,$$

k being a subfield of $\mathbb{C}$, and denote by $\mathcal{L}$ the local system of k-vector spaces $\mathcal{L}(\mathbf{q}^2)$ constructed above. We denote by $\mathcal{L}^{-1}$ the dual local system $\mathcal{L}(\mathbf{q}^{-2})$.

Let $j : \mathring{\mathbb{A}}_{\mathcal{H}} \longrightarrow \mathbb{A}$ denote an open embedding. For ? equal to one of the symbols !, $*$ or !$*$, let us consider perverse sheaves $\mathcal{L}_? := j_?\mathcal{L}[N]$. They belong to $\mathcal{M}(\mathbb{A}; \mathcal{S}_{\mathcal{H}})$; these sheaves will be called *standard extensions* of $\mathcal{L}$, or simply *standard sheaves*. Note that

$$D\mathcal{L}_! \cong \mathcal{L}_*^{-1};\ D\mathcal{L}_* \cong \mathcal{L}_!^{-1} \tag{29}$$

We have a canonical map

$$m : \mathcal{L}_! \longrightarrow \mathcal{L}_*, \tag{30}$$

and by definition $\mathcal{L}_{!*}$ coincides with its image.

Our aim in this section will be to compute explicitely the cochain complexes of standard sheaves.

4.6. **Orientations.** Let F be a facet which is a connected component of a real edge L_F. Consider a linear space $L_F^{\perp} = \mathbb{A}_{\mathbb{R}}/L_F$. Let us define

$$\lambda_F := H^0(L_F^{\perp}; \mathcal{O}r_{L_F^{\perp}}),$$

it is a free abelian group of rank 1. To choose an orientation of $L_F^{\perp}$ (as a real vector space) is the same as to choose a basis vector in λ_F. We will call an orientation of $L_F^{\perp}$ a *coorientation* of F.

We have an evident piecewise linear homeomorphism of $D_F = \bigcup D_{F<C}$ onto a closed disk in $L_F^{\perp}$; thus, a coorientation of F is the same as an orientation of D_F (as a C^0-manifold); it defines orientations of all cells $D_{F<C}$.

4.6.1. From now on until the end of the section, let us fix coorientations of all facets. Suppose we have a pair $E < F$, dim E = dim $F - 1$. The cell D_E is a part of the boundary of D_F. Let us define the sign $\mathrm{sgn}(F, E) = \pm 1$ as follows. Complete an orienting basis of D_E by a vector directed outside D_F; if we get the given orientation of D_F, set $\mathrm{sgn}(F, E) = 1$, otherwise set $\mathrm{sgn}(F, E) = -1$.

4.7. **Basis in $\Phi_F(\mathcal{L}_!)^*$.** Let F be a facet of dimension p. We have by definition

$$\Phi_F(\mathcal{L}_!) = H^{-p}(D_F, S_F; \mathcal{L}_!) = H^{N-p}(D_F, S_F; j_!\mathcal{L}) \cong H^{N-p}(D_F, S_F \cup ({}_{\mathcal{H}}H_{\mathbb{R}} \cap D_F); j_!\mathcal{L}).$$

By Poincaré duality,

$$H^{N-p}(D_F, S_F \cup ({}_{\mathcal{H}}H_{\mathbb{R}} \cap D_F); j_!\mathcal{L})^* \cong H^0(D_F - (S_F \cup_{\mathcal{H}} H_{\mathbb{R}}); \mathcal{L}^{-1})$$

(recall that we have fixed an orientation of D_F). The space $D_F - (S_F \cup_{\mathcal{H}} H_{\mathbb{R}})$ is a disjoint union

$$D_F - (S_F \cup_{\mathcal{H}} H_{\mathbb{R}}) = \bigcup_{C \in \mathrm{Ch}(F)} \overset{\circ}{D}_{F<C}.$$

Consequently,

$$H^0(D_F - (S_F \cup_{\mathcal{H}} H_{\mathbb{R}}); \mathcal{L}^{-1}) \cong \oplus_{C \in \mathrm{Ch}(F)} H^0(\overset{\circ}{D}_{F<C}; \mathcal{L}^{-1}).$$

By definition of $\mathcal{L}$, we have canonical identifications $H^0(\overset{\circ}{D}_{F<C}; \mathcal{L}^{-1}) = \mathsf{k}$. We will denote by $c(\mathcal{L}_!)_{F<C} \in \Phi_F(\mathcal{L})^*$ the image of $1 \in H^0(\overset{\circ}{D}_{F<C}; \mathcal{L}^{-1})$ with respect to the embedding

$$H^0(\overset{\circ}{D}_{F<C}; \mathcal{L}^{-1}) \hookrightarrow \Phi_F(\mathcal{L}_!)^*$$

following from the above.

Thus, classes $c(\mathcal{L}_!)_{F<C}$, $C \in \mathrm{Ch}(F)$, form a basis of $\Phi_F(\mathcal{L}_!)^*$.

4.8. Let us describe canonical maps for $\mathcal{L}_!$. If $F < E$, dim E = dim $F + 1$, let $u^* : \Phi_F(\mathcal{L}_!)^* \longrightarrow \Phi_E(\mathcal{L}_!)^*$ denote the map dual to $u_F^E(\mathcal{L}_!)$. Let C be a chamber adjacent to F. Then

$$u^*(c(\mathcal{L}_!)_{F<C}) = \begin{cases} \mathrm{sgn}(F, E) c(\mathcal{L}_!)_{E<C} & \text{if } E < C \\ 0 & \text{otherwise} \end{cases} \tag{31}$$

4.9. **Basis in $\Phi_F(\mathcal{L}_*)^*$.** We have isomorphisms

$$\Phi_F(\mathcal{L}_*) \cong \Phi_F(D\mathcal{L}_!^{-1}) \cong \Phi_F(\mathcal{L}_!^{-1})^* \tag{32}$$

Hence, the defined above basis $\{c(\mathcal{L}_!^{-1})_{F<C}\}_{C \in \mathrm{Ch}(F)}$ of $\Phi_F(\mathcal{L}_!^{-1})^*$, gives a basis in $\Phi_F(\mathcal{L}_*)$. We will denote by $\{c(\mathcal{L}_*)_{F<C}\}_{C \in \mathrm{Ch}(F)}$ the dual basis of $\Phi_F(\mathcal{L}_*)$.

4.10. **Example.** Let us describe our chains explicitely in the simplest one-dimensional case, in the setup 3.6. We choose a natural orientation on $\mathbb{A}_{\mathbb{R}}$. A local system $\mathcal{L} = \mathcal{L}(q^2)$ is uniquely determined by one nonzero complex number q. By definition, the upper (resp., lower) halfplane halfmonodromy from w_+ to w_- is equal to q (resp., q^{-1}).

4.10.1. *Basis in* $\Phi_F(\mathcal{L}_!)^*$. The space $\Phi_F(\mathcal{L}_!)^*$ admits a basis consisting of two chains $c_\pm = c(\mathcal{L}_!)_{F<E_\pm}$ shown below, see Fig. 3(a). By definition, a homology class is represented by a cell together with a section of a local system $\mathcal{L}^{-1}$ over it. The section of $\mathcal{L}^{-1}$ over c_+ (resp., c_-) takes value 1 over w_+ (resp., w_-).

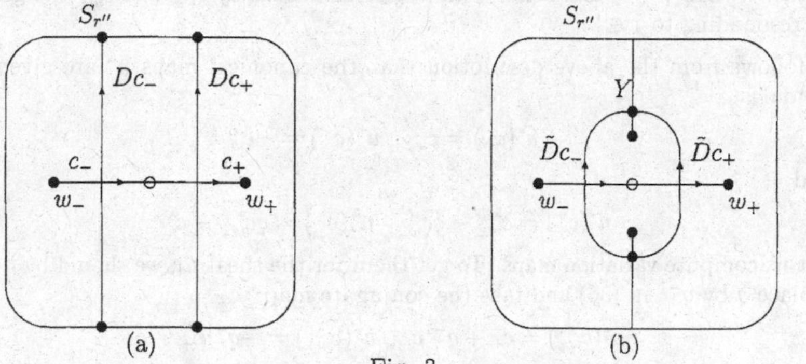

Fig. 3

4.10.2. *Basis in* $\Phi_F(\mathcal{L}_*)^*$. Let us adopt notations of 3.8, with $\mathcal{K} = \mathcal{L}_!^{-1}$. It is easy to find the basis $\{Dc_+, Dc_-\}$ of the dual space

$$\Phi_F(\mathcal{L}_!^{-1}) \cong H^0(\mathbb{A}, \mathbb{A}_{\geq r''}; j^*j_*D\mathcal{L}_!^{-1})^* = H^1(\mathbb{A} - \{w_+, w_-\}, \mathbb{A}_{\geq r''}; \mathcal{L})^*$$

dual to $\{c(\mathcal{L}^{-1})_{F<E_+}, c(\mathcal{L}^{-1})_{F<E_-}\}$. Namely, $Dc_\pm$ is represented by the relative 1-chain

$$\{\pm\frac{1}{2} + y \cdot i | -\sqrt{(r'')^2 - \frac{1}{4}} \leq y \leq \sqrt{(r'')^2 - \frac{1}{4}}\},$$

with evident sections of $\mathcal{L}^{-1}$ over them, see Fig. 3(a).

Next, one has to deform these chains to chains $\tilde{D}c_\pm$ with their ends on Y, as in Fig. 3(b). Finally, one has to make a clockwise rotation of the picture by $\pi/2$. As a result, we arrive at the following two chains $c_\pm^*$ forming a basis of $\Phi_F(\mathcal{L}_*)^*$:

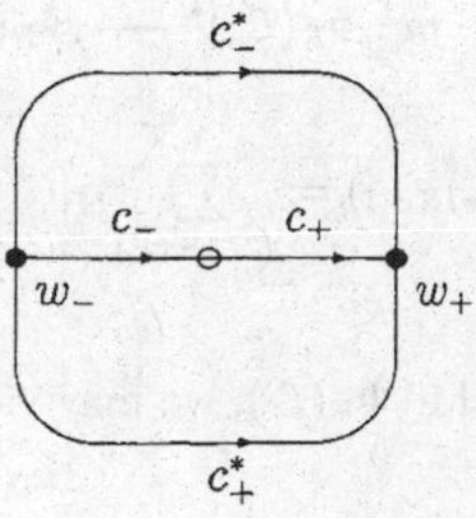

Fig. 4

The section of $\mathcal{L}^{-1}$ over c_+^* (resp., c_-) has value 1 at w_+ (resp., w_-).

It follows from this description that the natural map

$$m : \Phi_F(\mathcal{L}_*)^* \longrightarrow \Phi_F(\mathcal{L}_!)^* \tag{33}$$

is given by the formulas

$$m(c_+^*) = c_+ + qc_-;\ m(c_-^*) = qc_+ + c_- \tag{34}$$

By definition, spaces $\Phi_{E_\pm}(\mathcal{L}_?)^*$ may be identified with fibers $\mathcal{L}_{w_\pm}$ respectively, for both $? =!$ and $? = *$, and hence with k. Let us denote by $c_{w_\pm}$ and $c^*_{w_\pm}$ the generators corresponding to $1 \in$ k.

It follows from the above description that the canonical maps u^* are given by the formulas

$$u^*(c_+) = c_{w_+};\ u^*(c_-) = -c_{w_-}; \tag{35}$$

and

$$u^*(c_+^*) = c^*_{w_+} - qc^*_{w_-};\ u^*(c_-^*) = qc^*_{w_+} - c^*_{w_-}; \tag{36}$$

Let us compute variation maps. To get them for the sheaf $\mathcal{L}_!$, we should by definition replace q by q^{-1} in (36) and take the conjugate map:

$$v^*(c_{w_+}) = c_+ + q^{-1}c_-;\ v^*(c_{w_-}) = -q^{-1}c_+ - c_- \tag{37}$$

To compute v^* for $\mathcal{L}_*$, note that the basis in

$$H^0(\mathbb{A}, \{w_+, w_-\}; \mathcal{L}_!^{-1})^* = H_1(\mathbb{A}, \{w_+, w_-, 0\}; \mathcal{L})$$

dual to $\{c_+^*, c_-^*\}$, is $\{q^{-1}\tilde{c}_-, q^{-1}\tilde{c}_+\}$ *(sic!)* where $\tilde{c}_\pm$ denote the chains defined in the same way as $c_\pm$, with $\mathcal{L}$ replaced by $\mathcal{L}^{-1}$. From this remark it follows that

$$v^*(c^*_{w_+}) = q^{-1}c_-^*;\ v^*(c^*_{w_-}) = -q^{-1}c_+^* \tag{38}$$

4.11. Let us return to the case of an arbitrary arrangement. Let us say that a hyperplane $H \in \mathcal{H}$ *separates* two chambers C, C' if they lie in different halfspaces with respect to $H_{\mathbb{R}}$. Let us define numbers

$$\mathbf{q}(C, C') = \prod \mathbf{q}(H), \tag{39}$$

the product over all hyperplanes $H \in \mathcal{H}$ separating C and C'. In particular, $\mathbf{q}(C, C) = 1$.

4.12. **Lemma.** *Let F be a face, $C \in \mathrm{Ch}(F)$. The canonical mapping*

$$m : \Phi_F(\mathcal{L}_*)^* \longrightarrow \Phi_F(\mathcal{L}_!)^*$$

is given by the formula

$$m(c(\mathcal{L}_*)_{F<C}) = \sum_{C' \in \mathrm{Ch}(F)} \mathbf{q}(C, C')c(\mathcal{L}_!)_{F<C'} \tag{40}$$

4.12.1. Since $\Phi_F(\mathcal{L}_*)$ is dual to $\Phi_F(\mathcal{L}_!)$, we may view m as a *bilinear form* on $\Phi_F(\mathcal{L}_!)$. By (40) it is *symmetric*.

Proof of lemma. We generalize the argument of the previous example. First consider the case of zero-dimensional F. Given a chain $c(\mathcal{L}_!^{-1})_{F<C}$, the corresponding dual chain may be taken as

$$Dc_{F<C} = \epsilon \cdot {}^{C}w \oplus i \cdot \mathbb{A}_{\mathbb{R}},$$

were ϵ is a sufficiently small positive real. Next, to get the dual chain $c(\mathcal{L}_*)_{F<C}$, we should make a deformation similar to the above one, and a rotation by $\frac{\pi}{2}$. It is

convenient to make the rotation first. After the rotation, we get a chain $\mathbb{A}_{\mathbb{R}} - \epsilon i \cdot {}^{C}w$. The value of $m(c(\mathcal{L}_*)_{F<C})$ is given by the projection of this chain to $\mathbb{A}_{\mathbb{R}}$.

The coefficient at $c(\mathcal{L}_!)_{F<C'}$ is given by the monodromy of $\mathcal{L}^{-1}$ along the following path from C to C'. First, go "down" from ${}^{C}w$ to $-\epsilon i \cdot {}^{C}w$; next, travel in $\mathbb{A}_{\mathbb{R}} - \epsilon i \cdot {}^{C}w$ along the straight line from $-\epsilon i \cdot {}^{C}w$ to $-\epsilon i \cdot {}^{C'}.w$, and then go "up" to ${}^{C'}w$. Each time we are passing under a hyperplane $H_{\mathbb{R}}$ separating C and C', we gain a factor $\mathbf{q}(H)$. This gives desired coefficient for the case dim $F = 0$.

For an arbitrary F we use the same argument by considering the intersection of our picture with a transversal slice. □

4.13. Let E be a facet which is a component of a real edge $L_{E,\mathbb{R}}$; as usually L will denote the complexification. Let $\mathcal{H}_L \subset \mathcal{H}$ be a subset consisting of all hyperplanes containg L. If we assign to a chamber $C \in \mathrm{Ch}(E)$ a unique chamber of the subarrangement $\mathcal{H}_L$ comtaining C, we get a bijection of $\mathrm{Ch}(E)$ with the set of *all* chambers of $\mathcal{H}_L$.

Let $F < E$ be another facet. Each chamber $C \in \mathrm{Ch}(F)$ is contained in a unique chamber of $\mathcal{H}_L$. Taking into account a previous bijection, we get a mapping

$$\pi_E^F : \mathrm{Ch}(F) \longrightarrow \mathrm{Ch}(E) \tag{41}$$

4.14. **Lemma.** *Let F be a facet, $C \in \mathrm{Ch}(F)$. We have*

$$u^*(c(\mathcal{L}_*)_{F<C}) = \sum \mathrm{sgn}(F,E)\mathbf{q}(C, \pi_E^F(C))c(\mathcal{L}_*)_{E<\pi_E^F(C)}, \tag{42}$$

the summation over all facets E such that $F < E$ and dim $E =$ dim $F + 1$.

(Signs $\mathrm{sgn}(F, E)$ have been defined in 4.6.1.)

Proof. Again, the crucial case is dim $F = 0$ — the case of arbitrary dimension is treated using a transversal slice. So, let us suppose that F is zero-dimensional. In order to compute the coefficient of $u^*(c(\mathcal{L}_*)_{F<C})$ at $c(\mathcal{L}^*)_{E<C'}$ where E is a one-dimensional facet adjacent to F and $C' \in \mathrm{Ch}(E)$, we have to do the following.

Consider the intersection of a real affine subspace $\mathbb{A}_{\mathbb{R}} - \epsilon \cdot i \cdot {}^{C}w$ (as in the proof of the previous lemma) with a complex hyperplane M_E passing through ${}^{E}w$ and transversal to E. The intersection will be homotopic to a certain chain $c(\mathcal{L}_*)_{E<C''}$ where C'' is easily seen to be equal to $\pi_E^F(C)$, and the coefficient is obtained by the same rule as described in the previous proof. The sign will appear in accordance with compatibility of orientations of D_F and D_E. □

4.15. Let us collect our results. Let us denote by $\{b(\mathcal{L}_?)_{F<C}\}_{C\in\mathrm{Ch}(F)}$ the basis in $\Phi_F(\mathcal{L}_?)$ dual to $\{c(\mathcal{L}_?)_{F<C}\}$, where ? =! or $*$.

4.16. **Theorem.** *(i) The complex $C^\bullet_{\mathcal{H}}(\mathbb{A}; \mathcal{L}_!)$ is described as follows. For each p, $0 \le p \le N$, the space $C^{-p}_{\mathcal{H}}(\mathbb{A}; \mathcal{L}_!)$ admits a basis consisting of all cochains $b(\mathcal{L}_!)_{F<C}$ where F runs through all facets of $\mathcal{H}_{\mathbb{R}}$ of dimension p, and C through $\mathrm{Ch}(F)$. The differential*

$$d : C^{-p}_{\mathcal{H}}(\mathbb{A}; \mathcal{L}_!) \longrightarrow C^{-p+1}_{\mathcal{H}}(\mathbb{A}; \mathcal{L}_!)$$

is given by the formula

$$d(b(L_!)_{F<C}) = \sum_{E:E<F,\ \dim\ E=\dim\ F-1} \mathrm{sgn}(E,F)b(L_!)_{E<C} \tag{43}$$

(ii) The complex $C^\bullet_{\mathcal{H}}(\mathbb{A};\mathcal{L}_*)$ *is described as follows. For each* p, $0 \le p \le N$, *the space* $C^{-p}_{\mathcal{H}}(\mathbb{A};\mathcal{L}_*)$ *admits a basis consisting of all cochains* $b(\mathcal{L}_*)_{F<C}$ *where* F *runs through all facets of* $\mathcal{H}_{\mathbb{R}}$ *of dimension* p, *and* C *through* $\mathrm{Ch}(F)$. *The differential*

$$d : C^{-p}_{\mathcal{H}}(\mathbb{A};\mathcal{L}_*) \longrightarrow C^{-p+1}_{\mathcal{H}}(\mathbb{A};\mathcal{L}_*)$$

is given by the formula

$$d(b(L_*)_{F<C}) = \sum \mathrm{sgn}(E,F)\mathbf{q}(C,C')b(L_*)_{E<C'}, \tag{44}$$

the summation over all facets $E < F$ *such that* $\dim\ E = \dim\ F - 1$ *and all chambers* $C' \in \mathrm{Ch}(E)$ *such that* $\pi^E_F(C') = C$.

(iii) The natural map of complexes

$$m : C^\bullet_{\mathcal{H}}(\mathbb{A};\mathcal{L}_!) \longrightarrow C^\bullet_{\mathcal{H}}(\mathbb{A};\mathcal{L}_*) \tag{45}$$

induced by the canonical map $\mathcal{L}_! \longrightarrow \mathcal{L}_*$, *is given by the formula*

$$m(b(\mathcal{L}_!)_{F<C}) = \sum_{C'\in \mathrm{Ch}(F)} \mathbf{q}(C,C')b(\mathcal{L}_*)_{F<C'} \tag{46}$$

All statements have already been proven.

4.16.1. **Corollary.** *The complexes* $C^\bullet_{\mathcal{H}}(\mathbb{A};\mathcal{L}_!)$ *and* $C^\bullet_{\mathcal{H}}(\mathbb{A};\mathcal{L}_*)$ *described explicitely in the above theorem, compute the relative cohomology* $H^\bullet(\mathbb{A},\ _{\mathcal{H}}H;\mathcal{L})$ *and the cohomology of the open stratum* $H^\bullet(\overset{\circ}{\mathbb{A}},\mathcal{L})$ *respectively, and the map* m *induces the canonical map in cohomology.*

Proof. This follows immediately from 3.14. □

This corollary was proven in [V], Sec. 2, by a different argument.

4.17. **Theorem.** *The complex* $C^\bullet_{\mathcal{H}}(\mathbb{A};\mathcal{L}_{!*})$ *is canonically isomorphic to the image of (45).*

Proof. This follows from the previous theorem and the exactness of the functor $\mathcal{M} \mapsto C^\bullet_{\mathcal{H}}(\mathbb{A};\mathcal{M})$, cf. Thm. 3.14 (i). □

The above description of cohomology is analogous to [SV1], p. I, whose results may be considered as a "quasiclassical" version of the above computations.

Part II. CONFIGURATION SPACES AND QUANTUM GROUPS

1. Introduction

1.1. We are starting here the geometric study of the tensor category $\mathcal{C}$ associated with a quantum group (corresponding to a Cartan matrix of finite type) at a root of unity (see [AJS], 1.3 and the present part, 11.3 for the precise definitions).

The main results of this part are Theorems 8.18, 8.23, 12.7 and 12.8 which

— establish isomorphisms between homogeneous components of irreducible objects in $\mathcal{C}$ and spaces of vanishing cycles at the origin of certain Goresky-MacPherson sheaves on configuration spaces;

— establish isomorphisms of the stalks at the origin of the above GM sheaves with certain Hochschild complexes (which compute the Hochschild homology of a certain "triangular" subalgebra of our quantum group with coefficients in the coresponding irreducible representation);

— establish the analogous results for tensor products of irreducibles. In geometry, the tensor product of representations corresponds to a "fusion" of sheaves on configuration spaces — operation defined using the functor of nearby cycles, see Section 10.

We must mention that the assumption that we are dealing with a Cartan matrix of finite type and a root of unity appears only at the very end (see Chapter 4). We need these assumptions in order to compare our representations with the conventional definition of the category $\mathcal{C}$. All previous results are valid in more general assumptions. In particular a Cartan matrix could be arbitrary and a deformation parameter ζ not necessarily a root of unity.

1.2. Some of the results of this part constitute the description of the cohomology of certain "standard" local systems over configuration spaces in terms of quantum groups. These results, due to Varchenko and one of us, were announced several years ago in [SV2]. The proofs may be found in [V]. Our proof of these results uses completely different approach. Some close results were discussed in [S].

Certain results of a similar geometric spirit are discussed in [FW].

1.3. We are grateful to A.Shen who made our communication during the writing of this part possible.

1.4. *Notations.* We will use all the notations from the part I. References to *loc. cit.* will look like I.1.1. If a, b are two integers, we will denote by $[a, b]$ the set of all integers c such that $a \leq c \leq b$; $[1, a]$ will be denoted by $[a]$. $\mathbb{N}$ will denote the set of non-negative integers. For $r \in \mathbb{N}$, Σ_r will denote the group of all bijections $[r] \overset{\sim}{\longrightarrow} [r]$.

We suppose that our ground field k has characteristic 0, and fix an element $\zeta \in \mathsf{k}$, $\zeta \neq 0$. For $a \in \mathbb{Z}$ we will use the notation

$$[a]_\zeta = 1 - \zeta^{-2a} \tag{47}$$

The word "t-exact" will allways mean t-exactness with respect to the middle perversity.

Chapter 1. Algebraic discussion

2. Free algebras and bilinear forms

Most definitions of this section follow [L1] and [SV2] (with slight modifications). We also add some new definitions and computations important for the sequel. Cf. also [V], Section 4.

2.1. Until the end of this part, let us fix a finite set I and a symmetric $\mathbb{Z}$-valued bilinear form $\nu, \nu' \mapsto \nu \cdot \nu'$ on the free abelian group $\mathbb{Z}[I]$ (cf. [L1], 1.1). We will denote by X the dual abelian group $\mathrm{Hom}(\mathbb{Z}[I], \mathbb{Z})$. Its elements will be called *weights*. Given $\nu \in \mathbb{Z}[I]$, we will denote by $\lambda_\nu \in X$ the functional $i \mapsto i \cdot \nu$. Thus we have

$$\langle \lambda_\nu, \mu \rangle = \nu \cdot \mu \tag{48}$$

for all $\nu, \mu \in \mathbb{N}[I]$.

2.2. Let $'\mathfrak{f}$ denote a free associative k-algebra with 1 with generators θ_i, $i \in I$. Let $\mathbb{N}[I]$ be a submonoid of $\mathbb{Z}[I]$ consisting of all linear combinations of elements of I with coefficients in $\mathbb{N}$. For $\nu = \sum \nu_i i \in \mathbb{N}[I]$ we denote by $'\mathfrak{f}_\nu$ the k-subspace of $'\mathfrak{f}$ spanned by all monomials $\theta_{i_1}\theta_{i_2} \cdot \ldots \cdot \theta_{i_p}$ such that for any $i \in I$, the number of occurences of i in the sequence $i_1, \ldots, i_p$ is equal to ν_i.

We have a direct sum decomposition $'\mathfrak{f} = \oplus_{\nu \in \mathbb{N}[I]} {}'\mathfrak{f}_\nu$, all spaces $'\mathfrak{f}_\nu$ are finite dimensional, and we have $'\mathfrak{f}_0 = \mathsf{k} \cdot 1$, $'\mathfrak{f}_\nu \cdot {}'\mathfrak{f}_{\nu'} \subset {}'\mathfrak{f}_{\nu+\nu'}$.

Let $\epsilon :\ '\mathfrak{f} \longrightarrow \mathsf{k}$ denote the augmentation — a unique k-algebra map such that $\epsilon(1) = 1$ and $\epsilon(\theta_i) = 0$ for all i. Set $'\mathfrak{f}^+ := \mathrm{Ker}(\epsilon)$. We have $'\mathfrak{f}^+ = \oplus_{\nu \neq 0} {}'\mathfrak{f}_\nu$.

An element $x \in {}'\mathfrak{f}$ is called *homogeneous* if it belongs to $'\mathfrak{f}_\nu$ for some ν. We then set $|x| = \nu$. We will use the notation $\mathrm{depth}(x)$ for the number $\sum_i \nu_i$ if $\nu = \sum_i \nu_i i$; it will be called *the depth* of x.

2.3. Given a sequence $\vec{K} = (i_1, \ldots, i_N)$, $i_j \in I$, let us denote by $\theta_{\vec{K}}$ the monomial $\theta_{i_1} \cdot \ldots \cdot \theta_{i_N}$. For an empty sequence we set $\theta_\emptyset = 1$.

For $\tau \in \Sigma_N$ let us introduce the number

$$\zeta(\vec{K}; \tau) = \prod \zeta^{i_a \cdot i_b}, \tag{49}$$

the product over all a, b such that $1 \le a < b \le N$ and $\tau(a) > \tau(b)$.

We will call this number *the twisting number of the sequence $\vec{K}$ with respect to the permutation τ.*

We will use the notation

$$\tau(\vec{K}) = (i_{\tau(1)}, i_{\tau(2)}, \ldots, i_{\tau(N)}) \tag{50}$$

2.4. Let us regard the tensor product $'\mathfrak{f}\otimes{}'\mathfrak{f}$ (in the sequel $\otimes$ will mean $\otimes_{\mathsf{k}}$ unless specified otherwise) as a k-algebra with multiplication

$$(x_1\otimes x_2)\cdot(x_1'\otimes x_2')=\zeta^{|x_2|\cdot|x_1'|}x_1x_1'\otimes x_2x_2' \tag{51}$$

for homogeneous x_2, x_1'. Let us define a map

$$\Delta:{}'\mathfrak{f}\longrightarrow{}'\mathfrak{f}\otimes{}'\mathfrak{f} \tag{52}$$

as a unique algebra homomorphism carrying θ_i to $\theta_i\otimes 1+1\otimes\theta_i$.

2.5. Let us define a coalgebra structure on $'\mathfrak{f}\otimes{}'\mathfrak{f}$ as follows. Let us introduce the braiding isomorphism

$$r:{}'\mathfrak{f}\otimes{}'\mathfrak{f}\xrightarrow{\sim}{}'\mathfrak{f}\otimes{}'\mathfrak{f} \tag{53}$$

by the rule

$$r(x\otimes y)=\zeta^{|x|\cdot|y|}y\otimes x \tag{54}$$

for homogeneous x,y. By definition,

$$\Delta_{'\mathfrak{f}\otimes{}'\mathfrak{f}}:{}'\mathfrak{f}\otimes{}'\mathfrak{f}\longrightarrow({}'\mathfrak{f}\otimes{}'\mathfrak{f})\otimes({}'\mathfrak{f}\otimes{}'\mathfrak{f}) \tag{55}$$

coincides with the composition $(1_{'\mathfrak{f}}\otimes r\otimes 1_{'\mathfrak{f}})\circ(\Delta_{'\mathfrak{f}}\otimes\Delta_{'\mathfrak{f}})$.

The multiplication

$$'\mathfrak{f}\otimes{}'\mathfrak{f}\longrightarrow{}'\mathfrak{f} \tag{56}$$

is a coalgebra morphism.

2.6. Let us describe Δ more explicitely. Suppose a sequence $\vec{K}=(i_1,\dots,i_N)$, $i_j\in I$, is given. For a subset $A=\{j_1,\dots,j_a\}\subset[N]$, $j_1<\dots<j_a$, let $A'=[N]-A=\{k_1,\dots,k_{N-a}\}$, $k_1<\dots<k_{N-a}$. Define a permutation τ_A by the formula

$$(\tau(1),\dots,\tau(N))=(j_1,j_2,\dots,j_a,k_1,k_2,\dots,k_{N-a}) \tag{57}$$

Set $\vec{K}_A:=(i_{j_1},i_{j_2},\dots,i_{j_a})$, $\vec{K}_{A'}:=(i_{k_1},i_{k_2},\dots,i_{k_{N-a}})$.

2.6.1. **Lemma.**

$$\Delta(\theta_{\vec{K}})=\sum_{A\subset K}\zeta(\vec{K};\tau_A)\theta_{\vec{K}_A}\otimes\theta_{\vec{K}_{A'}},$$

the summation ranging over all subsets $A\subset[N]$.

Proof follows immediately from the definitions. □

2.7. Let us denote by

$$\Delta^{(N)}:{}'\mathfrak{f}\longrightarrow{}'\mathfrak{f}^{\otimes N} \tag{58}$$

iterated coproducts; by the coassociativity they are well defined.

Let us define a structure of an algebra on $'\mathfrak{f}^{\otimes N}$ as follows:

$$(x_1\otimes\dots\otimes x_N)\cdot(y_1\otimes\dots\otimes y_N)=\zeta^{\sum_{j<i}|x_i|\cdot|y_j|}x_1y_1\otimes\dots\otimes x_Ny_N \tag{59}$$

for homogeneous $x_1,\dots,x_N;y_1,\dots,y_N$. The map $\Delta^{(N)}$ is an algebra morphism.

2.8. Suppose we have a sequence $\vec{K}=(i_1,\dots,i_N)$. Let us consider an element $\Delta^{(N)}(\theta_{\vec{K}})$; let $\Delta^{(N)}(\theta_{\vec{K}})^+$ denote its projection to the subspace $'\mathfrak{f}^{+\otimes N}$.

2.8.1. **Lemma.**

$$\Delta^{(N)}(\theta_{\vec{K}})^+ = \sum_{\tau\in\Sigma_N} \zeta(\vec{K};\tau)\theta_{i_{\tau(1)}}\otimes\ldots\otimes\theta_{i_{\tau(N)}}$$

Proof follows from 2.6.1 by induction on N. □

2.9. For each component $'\mathfrak{f}_\nu$ consider the dual k-space $'\mathfrak{f}^*_\nu$, and set $'\mathfrak{f}^* := \oplus\, '\mathfrak{f}^*_\nu$. Graded components $\Delta_{\nu,\nu'}: {}'\mathfrak{f}_{\nu+\nu'} \longrightarrow {}'\mathfrak{f}_\nu \otimes {}'\mathfrak{f}_{\nu'}$ define dual maps $'\mathfrak{f}^*_\nu \otimes {}'\mathfrak{f}^*_{\nu'} \longrightarrow {}'\mathfrak{f}^*_{\nu+\nu'}$ which give rise to a multiplication

$$'\mathfrak{f}^* \otimes {}'\mathfrak{f}^* \longrightarrow {}'\mathfrak{f}^* \qquad (60)$$

making $'\mathfrak{f}^*$ a graded associative algebra with 1 (dual to the augmentation of $'\mathfrak{f}$). This follows from the coassociativity of Δ, cf. [L1], 1.2.2.

Here and in the sequel, we will use identifications $(V\otimes W)^* = V^*\otimes W^*$ (for finite dimensional spaces V, W) by the rule $\langle\phi\otimes\psi, x\otimes y\rangle = \langle\phi, x\rangle\cdot\langle\psi, y\rangle$.

The dual to (56) defines a comultiplication

$$\delta: {}'\mathfrak{f}^* \longrightarrow {}'\mathfrak{f}^* \otimes {}'\mathfrak{f}^* \qquad (61)$$

It makes $'\mathfrak{f}^*$ a graded coassociative coalgebra with a counit.

The constructions dual to 2.4 and 2.5 equip $'\mathfrak{f}^* \otimes {}'\mathfrak{f}^*$ with a structure of a coalgebra and an algebra. It follows from *loc. cit* that (60) is a coalgebra morphism, and δ is an algebra morphism.

By iterating δ we get maps

$$\delta^{(N)}: {}'\mathfrak{f}^* \longrightarrow {}'\mathfrak{f}^{*\otimes N} \qquad (62)$$

If we regard $'\mathfrak{f}^{*\otimes N}$ as an algebra by the same construction as in (59), $\delta^{(N)}$ is an algebra morphism.

2.10. **Lemma.** *There exists a unique bilinear form*

$$S(\ ,\ ): {}'\mathfrak{f}\otimes {}'\mathfrak{f} \longrightarrow \mathsf{k}$$

such that

(a) $S(1,1) = 1$ *and* $(\theta_i, \theta_j) = \delta_{i,j}$ *for all* $i, j\in I$;
(b) $S(x, y'y'') = S(\Delta(x), y'\otimes y'')$ *for all* $x, y', y'' \in {}'\mathfrak{f}$;
(c) $S(xx', y'') = S(x\otimes x', \Delta(y''))$ *for all* $x, x', y'' \in {}'\mathfrak{f}$.

(The bilinear form

$$({}'\mathfrak{f}\otimes {}'\mathfrak{f})\otimes({}'\mathfrak{f}\otimes {}'\mathfrak{f}) \longrightarrow \mathsf{k}$$

given by

$$(x_1\otimes x_2)\otimes(y_1\otimes y_2) \mapsto S(x_1, y_1)S(x_2, y_2)$$

is denoted again by $S(\ ,\ )$.*)*

The bilinear form $S(\ ,\ )$ *on* $'\mathfrak{f}$ *is symmetric. The different homogeneous components* $'\mathfrak{f}_\nu$ *are mutually orthogonal.*

Proof. See [L1], 1.2.3. Cf. also [SV2], (1.8)-(1.11). □

2.11. Following [L1], 1.2.13 and [SV2], (1.10)-(1.11), let us introduce operators $\delta_i : {}'\mathfrak{f} \longrightarrow {}'\mathfrak{f}$, $i \in I$, as unique linear mappings satisfying

$$\delta_i(1) = 0;\ \delta_i(\theta_j) = \delta_{i,j},\ j \in I;\ \delta_i(xy) = \delta_i(x)y + \zeta^{|x|\cdot i} x\delta_i(y) \tag{63}$$

for homogeneous x.

It follows from 2.10 (c) that

$$S(\theta_i x, y) = S(x, \delta_i(y)) \tag{64}$$

for all $i \in I$, $x, y \in {}'\mathfrak{f}$, and obviously S is determined uniquely by this property, together with the requirement $S(1,1) = 1$.

2.12. **Lemma.** *For any two sequences $\vec{K}$, $\vec{K}'$ of N elements from I we have*

$$S(\theta_{\vec{K}}, \theta_{\vec{K}'}) = \sum_{\tau\in\Sigma_N:\ \tau(\vec{K})=\vec{K}'} \zeta(\vec{K};\tau).$$

Proof follows from (64) by induction on N, or else from 2.8.1. □

2.13. Let us define elements $\theta_i^* \in {}'\mathfrak{f}_i^*$ by the rule $<\theta_i^*, \theta_i>= 1$. The form S defines a homomomorphism of graded algebras

$$S : {}'\mathfrak{f} \longrightarrow {}'\mathfrak{f}^* \tag{65}$$

carrying θ_i to θ_i^*. S is determined uniquely by this property.

2.14. **Lemma.** *The map S is a morphism of coalgebras.*

Proof. This follows from the symmetry of S. □

VERMA MODULES

2.15. Let us pick a weight Λ. Our aim now will be to define certain X-graded vector space $V(\Lambda)$ equipped with the following structures.

(i) A structure of left ${}'\mathfrak{f}$-module ${}'\mathfrak{f} \otimes V(\Lambda) \longrightarrow V(\Lambda)$;

(ii) a structure of left ${}'\mathfrak{f}$-comodule $V(\Lambda) \longrightarrow {}'\mathfrak{f} \otimes V(\Lambda)$;

(iii) a symmetric bilinear form S_Λ on $V(\Lambda)$.

As a vector space, we set $V(\Lambda) = {}'\mathfrak{f}$. We will define on $V(\Lambda)$ two gradings. The first one, $\mathbb{N}[I]$-grading coincides with the grading on ${}'\mathfrak{f}$. If $x \in V(\Lambda)$ is a homogeneous element, we will denote by $\mathrm{depth}(x)$ its depth as an element of ${}'\mathfrak{f}$.

The second grading — X-grading — is defined as follows. By definition, we set

$$V(\Lambda)_\lambda = \oplus_{\nu\in\mathbb{N}[I]|\Lambda-\lambda_\nu=\lambda}\, {}'\mathfrak{f}_\nu$$

for $\lambda \in X$. In particular, $V(\Lambda)_\Lambda = {}'\mathfrak{f}_0 = \mathsf{k}\cdot 1$. We will denote the element 1 in $V(\Lambda)$ by v_Λ.

By definition, multiplication

$$'\mathfrak{f} \otimes V(\Lambda) \longrightarrow V(\Lambda) \tag{66}$$

coincides with the multiplication in $'\mathfrak{f}$.

Let us define an X-grading in $'\mathfrak{f}$ by setting

$$'\mathfrak{f}_\lambda = \oplus_{\nu\in\mathbb{N}[I]|-\lambda_\nu=\lambda}\, '\mathfrak{f}_\nu$$

for $\lambda \in X$. The map (66) is compatible with both $\mathbb{N}[I]$ and X-gradings (we define gradings on the tensor product as usually as a sum of gradings of factors).

2.16. The form S_Λ. Let us define linear operators $\epsilon_i : V(\Lambda) \longrightarrow V(\Lambda),\ i \in I$, as unique operators such that $\epsilon_i(v_\Lambda) = 0$ and

$$\epsilon_i(\theta_j x) = [\langle\beta, i\rangle]_\zeta \delta_{i,j} x + \zeta^{i\cdot j}\theta_j\epsilon_i(x) \tag{67}$$

for $j \in I,\ x \in V(\Lambda)_\beta$.

We define $S_\Lambda : V(\Lambda) \otimes V(\Lambda) \longrightarrow \mathsf{k}$ as a unique linear map such that $S_\Lambda(v_\Lambda, v_\Lambda) = 1$, and

$$S_\Lambda(\theta_i x, y) = S_\Lambda(x, \epsilon_i(y)) \tag{68}$$

for all $x, y \in V(\Lambda),\ i \in I$. Let us list elementary properties of S_Λ.

2.16.1. *Different graded components $V(\Lambda)_\nu,\ \nu \in \mathbb{N}[I]$, are orthogonal with respect to S_Λ.*

This follows directly from the definition.

2.16.2. *The form S_Λ is symmetric.*

This is an immediate corollary of the formula

$$S_\Lambda(\epsilon_i(y), x) = S_\Lambda(y, \theta_i x) \tag{69}$$

which in turn is proved by an easy induction on depth(x).

2.16.3. *"Quasiclassical" limit.* Let us consider restriction of our form to the homogeneous component $V(\Lambda)_\lambda$ of depth N. If we divide our form by $(\zeta - 1)^N$ and formally pass to the limit $\zeta \longrightarrow 1$, we get the "Shapovalov" contravariant form as defined in [SV1], 6.4.1.

The next lemma is similar to 2.12.

2.17. **Lemma.** *For any $\vec{K},\ \vec{K'}$ as in 2.12 we have*

$$S_\Lambda(\theta_{\vec{K}} v_\Lambda, \theta_{\vec{K'}} v_\Lambda) = \sum_{\tau\in\Sigma_N:\ \tau(\vec{K})=\vec{K'}} \zeta(\vec{K};\tau)A(\vec{K},\Lambda;\tau)$$

where

$$A(\vec{K},\Lambda;\tau) = \prod_{a=1}^{N} [\langle\Lambda - \sum_{b:\ b<a,\tau(b)<\tau(a)} \lambda_{i_b}, i_a\rangle]_\zeta.$$

Proof. Induction on N, using definition of S_Λ. □

2.18. **Coaction.** Let us define a linear map

$$\Delta_\Lambda : V(\Lambda) \longrightarrow {}'\mathfrak{f} \otimes V(\Lambda) \tag{70}$$

as follows. Let us introduce linear operators $t_i : {}'\mathfrak{f}^+ \otimes V(\Lambda) \longrightarrow {}'\mathfrak{f}^+ \otimes V(\Lambda)$, $i \in I$, by the formula

$$t_i(x \otimes y) = \theta_i x \otimes y - \zeta^{i\cdot\nu - 2\langle\lambda,i\rangle} \cdot x\theta_i \otimes y + \zeta^{i\cdot\nu} x \otimes \theta_i y \tag{71}$$

for $x \in {}'\mathfrak{f}_\nu$ and $y \in V(\Lambda)_\lambda$.

By definition,

$$\Delta_\Lambda(\theta_{i_N} \cdot \ldots \cdot \theta_{i_1} v_\Lambda) = 1 \otimes \theta_{i_N} \cdot \ldots \cdot \theta_{i_1} v_\Lambda \tag{72}$$
$$+[\langle \Lambda - \lambda_{i_1} - \ldots - \lambda_{i_{N-1}}, i_N\rangle]_\zeta \cdot \theta_{i_N} \otimes \theta_{i_{N-1}} \cdot \ldots \cdot \theta_{i_1} v_\Lambda$$
$$+ \sum_{j=1}^{N-1} [\langle \Lambda - \lambda_{i_1} - \ldots - \lambda_{i_{j-1}}, i_j\rangle]_\zeta \cdot t_{i_N} \circ t_{i_{N-1}} \circ \ldots \circ t_{i_{j+1}}(\theta_{i_j} \otimes \theta_{i_{j-1}} \cdot \ldots \cdot \theta_{i_1} v_\Lambda)$$

2.19. Let us define linear operators

$$\mathbf{ad}_{\theta_i,\lambda} : {}'\mathfrak{f} \longrightarrow {}'\mathfrak{f},\ i \in I,\ \lambda \in X \tag{73}$$

by the formula

$$\mathbf{ad}_{\theta_i,\lambda}(x) = \theta_i x - \zeta^{i\cdot\nu - 2\langle\lambda,i\rangle} \cdot x\theta_i \tag{74}$$

for $x \in {}'\mathfrak{f}_\nu$.

Let us note the following relation

$$(\delta_i \circ \mathbf{ad}_{\theta_j,\lambda} - \zeta^{i\cdot j} \cdot \mathbf{ad}_{\theta_j,\lambda} \circ \delta_i)(x) = [\langle \lambda - \lambda_\nu, i\rangle]_\zeta \delta_{ij} x \tag{75}$$

for $x \in {}'\mathfrak{f}_\nu$, where δ_i are operators defined in 2.11, and δ_{ij} the Kronecker symbol.

2.20. **Formula for coaction.** Let us pick a sequence $\vec{I} = (i_N, i_{N-1}, \ldots, i_1)$. To shorten the notations, we set

$$\mathbf{ad}_{j,\lambda} := \mathbf{ad}_{\theta_{i_j},\lambda},\ j = 1, \ldots, N \tag{76}$$

2.20.1. *Quantum commutators.* For any non-empty subset $Q \subset [N]$, set $\theta_{\vec{I},Q} := \theta_{\vec{I}_Q}$ where $\vec{I}_Q$ denotes the sequence obtained from $\vec{I}$ by omitting all entries i_j, $j \in Q$. We will denote ${}'\mathfrak{f}_Q = {}'\mathfrak{f}_{\nu_Q}$ where $\nu_Q := \sum_{j\in Q} i_j$.

Let us define an element $[\theta_{\vec{I},Q,\Lambda}] \in {}'\mathfrak{f}_Q$ as follows. Set

$$[\theta_{\vec{I},\{j\},\Lambda}] = \zeta^{i_j\cdot(\sum_{k>j} i_k)} \theta_{i_j} \tag{77}$$

for all $j \in [N]$.

Suppose now that $\mathrm{card}(Q) = l + 1 \geq 2$. Let $Q = \{j_0, j_1, \ldots, j_l\}$, $j_0 < j_1 < \ldots < j_l$. Define the weights

$$\lambda_a = \Lambda - \lambda_{\sum i_k},\ a = 1, \ldots, l,$$

where the summation is over k from 1 to $j_a - 1$, $k \neq j_1, j_2, \ldots, j_{a-1}$.

Let us define sequences $\vec{N} := (N, N-1, \ldots, 1)$, $\vec{Q} = (j_l, j_{l-1}, \ldots, j_0)$ and $\vec{N}_Q$ obtained from $\vec{N}$ by omitting all entries $j \in Q$. Define the permutation $\tau_Q \in \Sigma_N$ by the requirement

$$\tau_Q(\vec{N}) = \vec{Q} || \vec{N}_Q$$

where || denotes concatenation.

Set by definition

$$[\theta_{\vec{I},Q,\Lambda}] = \zeta(\vec{I},\tau_Q)\cdot \mathbf{ad}_{j_l,\lambda_l}\circ \mathbf{ad}_{j_{l-1},\lambda_{l-1}}\circ\ldots\circ \mathbf{ad}_{j_1,\lambda_1}(\theta_{i_{j_0}}) \tag{78}$$

2.20.2. **Lemma.** *We have*

$$\Delta_\Lambda(\theta_{\vec{I}}v_\Lambda) = 1\otimes\theta_{\vec{I}}v_\Lambda + \sum_Q [(\Lambda-\lambda_{i_1}-\lambda_{i_2}-\ldots-\lambda_{i_{j(Q)-1}}, i_{j(Q)})]_\zeta\cdot[\theta_{\vec{I},Q,\Lambda}]\otimes\theta_{\vec{I},Q}v_\Lambda, \tag{79}$$

the summation over all non-empty subsets $Q\subset[N]$, $j(Q)$ denotes the minimal element of Q.

Proof. The statement of the lemma follows at once from the inspection of definition (72), after rearranging the summands. □

Several remarks are in order.

2.20.3. Formula (79) as similar to [S], 2.5.4.

2.20.4. If all elements i_j are distinct then the part of the sum in the rhs of (79) corresponding to one-element subsets Q is equal to $\sum_{j=1}^N \theta_{i_j}\otimes\epsilon_{i_j}(\theta_{\vec{I}}v_\Lambda)$.

2.20.5. *"Quasiclassical" limit.* It follows from the definition of quantum commutators that if we divide the rhs of (79) by $(\zeta-1)^N)$ and formally pass to the limit $\zeta\longrightarrow 1$, we get the expression for the coaction obtained in [SV1], 6.15.3.2.

2.21. Let us define the space $V(\Lambda)^*$ as the direct sum $\oplus_\nu V(\Lambda)^*_\nu$. We define an $\mathbb{N}[I]$-grading on it as $(V(\Lambda)^*)_\nu = V(\Lambda)^*_\nu$, and an X-grading as $V(\Lambda)^*_\lambda = \oplus_{\nu:\ \Lambda-\lambda_\nu=\lambda}V(\Lambda)^*_\nu$. The form S_Λ induces the map

$$S_\Lambda : V(\Lambda)\longrightarrow V(\Lambda)^* \tag{80}$$

compatible with both gradings.

2.22. **Tensor products.** Suppose we are given n weights $\Lambda_0,\ldots,\Lambda_{n-1}$.

2.22.1. For every $m\in\mathbb{N}$ we introduce a bilinear form $S = S_{m;\Lambda_0,\ldots,\Lambda_{n-1}}$ on the tensor product $'\mathfrak{f}^{\otimes m}\otimes V(\Lambda_0)\otimes\ldots\otimes V(\Lambda_{n-1})$ by the formula

$$S(x_1\otimes\ldots\otimes x_m\otimes y_0\otimes\ldots\otimes y_{n-1}, x'_1\otimes\ldots\otimes x'_m\otimes y'_0\otimes\ldots\otimes y'_{n-1}) = \prod_{i=1}^m S(x_i,x'_i)\prod_{j=0}^{n-1} S_{\Lambda_j}(y_j,y'_j)$$

(in the evident notations). This form defines mappings

$$S:\ '\mathfrak{f}^{\otimes m}\otimes V(\Lambda_0)\otimes\ldots\otimes V(\Lambda_{n-1})\longrightarrow\ '\mathfrak{f}^{*\otimes m}\otimes V(\Lambda_0)^*\otimes\ldots\otimes V(\Lambda_{n-1})^* \tag{81}$$

2.22.2. We will regard $V(\Lambda_0)\otimes\ldots\otimes V(\Lambda_{n-1})$ as an $'\mathfrak{f}^{\otimes n}$-module with an action

$$(u_0\otimes\ldots\otimes u_{n-1})\cdot(x_1\otimes\ldots\otimes x_{n-1})=\zeta^{-\sum_{j<i}\langle\lambda_j,\nu_i\rangle}u_0x_0\otimes\ldots\otimes u_{n-1}x_{n-1} \tag{82}$$

for $u_i\in\ '\mathfrak{f}_{\nu_i}$, $x_j\in V(\Lambda_j)_{\lambda_j}$, cf. (59). Here we regard $'\mathfrak{f}^{\otimes n}$ as an algebra according to the rule of *loc. cit.*; one checks easily using (48) that we really get a module structure.

Using the iterated comultiplication $\Delta^{(n)}$, we get a structure of an $'\mathfrak{f}$-module on $V(\Lambda_0)\otimes\ldots\otimes V(\Lambda_{n-1})$.

2.23. **Theorem.** *We have an identity*

$$S_\Lambda(xy,z)=S_{1;\Lambda}(x\otimes y,\Delta_\Lambda(z)) \tag{83}$$

for any $x\in\ '\mathfrak{f}$, $y,z\in V(\Lambda)$ and any weight Λ.

2.24. **Proof.** We may suppose that x,y and z are monomials. Let $z=\theta_{\vec{I}}v_\Lambda$ where $\vec{I}=(i_N,\ldots,i_1)$.

(a) Let us suppose first that all indices i_j are distinct. We will use the notations and computations from 2.20. The sides of (83) are non-zero only if y is equal to $z_Q:=\theta_{\vec{I},Q}v_\Lambda$ for some subset $Q\subset[N]$.

Therefore, it follows from Lemma 2.20.2 that it is enough to prove

2.24.1. **Lemma.** *For every non-empty $Q\subset[N]$ and $x\in\ '\mathfrak{f}_Q$ we have*

$$S_\Lambda(xz_Q,z)=[\langle\Lambda,i_{j(Q)}\rangle-\mu_Q\cdot i_{j(Q)}]_\zeta\cdot S(x,[\theta_{\vec{I},Q,\Lambda}])\cdot S_\Lambda(z_Q,z_Q) \tag{84}$$

where $j(Q)$ denotes the minimal element of Q, and

$$\mu_Q:=\sum_{a=1}^{j(Q)-1}i_a.$$

Proof. If $\mathrm{card}(Q)=1$ the statement follows from the definiton (77). The proof will proceed by the simultaneous induction by l and N. Suppose that $x=\theta_{i_p}\cdot x'$, so $x'\in\ '\mathfrak{f}_{Q'}$ where $Q'=Q-\{i_p\}$, $p=j_a$ for some $a\in[0,l]$. Let us set $\vec{I}'=\vec{I}-\{i_p\}$, $z'=z_{\{i_p\}}$, so that $z_Q=z'_{Q'}$.

We have

$$S_\Lambda(\theta_{i_p}x'\cdot z_Q,z)=S_\Lambda(x'\cdot z_Q,\epsilon_{i_p}(z))= \tag{85}$$

$$=[\langle\Lambda,i_p\rangle-(\sum_{k<p}i_k)\cdot i_p]_\zeta\cdot\zeta^{(\sum_{k>p}i_k)\cdot i_p}\cdot S(x'\cdot z_Q,z')=$$

$$=[\langle\Lambda,i_p\rangle-(\sum_{k<p}i_k)\cdot i_p]_\zeta\cdot[\langle\Lambda,i_{j(Q')}\rangle-\mu_{Q'}\cdot i_{j(Q')}]_\zeta\cdot\zeta^{(\sum_{k>p}i_k)\cdot i_p}\cdot$$

$$\cdot S(x,[\theta_{\vec{I},Q,\Lambda}])\cdot S_\Lambda(z'_{Q'},z'_{Q'})$$

by induction hypothesis. On the other hand,

$$S(\theta_{i_p}\cdot x',[\theta_{\vec{I},Q,\Lambda}])=S(x',\delta_{i_p}([\theta_{\vec{I},Q,\Lambda}])).$$

Therefore, to complete the induction step it is enough to prove that

$$[\langle\Lambda, i_{j(Q)}\rangle - \mu_Q \cdot i_{j(Q)}]_\zeta \cdot \delta_{i_p}([\theta_{\vec{I},Q,\Lambda}]) = \qquad (86)$$
$$= [\langle\Lambda, i_p\rangle - (\sum_{k<p} i_k)\cdot i_p]_\zeta \cdot [\langle\Lambda, i_{j(Q')}\rangle - \mu_{Q'} \cdot i_{j(Q')}]_\zeta \cdot \zeta^{(\sum_{k>p} i_k)\cdot i_p}[\theta_{\vec{I}',Q',\Lambda}]$$

This formula follows directly from the definition of quantum commutators (78) and formula (75). One has to treat separately two cases: $a > 0$, in which case $j(Q) = j(Q') = j_0$ and $a = 0$, in which case $j(Q) = j_0,\ j(Q') = j_1$. Lemma is proven. □

This completes the proof of case (a).

(b) There are repeating indices in the sequence $\vec{I}$. Suppose that $\theta_{\vec{I}} \in {}'\mathfrak{f}_\nu$. At this point we will use symmetrization constructions (and simple facts) from Section 4 below. The reader will readily see that there is no vicious circle. So, this part of the proof must be read after *loc.cit.*

There exists a finite set J and a map $\pi : J \longrightarrow I$ such that $\nu = \nu_\pi$. Using compatibility of the coaction and the forms S with symmetrization — cf. Lemmata 4.5 and 4.8 below — our claim is immediately reduced to the analogous claim for the algebra ${}^\pi\,{}'\mathfrak{f}$, the module $V({}^\pi\Lambda)$ and homogeneous weight χ_J which does not contain multiple indices and therefore follows from (a) above.

This completes the proof of the theorem. □

2.25. Let us pick a weight Λ. We can consider numbers $q_{ij} := \zeta^{i\cdot j}$ and $r_i := \langle\Lambda, i\rangle$, $i, j \in I$ as parameters of our bilinear forms.

More precisely, for a given $\nu \in \mathbb{N}[I]$ the matrix elements of the form S (resp., S_Λ) on ${}'\mathfrak{f}_\nu$ (resp., on $V(\Lambda)_{\Lambda-\lambda_\nu}$) in the standard bases of these spaces are certain universal polynomials of q_{ij} (resp., q_{ij} and r_i). Let us denote their determinants by $\det(S_\nu)(\mathbf{q})$ and $\det(S_{\Lambda,\nu})(\mathbf{q};\mathbf{r})$ respectively. These determinants are polynomials of corresponding variables with integer coefficients.

2.25.1. **Lemma.** *Polynomials* $\det(S_\nu)(\mathbf{q})$ *and* $\det(S_{\Lambda,\nu})(\mathbf{q};\mathbf{r})$ *are not identically zero.*

In other words, bilinear forms S and S_Λ are non-degenerate for generic values of parameters — "Cartan matrix" (q_{ij}) and "weight" (r_i).

Proof. Let us consider the form S_Λ first. The specialization of the matrix of $S_{\Lambda,\nu}$ at $\zeta = 1$ is the identity matrix. It follows easily that $\det(S_{V,\nu})(\mathbf{q};\mathbf{r}) \neq 0$.

Similarly, the matrix of S_ν becomes identity at $\zeta = 0$, which implies the generic non-degeneracy. □

2.26. **Theorem.** *Coaction Δ_Λ is coassociative, i.e.*

$$(1_{'\mathfrak{f}} \otimes \Delta_\Lambda) \circ \Delta_\Lambda = (\Delta \otimes 1_{V(\Lambda)}) \circ \Delta_\Lambda. \qquad (87)$$

Proof. The equality (87) is a polynomial identity depending on parameters q_{ij} and r_i of the preceding subsection. For generic values of these parameters it is true due to associativity of the action of ${}'\mathfrak{f}$ an $V(\Lambda)$, Theorem 2.23 and Lemma 2.25.1. Therefore it is true for all values of parameters. □

2.27. The results Chapter 2 below provide a different, geometric proof of Theorems 2.23 and 2.26. Namely, the results of Section 8 summarized in Theorem 8.21 provide an isomorphism of our algebraic picture with a geometric one, and in the geometrical language the above theorems are obvious: they are nothing but the naturality of the canonical morphism between the extension by zero and the extension by star, and the claim that a Cousin complex is a complex. Lemma 2.25.1 also follows from geometric considerations: the extensions by zero and by star coincide for generic values of monodromy.

2.28. By Theorem 2.26 the dual maps

$$\Delta_\Lambda^* : {}'\mathfrak{f}\otimes V(\Lambda)^* \longrightarrow V(\Lambda)^* \tag{88}$$

give rise to a structure of a ${}'\mathfrak{f}^*$-module on $V(\Lambda)^*$.

More generally, suppose we are given n modules $V(\Lambda_0),\ldots,V(\Lambda_{n-1})$. We regard the tensor product $V(\Lambda_0)^*\otimes\ldots\otimes V(\Lambda_{n-1})^*$ as a ${}'\mathfrak{f}^{*\otimes n}$-module according to the "sign" rule (82). Using iterated comultiplication (62) we get a structure of a ${}'\mathfrak{f}^*$-module on $V(\Lambda_0)^*\otimes\ldots\otimes V(\Lambda_{n-1})^*$.

2.28.1. The square

$$\begin{array}{ccc} {}'\mathfrak{f}\otimes V(\Lambda_0)\otimes\ldots\otimes V(\Lambda_{n-1}) & \longrightarrow & V(\Lambda_0)\otimes\ldots\otimes V(\Lambda_{n-1}) \\ S\downarrow & & \downarrow S \\ {}'\mathfrak{f}^*\otimes V(\Lambda_0)^*\otimes\ldots\otimes V(\Lambda_{n-1})^* & \longrightarrow & V(\Lambda_0)^*\otimes\ldots\otimes V(\Lambda_{n-1})^* \end{array}$$

commutes.

This follows from 2.23 and 2.14.

3. Hochschild complexes

3.1. If A is an augmented k-algebra, A^+ — the kernel of the augmentation, M an A-module, let $C_A^\bullet(M)$ denote the following complex. By definition, $C_A^\bullet(M)$ is concentrated in non-positive degrees. For $r\geq 0$

$$C_A^{-r}(M) = A^{+\otimes r}\otimes M.$$

We will use a notation $a_r|\ldots|a_1|m$ for $a_r\otimes\ldots a_1\otimes m$.

The differential $d: C_A^{-r}(M)\longrightarrow C_A^{-r+1}(M)$ acts as

$$d(a_r|\ldots|a_1|m) = \sum_{p=1}^{r-1}(-1)^p a_r|\ldots|a_{p+1}a_p|\ldots a_1|m + a_r|\ldots a_2|a_1 m.$$

We have canonically $H^{-r}(C_A^\bullet(M))\cong \mathrm{Tor}_r^A(\mathsf{k},M)$ where k is considered as an A-module by means of the augmentation, cf. [M], Ch. X, §2.

We will be interested in the algebras ${}'\mathfrak{f}$ and ${}'\mathfrak{f}^*$. We define the augmentation ${}'\mathfrak{f}\longrightarrow\mathsf{k}$ as being zero on all ${}'\mathfrak{f}_\nu$, $\nu\in\mathbb{N}[I]$, $\nu\neq 0$, and identity on ${}'\mathfrak{f}_0$; in the same way it is defined on ${}'\mathfrak{f}^*$.

3.2. Let M be a $\mathbb{N}[I]$-graded $'\mathfrak{f}$-module. Each term $C^{-r}_{'\mathfrak{f}}(M)$ is $\mathbb{N}[I]$-graded by the sum of gradings of tensor factors. We will denote ${}_\nu C^{-r}_{'\mathfrak{f}}(M)$ the weight ν component. For $\boldsymbol{\nu} = (\nu_0, \ldots, \nu_r) \in \mathbb{N}[I]^{r+1}$ we set

$$_{\boldsymbol{\nu}}C^{-r}_{'\mathfrak{f}}(M) = {}_{\nu_r,\ldots,\nu_0}C^{-r}_{'\mathfrak{f}}(M) = {}'\mathfrak{f}_{\nu_r} \otimes \ldots \otimes {}'\mathfrak{f}_{\nu_1} \otimes M_{\nu_0}.$$

Thus,

$$_{\nu}C^{-r}_{'\mathfrak{f}}(M) = \oplus_{\nu_0+\ldots\nu_r=\nu}\ {}_{\nu_r,\ldots,\nu_0}C^{-r}_{'\mathfrak{f}}(M).$$

Note that all ν_p must be > 0 for $p > 0$ since tensor factors lie in $'\mathfrak{f}^+$.

The differential d clearly respects the $\mathbb{N}[I]$-grading; thus the whole complex is $\mathbb{N}[I]$-graded:

$$C^\bullet_{'\mathfrak{f}}(M) = \oplus_{\nu\in\mathbb{N}[I]}\ {}_\nu C^\bullet_{'\mathfrak{f}}(M).$$

The same discussion applies to $\mathbb{N}[I]$-graded $'\mathfrak{f}^*$-modules.

3.3. Let us fix weights $\Lambda_0, \ldots, \Lambda_{n-1}$, $n \geq 1$. We will consider the Hochschild complex $C^\bullet_{'\mathfrak{f}}(V(\Lambda_0) \otimes \ldots \otimes V(\Lambda_{n-1}))$ where the structure of an $'\mathfrak{f}$-module on $V(\Lambda_0) \otimes \ldots \otimes V(\Lambda_{n-1})$ has been introduced in 2.22.

3.3.1. In the sequel we will use the following notation. If $K \subset I$ is a subset, we will denote by $\chi_K := \sum_{i\in K} i \in \mathbb{N}[I]$.

3.3.2. Suppose we have a map

$$\varrho : I \longrightarrow [-n+1, r] \tag{89}$$

where r is some non-negative integer. Let us introduce the elements

$$\nu_a(\varrho) = \chi_{\varrho^{-1}(a)}, \tag{90}$$

$a \in [-n+1, r]$. Let us denote by $\mathcal{P}_r(I; n)$ the set of all maps (89) such that $\varrho^{-1}(a) \neq \emptyset$ for all $a \in [r]$. It is easy to see that this set is not empty iff $0 \leq r \leq N$.

Let us assign to such a ϱ the space

$$_\varrho C^{-r}_{'\mathfrak{f}}(V(\Lambda_0) \otimes \ldots \otimes V(\Lambda_{n-1})) := {}'\mathfrak{f}_{\nu_r(\varrho)} \otimes \ldots \otimes {}'\mathfrak{f}_{\nu_1(\varrho)} \otimes V(\Lambda_0)_{\nu_0(\varrho)} \otimes \ldots \otimes V(\Lambda_{n-1})_{\nu_{-n+1}(\varrho)} \tag{91}$$

For each $\varrho \in \mathcal{P}_r(I; n)$ this space is non-zero, and we have

$$_{\chi_I}C^{-r}_{'\mathfrak{f}}(V(\Lambda_0) \otimes \ldots \otimes V(\Lambda_{n-1})) = \oplus_{\varrho\in\mathcal{P}_r(I;n)}\ {}_\varrho C^{-r}_{'\mathfrak{f}}(V(\Lambda_0) \otimes \ldots \otimes V(\Lambda_{n-1})) \tag{92}$$

3.4. **Bases.** Let us consider the set $\mathcal{P}_N(I; n)$. Obviously, if $\varrho \in \mathcal{P}_N(I; n)$ then $\varrho(I) = [N]$, and the induced map $I \longrightarrow [N]$ is a bijection; this way we get an isomorphism between $\mathcal{P}_N(I; n)$ and the set of all bijections $I \stackrel{\sim}{\longrightarrow} [N]$ or, to put it differently, with the set of all total orders on I.

For an arbitrary r, let $\varrho \in \mathcal{P}_r(I; n)$ and $\tau \in \mathcal{P}_N(I; n)$. Let us say that τ is *a refinement of* ϱ, and write $\varrho \leq \tau$, if $\varrho(i) < \varrho(j)$ implies $\tau(i) < \tau(j)$ for each $i, j \in I$. The map τ induces total orders on all subsets $\varrho^{-1}(a)$. We will denote by $\mathcal{O}rd(\varrho)$ the set of all refinements of a given ϱ.

Given $\varrho \leq \tau$ as above, and $a \in [-n+1, r]$, suppose that $\varrho^{-1}(a) = \{i_1, \ldots, i_p\}$ and $\tau(i_1) < \tau(i_2) < \ldots < \tau(i_p)$. Let us define a monomial

$$\theta_{\varrho\leq\tau;a} = \theta_{i_p}\theta_{i_{p-1}} \cdot \ldots \cdot \theta_{i_1} \in {}'\mathfrak{f}_{\nu_a(\varrho)}$$

If $\varrho^{-1}(a)=\emptyset$, we set $\theta_{\varrho\le\tau;a}=1$. This defines a monomial

$$\theta_{\varrho\le\tau}=\theta_{\varrho\le\tau;r}\otimes\ldots\otimes\theta_{\varrho\le\tau;1}\otimes\theta_{\varrho\le\tau;0}v_{\Lambda_0}\otimes\ldots\otimes\theta_{\varrho\le\tau;-n+1}v_{\Lambda_{n-1}}\in{}_{\varrho}C^{-r}_{'\mathfrak{f}}(V(\Lambda_0)\otimes\ldots\otimes V(\Lambda_{n-1})) \tag{93}$$

3.4.1. **Lemma.** *The set* $\{\theta_{\varrho\le\tau}|\tau\in\mathcal{O}rd(\varrho)\}$ *forms a basis of the space* ${}_{\varrho}C^{-r}_{'\mathfrak{f}}(V(\Lambda_0)\otimes\ldots\otimes V(\Lambda_{n-1}))$.

Proof is obvious. □

3.4.2. **Corollary.** *The set* $\{\theta_{\varrho\le\tau}|\varrho\in\mathcal{P}_r(I;n),\ \tau\in\mathcal{O}rd(\varrho)\}$ *forms a basis of the space* ${}_{\chi_I}C^{-r}_{'\mathfrak{f}}(V(\Lambda_0)\otimes\ldots\otimes V(\Lambda_{n-1}))$. □

3.5. We will also consider dual Hochschild complexes $C^{\bullet}_{'\mathfrak{f}^*}(V(\Lambda_0)^*\otimes\ldots\otimes V(\Lambda_{n-1})^*)$ where $V(\Lambda_0)^*\otimes\ldots\otimes V(\Lambda_{n-1})^*$ is regarded as an $'\mathfrak{f}^*$-module as in 2.28.1.

We have obvious isomorphisms

$$C^{-r}_{'\mathfrak{f}^*}(V(\Lambda_0)^*\otimes\ldots\otimes V(\Lambda_{n-1})^*)\cong C^{-r}_{'\mathfrak{f}}(V(\Lambda_0)\otimes\ldots\otimes V(\Lambda_{n-1}))^*$$

We define graded components

$${}_{\varrho}C^{-r}_{'\mathfrak{f}^*}(V(\Lambda_0)^*\otimes\ldots\otimes V(\Lambda_{n-1})^*),\ \varrho\in\mathcal{P}_r(I;n),$$

as duals to ${}_{\varrho}C^{-r}_{'\mathfrak{f}}(V(\Lambda_0)\otimes\ldots\otimes V(\Lambda_{n-1}))$.

We will denote by $\{\theta^*_{\varrho\le\tau}|\varrho\in\mathcal{P}_r(I;n),\ \tau\in\mathcal{O}rd(\varrho)\}$ the basis of ${}_{\chi_I}C^{-r}_{'\mathfrak{f}^*}(V(\Lambda_0)^*\otimes\ldots\otimes V(\Lambda_{n-1})^*)$ dual to the basis $\{\theta_{\varrho\le\tau}|\varrho\in\mathcal{P}_r(I;n),\ \tau\in\mathcal{O}rd(\varrho)\}$, 3.4.2.

3.6. The maps $S_{r;\Lambda_0,\ldots,\Lambda_{n-1}}$, cf. (81), for different r are compatible with differentials in Hochschild complexes, and therefore induce morphism of complexes

$$S:C^{\bullet}_{'\mathfrak{f}}(V(\Lambda_0)\otimes\ldots\otimes V(\Lambda_{n-1}))\longrightarrow C^{\bullet}_{'\mathfrak{f}^*}(V(\Lambda_0)^*\otimes\ldots\otimes V(\Lambda_{n-1})^*) \tag{94}$$

This follows from 2.28.1 and 2.10 (b).

4. Symmetrization

4.1. Let us fix a finite set J and a map $\pi:J\longrightarrow I$. We set $\nu_\pi:=\sum_i N_i i\in\mathbb{Z}[I]$ where $N_i:=\mathrm{card}(\pi^{-1}(i))$. The map π induces a map $\mathbb{Z}[J]\longrightarrow\mathbb{Z}[I]$ also to be denoted by π. We will use the notation $\chi_K:=\sum_{j\in K}j\in\mathbb{N}[J]$ for $K\subset J$. Thus, $\pi(\chi_J)=\nu_\pi$.

We will denote also by $\mu,\mu'\mapsto\mu\cdot\mu':=\pi(\mu)\cdot\pi(\mu')$ the bilinear form on $\mathbb{Z}[J]$ induced by the form on $\mathbb{Z}[I]$.

We will denote by Σ_π the group of all bijections $\sigma:J\longrightarrow J$ preserving fibers of π.

Let ${}^{\pi}\,'\mathfrak{f}$ be a free associative k-algebra with 1 with generators $\tilde\theta_j,\ j\in J$. It is evidently $\mathbb{N}[J]$-graded. For $\nu\in\mathbb{N}[J]$ the corresponding homogeneous component will be denoted ${}^{\pi}\,'\mathfrak{f}_\nu$. The degree of a homogeneous element $x\in{}^{\pi}\,'\mathfrak{f}$ will be denoted by $|x|\in\mathbb{N}[J]$. The group Σ_π acts on algebras ${}^{\pi}\,'\mathfrak{f}$, ${}^{\pi}\,'\mathfrak{f}^*$ by permutation of generators.

4.2. In the sequel, if G is a group and M is a G-module, M^G will denote the subset of G-invariants in M.

Let us define a k-linear "averaging" mapping

$$^{\pi}a : {}'\mathfrak{f}_{\nu_\pi} \longrightarrow ({}^{\pi}\,{}'\mathfrak{f}_{\chi_J})^{\Sigma_\pi} \tag{95}$$

by the rule

$$^{\pi}a(\theta_{i_1}\cdot\ldots\cdot\theta_{i_N}) = \sum \tilde{\theta}_{j_1}\cdot\ldots\cdot\tilde{\theta}_{j_N}, \tag{96}$$

the sum being taken over the set of all sequences $(j_1,\ldots,j_N)$ such that $\pi(j_p)=i_p$ for any p. Note that this set is naturally a Σ_π-torsor. Alternatively, $^{\pi}a$ may be defined as follows. Pick some sequence $(j_1,\ldots,j_N)$ as above, and consider an element

$$\sum_{\sigma\in\Sigma_\pi}\sigma(\tilde{\theta}_{j_1}\cdot\ldots\cdot\tilde{\theta}_{j_N});$$

this element obviously lies in $({}^{\pi}\,{}'\mathfrak{f}_{\chi_J})^{\Sigma_\pi}$ and is equal to $^{\pi}a(\theta_{i_1}\cdot\ldots\cdot\theta_{i_N})$.

The map π induces the map between homogeneous components

$$\pi : {}^{\pi}\,{}'\mathfrak{f}_{\chi_J} \longrightarrow {}'\mathfrak{f}_{\nu_\pi}. \tag{97}$$

It is clear that the composition $\pi\circ{}^{\pi}a$ is equal to the multiplication by $\mathrm{card}(\Sigma_\pi)$, and $^{\pi}a\circ\pi$ — to the action of operator $\sum_{\sigma\in\Sigma_\pi}\sigma$. As a consequence, we get

4.2.1. **Lemma,** [SV1], 5.11. *The map $^{\pi}a$ is an isomorphism.* □

4.3. Let us consider the dual to the map (97): ${}'\mathfrak{f}^*_{\nu_\pi} \longrightarrow {}^{\pi}\,{}'\mathfrak{f}^*_{\chi_J}$; it is obvious that it lands in the subspace of Σ_π-invariant functionals. Let us consider the induced map

$$^{\pi}a^* : {}'\mathfrak{f}^*_{\nu_\pi} \xrightarrow{\sim} ({}^{\pi}\,{}'\mathfrak{f}^*_{\chi_J})^{\Sigma_\pi} \tag{98}$$

It follows from the above discussion that $^{\pi}a^*$ is an isomorphism.

4.4. Given a weight $\Lambda\in X=\mathrm{Hom}(\mathbb{Z}[I],\mathbb{Z})$, we will denote by $^{\pi}\Lambda$ the composition $\mathbb{Z}[J]\xrightarrow{\pi}\mathbb{Z}[I]\xrightarrow{\Lambda}\mathbb{Z}$, and by $V(^{\pi}\Lambda)$ the corresponding Verma module over $^{\pi}\,{}'\mathfrak{f}$.

Suppose we are given n weights $\Lambda_0,\ldots,\Lambda_{n-1}$. Let us consider the Hochschild complex $C^{\bullet}_{\pi\,'\mathfrak{f}}(V(^{\pi}\Lambda_0)\otimes\ldots\otimes V(^{\pi}\Lambda_{n-1}))$. By definition, its $(-r)$-th term coincides with the tensor power $^{\pi}\,{}'\mathfrak{f}^{\otimes n+r}$. Therefore we can identify the homogeneous component ${}_{\chi_J}C^{-r}_{\pi\,'\mathfrak{f}}(V(^{\pi}\Lambda_0)\otimes\ldots\otimes V(^{\pi}\Lambda_{n-1}))$ with $(^{\pi}\,{}'\mathfrak{f}^{\otimes n+r})_{\chi_J}$ which in turn is isomorphic to $^{\pi}\,{}'\mathfrak{f}_{\chi_J}$, by means of the multiplication map $^{\pi}\,{}'\mathfrak{f}^{\otimes n+r}\longrightarrow{}^{\pi}\,{}'\mathfrak{f}$. This defines a map

$${}_{\chi_J}C^{-r}_{\pi\,'\mathfrak{f}}(V(^{\pi}\Lambda_0)\otimes\ldots\otimes V(^{\pi}\Lambda_{n-1}))\longrightarrow{}^{\pi}\,{}'\mathfrak{f}_{\chi_J} \tag{99}$$

which is an embedding when restricted to polygraded components. The Σ_π-action on ${}'\mathfrak{f}$ induces the Σ_π-action on ${}_{\chi_J}C^{-r}_{\pi\,'\mathfrak{f}}(V(^{\pi}\Lambda_0)\otimes\ldots\otimes V(^{\pi}\Lambda_{n-1}))$.

In the same manner we define a map

$${}_{\nu_\pi}C^{-r}_{'\mathfrak{f}}(V(\Lambda_0)\otimes\ldots\otimes V(\Lambda_{n-1}))\longrightarrow{}'\mathfrak{f}_{\nu_\pi} \tag{100}$$

Let us define an averaging map

$$^{\pi}a : {}_{\nu_\pi}C^{-r}_{'\mathfrak{f}}(V(\Lambda_0)\otimes\ldots\otimes V(\Lambda_{n-1}))\longrightarrow{}_{\chi_J}C^{-r}_{\pi\,'\mathfrak{f}}(V(^{\pi}\Lambda_0)\otimes\ldots\otimes V(^{\pi}\Lambda_{n-1}))^{\Sigma_\pi} \tag{101}$$

as the map induced by (95). It follows at once that this map is is an isomorphism.

These maps for different r are by definition compatible with differentials in Hochschild complexes. Therefore we get

4.4.1. **Lemma.** *The maps (101) induce isomorphism of complexes*

$$^{\pi}a:\ _{\nu_\pi}C^{\bullet}_{'\mathfrak{f}}(V(\Lambda_0)\otimes\ldots\otimes V(\Lambda_{n-1}))\overset{\sim}{\longrightarrow}\ _{\chi_J}C^{\bullet}_{\pi\ '\mathfrak{f}}(V(^{\pi}\Lambda_0)\otimes\ldots\otimes V(^{\pi}\Lambda_{n-1}))^{\Sigma_\pi}.\ \square \qquad (102)$$

4.5. **Lemma.** *The averaging is compatible with coaction. In other words, for any $\Lambda\in X$ the square*

$$\begin{array}{ccc} V(\Lambda) & \xrightarrow{\Delta_\Lambda} & '\mathfrak{f}\otimes V(\Lambda) \\ {}^{\pi}a\downarrow & & \downarrow {}^{\pi}a \\ V(^{\pi}\Lambda) & \xrightarrow{\Delta_{^{\pi}\Lambda}} & {}^{\pi}\ '\mathfrak{f}\otimes V(^{\pi}\Lambda) \end{array}$$

commutes.

Proof follows at once by inspection of the definition (72). □

4.6. Consider the dual Hochschild complexes. We have an obvious isomorphism

$$_{\chi_J}C^{-r}_{\pi\ '\mathfrak{f}^*}(V(^{\pi}\Lambda_0)^*\otimes\ldots\otimes V(^{\pi}\Lambda_{n-1})^*)\cong\ _{\chi_J}C^{-r}_{\pi\ '\mathfrak{f}}(V(^{\pi}\Lambda_0)\otimes\ldots\otimes V(^{\pi}\Lambda_{n-1}))^*;$$

using it, we define the isomorphism

$$_{\chi_J}C^{-r}_{\pi\ '\mathfrak{f}^*}(V(^{\pi}\Lambda_0)^*\otimes\ldots\otimes V(^{\pi}\Lambda_{n-1})^*)\overset{\sim}{\longrightarrow}{}^{\pi}\ '\mathfrak{f}^*_{\chi_J}$$

as the dual to (99). The Σ_π-action on the target induces the action on $_{\chi_J}C^{-r}_{\pi\ '\mathfrak{f}^*}(V(^{\pi}\Lambda_0)^*\otimes\ldots\otimes V(^{\pi}\Lambda_{n-1})^*)$. Similarly, the isomorphism

$$_{\nu_\pi}C^{-r}_{'\mathfrak{f}^*}(V(\Lambda_0)^*\otimes\ldots\otimes V(\Lambda_{n-1})^*)\overset{\sim}{\longrightarrow}\ '\mathfrak{f}^*_{\nu_\pi}$$

is defined. We define the averaging map

$$^{\pi}a^*:_{\nu_\pi}C^{-r}_{'\mathfrak{f}^*}(V(\Lambda_0)^*\otimes\ldots\otimes V(\Lambda_{n-1})^*)\longrightarrow_{\chi_J}C^{-r}_{\pi\ '\mathfrak{f}^*}(V(^{\pi}\Lambda_0)^*\otimes\ldots\otimes V(^{\pi}\Lambda_{n-1})^*)^{\Sigma_\pi} \qquad (103)$$

as the map which coincides with (98) modulo the above identifications. Again, this map is an isomorphism.

Due to Lemma 4.5 these maps for different r are compatible with the differentials in Hochschild complexes. Therefore we get

4.6.1. **Lemma.** *The maps (103) induce isomorphism of complexes*

$$^{\pi}a^*:_{\nu_\pi}C^{\bullet}_{'\mathfrak{f}^*}(V(\Lambda_0)^*\otimes\ldots\otimes V(\Lambda_{n-1})^*)\overset{\sim}{\longrightarrow}\ _{\chi_J}C^{\bullet}_{\pi\ '\mathfrak{f}^*}(V(^{\pi}\Lambda_0)^*\otimes\ldots\otimes V(^{\pi}\Lambda_{n-1})^*)^{\Sigma_\pi}.\ \square \qquad (104)$$

BILINEAR FORMS

4.7. Using the bilinear form on $\mathbb{Z}[J]$ introduced above, we define the symmetric bilinear form $S(\ ,\ )$ on ${}^{\pi}\,'\mathfrak{f}$ exactly in the same way as the form S on $'\mathfrak{f}$. Similarly, given $\Lambda \in X$, we define the bilinear form $S_{{}^{\pi}\Lambda}$ on $V({}^{\pi}\Lambda)$ as in 2.16, with I replaced by J.

4.7.1. **Lemma.** (i) *The square*

$$\begin{array}{ccc} '\mathfrak{f}_{\nu_\pi} & \xrightarrow{S} & '\mathfrak{f}^*_{\nu_\pi} \\ {}^{\pi}a \downarrow & & \downarrow {}^{\pi}a^* \\ {}^{\pi}\,'\mathfrak{f}_{\chi_J} & \xrightarrow{S} & {}^{\pi}\,'\mathfrak{f}^*_{\chi_J} \end{array}$$

commutes.

(ii) *For any $\Lambda \in X$ the square*

$$\begin{array}{ccc} V(\Lambda)_{\nu_\pi} & \xrightarrow{S_\Lambda} & V(\Lambda)^*_{\nu_\pi} \\ {}^{\pi}a \downarrow & & \downarrow {}^{\pi}a^* \\ V({}^{\pi}\Lambda)_{\chi_J} & \xrightarrow{S_{{}^{\pi}\Lambda}} & V({}^{\pi}\Lambda)^*_{\chi_J} \end{array}$$

commutes.

Proof. (i) Let us consider an element $\theta_{\vec{I}} = \theta_{i_1} \cdot \ldots \cdot \theta_{i_N} \in\ '\mathfrak{f}_{\nu_\pi}$ (we assume that $N = \mathrm{card}(J)$). The functional ${}^{\pi}a^* \circ S(\theta_{\vec{I}})$ carries a monomial $\tilde{\theta}_{j_1} \cdot \ldots \cdot \tilde{\theta}_{j_N}$ to

$$S(\theta_{i_1} \cdot \ldots \cdot \theta_{i_N}, \theta_{\pi(j_1)} \cdot \ldots \cdot \theta_{\pi(j_N)}).$$

On the other hand,

$$S \circ {}^{\pi}a(\theta_{\vec{I}})(\tilde{\theta}_{j_1} \cdot \ldots \cdot \tilde{\theta}_{j_N}) = \sum S(\tilde{\theta}_{k_1} \cdot \ldots \cdot \tilde{\theta}_{k_N}, \tilde{\theta}_{j_1} \cdot \ldots \cdot \tilde{\theta}_{j_N}),$$

the summation ranging over all sequences $\vec{K} = (k_1, \ldots, k_N)$ such that $\pi(\vec{K}) = \vec{I}$. It follows from Lemma 2.12 that both expressions are equal.

(ii) The same argument as in (i), using Lemma 2.17 instead of 2.12. □

More generally, we have

4.8. **Lemma.** *For every $m \geq 0$ and weights $\Lambda_0, \ldots, \Lambda_{n-1} \in X$ the square*

$$\begin{array}{ccc} ('\mathfrak{f}^{\otimes m} \otimes V(\Lambda_0) \otimes \ldots \otimes V(\Lambda_{n-1}))_{\nu_\pi} & \xrightarrow{S_{m;\Lambda_0,\ldots,\Lambda_{n-1}}} & ('\mathfrak{f}^{*\otimes m} \otimes V(\Lambda_0)^* \otimes \ldots \otimes V(\Lambda_{n-1})^*)_{\nu_\pi} \\ {}^{\pi}a \downarrow & & \downarrow {}^{\pi}a^* \\ ({}^{\pi}\,'\mathfrak{f}^{\otimes m} \otimes V({}^{\pi}\Lambda_0) \otimes \ldots \otimes V({}^{\pi}\Lambda_{n-1}))_{\chi_J} & \xrightarrow{S_{m;{}^{\pi}\Lambda_0,\ldots,{}^{\pi}\Lambda_{n-1}}} & ({}^{\pi}\,'\mathfrak{f}^{*\otimes m} \otimes V({}^{\pi}\Lambda_0)^* \otimes \ldots \otimes V({}^{\pi}\Lambda_{n-1})^*)_{\chi_J} \end{array}$$

commutes.

Proof is quite similar to the proof of the previous lemma. We leave it to the reader. □

5. QUOTIENT ALGEBRAS

5.1. Let us consider the map (65) $S :\ '\mathfrak{f} \longrightarrow\ '\mathfrak{f}^*$. Let us consider its kernel $\mathrm{Ker}(S)$. It follows at once from (64) that $\mathrm{Ker}(S)$ is a left ideal in $'\mathfrak{f}$. In the same manner, it is easy to see that it is also a right ideal, cf. [L1], 1.2.4.

We will denote by $\mathfrak{f}$ the quotient algebra $'\mathfrak{f}/\mathrm{Ker}(S)$. It inherits the $\mathbb{N}[I]$-grading and the coalgebra structure from $'\mathfrak{f}$, cf. *loc.cit.* 1.2.5, 1.2.6.

5.2. In the same manner, given a weight Λ, consider the kernel of $S_\Lambda : V(\Lambda) \longrightarrow V(\Lambda)^*$. Let us denote by $L(\Lambda)$ the quotient space $V(\Lambda)/\mathrm{Ker}(S_\Lambda)$. It inherits $\mathbb{N}[I]$- and X-gradings from $V(\Lambda)$. Due to Theorem 2.23 the structure of $'\mathfrak{f}$-module on $V(\Lambda)$ induces the structure of $\mathfrak{f}$-module on $L(\Lambda)$.

More generally, due to the structure of a coalgebra on $\mathfrak{f}$, all tensor products $L(\Lambda_0) \otimes \ldots \otimes L(\Lambda_{n-1})$ become $\mathfrak{f}$-modules (one should take into account the "sign rule" (82)).

5.3. We can consider Hochschild complexes $C^\bullet_{\mathfrak{f}}(L(\Lambda_0) \otimes \ldots \otimes L(\Lambda_{n-1}))$.

5.3.1. **Lemma.** *The map*

$$S : C^\bullet{}_{'\mathfrak{f}}(V(\Lambda_0) \otimes \ldots \otimes V(\Lambda_{n-1})) \longrightarrow C^\bullet{}_{'\mathfrak{f}^*}(V(\Lambda_0)^* \otimes \ldots \otimes V(\Lambda_{n-1})^*)$$

factors through the isomorphism

$$\mathrm{Im}(S) \xrightarrow{\sim} C^\bullet_{\mathfrak{f}}(L(\Lambda_0) \otimes \ldots \otimes L(\Lambda_{n-1})) \qquad (105)$$

Proof. This follows at once from the definitions. □

Chapter 2. Geometric discussion

6. Diagonal stratification and related algebras

6.1. Let us adopt notations of 4.1. We set $N := \mathrm{card}(J)$. Let ${}^{\pi}\mathbb{A}_{\mathbb{R}}$ denote a real affine space with coordinates t_j, $j \in J$, and ${}^{\pi}\mathbb{A}$ its complexification. Let us consider an arrangement $\mathcal{H}_\emptyset$ consisting of all diagonals Δ_{ij}, $i, j \in J$. Let us denote by $\mathcal{S}_\emptyset$ the corresponding stratification; $\mathcal{S}_{\emptyset,\mathbb{R}}$ will denote the corresponding real stratification of $\mathbb{A}_{\mathbb{R}}$.

The stratification $\mathcal{S}_\emptyset$ has a unique minimal stratum

$$\Delta = \bigcap \Delta_{ij} \tag{106}$$

— main diagonal; it is one-dimensional. We will denote by ${}^{\pi}\mathring{\mathbb{A}}_\emptyset$ (resp., ${}^{\pi}\mathring{\mathbb{A}}_{\emptyset,\mathbb{R}}$) the open stratum of $\mathcal{S}_\emptyset$ (resp., of $\mathcal{S}_{\emptyset,\mathbb{R}}$).

6.2. Let us describe the chambers of $\mathcal{S}_{\emptyset,\mathbb{R}}$. If C is a chamber and $\mathbf{x} = (x_j) \in C$, i.e. the embedding $J \hookrightarrow \mathbb{R}$, $j \mapsto x_j$, it induces an obvious total order on J, i.e. a bijection

$$\tau_C : J \xrightarrow{\sim} [N] \tag{107}$$

Namely, τ_C is determined uniquely by the requirement $\tau_C(i) < \tau_C(j)$ iff $x_i < x_j$; it does not depend on the choice of $\mathbf{x}$. This way we get a one-to-one correspondence between the set of chambers of $\mathcal{S}_\emptyset$ and the set of all bijections (107). We will denote by C_τ the chamber corresponding to τ.

Given C and $\mathbf{x}$ as above, suppose that we have $i, j \in J$ such that $x_i < x_j$ and there is no $k \in J$ such that $x_i < x_k < x_j$. We will say that i, j are *neighbours* in C, more precisely that *i is a left neighbour of j*.

Let $\mathbf{x}' = (x'_j)$ be a point with $x'_p = x_p$ for all $p \neq j$, and x'_j equal to some number smaller than x_i but greater than any x_k such that $x_k < x_i$. Let ${}^{ji}C$ denote the chamber containing $\mathbf{x}'$. Let us introduce a homotopy class of paths ${}^{C}\gamma_{ij}$ connecting $\mathbf{x}$ and $\mathbf{x}'$ as shown on Fig. 1 below.

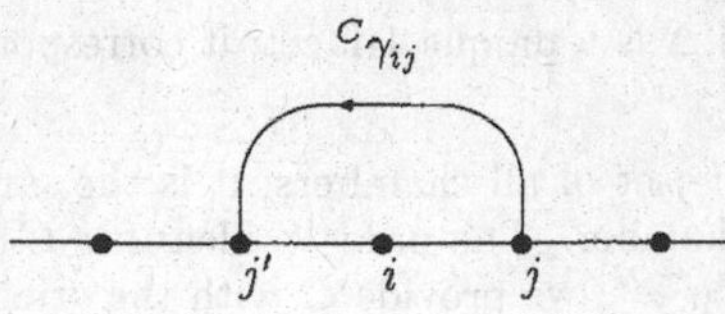

Fig. 1.

We can apply the discussion I.4.1 and consider the groupoid $\pi_1({}^{\pi}\overset{\circ}{\mathbb{A}}_{\emptyset}, {}^{\pi}\overset{\circ}{\mathbb{A}}_{\emptyset,\mathbb{R}})$. It has as the set of objects the set of all chambers. The set of morphisms is generated by all morphisms ${}^{C}\gamma_{ij}$ subject to certain evident braiding relations. We will need only the following particular case.

To define a *one-dimensional* local system $\mathcal{L}$ over ${}^{\pi}\overset{\circ}{\mathbb{A}}_{\emptyset}$ is the same as to give a set of one-dimensional vector spaces $\mathcal{L}_C$, $C \in \pi_0({}^{\pi}\overset{\circ}{\mathbb{A}}_{\emptyset,\mathbb{R}})$, together with arbitrary invertible linear operators

$$^{C}T_{ij} : \mathcal{L}_C \longrightarrow \mathcal{L}_{j i C} \tag{108}$$

("half-monodromies") defined for chambers having i as a left neighbour of j.

6.3. We define a one-dimensional local system ${}^{\pi}\mathcal{I}$ over ${}^{\pi}\overset{\circ}{\mathbb{A}}_{\emptyset}$ as follows. Its fibers ${}^{\pi}\mathcal{I}_C$ are one-dimensional linear spaces with fixed basis vectors; they will be identified with k.

Half-monodromies are defined as

$$^{C}T_{ij} = \zeta^{i \cdot j}, \ i, j \in J$$

6.4. Let $j : {}^{\pi}\overset{\circ}{\mathbb{A}}_{\emptyset} \longrightarrow {}^{\pi}\mathbb{A}$ denote an open embedding. We will study the following objects of $\mathcal{M}({}^{\pi}\mathbb{A}; \mathcal{S}_{\emptyset})$:

$$^{\pi}\mathcal{I}_? = j_?^{\pi}\mathcal{I}[N],$$

where $? = !, *$. We have a canonical map

$$m : {}^{\pi}\mathcal{I}_! \longrightarrow {}^{\pi}\mathcal{I}_* \tag{109}$$

and by definition ${}^{\pi}\mathcal{I}_{!*}$ is its image, cf. I.4.5.

6.5. For an integer r let us denote by $\mathcal{P}_r(J)$ the set of all surjective mappings $J \longrightarrow [r]$. It is evident that $\mathcal{P}_r(J) \neq \emptyset$ if and only if $1 \leq r \leq N$. To each $\rho \in \mathcal{P}_r(J)$ let us assign a point $w_\rho = (\rho(j)) \in {}^{\pi}\mathbb{A}_{\mathbb{R}}$. Let F_ρ denote the facet containing w_ρ. This way we get a bijection between $\mathcal{P}_r(J)$ and the set of r-dimensional facets. For $r = N$ we get the bijection from 6.2.

At the same time we have defined a marking of $\mathcal{H}_{\emptyset}$: by definition, ${}^{F_\rho}w = w_\rho$. This defines cells D_F, S_F.

6.6. The main diagonal Δ is a unique 1-facet; it corresponds to the unique element $\rho_0 \in \mathcal{P}_1(J)$.

We will denote by Ch the set of all chambers; it is the same as $\mathrm{Ch}(\Delta)$ in notations of Part I. Let C_τ be a chamber. The order τ identifies C with an open cone in the standard coordinate space $\mathbb{R}^N$; we provide C with the orientation induced from $\mathbb{R}^N$.

6.7. **Basis in $\Phi_\Delta({}^{\pi}\mathcal{I}_*)$.** The construction I.4.7 gives us the basis $\{c_{\Delta<C}\}$ in $\Phi_\Delta({}^{\pi}\mathcal{I}_!)^*$ indexed by $C \in \mathrm{Ch}$. We will use notation $c_{\tau,!} := c_{\Delta<C_\tau}$.

A chain $c_{\tau,!}$ looks as follows.

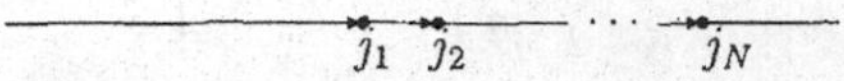

Fig. 2. A chain $c_{\tau,!}$.

Here $\tau(j_i) = i$. We will denote by $\{b_{\tau,!}\}$ the dual basis in $\Phi_\Delta(\mathcal{I}_!)$.

6.8. **Basis in $\Phi_\Delta(\mathcal{I}_*)$.** Similarly, the definition I.4.9 gives us the basis $\{c_{\Delta<C}\}$, $C \in$ Ch in $\Phi_\Delta(\mathcal{I}_*)^*$. We will use the notations $c_{\tau,*} := c_{\Delta<C_\tau}$.

If we specify the definition I.4.9 and its explanation I.4.12 to our arrangement, we get the following picture for a dual chain $c_{\tau,*}$.

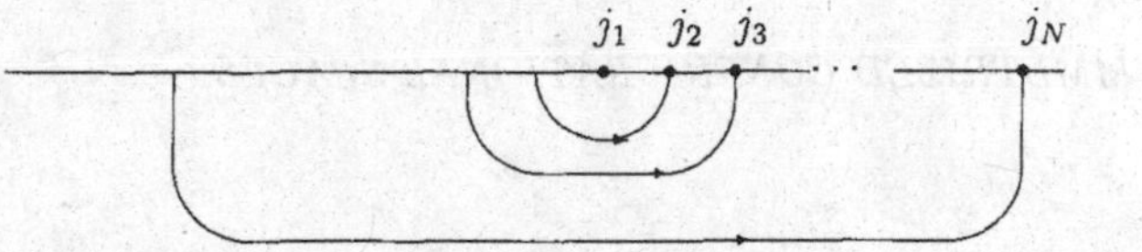

Fig. 3. A chain $c_{\tau,*}$.

This chain is represented by the section of a local system $\mathcal{I}^{-1}$ over the cell in ${}^\pi\mathring{A}_\emptyset$ shown above, which takes value 1 at the point corresponding to the end of the travel (direction of travel is shown by arrows).

To understand what is going on, it is instructive to treat the case $N = 2$ first, which essentially coincides with the Example I.4.10.

We will denote by $\{b_{\tau,*}\}$ the dual basis in $\Phi_\Delta(\mathcal{I}_*)$.

6.9. Obviously, all maps $\tau : J \longrightarrow [N]$ from $\mathcal{P}_N(J)$ are bijections. Given two such maps τ_1, τ_2, define the sign $\mathrm{sgn}(\tau_1, \tau_2) = \pm 1$ as the sign of the permutation $\tau_1\tau_2^{-1} \in \Sigma_N$.

For any $\tau \in \mathcal{P}_N(J)$ let us denote by $\vec{J}_\tau$ the sequence $(\tau^{-1}(N), \tau^{-1}(N-1), \dots, \tau^{-1}(1))$.

6.10. Let us pick $\eta \in \mathcal{P}_N(J)$. Let us define the following maps:

$$ {}^\pi\phi^{(\eta)}_{\Delta,!} : \Phi_\Delta({}^\pi\mathcal{I}_!) \longrightarrow {}^\pi{}'\mathfrak{f}_{\chi_J} \tag{110} $$

which carries $b_{\tau,!}$ to $\mathrm{sgn}(\tau, \eta) \cdot \theta_{\vec{J}_\tau}$, and

$$ {}^\pi\phi^{(\eta)}_{\Delta,*} : \Phi_\Delta({}^\pi\mathcal{I}_*) \longrightarrow {}^\pi{}'\mathfrak{f}^*_{\chi_J} \tag{111} $$

which carries $b_{\tau,*}$ to $\mathrm{sgn}(\tau, \eta) \cdot \theta^*_{\vec{J}_\tau}$.

6.11. **Theorem.** *(i) The maps* ${}^\pi\phi^{(\eta)}_{\Delta,!}$ *and* ${}^\pi\phi^{(\eta)}_{\Delta,*}$ *are isomorphisms. The square*

$$\begin{array}{ccc} \Phi_\Delta({}^\pi\mathcal{I}_!) & \overset{{}^\pi\phi^{(\eta)}_{\Delta,!}}{\underset{\sim}{\longrightarrow}} & {}^\pi\,{}'\mathfrak{f}_{\chi_J} \\ m\downarrow & & \downarrow S \\ \Phi_\Delta({}^\pi\mathcal{I}_*) & \overset{{}^\pi\phi^{(\eta)}_{\Delta,*}}{\underset{\sim}{\longrightarrow}} & {}^\pi\,{}'\mathfrak{f}^*_{\chi_J} \end{array}$$

commutes.

(ii) The map ${}^\pi\phi^{(\eta)}_{\Delta,!}$ *induces an isomorphism*

$$ {}^\pi\phi^{(\eta)}_{\Delta,!*} : \Phi_\Delta({}^\pi\mathcal{I}_{!*}) \overset{\sim}{\longrightarrow} {}^\pi\mathfrak{f}_{\chi_J} \ \square \tag{112}$$

Proof. This theorem is particular case of I.14.16, I.4.17. The claim about isomorphisms in (i) is clear. To prove the commutativity of the square, we have to compute the action of the canonical map m on our standard chains. The claim follows at once from their geometric description given above. Note that here the sign in the definition of morphisms ϕ is essential, due to orientations of our chains. (ii) is a direct corollary of (i) □

SYMMETRIZED CONFIGURATIONAL SPACES

6.12. **Colored configuration spaces.** Let us fix $\nu = \sum \nu_i i \in \mathbb{N}[I]$, $\sum_i \nu_i = N$. There exists a finite set J and a morphism $\pi : J \longrightarrow I$ such that $\mathrm{card}(\pi^{-1}(i)) = \nu_i$ for all $i \in I$. Let us call such π *an unfolding of* ν. It is unique up to a non-unique isomorphism; the automorphism group of π is precisely Σ_π, and $\nu = \nu_\pi$ in our previous notations.

Let us pick an unfolding π. As in the above discussion, we define ${}^\pi\mathbb{A}$ as a complex affine space with coordinates t_j, $j \in J$. Thus, $\dim {}^\pi\mathbb{A} = N$. The group Σ_π acts on the space ${}^\pi\mathbb{A}$ by permutations of coordinates.

Let us denote by $\mathcal{A}_\nu$ the quotient manifold ${}^\pi\mathbb{A}/\Sigma_\pi$. As an algebraic manifold, $\mathcal{A}_\nu$ is also a complex N-dimensional affine space. We have a canonical projection

$$\pi : {}^\pi\mathbb{A} \longrightarrow \mathcal{A}_\nu \tag{113}$$

The space $\mathcal{A}_\nu$ does not depend on the choice of an unfolding π. It will be called *the configuration space of* ν*-colored points on the affine line* $\mathbb{A}^1$.

We will consider the stratification on $\mathcal{A}_\nu$ whose strata are $\pi(S)$, $S \in \mathcal{S}_\emptyset$; we will denote this stratification also by $\mathcal{S}_\emptyset$; this definition does not depend on the choice of π. We will study the category $\mathcal{M}(\mathcal{A}_\nu; \mathcal{S}_\emptyset)$.

We will denote by $\overset{\circ}{\mathcal{A}}_{\nu,\emptyset}$ the open stratum. It is clear that $\pi^{-1}(\overset{\circ}{\mathcal{A}}_{\nu,\emptyset}) = {}^\pi\overset{\circ}{\mathbb{A}}_\emptyset$. The morphism π is unramified over $\overset{\circ}{\mathcal{A}}_{\nu,\emptyset}$.

The action of Σ_π on ${}^\pi\mathbb{A}$ may be extended in the evident way to the local system ${}^\pi\mathcal{I}$, hence all our spaces of geometric origin — like $\Phi_\Delta({}^\pi\mathcal{I}_!)$, etc. — get an action of Σ_π.

6.12.1. If M is an object with a Σ_π-action (for example a vector space or a sheaf), we will denote by $M^{\Sigma_\pi,-}$ the subobject $\{x \in M|$ for every $\sigma \in \Sigma_\pi$ $\sigma x = \mathrm{sgn}(\sigma)x\}$ where $\mathrm{sgn}(\sigma) = \pm 1$ is the sign of a permutation.

A morphism $f : M \longrightarrow N$ between two objects with Σ_π-action will be called *skew* *(Σ_π)-equivariant* if for any $x \in M$, $\sigma \in \Sigma_\pi$, $f(\sigma x) = \mathrm{sgn}(\sigma)\sigma f(x)$.

Let us define a local system over $\mathcal{A}_\nu$

$$\mathcal{I}_\nu = (\pi_* {}^\pi\mathcal{I})^{\Sigma_\pi,-} \tag{114}$$

6.13. Let $j : {}^\pi\mathring{\mathbb{A}}_\emptyset \hookrightarrow {}^\pi\mathbb{A}$, $j_\mathcal{A} : \mathring{\mathcal{A}}_{\nu,\emptyset} \hookrightarrow \mathcal{A}_\nu$ be the open embeddings. Let us define the following objects of $\mathcal{M}(\mathcal{A}_\nu; \mathcal{S}_\emptyset)$:

$$\mathcal{I}_{\nu ?} := j_{\mathcal{A}?}\mathcal{I}_\nu[N] \tag{115}$$

where $? =!, *$ or $!*$. We have by definition

$$\mathcal{I}_{\nu !} = (\pi_* {}^\pi\mathcal{I}_!)^{\Sigma_\pi,-} \tag{116}$$

The morphism π is finite; consequently π_* is t-exact (see [BBD], 4.1.3) and commutes with the Verdier duality. Therefore,

$$\mathcal{I}_{\nu *} = (\pi_* {}^\pi\mathcal{I}_*)^{\Sigma_\pi,-};\ \mathcal{I}_{\nu !*} = (\pi_* {}^\pi\mathcal{I}_{!*})^{\Sigma_\pi,-} \tag{117}$$

6.14. Let us define vector spaces

$$\Phi_\Delta(\mathcal{I}_{\nu ?}) := (\Phi_\Delta({}^\pi\mathcal{I}_?))^{\Sigma_\pi,-} \tag{118}$$

where $? =!, *$ or $!*$.

Let us pick a Σ_π-equivariant marking of $\mathcal{H}_\emptyset$, for example the one from 6.5; consider the corresponding cells D_Δ, S_Δ. It follows from (116) and (117) that

$$\Phi_\Delta(\mathcal{I}_{\nu ?}) = R\Gamma(\pi(D_\Delta), \pi(S_\Delta); \mathcal{I}_{\nu ?})[-1] \tag{119}$$

where $? =!, *$ or $!*$, cf. I.3.3.

6.15. The group Σ_π is acting on on ${}^\pi\,'\mathfrak{f}$. Let us pick $\eta \in \mathcal{P}_N(J)$. It follows from the definitions that the isomorphisms ${}^\pi\phi^{(\eta)}_{\Delta,!}$, ${}^\pi\phi^{(\eta)}_{\Delta,*}$ are skew Σ_π-equivariant. Therefore, passing to invariants in Theorem 6.11 we get

6.16. **Theorem.** *The maps* ${}^\pi\phi^{(\eta)}_{\Delta,!}, {}^\pi\phi^{(\eta)}_{\Delta,*}$ *induce isomorphisms included into a commutative square*

$$\begin{array}{ccc} \Phi_\Delta(\mathcal{I}_{\nu !}) & \overset{\phi^{(\eta)}_{\nu,!}}{\xrightarrow{\sim}} & '\mathfrak{f}_\nu \\ m \downarrow & & \downarrow S \\ \Phi_\Delta(\mathcal{I}_{\nu *}) & \overset{\phi^{(\eta)}_{\nu,*}}{\xrightarrow{\sim}} & '\mathfrak{f}^*_\nu \end{array}$$

and

$$\phi^{(\eta)}_{\nu,!*} : \Phi_\Delta(\mathcal{I}_{\nu !*}) \xrightarrow{\sim} \mathfrak{f}_\nu \quad \Box \tag{120}$$

7. Principal stratification

The contents of this section is parallel to I, Section 3. However, we present here certain modification of general constructions from *loc. cit.*

7.1. Let us fix a finite set J of cardinality N. In this section we will denote by $\mathbb{A}_{\mathbb{R}}$ a real affine space with fixed coordinates $t_j : \mathbb{A}_{\mathbb{R}} \longrightarrow \mathbb{R}$, $j \in J$, and by $\mathbb{A}$ its complexification. For $z \in \mathbb{C}$, $i, j \in J$ denote by $H_j(z) \subset \mathbb{A}$ a hyperplane $t_j = z$, and by Δ_{ij} a hyperplane $t_i = t_j$.

Let us consider an arrangement $\mathcal{H}$ in $\mathbb{A}$ consisting of hyperplanes $H_i(0)$ and Δ_{ij}, $i, j \in J$, $i \neq j$. It is a complexification of an evident real arrangement $\mathcal{H}_{\mathbb{R}}$ in $\mathbb{A}_{\mathbb{R}}$. As usual, the subscript $_{\mathbb{R}}$ will denote real points.

Denote by $\mathcal{S}$ the corresponding stratification of $\mathbb{A}$. To distinguish this stratification from the diagonal stratification of the previous section, we will call it *the principal stratification.* To shorten the notation, we will denote in this part by $\mathcal{D}(\mathcal{A}, \mathcal{S})$ a category which would be denoted $\mathcal{D}^b(\mathbb{A}; \mathcal{S})$ in I. In this section we will study the category $\mathcal{M}(\mathbb{A}; \mathcal{S})$.

7.1.1. Let us consider a positive cone

$$\mathbb{A}_{\mathbb{R}}^{+} = \{(t_j) | \text{ all } t_j \geq 0\} \subset \mathbb{A}_{\mathbb{R}}$$

A facet will be called *positive* if it lies inside $\mathbb{A}_{\mathbb{R}}^{+}$.

A flag $\mathbf{F}$ is called *positive* if all its facets are positive.

7.2. Let us fix a marking $\mathbf{w} = \{ {}^{F}w\}$ of $\mathcal{H}_{\mathbb{R}}$ (cf. I.3.2). For a positive facet F define

$$D_F^{+} = D_F \cap \mathbb{A}_{\mathbb{R}}^{+};\; S_F^{+} = S_F \cap \mathbb{A}_{\mathbb{R}}^{+};\; \overset{\circ}{D}{}_F^{+} = D_F^{+} - S_F^{+}.$$

Note that D_F^{+} coincides with the union of ${}^{\mathbf{F}}\Delta$ over all positive flags beginning at F, and S_F^{+} coincides with the union of ${}^{\mathbf{F}}\Delta$ as above with dim ${}^{\mathbf{F}}\Delta <$ codim F. It follows that only marking points ${}^{F}w$ for positive facets F take part in the definition of cells D_F^{+}, S_F^{+}.

7.3. Let $\mathcal{K}$ be an object of $\mathcal{D}(\mathbb{A}; \mathcal{S})$, F a positive facet of dimension p. Let us introduce a notation

$$\Phi_F^{+}(\mathcal{K}) = \Gamma(D_F^{+}, S_F^{+}; \mathcal{K})[-p].$$

This way we get a functor

$$\Phi_F^{+} : \mathcal{D}(\mathbb{A};\; \mathcal{S}) \longrightarrow \mathcal{D}^b(pt) \tag{121}$$

7.4. **Theorem.** *Functors Φ_F^{+} commute with Verdier duality. More precisely, we have canonical natural isomorphisms*

$$D\Phi_F^{+}(\mathcal{K}) \xrightarrow{\sim} \Phi_F^{+}(D\mathcal{K}). \tag{122}$$

Proof goes along the same lines as the proof of Theorem I.3.5.

7.5. First let us consider the case $N = 1$, cf. I.3.6. We will adopt notations from there and from I, Fig. 1. Our arrangement has one positive 1-dimensional facet $E = \mathbb{R}_{>0}$, let $w \in E$ be a marking.

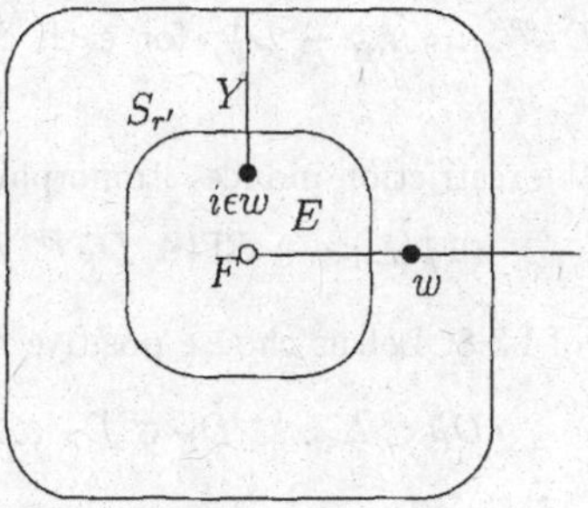

Fig. 4

We have an isomorphism

$$\Phi_F^+(\mathcal{K}) \xrightarrow{\sim} R\Gamma(\mathbb{A}, \{w\}; \mathcal{K}) \tag{123}$$

Denote $j := j_{\mathbb{A}-\{w\}}$. We have by Poincaré duality

$$D\Phi_F^+(\mathcal{K}) \xrightarrow{\sim} R\Gamma_c(\mathbb{A}; j_*j^*D\mathcal{K}) \xrightarrow{\sim} R\Gamma(\mathbb{A}, \mathbb{A}_{\geq r''}; j_*j^*D\mathcal{K}) \tag{124}$$

Let us denote $D_F^{+opp} := \mathbb{R}_{\geq w}$, and $Y = \epsilon i \cdot D_F^{+opp}$. We have

$$R\Gamma(\mathbb{A}, \mathbb{A}_{\geq r''}; j_*j^*D\mathcal{K}) \xrightarrow{\sim} R\Gamma(\mathbb{A}, Y \cup \mathbb{A}_{\geq r''}; j_*j^*D\mathcal{K}) \tag{125}$$

by homotopy. Consider the restriction map

$$res : R\Gamma(\mathbb{A}, Y \cup \mathbb{A}_{\geq r''}; j_*j^*D\mathcal{K}) \longrightarrow R\Gamma(\mathbb{A}_{\leq r'}, Y \cap \mathbb{A}_{\leq r'}; D\mathcal{K}) \tag{126}$$

7.5.1. **Claim.** *res is an isomorphism.*

In fact, Cone(res) is isomorphic to

$$R\Gamma(\mathbb{A}, \mathbb{A}_{\leq r'} \cup \mathbb{A}_{\geq r''} \cup Y; j_*j^*D\mathcal{K}) = R\Gamma_c(\mathbb{A}_{<r''}, \mathbb{A}_{\leq r'} \cup Y; j_*j^*D\mathcal{K}) \cong$$
$$\cong DR\Gamma(\mathbb{A}_{<r''} - (\mathbb{A}_{\leq r'} \cup Y); j_!j^*\mathcal{K})$$

But

$$R\Gamma(\mathbb{A}_{<r''} - (\mathbb{A}_{\leq r'} \cup Y); j_!j^*\mathcal{K}) = R\Gamma(\mathbb{A}_{<r''} - (\mathbb{A}_{\leq r'} \cup Y), \{w\}; \mathcal{K}) = 0 \tag{127}$$

evidently. This proves the claim. □

A clockwise rotation by $\pi/2$ induces an isomorphism

$$R\Gamma(\mathbb{A}_{\leq r'}, Y \cap \mathbb{A}_{\leq r'}; D\mathcal{K}) \cong R\Gamma(\mathbb{A}_{\leq r'}, \epsilon \cdot D_F^{opp}; D\mathcal{K}),$$

and the last complex is isomorphic to $\Phi_F^+(D\mathcal{K})$. This proves the theorem for $N = 1$.

7.6. Now let us return to an arbitrary J. Let us prove the theorem for F equal to the unique 0-dimensional facet.

Let us introduce the following subspaces of $\mathbb{A}_{\mathbb{R}}$ (as usually, a circle on the top will denote the interior).

$D_F^{opp} := \mathbb{A}_{\mathbb{R}} - \overset{\circ}{D}_F$; $D_F^{+opp} := \mathbb{A}_{\mathbb{R}}^+ - \overset{\circ}{D}{}_F^+$; for each cell $D_{F<C}$, $C \in \mathrm{Ch}(F)$ define $D_{F<C}^{opp} := C - \overset{\circ}{D}_{F<C}$.

It is easy to see that the restriction induces isomorphism

$$\Phi_F^+(\mathcal{K}) \xrightarrow{\sim} R\Gamma(\mathbb{A}, D_F^{+opp}; \mathcal{K}).$$

We use the notations of I.3.8. Let us choose positive numbers $r' < r''$, ϵ, such that

$$\epsilon D_F \subset \mathbb{A}_{<r'} \subset \overset{\circ}{D}_F \subset D_F \subset \mathbb{A}_{<r''}$$

Define the subspace

$$Y^+ = \epsilon i \cdot D_F^{+opp};$$

denote $j := j_{\mathbb{A}-S_F^+}$. We have isomorphisms

$$D\Phi_F^+(\mathcal{K}) \cong DR\Gamma(\mathbb{A}, S_F^+; \mathcal{K}) \cong R\Gamma_c(\mathbb{A}; j_*j^*D\mathcal{K}) \cong \tag{128}$$
$$\cong R\Gamma(\mathbb{A}, \mathbb{A}_{\geq r''}; j_*j^*D\mathcal{K}) \cong R\Gamma(\mathbb{A}, Y^+ \cup \mathbb{A}_{\geq r''}; j_*j^*D\mathcal{K})$$

Consider the restriction map

$$res : R\Gamma(\mathbb{A}, Y^+ \cup \mathbb{A}_{\geq r''}; j_*j^*D\mathcal{K}) \longrightarrow R\Gamma(\mathbb{A}_{\leq r'}, Y^+ \cap \mathbb{A}_{\leq r'}; D\mathcal{K}) \tag{129}$$

Cone(res) is isomorphic to

$$R\Gamma(\mathbb{A}, \mathbb{A}_{\leq r'} \cup \mathbb{A}_{\geq r''} \cup Y^+; j_*j^*D\mathcal{K}) = R\Gamma_c(\mathbb{A}_{<r''}, \mathbb{A}_{\leq r'} \cup Y^+; j_*j^*D\mathcal{K}) \cong$$
$$\cong DR\Gamma(\mathbb{A}_{<r''} - (\mathbb{A}_{\leq r'} \cup Y^+); j_!j^*\mathcal{K}) = R\Gamma(\mathbb{A}_{(r',r'')} - Y^+, S_F^+; \mathcal{K})$$

7.6.1. **Lemma.** (Cf. I.3.8.1.) $R\Gamma(\mathbb{A}_{(r',r'')} - Y^+, S_F^+; \mathcal{K}) = 0$.

Proof. Let us define the following subspaces of $\mathbb{A}$.

$A := \{(t_j) |$ for all j $|t_j| < 1$; there exists $j : t_j \neq 0\}$; $A_{\mathbb{R}}^+ := A \cap \mathbb{A}_{\mathbb{R}}^+$. Note that $A_{\mathbb{R}}^+ \cap i \cdot A_{\mathbb{R}}^+ = \emptyset$. Due to monodromicity, it is easy to see that

$$R\Gamma(\mathbb{A}_{(r',r'')} - Y^+, S_F^+; \mathcal{K}) \cong R\Gamma(A - i \cdot A_{\mathbb{R}}^+, A_{\mathbb{R}}^+; \mathcal{K}).$$

Therefore, it is enough to prove the following

7.6.2. **Claim.** *The restriction map*

$$R\Gamma(A - i \cdot A_{\mathbb{R}}^+; \mathcal{K}) \longrightarrow R\Gamma(A_{\mathbb{R}}^+; \mathcal{K}) \tag{130}$$

is an isomorphism.

Proof of the Claim. Let us introduce for each $k \in J$ open subspaces

$$A_k = \{(t_j) \in A | t_k \notin i \cdot \mathbb{R}_{\geq 0}\} \subset A - i \cdot A_{\mathbb{R}}^+$$

and

$$A'_k = \{(t_j) \in A_{\mathbb{R}}^+ | t_k > 0\} \subset A_{\mathbb{R}}^+$$

Obviously $A'_k = A_k \cap A_{\mathbb{R}}^+$. For each subset $M \subset J$ set $A_M := \bigcap_{k \in M} A_k$; $A'_M := \bigcap_{k \in M} A'_k$.

For each non-empty M define the spaces $B_M := \{(t_j)_{j\in M}|$ for all $j\ |t_j| < 1, t_j \notin i\cdot\mathbb{R}_{\geq 0}\}$ and $B'_M := \{(t_j)_{j\in M}|$ for all $j\ t_j \in \mathbb{R},\ 0 < t_j < 1\}$. We have obvious projections $f_M : A_M \longrightarrow B_M$, $f'_M : A'_M \longrightarrow B'_M$.

Let us look at fibers of f_M and f'_M. Given $b =\cdot (t_j)_{j\in M} \in B_M$, the fiber $f_M^{-1}(b)$ is by definition $\{(t_k)_{k\in J-M}||t_k| < 1\}$, the possible singularities of our sheaf $\mathcal{K}$ are at the hyperplanes $t_k = t_j$ and $t_k = 0$. It is easy to see that f_M is "lisse" with respect to $\mathcal{K}$ which means in particular that a stalk $(f_{M*}\mathcal{K})_b$ is equal to $R\Gamma(f_M^{-1}(b);\mathcal{K})$. The same considerations apply to f'_M. Moreover, it follows from I.2.12 that the restriction maps

$$R\Gamma(f_M^{-1}(b);\mathcal{K}) \longrightarrow R\Gamma((f'_M)^{-1}(b);\mathcal{K})$$

are isomorphisms for every $b \in B'_M$. This implies that $f'_{M*}\mathcal{K}$ is equal to the restriction of $f_{M*}\mathcal{K}$ to B'_M.

The sheaf $f_{M*}\mathcal{K}$ is smooth along the diagonal stratification. For a small $\delta > 0$ let $U_\delta \subset B_M$ denote an open subset $\{(t_j) \in B_M|\ |\arg(t_j)| < \delta$ for all $j\}$. The restriction maps $R\Gamma(B_M; f_{M*}\mathcal{K}) \longrightarrow R\Gamma(U_\delta; f_{M*}\mathcal{K})$ are isomorphisms. We have $B'_M = \bigcap_\delta U_\delta$, therefore by I.2.12 the restriction $R\Gamma(B_M; f_{M*}\mathcal{K}) \longrightarrow R\Gamma(B'_M; f_{M*}\mathcal{K})$ is an isomorphism. This implies, by Leray, that restriction maps

$$R\Gamma(A_M;\mathcal{K}) \longrightarrow R\Gamma(A'_M;\mathcal{K})$$

are isomorphisms for every non-empty M.

Obviously $A - i\cdot A^+_{\mathbb{R}} = \bigcup_{k\in J} A_k$ and $A^+_{\mathbb{R}} = \bigcup_{k\in J} A'_k$. Therefore, by Mayer-Vietoris the map (130) is an isomorphism. This proves the claim, together with the lemma. □

A clockwise rotation by $\pi/2$ induces an isomorphism

$$R\Gamma(\mathbb{A}_{\leq r'}, Y^+ \cap \mathbb{A}_{\leq r'}; D\mathcal{K}) \cong R\Gamma(\mathbb{A}_{\leq r'}, \epsilon\cdot D_F^{+opp}; D\mathcal{K}) \cong R\Gamma(\mathbb{A}_{\leq r'}, \epsilon\cdot S_F^+; D\mathcal{K})$$

and the last complex is isomorphic to $\Phi_F^+(D\mathcal{K})$ in view of 7.6. This proves the theorem for the case of the 0-facet F.

7.7. Suppose that F is an arbitrary positive facet. From the description of positive facets (see *infra*, 8.4) one sees that the cell D_F^+ is homeomorphic to a cartesian product of the form

$$D_{F_0}^+ \times D_{F_1}^+ \times \ldots \times D_{F_a}^+$$

where F_0 is a 0-facet of the principal stratification in some affine space of smaller dimension, and $D_{F_i}^+$ are the cells of *the diagonal* stratification discussed in the previous section.

Using this remark, we apply a combination of the arguments of the previous subsection (to the first factor) and of I.3.8 (to the remaining factors) to get the required isomorphism. We leave details to the reader.

The theorem is proved. □

7.8. **Theorem.** *All functors Φ_F^+ are t-exact. In other words, for all positive facets F,*

$$\Phi_F^+(\mathcal{M}(\mathbb{A};\mathcal{S})) \subset \mathcal{V}ect \subset \mathcal{D}^b(pt).$$

Proof. The same as that of I.3.9. □.

7.9. Thus we get exact functors

$$\Phi^{+} : \mathcal{M}(\mathcal{A};\mathcal{S}) \longrightarrow \mathcal{V}ect \tag{131}$$

commuting with Verdier duality.

We will also denote vector spaces $\Phi^{+}_{F}(\mathcal{M})$ by $\mathcal{M}^{+}_{F}$.

7.10. **Canonical and variation maps.** Suppose we have a positive facet E. Let us denote by $^{+}\mathcal{F}ac^{1}(E)$ the set of all positive facets F such that $E < F$, dim $F =$ dim $E + 1$. We have

$$S^{+}_{E} = \bigcup_{F\in\ ^{+}\mathcal{F}ac^{1}(E)} D^{+}_{F} \tag{132}$$

Suppose we have $\mathcal{K} \in \mathcal{D}(\mathbb{A},\mathcal{S})$.

7.10.1. **Lemma.** *We have a natural isomorphism*

$$R\Gamma(S^{+}_{E}, \bigcup_{F\in\ ^{+}\mathcal{F}ac^{1}(E)} S^{+}_{F};\mathcal{K}) \cong \oplus_{F\in\ ^{+}\mathcal{F}ac^{1}(E)} R\Gamma(D^{+}_{F}, S^{+}_{F};\mathcal{K})$$

Proof. Note that $S^{+}_{E} - \bigcup_{F\in\ ^{+}\mathcal{F}ac^{1}(E)} S^{+}_{F} = \bigcup_{F\in\ ^{+}\mathcal{F}ac^{1}(E)} \overset{\circ}{D}{}^{+}_{F}$ (disjoint union). The claim follows now from the Poincaré duality. □

Therefore, for any $F \in\ ^{+}\mathcal{F}ac^{1}(E)$ we get a natural inclusion map

$$i^{F}_{E} : R\Gamma(D^{+}_{F}, S^{+}_{F};\mathcal{K}) \hookrightarrow R\Gamma(S^{+}_{E}, \bigcup_{F'\in\mathcal{F}ac^{1}(E)} S^{+}_{F'};\mathcal{K}) \tag{133}$$

Let us define a map

$$u^{F}_{E}(\mathcal{K}) : \Phi^{+}_{F}(\mathcal{K}) \longrightarrow \Phi^{+}_{E}(\mathcal{K})$$

as a composition

$$R\Gamma(D^{+}_{F}, S^{+}_{F};\mathcal{K})[-p] \xrightarrow{i^{F}_{E}} R\Gamma(S^{+}_{E}, \bigcup_{F'\in\ ^{+}\mathcal{F}ac^{1}(E)} S^{+}_{F'};\mathcal{K})[-p] \longrightarrow \tag{134}$$

$$R\Gamma(S^{+}_{E};\mathcal{K})[-p] \longrightarrow R\Gamma(D^{+}_{E}, S^{+}_{E})[-p+1]$$

where the last arrow is the coboundary map for the couple (S^{+}_{E}, D^{+}_{E}), and the second one is evident.

This way we get a natural transormation

$$^{+}u^{F}_{E} : \Phi^{+}_{F} \longrightarrow \Phi^{+}_{E} \tag{135}$$

which will be called a *canonical map*.

We define a *variation morphism*

$$^{+}v^{E}_{F} : \Phi^{+}_{E} \longrightarrow \Phi^{+}_{F} \tag{136}$$

as follows. By definition, $^{+}v^{E}_{F}(\mathcal{K})$ is the map dual to the composition

$$D\Phi^{+}_{F}(\mathcal{K}) \xrightarrow{\sim} \Phi^{+}_{F}(D\mathcal{K}) \overset{^{+}u^{F}_{E}(D\mathcal{K})}{\longrightarrow} \Phi^{+}_{E}(D\mathcal{K}) \xrightarrow{\sim} D\Phi^{+}_{E}(\mathcal{K}).$$

7.11. **Cochain complexes.** For each $r \in [0, N]$ and $\mathcal{M} \in \mathcal{M}(\mathbb{A}, \mathcal{S})$ introduce vector spaces

$$^{+}C^{-r}(\mathbb{A}; \mathcal{M}) = \oplus_{F:F \text{ positive, } \dim F = r} \mathcal{M}_F^{+} \qquad (137)$$

For $r > 0$ or $r < -N$ set $^{+}C^{r}(\mathbb{A}; \mathcal{M}) = 0$.

Define operators

$$d : {}^{+}C^{-r}(\mathbb{A}; \mathcal{M}) \longrightarrow {}^{+}C^{-r+1}(\mathbb{A}; \mathcal{M})$$

having components $^{+}u_E^F$.

7.11.1. **Lemma.** $d^2 = 0$.

Proof. The same as that of I.3.13.1. □

This way we get a complex $^{+}C^{\bullet}(\mathbb{A}; \mathcal{M})$ lying in degrees from $-N$ to 0. It will be called the *complex of positive cochains* of the sheaf $\mathcal{M}$.

7.12. **Theorem.** *(i) A functor*

$$\mathcal{M} \mapsto {}^{+}C^{\bullet}(\mathbb{A}; \mathcal{M})$$

is an exact functor from $\mathcal{M}(\mathbb{A}; \mathcal{S})$ to the category of complexes of vector spaces.

(ii) We have a canonical natural isomorphism in $\mathcal{D}(\{pt\})$

$$^{+}C^{\bullet}(\mathbb{A}; \mathcal{M}) \xrightarrow{\sim} R\Gamma(\mathbb{A}; \mathcal{M})$$

Proof. One sees easily that restriction maps

$$R\Gamma(\mathbb{A}; \mathcal{M}) \longrightarrow R\Gamma(\mathbb{A}_{\mathbb{R}}^{+}; \mathcal{M})$$

are isomorphisms. The rest of the argument is the same as in I.3.14. □

7.13. Let us consider a function $\sum_j t_j : \mathbb{A} \longrightarrow \mathbb{A}^1$, and the corresponding vanishing cycles functor

$$\Phi_{\Sigma\, t_j} : \mathcal{D}^b(\mathbb{A}) \longrightarrow \mathcal{D}^b(\mathbb{A}_{(0)}) \qquad (138)$$

where $\mathbb{A}_{(0)} = \{(t_j) | \sum_j t_j = 0\}$, cf. [KS], 8.6.2.

If $\mathcal{K} \in \mathcal{D}(\mathbb{A}; \mathcal{S})$, it is easy to see that the complex $\Phi_{\Sigma\, t_j}(\mathcal{K})$ has the support at the origin. Let us denote by the same letter its stalk at the origin — it is a complex of vector spaces.

7.13.1. **Lemma.** *We have a natural isomorphism*

$$\Phi_{\Sigma\, t_j}(\mathcal{K}) \xrightarrow{\sim} \Phi_{\{0\}}^{+}(\mathcal{K}) \qquad (139)$$

Proof is left to the reader. □

7.14. Let us consider the setup of 6.12. Let us denote by the same letter $\mathcal{S}$ the stratification of $\mathcal{A}_\nu$ whose strata are subspaces $\pi(S)$, S being a stratum of the stratification $\mathcal{S}$ on ${}^\pi\mathbb{A}$. This stratification will be called *the principal stratification of* $\mathcal{A}_\nu$.

The function $\Sigma\ t_j$ is obviously Σ_π-equivariant, therefore it induces the function for which we will use the same notation,

$$\Sigma\ t_j : \mathcal{A}_\nu \longrightarrow \mathbb{A}^1 \tag{140}$$

Again, it is easy to see that for $\mathcal{K} \in \mathcal{D}^b(\mathcal{A}_\nu; \mathcal{S})$ the complex $\Phi_{\Sigma t_j}(\mathcal{K})$ has the support at the origin. Let us denote by $\Phi_\nu(\mathcal{K})$ its stalk at the origin.

It is known that the functor of vanishing cycles is t-exact with respect to the middle perversity; whence we get an exact functor

$$\Phi_\nu : \mathcal{M}(\mathcal{A}_\nu; \mathcal{S}) \longrightarrow \mathcal{V}ect \tag{141}$$

7.14.1. **Lemma.** *Suppose that $\mathcal{N}$ is a Σ_π- equivariant complex of sheaves over ${}^\pi\mathbb{A}$ which belongs to to $\mathcal{D}^b({}^\pi\mathbb{A}; \mathcal{S})$ (after forgetting Σ_π-action). We have a natural isomorphism*

$$\Phi_\nu((\pi_*\mathcal{N})^{\Sigma_\pi,-}) \xrightarrow{\sim} (\Phi_{\Sigma\ t_j}(\mathcal{N}))^{\Sigma_\pi,-} \tag{142}$$

Proof follows from the proper base change for vanishing cycles (see [D3], 2.1.7.1) and the exactness of the functor $(\cdot)^{\Sigma_\pi,-}$. We leave details to the reader. □

7.14.2. **Corollary.** *For a Σ_π-equivariant sheaf $\mathcal{N} \in \mathcal{M}({}^\pi\mathbb{A}_\nu; \mathcal{S})$ we have a natural isomorphism of vector spaces*

$$\Phi_\nu(\mathcal{M}) \xrightarrow{\sim} (\Phi^+_{\{0\}}(\mathcal{N}))^{\Sigma_\pi,-} \tag{143}$$

where $\mathcal{M} = (\pi_*\mathcal{N})^{\Sigma_\pi,-}$. □

8. Standard sheaves

Let us keep assumptions and notations of 4.1.

8.1. Let us denote by ${}^\pi\mathbb{A}$ the complex affine space with coordinates t_j, $j \in J$. We will consider its principal stratification as in 7.1.

Suppose we are given a positive chamber C and a point $\mathbf{x} = (x_j)_{j\in J} \in C$. There extists a unique bijection

$$\sigma_C : J \xrightarrow{\sim} [N] \tag{144}$$

such that for any $i, j \in J$, $\sigma_C(i) < \sigma_C(j)$ iff $x_i < x_j$. This bijection does not depend upon the choice of $\mathbf{x}$. This way we can identify the set of all positive chambers with the set of all isomorphisms (144), or, in other words, with the set of all total orderings of J.

Given C and $\mathbf{x}$ as above, suppose that we have $i, j \in J$ such that $x_i < x_j$ and there is no $k \in J$ such that $x_i < x_k < x_j$. We will say that i, j are *neighbours* in C, more precisely that i is a left neighbour of j.

Let $\mathbf{x}' = (x'_j)$ be a point with $x'_p = x_p$ for all $p \neq j$, and x'_j equal to some number smaller than x_i but greater than any x_k such that $x_k < x_i$. Let ${}^{ji}C$ denote the chamber containing $\mathbf{x}'$. Let us introduce a homotopy class of paths ${}^C\gamma_{ij}$ connecting $\mathbf{x}$ and $\mathbf{x}'$ as shown on Fig. 5(a) below.

On the other hand, suppose that i and 0 are neighbours in C, there is no x_j between 0 and x_i. Then we introduce the homotopy class of paths from $\mathbf{x}$ to itself as shown on Fig. 5 (b).

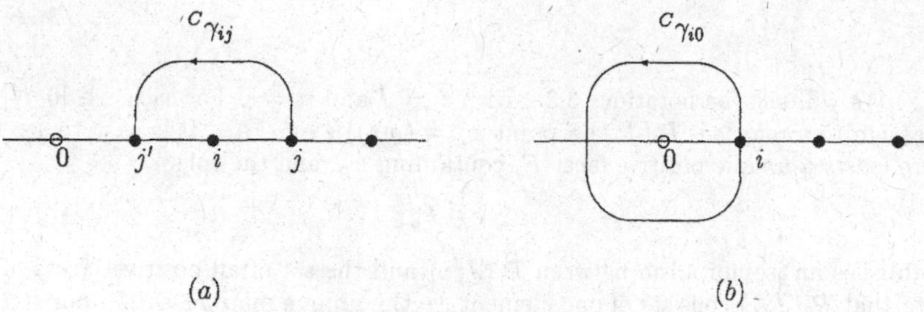

Fig. 5

All chambers are contractible. Let us denote by ${}^\pi\mathring{\mathbb{A}}^+_{\mathbb{R}}$ the union of all positive chambers. We can apply the discussion I.4.1 and consider the groupoid $\pi_1({}^\pi\mathring{\mathbb{A}}, {}^\pi\mathring{\mathbb{A}}^+_{\mathbb{R}})$. It has as the set of objects the set of all positive chambers. The set of morphisms is generated by all morphisms ${}^C\gamma_{ij}$ and ${}^C\gamma_{i0}$ subject to certain evident braiding relations. We will need only the following particular case.

To define a *one-dimensional* local system $\mathcal{L}$ over ${}^\pi\mathring{\mathbb{A}}$ is the same as to give a set of one-dimensional vector spaces $\mathcal{L}_C$, $C \in \pi_0({}^\pi\mathring{\mathbb{A}}^+_{\mathbb{R}})$, together with arbitrary invertible linear operators

$$ {}^CT_{ij} : \mathcal{L}_C \longrightarrow \mathcal{L}_{ji C} \tag{145}$$

("half-monodromies") defined for chambers where i is a left neighbour of j and

$$ {}^CT_{i0} : \mathcal{L}_C \longrightarrow \mathcal{L}_C \tag{146}$$

defined for chambers with neighbouring i and 0.

8.2. Let us fix a weight $\Lambda \in X$. We define a one-dimensional local system $\mathcal{I}({}^\pi\Lambda)$ over ${}^\pi\mathring{\mathbb{A}}$ as follows. Its fibers $\mathcal{I}({}^\pi\Lambda)_C$ are one-dimensional linear spaces with fixed basis vectors; they will be identified with k.

Monodromies are defined as

$$ {}^CT_{ij} = \zeta^{i\cdot j},\ {}^CT_{i0} = \zeta^{-2\langle\Lambda,\pi(i)\rangle}, $$

for $i, j \in J$.

8.3. Let $j: {}^\pi\mathring{\mathbb{A}} \longrightarrow {}^\pi\mathbb{A}$ denote an open embedding. We will study the following objects of $\mathcal{M}({}^\pi\mathbb{A}; \mathcal{S})$:

$$ \mathcal{I}({}^\pi\Lambda)_? = j_?\mathcal{I}({}^\pi\Lambda)[N], $$

where ? =!, $*$. We have a canonical map

$$m : \mathcal{I}(^{\pi}\Lambda)_! \longrightarrow \mathcal{I}(^{\pi}\Lambda)_* \tag{147}$$

and by definition $\mathcal{I}(^{\pi}\Lambda)_{!*}$ is its image, cf. I.4.5.

COMPUTATIONS FOR $\mathcal{I}(^{\pi}\Lambda)_!$

8.4. We will use the notations 3.3.2 with $I = J$ and $n = 1$. For each $r \in [0, N]$, let us assign to a map $\varrho \in \mathcal{P}_r(J;1)$ a point $w_\varrho = (\varrho(j))_{j\in J} \in {}^{\pi}\mathbb{A}_{\mathbb{R}}$. It is easy to see that there exists a unique positive facet F_ϱ containing w_ϱ, and the rule

$$\varrho \mapsto F_\varrho \tag{148}$$

establishes an isomorphism between $\mathcal{P}_r(J;1)$ and the set of all positive facets of $\mathcal{S}_{\mathbb{R}}$. Note that $\mathcal{P}_0(J;1)$ consists of one element — the unique map $J \longrightarrow [0]$; our stratification has one zero-dimensional facet.

At the same time we have picked a point ${}^{F_\varrho}w := w_\varrho$ on each positive facet F_ϱ; this defines cells D_F^+, S_F^+ (cf. the last remark in 7.2).

8.5. Given $\varrho \in \mathcal{P}_r(J;1)$ and $\tau \in \mathcal{P}_N(J;1)$, it is easy to see that the chamber $C = F_\tau$ belongs to $\mathrm{Ch}(F_\varrho)$ iff τ is a refinement of ϱ in the sense of 3.4. This defines a bijection between the set of all refinements of ϱ and the set of all positive chambers containing F_ϱ.

We will denote the last set by $\mathrm{Ch}^+(F_\varrho)$.

8.5.1. *Orientations.* Let $F = F_\varrho$ be a positive facet and $C = F_\tau \in \mathrm{Ch}^+(F)$. The map τ defines an isomorphism denoted by the same letter

$$\tau : J \xrightarrow{\sim} [N] \tag{149}$$

Using τ, the natural order on $[N]$ induces a total order on J. For each $a \in [r]$, let m_a denote the minimal element of $\varrho^{-1}(a)$, and set

$$J' = J_{\varrho \le \tau} := J - \{m_1, \ldots, m_r\} \subset J$$

Let us consider the map

$$(x_j) \in D_{F<C} \mapsto \{x_j - m_{\varrho(j)} | j \in J'\} \in \mathbb{R}^{J'}.$$

It is easy to see that this mapping establishes an isomorphism of the germ of the cell $D_{F<C}$ near the point ${}^{F}w$ onto a germ of the cone

$$\{0 \le u_1 \le \ldots \le u_{N-r}\}$$

in $\mathbb{R}^{J'}$ where we have denoted for a moment by (u_i) the coordinates in $\mathbb{R}^{J'}$ ordered by the order induced from J.

This isomorphism together with the above order defines an orientation on $D_{F<C}$.

8.6. **Basis in $\Phi_F^+(\mathcal{I}(^\pi\Lambda)_!)^*$.** We follow the pattern of I.4.7. Let F be a positive facet of dimension r. By definition,

$$\Phi_F^+(\mathcal{I}(^\pi\Lambda)_!) = H^{-r}(D_F^+, S_F^+; \mathcal{I}(^\pi\Lambda)_!) = H^{N-r}(D_F^+, S_F^+; j_!\mathcal{I}(^\pi\Lambda)) \cong H^{N-r}(D_F^+, S_F^+ \cup ({}_{\mathcal{H}}H_{\mathbb{R}} \cap D_F^+); j_!\mathcal{I}(^\pi$$

Note that we have

$$D_F^+ - (S_F^+ \cup {}_{\mathcal{H}}H_{\mathbb{R}}) = \bigcup_{C \in \mathrm{Ch}^+(F)} \overset{\circ}{D}_{F<C}$$

(disjoint union), therefore by additivity

$$H^{N-r}(D_F^+, S_F^+ \cup ({}_{\mathcal{H}}H_{\mathbb{R}} \cap D_F^+); j_!\mathcal{I}(^\pi\Lambda)) \cong \oplus_{C \in \mathrm{Ch}^+(F)} H^{N-r}(D_{F<C}, S_{F<C}; j_!\mathcal{I}(^\pi\Lambda)).$$

By Poincaré duality,

$$H^{N-r}(D_{F<C}, S_{F<C}; j_!\mathcal{I}(^\pi\Lambda))^* \cong H^0(\overset{\circ}{D}_{F<C}; \mathcal{I}(^\pi\Lambda)^{-1})$$

— here we have used the orientations of cells $D_{F<C}$ introduced above. By definition of the local system $\mathcal{I}$, the last space is canonically identified with k.

If $F = F_\varrho$, $C = F_\tau$, we will denote by $c_{\varrho\le\tau} \in \Phi_F^+(\mathcal{I}(^\pi\Lambda)_!)^*$ the image of $1 \in H^0(\overset{\circ}{D}_{F<C}; \mathcal{I}(^\pi\Lambda)^{-1})$. Thus the chains $c_{\varrho\le\tau}$, $\tau \in \mathcal{O}rd(\varrho)$, form a basis of the space $\Phi_F^+(\mathcal{I}(^\pi\Lambda)_!)^*$.

8.7. **Diagrams.** It is convenient to use the following diagram notations for chains $c_{\varrho\le\tau}$.

Let us denote elements of J by letters $a, b, c, \ldots$. An r-dimensional chain $c_{\varrho\le\tau}$ where $\varrho : J \longrightarrow [0, r]$, is represented by a picture:

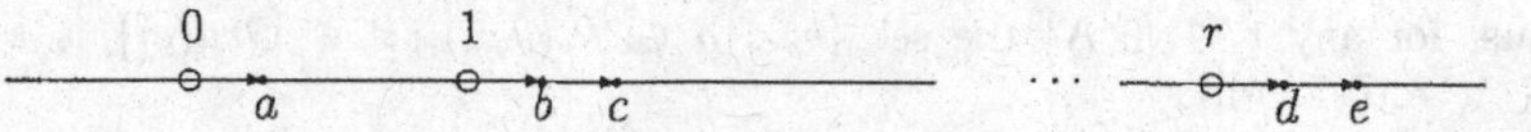

Fig. 6. Chain $c_{\varrho\le\tau}$.

A picture consists of $r+1$ fragments:

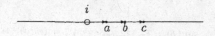

Fig. 7. Set $\varrho^{-1}(i)$.

where $i = 0, \ldots, r$, the i-th fragment being a blank circle with a number of small vectors going from it. These vectors are in one-to-one correspondence with the set

$\varrho^{-1}(i)$; their ends are labeled by elements of this set. Their order (from left to right) is determined by the order on $\varrho^{-1}(i)$ induced by τ. The point 0 may have no vectors (when $\varrho^{-1}(0) = \emptyset$); all other points have at least one vector.

8.8. Suppose we have $\varrho \in \mathcal{P}_r(J;1)$, $\varrho' \in \mathcal{P}_{r+1}(J;1)$. It is easy to see that $F_\varrho < F_{\varrho'}$ if and only if there exists $i \in [0,r]$ such that $\varrho = \delta_i \circ \varrho'$ where $\delta_i : [0, r+1] \longrightarrow [0,r]$ carries a to a if $a \leq i$ and to $a-1$ if $a \geq i+1$. We will write in this case that $\varrho < \varrho'$.

Let us compute the dual to the canonical map

$$ {}^+u^* : \Phi^+_{F_\varrho}(\mathcal{I}({}^\pi\Lambda)_!)^* \longrightarrow \Phi^+_{F_{\varrho'}}(\mathcal{I}({}^\pi\Lambda)_!)^*. \tag{150} $$

Suppose we have $\tau \in \mathcal{O}rd(\varrho')$; set $C = F_\tau$, thus $F_\varrho < F_{\varrho'} < C$. Let us define the sign

$$ \operatorname{sgn}(\varrho < \varrho' \leq \tau) = (-1)^{\sum_{j=i+1}^{r+1} \operatorname{card}((\varrho')^{-1}(j)-1)} \tag{151} $$

This sign has the following geometrical meaning. The cell $D_{F_{\varrho'<C}}$ lies in the boundary of $D_{F_{\varrho<C}}$. We have oriented these cells above. Let us define the compatibility of these orientations as follows. Complete an orienting basis of the smaller cell by a vector directed outside the larger cell — if we get the orientation of the larger cell, we say that the orientations are compatible, cf. I.4.6.1.

It is easy to see from the definitions that the sign (151) is equal $+1$ iff the orientations of $D_{F_{\varrho'<C}}$ and $D_{F_{\varrho<C}}$ are compatible. As a consequence, we get

8.8.1. **Lemma.** *The map (150) has the following matrix:*

$$ {}^+u^*(c_{\varrho\leq\tau}) = \sum \operatorname{sgn}(\varrho < \varrho' \leq \tau) c_{\varrho'\leq\tau}, $$

the summation over all ϱ' such that $F_\varrho < F_{\varrho'} < F_\tau$ and $\dim F_{\varrho'} = \dim F_\varrho + 1$. □

8.9. **Isomorphisms ${}^\pi\phi_!$.** We will use notations of 3.4 with I replaced by J, $'\mathfrak{f}$ by ${}^\pi{}'\mathfrak{f}$, with $n = 1$ and $\Lambda_0 = {}^\pi\Lambda$.

Thus, for any $r \in [0, N]$ the set $\{b_{\varrho\leq\tau} | \varrho \in \mathcal{P}_r(J;1),\ \tau \in \mathcal{O}rd(\varrho)\}$ is a basis of ${}^+C^{-r}({}^\pi\mathbb{A}; \mathcal{I}({}^\pi\Lambda)_!)$.

Let us pick $\eta \in \mathcal{P}_N(J)$. Any $\tau \in \mathcal{P}_N(J;1)$ induces the bijection $\tau' : J \xrightarrow{\sim} [N]$. We will denote by $\operatorname{sgn}(\tau, \eta) = \pm 1$ the sign of the permutation $\tau'\eta^{-1}$.

Let us denote by $\{b_{\varrho\leq\tau} | \tau \in \mathcal{O}rd(\varrho)\}$ the basis in $\Phi^+_{F_\varrho}(\mathcal{I}({}^\pi\Lambda)_!)$ dual to $\{c_{\varrho\leq\tau}\}$.

Let us define isomorphisms

$$ {}^\pi\phi^{(\eta)}_{\varrho,!} : \Phi^+_{F_\varrho}(\mathcal{I}({}^\pi\Lambda)_!) \xrightarrow{\sim} {}_\varrho C^{-r}_{{}^\pi{}'\mathfrak{f}}(V({}^\pi\Lambda)) \tag{152} $$

by the formula

$$ {}^\pi\phi^{(\eta)}_{\varrho,!}(b_{\varrho\leq\tau}) = \operatorname{sgn}(\tau,\eta)\operatorname{sgn}(\varrho)\theta_{\varrho\leq\tau} \tag{153} $$

(see (93)) where

$$ \operatorname{sgn}(\varrho) = (-1)^{\sum_{i=1}^{r}(r-i+1)\cdot(\operatorname{card}(\varrho^{-1}(i))-1)} \tag{154} $$

for $\varrho \in \mathcal{P}_r(J;1)$. Taking the direct sum of ${}^\pi\phi^{(\eta)}_{\varrho,!}$, $\varrho \in \mathcal{P}_r(J;1)$, we get isomorphisms

$$ {}^\pi\phi^{(\eta)}_{r,!} : {}^+C^{-r}({}^\pi\mathbb{A}; \mathcal{I}({}^\pi\Lambda)_!) \xrightarrow{\sim} {}_{\chi_J}C^{-r}_{{}^\pi{}'\mathfrak{f}}(V({}^\pi\Lambda)) \tag{155} $$

A direct computation using 8.8.1 shows that the maps ${}^\pi\phi^{(\eta)}_{r,!}$ are compatible with differentials. Therefore, we arrive at

8.10. **Theorem.** *The maps* ${}^{\pi}\phi^{(\eta)}_{\tau,!}$ *induce an isomorphism of complexes*

$$ {}^{\pi}\phi^{(\eta)}_{!} : {}^{+}C^{\bullet}({}^{\pi}\mathbb{A}; \mathcal{I}({}^{\pi}\Lambda)_!) \xrightarrow{\sim} {}_{\chi_J}C^{\bullet}_{{}^{\pi}\; ' \mathfrak{f}}(V({}^{\pi}\Lambda)) \;\Box \tag{156}$$

COMPUTATIONS FOR $\mathcal{I}({}^{\pi}\Lambda)_*$

8.11. **Bases.** The Verdier dual to $\mathcal{I}({}^{\pi}\Lambda)_*$ is canonically isomorphic to $\mathcal{I}({}^{\pi}\Lambda)_!^{-1}$. Therefore, by Theorem 7.4 for each positive facet F we have natural isomorphisms

$$\Phi_F^+(\mathcal{I}({}^{\pi}\Lambda)_*)^* \cong \Phi_F^+(\mathcal{I}({}^{\pi}\Lambda)_!^{-1}) \tag{157}$$

Let $\{\tilde{c}_{\varrho\leq\tau} | \tau \in \mathcal{O}rd(\varrho)\}$ be the basis in $\Phi_F^+(\mathcal{I}({}^{\pi}\Lambda)_!^{-1})^*$ defined in 8.6, with $\mathcal{I}({}^{\pi}\Lambda)$ replaced by $\mathcal{I}({}^{\pi}\Lambda)^{-1}$. We will denote by $\{c_{\varrho\leq\tau,*} | \tau \in \mathcal{O}rd(\varrho)\}$ the dual basis in $\Phi_F^+(\mathcal{I}({}^{\pi}\Lambda)_*)^*$. Finally, we will denote by $\{b_{\varrho\leq\tau,*} | \tau \in \mathcal{O}rd(\varrho)\}$ the basis in $\Phi_F^+(\mathcal{I}({}^{\pi}\Lambda)_*)$ dual to the previous one.

Our aim in the next subsections will be the description of canonical morphisms $m : \Phi_F^+(\mathcal{I}({}^{\pi}\Lambda)_!) \longrightarrow \Phi_F^+(\mathcal{I}({}^{\pi}\Lambda)_*)$ and of the cochain complex ${}^{+}C^{\bullet}({}^{\pi}\mathbb{A}; \mathcal{I}({}^{\pi}\Lambda)_*)$ in terms of our bases.

8.12. **Example.** Let us pick an element, $i \in I$ and let $\pi : J := \{i\} \hookrightarrow I$. Then we are in a one-dimensional situation, cf. 7.5. The space ${}^{\pi}\mathbb{A}$ has one coordinate t_i. By definition, the local system $\mathcal{I}({}^{\pi}\Lambda)$ over ${}^{\pi}\mathring{\mathbb{A}} = {}^{\pi}\mathbb{A} - \{0\}$ with the base point $w : t_i = 1$ has the fiber $\mathcal{I}_w = \mathsf{k}$ and monodromy equal to $\zeta^{-2\langle\Lambda,i\rangle}$ along a counterclockwise loop.

The stratification $\mathcal{S}_{\mathbb{R}}$ has a unique 0-dimensional facet $F = F_\varrho$ corresponding to the unique element $\varrho \in \mathcal{P}_0(J;1)$, and a unique 1-dimensional positive facet $E = F_\tau$ corresponding to the unique element $\tau \in \mathcal{P}_1(J;1)$.

Let us construct the dual chain $c_1^* := c_{\varrho\leq\tau,*}$. We adopt the notations of 7.5, in particular of Fig.1. The chain $\tilde{c}_1 := \tilde{c}_{\varrho\leq\tau} \in H^1({}^{\pi}\mathbb{A}, \{0,w\}; j_!\mathcal{I}({}^{\pi}\Lambda)^{-1})^*$ is shown on Fig. 8(a) below. As a first step, we define the dual chain $c_1^0 \in H^1({}^{\pi}\mathbb{A} - \{0,w\}, {}^{\pi}\mathbb{A}_{\geq r''}; \mathcal{I}({}^{\pi}\Lambda))^*$ — it is also shown on Fig. 8(a).

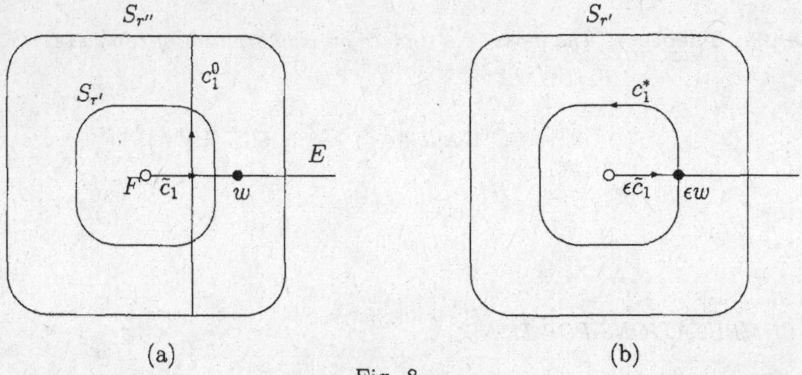

Fig. 8.

Next, we make a clockwise rotation of c_1^0 on $\pi/2$, and make a homotopy inside the disk ${}^\pi\mathbb{A}_{\leq r'}$ to a chain $c_1^* \in H^1({}^\pi\mathbb{A}_{\leq r'} - \{0\}, \epsilon \cdot w; \mathcal{I}({}^\pi\Lambda))^*$ beginning and ending at $\epsilon \cdot w$. This chain is shown on Fig. 8(b) (in a bigger scale). Modulo "homothety" identification

$$H^1({}^\pi\mathbb{A}_{\leq r'} - \{0\}, \epsilon \cdot w; \mathcal{I}({}^\pi\Lambda))^* \cong H^1({}^\pi\mathbb{A} - \{0\}, w; \mathcal{I}({}^\pi\Lambda))^* = \Phi_F(\mathcal{I}({}^\pi\Lambda)_*)^*$$

this chain coincides with $c_{\varrho\leq\tau,*}$.

The canonical map

$$m^* : \Phi_F^+(\mathcal{I}({}^\pi\Lambda)_*)^* \longrightarrow \Phi_F^+(\mathcal{I}({}^\pi\Lambda)_!)^*$$

carries $c_{\varrho\leq\tau,*}$ to $[\langle\Lambda, i\rangle]_\zeta \cdot c_{\varrho\leq\tau}$. The boundary map

$$d^* : \Phi_F^+(\mathcal{I}({}^\pi\Lambda)_*)^* \longrightarrow \Phi_E^+(\mathcal{I}({}^\pi\Lambda)_*)^*$$

carries $c_{\varrho\leq\tau,*}$ to $[\langle\Lambda, i\rangle]_\zeta \cdot c_{\tau\leq\tau,*}$.

8.13. **Vanishing cycles at the origin.** Let us return to the general situation. First let us consider an important case of the unique 0-dimensional facet — the origin. Let 0 denote the unique element of $\mathcal{P}_0(J;1)$. To shorten the notation, let us denote $\Phi_{F_0}^+$ by Φ_0^+.

The bases in $\Phi_0^+(\mathcal{I}({}^\pi\Lambda)_!)$, etc. are numbered by all bijections $\tau : J \xrightarrow{\sim} [N]$. Let us pick such a bijection. Let $\tau^{-1}(i) = \{j_i\}$, $i = 1, \ldots, N$. The chain $c_{0\leq\tau} \in \Phi_0^+(\mathcal{I}({}^\pi\Lambda)_!)^*$ is depicted as follows:

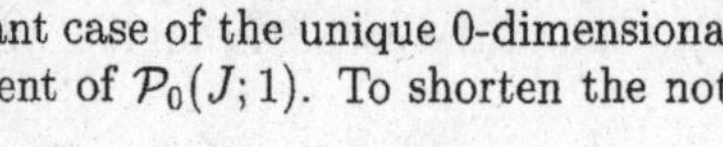

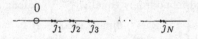

Fig. 9. $c_{0\leq\tau}$

Using considerations completely analogous to the one-dimensional case above, we see that the dual chain $c_{0\leq\tau,*} \in \Phi_0^+(\mathcal{I}(^\pi\Lambda)_*)^*$ is portayed as follows:

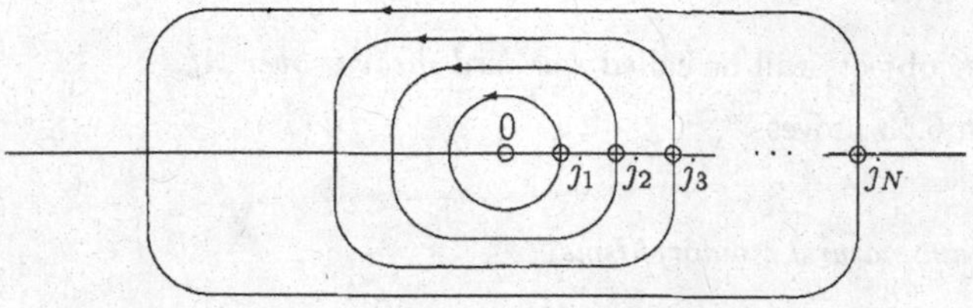

Fig. 10. $c_{0\leq\tau,*}$

The points t_{j_i} are travelling independently along the corresponding loops, in the indicated directions. The section of $\mathcal{I}(^\pi\Lambda)^{-1}$ over this cell is determined by the requierement to be equal to 1 when all points are equal to marking points, at the end of their the travel (coming from below).

8.14. **Isomorphisms** $\phi_{0,*}^{(\eta)}$. We will use notations of 3.4 and 3.5 with I replaced by J, $'\mathfrak{f}$ — by $^\pi\,'\mathfrak{f}$, with $n=1$ and $\Lambda_0 = {}^\pi\Lambda$. By definition, $C^0_{\pi\ '\mathfrak{f}}(V(^\pi\Lambda)) = V(^\pi\Lambda)$. The space $V(^\pi\Lambda)_{\chi_J}$ admits as a basis (of cardinality $N!$) the set consisting of all monomials

$$\theta_{0\leq\tau} = \theta_{\tau^{-1}(N)}\theta_{\tau^{-1}(N-1)}\cdot\ldots\cdot\theta_{\tau^{-1}(1)}v_{^\pi\Lambda},$$

where τ ranges through the set of all bijections $J \xrightarrow{\sim} [N]$. By definition, $\{\theta^*_{0\leq\tau}\}$ is the dual basis of $V(^\pi\Lambda)^*_{\chi_J}$.

Let us pick $\eta \in \mathcal{P}_N(J)$. Let us define an isomorphism

$$^\pi\phi_{0,*}^{(\eta)} : \Phi_0^+(\mathcal{I}(^\pi\Lambda)_*) \xrightarrow{\sim} V(^\pi\Lambda)^*_{\chi_J} \tag{158}$$

by the formula

$$^\pi\phi_{0,*}^{(\eta)}(b_{0\leq\tau}) = \mathrm{sgn}(\tau,\eta)\theta^*_{0\leq\tau} \tag{159}$$

8.15. **Theorem.** *The square*

$$\begin{array}{ccc} \Phi_0^+(\mathcal{I}(^\pi\Lambda)_!) & \overset{^\pi\phi_{0,!}^{(\eta)}}{\underset{\sim}{\longrightarrow}} & V(^\pi\Lambda)_{\chi_J} \\ m\downarrow & & \downarrow S_{^\pi\Lambda} \\ \Phi_0^+(\mathcal{I}(^\pi\Lambda)_*) & \overset{^\pi\phi_{0,*}^{(\eta)}}{\underset{\sim}{\longrightarrow}} & V(^\pi\Lambda)^*_{\chi_J} \end{array}$$

commutes.

Proof. This follows directly from the discussion of 8.13 and the definition of the form $S_{^\pi\Lambda}$. □

8.16. Let us pass to the setup of 6.12 and 7.14. Let $j_\nu : \overset{\circ}{\mathcal{A}}_\nu \hookrightarrow \mathcal{A}_\nu$ denote the embedding of the open stratum of the principal stratification.

Set

$$\mathcal{I}_\nu(\Lambda) = (\pi_*\mathcal{I}(^\pi\Lambda))^{\Sigma_\pi,-} \tag{160}$$

It is a local system over $\overset{\circ}{\mathcal{A}}_\nu$.

Let us define objects

$$\mathcal{I}_\nu(\Lambda)_? := j_{\nu ?}\mathcal{I}_\nu(\Lambda)[N] \in \mathcal{M}(\mathcal{A}_\nu; \mathcal{S}) \tag{161}$$

where ? =!, * or !*. These objects will be called *standard sheaves over* $\mathcal{A}_\nu$.

The same reasoning as in 6.13 proves

8.16.1. **Lemma.** *We have natural isomorphisms*

$$\mathcal{I}_\nu(\Lambda)_? \cong (\pi_* \mathcal{I}(^\pi\Lambda)_?)^{\Sigma_\pi,-} \tag{162}$$

for ? =!, * *or* !*. □

8.17. For a given $\eta \in \mathcal{P}_N(J)$ the isomorphisms $^\pi\phi^{(\eta)}_{0,*}$ and $^\pi\phi^{(\eta)}_{0,!}$ are skew Σ_π- equivariant. Therefore, after passing to invariants in Theorem 8.15 we get

8.18. **Theorem.** *The maps* $^\pi\phi^{(\eta)}_{0,*}$ *and* $^\pi\phi^{(\eta)}_{0,!}$ *induce isomorphisms included into a commutative square*

$$\begin{array}{ccc} \Phi_\nu(\mathcal{I}_\nu(\Lambda)_!) & \overset{\phi^{(\eta)}_{\nu,\Lambda,!}}{\underset{\sim}{\longrightarrow}} & V(\Lambda)_\nu \\ m \downarrow & & \downarrow S_\Lambda \\ \Phi_\nu(\mathcal{I}_\nu(\Lambda)_*) & \overset{\phi^{(\eta)}_{\nu,\Lambda,*}}{\underset{\sim}{\longrightarrow}} & V(\Lambda)^*_\nu \end{array}$$

and

$$\phi^{(\eta)}_{\nu,\Lambda,!*} : \Phi_\nu(\mathcal{I}_\nu(\Lambda)_{!*}) \overset{\sim}{\longrightarrow} L(\Lambda)_\nu \tag{163}$$

Proof follows from the previous theorem and Lemma 4.7.1(ii). □

8.19. Now suppose we are given an arbitrary r, $\varrho \in \mathcal{P}_r(J;1)$ and $\tau \in \mathcal{O}rd(\varrho)$. The picture of the dual chain $c_{\varrho\leq\tau,*}$ is a combination of Figures 10 and 3. For example, the chain dual to the one on Fig. 6 is portayed as follows:

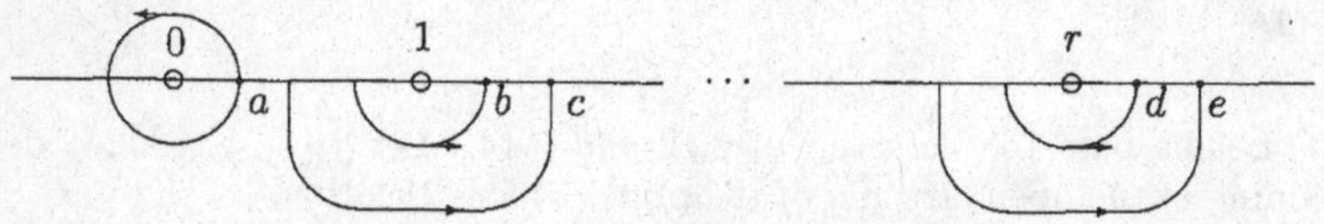

Fig. 11. Chain $c_{\varrho\leq\tau,*}$.

8.20. **Isomorphisms** ${}^{\pi}\Phi_*$. Let us pick $\eta \in \mathcal{P}_N(J)$. Let us define isomorphisms

$$^{\pi}\phi^{(\eta)}_{\varrho,*} : \Phi^+_{F_\varrho}(\mathcal{I}(^{\pi}\Lambda)_*) \xrightarrow{\sim} {}_{\varrho}C^{-r}_{\pi}{}'_{\mathfrak{f}^*}(V(^{\pi}\Lambda)^*) \tag{164}$$

by the formula

$$^{\pi}\phi^{(\eta)}_{\varrho,*}(b_{\varrho\leq\tau,*}) = \mathrm{sgn}(\tau,\eta)\mathrm{sgn}(\varrho)\theta^*_{\varrho\leq\tau} \tag{165}$$

where $\mathrm{sgn}(\varrho)$ is defined in (154).

Taking the direct sum of ${}^{\pi}\phi^{(\eta)}_{\varrho,*}$, $\varrho \in \mathcal{P}_r(J;1)$, we get isomorphisms

$$^{\pi}\phi^{(\eta)}_{r,*} : {}^{+}C^{-r}(^{\pi}\mathbb{A};\mathcal{I}(^{\pi}\Lambda)_*) \xrightarrow{\sim} {}_{\chi_J}C^{-r}_{\pi}{}'_{\mathfrak{f}^*}(V(^{\pi}\Lambda)^*) \tag{166}$$

8.21. **Theorem.** *The maps* ${}^{\pi}\phi^{(\eta)}_{r,*}$ *induce an isomorphism of complexes*

$$^{\pi}\phi^{(\eta)}_{*} : {}^{+}C^{\bullet}(^{\pi}\mathbb{A};\mathcal{I}(^{\pi}\Lambda)_*) \xrightarrow{\sim} {}_{\chi_J}C^{\bullet}_{\pi}{}'_{\mathfrak{f}^*}(V(^{\pi}\Lambda)^*) \tag{167}$$

which makes the square

$$\begin{array}{ccc} {}^{+}C^{\bullet}(^{\pi}\mathbb{A};\mathcal{I}(^{\pi}\Lambda)_!) & \overset{^{\pi}\phi^{(\eta)}_!}{\xrightarrow{\sim}} & {}_{\chi_J}C^{\bullet}_{\pi}{}'_{\mathfrak{f}}(V(^{\pi}\Lambda)) \\ m\downarrow & & \downarrow S \\ {}^{+}C^{\bullet}(^{\pi}\mathbb{A};\mathcal{I}(^{\pi}\Lambda)_*) & \overset{^{\pi}\phi^{(\eta)}_*}{\xrightarrow{\sim}} & {}_{\chi_J}C^{\bullet}_{\pi}{}'_{\mathfrak{f}^*}(V(^{\pi}\Lambda)^*) \end{array}$$

commutative.

Proof. Compatibility with differentials is verified directly and commutativity of the square are checked directly from the geometric description of our chains (actually, it is sufficient to check one of these claims — the other one follows formally).

Note the geometric meaning of operators t_i from (71) — they correspond to the deletion of the i-th loop on Fig. 10. □

8.22. Now let us pass to the situation 8.16. It follows from Theorem 7.12 (after passing to skew Σ_π- invariants) that the complexes ${}^{+}C^{\bullet}(^{\pi}\mathbb{A};\mathcal{I}_{\nu?})^{\Sigma_\pi,-}$ compute the stalk of $\mathcal{I}_{\nu?}$ at the origin. Let us denote this stalk by $\mathcal{I}_{\nu?,0}$.

Therefore, passing to Σ_π-invariants in the previous theorem, we get

8.23. **Theorem.** *The isomorphisms* ${}^{\pi}\phi^{(\eta)}_?$ *where* $? =!, *$ *or* $!*$, *induce isomorphisms in* $\mathcal{D}^b(pt)$ *included into a commutative square*

$$\begin{array}{ccc} \mathcal{I}_\nu(\Lambda)_{!,0} & \overset{_{\Lambda}\phi^{(\eta)}_{\nu,!,0}}{\xrightarrow{\sim}} & {}_{\nu}C^{\bullet}{}'_{\mathfrak{f}}(V(\Lambda)) \\ m\downarrow & & \downarrow S \\ \mathcal{I}_\nu(\Lambda)_{*,0} & \overset{_{\Lambda}\phi^{(\eta)}_{\nu,*,0}}{\xrightarrow{\sim}} & {}_{\nu}C^{\bullet}{}'_{\mathfrak{f}^*}(V(\Lambda)^*) \end{array}$$

and

$$_{\Lambda}\phi^{(\eta)}_{\nu,!*,0} : \mathcal{I}_\nu(\Lambda)_{!*,0} \xrightarrow{\sim} {}_{\nu}C^{\bullet}_{\mathfrak{f}}(L(\Lambda)) \quad \square \tag{168}$$

Chapter 3. Fusion

9. Additivity theorem

9.1. Let us start with the setup of 7.1. For a non-negative integer n let us denote by (n) the set $[-n, 0]$. Let us introduce the following spaces. $\mathbb{A}^{(n)}$ - a complex affine space with a fixed system of coordinates (t_i), $i \in (n)$. Let ${}^nJ$ denote the disjoint union $(n) \cup J$; ${}^n\mathbb{A}$ — a complex affine space with coordinates t_j, $j \in {}^nJ$.

In general, for an affine space with a distinguished coordinate system of, we will denote by $\mathcal{S}_\Delta$ its diagonal stratification as in 6.1.

Let ${}^n\overset{\circ}{\mathbb{A}} \subset {}^n\mathbb{A}$, $\overset{\circ}{\mathbb{A}}{}^{(n)} \subset \mathbb{A}^{(n)}$ be the open strata of $\mathcal{S}_\Delta$.

Let ${}^np: {}^n\mathbb{A} \longrightarrow \mathbb{A}^{(n)}$ be the evident projection; ${}^n\mathbb{B} = {}^np^{-1}(\overset{\circ}{\mathbb{A}}{}^{(n)})$. Given a point $\mathbf{z} = (z_i) \in \mathbb{A}^{(n)}$, let us denote by ${}^\mathbf{z}\mathbb{A}$ the fiber ${}^np^{-1}(\mathbf{z})$ and by ${}^\mathbf{z}\mathcal{S}$ the stratification induced by $\mathcal{S}_\Delta$. We will consider t_j, $j \in J$, as coordinates on ${}^\mathbf{z}\mathbb{A}$.

The subscript $(.)_\mathbb{R}$ will mean as usually "real points".

9.2. Let us fix a point $\mathbf{z} = (z_{-1}, z_0) \in \mathbb{A}^{(2)}_\mathbb{R}$ such that $z_{-1} < z_0$. Let us concentrate on the fiber ${}^\mathbf{z}\mathbb{A}$. As an abstract variety it is canonically isomorphic to $\mathbb{A}$ — a complex affine space with coordinates t_j, $j \in J$; so we will suppress $\mathbf{z}$ from its notation, keeping it in the notation for the stratification ${}^\mathbf{z}\mathcal{S}$ where the dependence on $\mathbf{z}$ really takes place.

Let us fix a real $w > z_0$. Let us pick two open non-intersecting disks $D_i \subset \mathbb{C}$ with centra at z_i and not containing w. Let us pick two real numbers $w_i > z_i$ such that $w_i \in D_i$, and paths P_i connecting w with w_i as shown on Fig. 12 below.

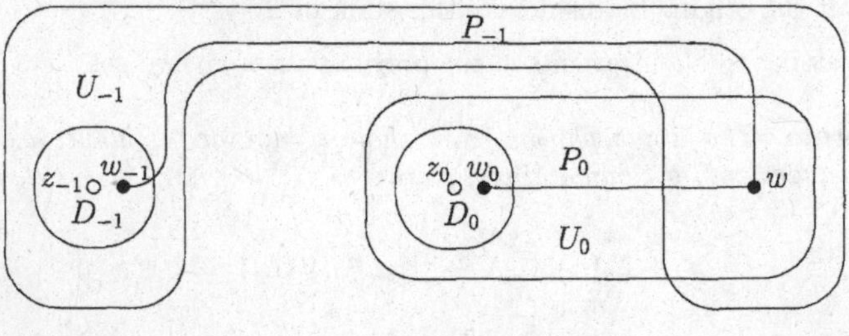

Fig. 12.

Let us denote by $\mathcal{Q}_2(J)$ the set of all maps $\rho: J \longrightarrow [-1, 0]$. Given such a map, let us denote by $\mathbb{A}_\rho \subset \mathbb{A}$ an open subvariety consisting of points $(t_j)_{j \in J}$ such that $t_j \in D_{\rho(j)}$ for all j. We will denote by the same letter ${}^\mathbf{z}\mathcal{S}$ the stratification of this space induced by ${}^\mathbf{z}\mathcal{S}$.

Set $H_w := \cup_j H_j(w)$; $P = P_{-1} \cup P_0$; $\tilde{P} = \{(t_j) \in \mathbb{A} |$ there exists j such that $t_j \in P\}$.

Given $\mathcal{K} \in \mathcal{D}(\mathbb{A};\ {}^\mathbf{z}\mathcal{S})$, the restriction map

$$R\Gamma(\mathbb{A}, H_w; \mathcal{K}) \longrightarrow R\Gamma(\mathbb{A}, \tilde{P}; \mathcal{K})$$

is an isomorphism by homotopy. On the other hand, we have restriction maps

$$R\Gamma(\mathbb{A}, \tilde{P}; \mathcal{K}) \longrightarrow R\Gamma(\mathbb{A}_\rho, \tilde{P}_\rho; \mathcal{K})$$

where $\tilde{P}_\rho := \tilde{P} \cap \mathbb{A}_\rho$. Therefore we have canonical maps

$$r_\rho : R\Gamma(\mathbb{A}, H_w; \mathcal{K}) \longrightarrow R\Gamma(\mathbb{A}_\rho, \tilde{P}_\rho; \mathcal{K}) \tag{169}$$

9.3. **Theorem.** *For every $\mathcal{K} \in \mathcal{D}(\mathbb{A};\ ^{\mathbf{z}}\mathcal{S})$ the canonical map*

$$r = \sum r_\rho : R\Gamma(\mathbb{A}, H_w; \mathcal{K}) \longrightarrow \oplus_{\rho \in \mathcal{Q}_2(J)} R\Gamma(\mathbb{A}_\rho, \tilde{P}_\rho; \mathcal{K}) \tag{170}$$

is an isomorphism.

9.4. **Proof.** Let us pick two open subsets $U_{-1}, U_0 \subset \mathbb{C}$ as shown on Fig. 12. Set $U = U_{-1} \cup U_0$, $\mathbb{A}_U = \{(t_j) \in \mathbb{A} | t_j \in U \text{ for all } j\}$. It is clear that the restriction morphism

$$R\Gamma(\mathbb{A}, \tilde{P}; \mathcal{K}) \longrightarrow R\Gamma(\mathbb{A}_U, \tilde{P}_U; \mathcal{K})$$

where $\tilde{P}_U := \tilde{P} \cap \mathbb{A}_U$, is an isomorphism.

For each $\rho \in \mathcal{Q}_2(J)$ set

$$\mathbb{A}_{U,\rho} := \{(t_j) \in \mathbb{A}_U | t_j \in U_{\rho(j)} \text{ for all } j\};\ \tilde{P}_{U,\rho} := \tilde{P} \cap \mathbb{A}_{U,\rho} \tag{171}$$

We have $\mathbb{A}_U = \bigcup_\rho \mathbb{A}_{U,\rho}$.

9.4.1. **Lemma.** *For every $\mathcal{K} \in \mathcal{D}(\mathbb{A};^{\mathbf{z}} \mathcal{S})$ the sum of restriction maps*

$$q : R\Gamma(\mathbb{A}_U, \tilde{P}_U; \mathcal{K}) \longrightarrow \sum_{\rho \in \mathcal{Q}_2(J)} R\Gamma(\mathbb{A}_{U,\rho}, \tilde{P}_{U,\rho}; \mathcal{K}) \tag{172}$$

is an isomorphism.

Proof. Suppose we have distinct $\rho_1, \dots, \rho_m$ such that $\mathbb{A}_{U;\rho_1,\dots,\rho_m} := \mathbb{A}_{U,\rho_1} \cap \dots \cap \mathbb{A}_{U,\rho_m} \neq \emptyset$; set $\tilde{P}_{U;\rho_1,\dots,\rho_m} := \tilde{P} \cap \mathbb{A}_{U;\rho_1,\dots,\rho_m}$.

Our lemma follows at once by Mayer-Vietoris argument from the following

9.4.2. **Claim.** *For every $m \geq 2$*

$$R\Gamma(\mathbb{A}_{U;\rho_1,\dots,\rho_m}, \tilde{P}_{U;\rho_1,\dots,\rho_m}; \mathcal{K}) = 0.$$

Proof of the claim. It is convenient to use the following notations. If $J = A \cup B$ is a disjoint union, we will denote by $\rho_{A,B}$ the map $J \longrightarrow [-1, 0]$ such that $\rho^{-1}(-1) = A$, $\rho^{-1}(0) = B$, and by $U_{A;B}$ the subspace $\mathbb{A}_{U,\rho_{A,B}}$.

Let us prove the claim for the case $N = 2$. Let $J = \{i, j\}$. In this case it is easy to see that the only non-trivial intersections are $U^{(1)} = U_{ij;\emptyset} \cap U_{j;i}$ and $U^{(2)} = U_{ij;\emptyset} \cap U_{\emptyset;ij}$.

To prove our claim for $U^{(1)}$ we will use *a shrinking neighbourhood argument* based on Lemma I.2.12. Let $\mathcal{K}'$ denote the sheaf on $U^{(1)}$ obtained by extension by zero of $\mathcal{K}|_{U^{(1)}-\tilde{P}}$. For each $\epsilon > 0$ let us denote by $U_{-1,\epsilon} \subset \mathbb{C}$ an open domain consisting of points having distance $< \epsilon$ from $D_{-1} \cup P$. Set

$$U^{(1)}_\epsilon = \{(t_i, t_j) |\ t_i \in U_{-1,\epsilon},\ t_j \in U_0 \cap U_{-1,\epsilon}\}.$$

It is clear that restriction maps $R\Gamma(U^{(1)};\mathcal{K}') \longrightarrow R\Gamma(U^{(1)}_\epsilon;\mathcal{K}')$ are isomorphisms. On the other hand, it follows from I.2.12 that

$$\lim_{\to\ \epsilon} R\Gamma(U^{(1)}_\epsilon;\mathcal{K}') \cong R\Gamma(\bigcap_\epsilon U^{(1)}_\epsilon;\mathcal{K}'),$$

and the last complex is zero by the definition of $\mathcal{K}'$ (the point t_j is confined to P in $\bigcap_\epsilon U^{(1)}_\epsilon$).

The subspace $U^{(2)}$ consists of (t_i, t_j) such that both t_i and t_j lie in $U_{-1}\cap U_0$. This case is even simpler. The picture is homeomorphic to an affine plane with a sheaf smooth along the diagonal stratification; and we are interested in its cohomology modulo the coordinate cross. This is clearly equal to zero, i.e. $R\Gamma(U^{(2)}, \tilde{P}\cap U^{(2)};\mathcal{K}) = 0$.

This proves the claim for $N = 2$. The case of an arbitrary N is treated in a similar manner, and we leave it to the reader. This completes the proof of the claim and of the lemma. □

9.4.3. **Lemma.** *For every $\rho \in \mathcal{Q}_2(J)$ the restriction map*

$$R\Gamma(\mathbb{A}_{U,\rho}, \tilde{P}_{U,\rho};\mathcal{K}) \longrightarrow R\Gamma(\mathbb{A}_\rho, \tilde{P}_\rho;\mathcal{K})$$

is an isomorphism.

Proof. Again let us consider the case $J = \{i, j\}$. If $\rho = \rho_{ij;\emptyset}$ or $\rho_{\emptyset;ij}$ the statement is obvious. Suppose $\rho = \rho_{i;j}$. Let us denote by $\mathbb{A}'_{U,\rho} \subset \mathbb{A}_{U,\rho}$ the subspace $\{(t_i, t_j) | t_i \in D_{-1},\ t_j \in U_0\}$. It is clear that the restriction map

$$R\Gamma(\mathbb{A}'_{U,\rho}, \tilde{P}'_{U,\rho};\mathcal{K}) \longrightarrow R\Gamma(\mathbb{A}_\rho, \tilde{P}_\rho;\mathcal{K})$$

where $\tilde{P}'_{U,\rho} := \tilde{P}\cap \mathbb{A}'_{U,\rho}$, is an isomorphism. Let us consider the restriction

$$R\Gamma(\mathbb{A}_{U,\rho}, \tilde{P}_{U,\rho};\mathcal{K}) \longrightarrow R\Gamma(\mathbb{A}'_{U,\rho}, \tilde{P}'_{U,\rho};\mathcal{K}) \qquad (173)$$

The cone of this map is isomorphic to $R\Gamma(\mathbb{A}_{U,\rho}, \tilde{P}_{U,\rho};\mathcal{M})$ where the sheaf $\mathcal{M}$ has the same singularities as $\mathcal{K}$ and in addition is 0 over the closure $\bar{D}_{-1}$. Now, consider a system of shrinking neighbourhoods of $P_{-1}\cup \bar{D}_{-1}$ as in the proof af the claim above, we see that $R\Gamma(\mathbb{A}_{U,\rho}, \tilde{P}_{U,\rho};\mathcal{M}) = 0$, i.e. (173) is an isomorphism. This implies the lemma for this case.

The case of arbitrary J is treated exactly in the same way. □

Our theorem is an obvious consequence of two previous lemmas. □

10. Fusion and tensor products

10.1. **Fusion functors.** The constructions below were inspired by [Dr].

For each integer $n \geq 1$ and $i \in [n]$, let us define functors

$$^n\psi_i : \mathcal{D}(^n\mathbb{A};\mathcal{S}_\Delta) \longrightarrow \mathcal{D}(^{n-1}\mathbb{A};\mathcal{S}_\Delta) \qquad (174)$$

as follows. We have the t-exact nearby cycles functors (see [D3] or [KS], 8.6, but note the shift by [-1]!)

$$\Psi_{t_{-i}-t_{-i+1}}[-1] : \mathcal{D}(^n\mathbb{A};\mathcal{S}_\Delta) \longrightarrow \mathcal{D}(\mathbb{A}';\mathcal{S}_\Delta)$$

where $\mathbb{A}'$ denotes (for a moment) an affine space with coordinates t_j, $j \in ((n)\cup J) - \{-i\}$. We can identify the last space with $^{n-1}\mathbb{A}$ simply by renaming coordinates t_j to

t_{j+1} for $-n \leq j \leq -i-1$. By definition, ${}^n\psi_i$ is equal to $\Psi_{t_{-i}-t_{-i+1}}[-1]$ followed by this identification.

10.2. **Lemma.** *(i) For each $n \geq 2$, $i \in [n]$ have canonical isomorphisms*

$$ {}^n\alpha_i : {}^{n-1}\psi_i \circ {}^n\psi_i \xrightarrow{\sim} {}^{n-1}\psi_i \circ {}^n\psi_{i+1} \tag{175}$$

and equalities

$$ {}^{n-1}\psi_j \circ {}^n\psi_i = {}^{n-1}\psi_i \circ {}^n\psi_{j+1} $$

for $j > i$, such that
(ii) ("Stasheff pentagon" identity*) the diagram below commutes:*

$$\begin{array}{ccc}
 & {}^{n-2}\psi_i \circ {}^{n-1}\psi_i \circ {}^n\psi_i & \\
\swarrow {}^{n-1}\alpha_i \circ {}^n\psi_i & & {}^{n-2}\psi_i \circ {}^n\alpha_i \searrow \\
{}^{n-2}\psi_i \circ {}^{n-1}\psi_{i+1} \circ {}^n\psi_i & & {}^{n-2}\psi_i \circ {}^{n-1}\psi_i \circ {}^n\psi_{i+1} \\
\| & & \downarrow {}^{n-1}\alpha_i \circ {}^n\psi_{i+1} \\
{}^{n-2}\psi_i \circ {}^{n-1}\psi_i \circ {}^n\psi_{i+2} & & {}^{n-2}\psi_i \circ {}^{n-1}\psi_{i+1} \circ {}^n\psi_{i+1} \\
{}^{n-1}\alpha_i \circ {}^n\psi_{i+2} \searrow & & \swarrow {}^{n-2}\psi_i \circ {}^n\alpha_{i+1} \\
 & {}^{n-2}\psi_i \circ {}^{n-1}\psi_{i+1} \circ {}^n\psi_{i+2} &
\end{array}$$

□

10.3. Let us define a t-exact functor

$$\psi : \mathcal{D}({}^n\mathbb{A}; \mathcal{S}_\Delta) \longrightarrow \mathcal{D}(\mathbb{A}; \mathcal{S}) \tag{176}$$

as a composition $i_0^*[-1] \circ {}^n\psi_1 \circ {}^{n-1}\psi_1 \circ \ldots \circ {}^1\psi_1$, where

$$i_0^* : \mathcal{D}({}^0\mathbb{A}; \mathcal{S}_\Delta) \longrightarrow \mathcal{D}(\mathbb{A}; \mathcal{S})$$

denotes the restriction to the subspace $t_0 = 0$. Note that $i_0^*[-1]$ is a t-exact equivalence. (Recall that $\mathbb{A}$ and $\mathcal{S}$ denote the same as in 7.1).

STANDARD SHEAVES

The constructions and computations below generalize Section 8.

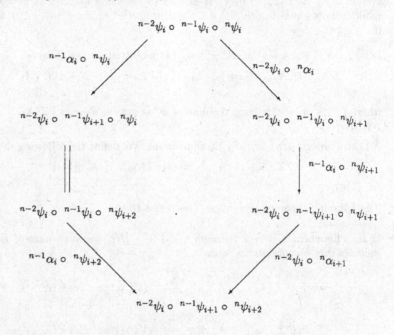

10.4. Let us make the following assumptions. Let us denote by A the $I \times I$-matrix $(i \cdot j)$. Let us suppose that det $A \neq 0$. There exists a unique $\mathbb{Z}[\frac{1}{\det A}]$-valued symmetric bilinear form on X (to be denoted by $\lambda, \mu \mapsto \lambda \cdot \mu$) such that the map $\mathbb{Z}[I] \longrightarrow X$, $\nu \mapsto \lambda_\nu$ respects scalar products.

Let us suppose that our field k contains an element ζ' such that $(\zeta')^{\det A} = \zeta$, and fix such ζ'. For $a = \frac{c}{\det A}$, $c \in \mathbb{Z}$, we set by definition $\zeta^a := (\zeta')^c$.

10.5. Let us fix $\nu = \sum \nu_i i \in \mathbb{N}[I]$ and its unfolding $\pi : J \longrightarrow I$ as in 6.12, and an integer $n \geq 1$. We will use the preceding notations with this J.

Let us fix $n+1$ weights $\Lambda_0, \Lambda_{-1}, \ldots, \Lambda_{-n} \in X$. We define a one-dimensional local system $\mathcal{I}(\Lambda_0, \ldots, \Lambda_{-n}; \nu)$ over ${}^n\overset{\circ}{\mathbb{A}}$ exactly in the same manner as in 6.3, with half-monodromies equal to $\zeta^{\pi(i) \cdot \pi(j)}$ if $i, j \in J$, to $\zeta^{-\langle \Lambda_i, \pi(j) \rangle}$ if $i \in (n), j \in J$ and to $\zeta^{\Lambda_i \cdot \Lambda_j}$ if $i, j \in (n)$.

Let $j : {}^n\overset{\circ}{\mathbb{A}} \longrightarrow {}^n\mathbb{A}$ be the embedding. Let us introduce the sheaves

$$\mathcal{I}(\Lambda_0, \ldots, \Lambda_{-n}; \nu)_? := j_? \mathcal{I}(\Lambda_0, \ldots, \Lambda_{-n}; \nu)[-n-N-1] \in \mathcal{M}({}^n\mathbb{A}; \mathcal{S}_\Delta) \tag{177}$$

where ? =!, $*$ or !$*$. Applying the functor ψ, we get the sheaves $\psi\mathcal{I}(\Lambda_0, \ldots, \Lambda_{-n}; \nu)_? \in \mathcal{M}(\mathbb{A}; \mathcal{S})$.

All these objects are naturally Σ_π-equivariant. We define the following sheaves on $\mathcal{A}_\nu$:

$${}^\psi\mathcal{I}_\nu(\Lambda_0, \ldots, \Lambda_{-n})_? := (\pi_* \psi\mathcal{I}(\Lambda_0, \ldots, \Lambda_{-n}; \nu)_?)^{\Sigma_\pi, -} \tag{178}$$

(cf. 8.16).

The following theorem generalizes Theorem 8.18.

10.6. **Theorem.** *Given a bijection $\eta : J \overset{\sim}{\longrightarrow} [N]$, we have natural isomorphisms included into a commutative square*

$$\begin{array}{ccc} \Phi_\nu({}^\psi\mathcal{I}_\nu(\Lambda_0, \ldots, \Lambda_{-n})_!) & \overset{\phi_!^{(\eta)}}{\underset{\sim}{\longrightarrow}} & (V(\Lambda_0) \otimes \ldots \otimes V(\Lambda_{-n}))_\nu \\ m \downarrow & & \downarrow S_\Lambda \\ \Phi_\nu({}^\psi\mathcal{I}_\nu(\Lambda_0, \ldots, \Lambda_{-n})_*) & \overset{\phi_*^{(\eta)}}{\underset{\sim}{\longrightarrow}} & (V(\Lambda_0)^* \otimes \ldots \otimes V(\Lambda_{-n})^*)_\nu \end{array}$$

and

$$\phi_{!*}^{(\eta)} : \Phi_\nu({}^\psi\mathcal{I}_\nu(\Lambda_0, \ldots, \Lambda_{-n})_{!*}) \overset{\sim}{\longrightarrow} (L(\Lambda_0) \otimes \ldots \otimes L(\Lambda_{-n}))_\nu \tag{179}$$

A change of η multiplies these isomorphisms by the sign of the corresponding permutation of $[N]$.

Proof. We may suppose that π is injective, i.e. all $\nu_i = 0$ or 1. The general case immediately follows from this one after passing to Σ_π-skew invariants.

Let us consider the case $n = 1$. In this case one sees easily from the definitions that we have a canonical isomorphism

$$\Phi_\nu({}^\psi\mathcal{I}_\nu(\Lambda_0, \Lambda_{-1})_?) \cong R\Gamma(\mathbb{A}, H_w; \mathcal{I}(\Lambda_0, \Lambda_{-1}; \nu)_?)$$

in the notations of Additivity Theorem 9.3. On the other hand, the set $\mathcal{Q}_2(J)$ is in one-to-one correspondence with the set of all decompositions $\nu = \nu_0 + \nu_{-1}$, $\nu_i \in \mathbb{N}[I]$, and if ρ corresponds to such a decomposition, we have a natural isomorphism

$$R\Gamma(\mathbb{A}_\rho, \tilde{P} \cap \mathbb{A}_\rho; \mathcal{I}_\nu(\Lambda_0, \Lambda_{-1})_?) \cong \Phi_{\nu_0}({}^\psi\mathcal{I}(\Lambda_0; \nu_0)_?) \otimes \Phi_{\nu_{-1}}({}^\psi\mathcal{I}(\Lambda_{-1}; \nu_{-1})_?) \tag{180}$$

by the Künneth formula. Therefore, Additivity Theorem implies isomorphisms

$$\Phi_\nu({}^\psi\mathcal{I}_\nu(\Lambda_0, \Lambda_{-1})_?) \cong \oplus_{\nu_0+\nu_{-1}=\nu}\Phi_{\nu_0}({}^\psi\mathcal{I}_{\nu_0}(\Lambda_0)_?) \otimes \Phi_{\nu_{-1}}({}^\psi\mathcal{I}_{\nu_{-1}}(\Lambda_{-1})_?)$$

which are the claim of our theorem.

The case $n > 2$ is obtained similarly by the iterated use of the Additivity Theorem. □

10.7. Next we will consider the stalks $(?)_0$ of our sheaves at 0, or what is the same (since they are $\mathbb{R}^{+*}$-homogeneous), the complexes $R\Gamma(\mathcal{A}_\nu; ?)$.

The next theorem generalizes Theorem 8.23.

10.8. **Theorem.** *Given a bijection $\eta : J \xrightarrow{\sim} [N]$, we have natural isomorphisms included into a commutative square*

$$\begin{array}{ccc} {}^\psi\mathcal{I}_\nu(\Lambda_0, \dots, \Lambda_{-n})_{!0} & \overset{\phi^{(\eta)}_{!,0}}{\xrightarrow{\sim}} & {}_\nu C^\bullet_{\cdot,\mathfrak{f}}(V(\Lambda_0) \otimes \dots \otimes V(\Lambda_{-n})) \\ m\downarrow & & \downarrow S_\Lambda \\ {}^\psi\mathcal{I}_\nu(\Lambda_0, \dots, \Lambda_{-n})_{*0} & \overset{\phi^{(\eta)}_{*,0}}{\xrightarrow{\sim}} & {}_\nu C^\bullet_{\cdot,\mathfrak{f}^*}(V(\Lambda_0)^* \otimes \dots \otimes V(\Lambda_{-n})^*) \end{array}$$

and

$$\phi^{(\eta)}_{!*,0} : {}^\psi\mathcal{I}_\nu(\Lambda_0, \dots, \Lambda_{-n})_{!*0} \xrightarrow{\sim} {}_\nu C^\bullet_{\mathfrak{f}}(L(\Lambda_0) \otimes \dots \otimes L(\Lambda_{-n})) \tag{181}$$

A change of η multiplies these isomorphisms by the sign of the corresponding permutation of $[N]$.

Proof. It is not hard to deduce from the previous theorem that we have natural isomorphisms of complexes included into a commutative square

$$\begin{array}{ccc} {}^+C(\mathbb{A}; \psi\mathcal{I}(\Lambda_0, \dots, \Lambda_{-n}; \nu)_!) & \xrightarrow{\sim} & {}_{\chi_J}C^\bullet_\pi{}_{,\mathfrak{f}}(V({}^\pi\Lambda_0) \otimes \dots \otimes V({}^\pi\Lambda_{-n})) \\ m\downarrow & & \downarrow S_\Lambda \\ {}^+C(\mathbb{A}; \psi\mathcal{I}(\Lambda_0, \dots, \Lambda_{-n}; \nu)_*) & \xrightarrow{\sim} & {}_{\chi_J}C^\bullet_\pi{}_{,\mathfrak{f}^*}(V({}^\pi\Lambda_0)^* \otimes \dots \otimes V({}^\pi\Lambda_{-n})^*) \end{array}$$

This implies our claim after passing to Σ_π-(skew) invariants. □

Chapter 4. Category $\mathcal{C}$

11. Simply laced case

11.1. From now on until the end of this part we will assume, in addition to the assumptions of 1.4, that ζ is a primitive l-th root of unity, where l is a fixed integer $l > 3$ prime to 2 and 3.

11.2. We will use notations of [L1], Chapters 1, 2, which we briefly recall.

11.2.1. Let $(I, \cdot)$ be *a simply laced Cartan datum of finite type* (cf. *loc.cit.*, 1.1.1, 2.1.3), that is, a finite set I and a nondegenerate symmetric bilinear form $\alpha, \beta \mapsto \alpha \cdot \beta$ on the free abelian group $\mathbb{Z}[I]$. This form satisfies conditions

(a) $i \cdot i = 2$ for any $i \in I$;

(b) $i \cdot j \in \{0, -1\}$ for any $i \neq j$ in I.

11.2.2. We will consider *the simply connected root datum of type* $(I, \cdot)$, that is (see *loc. cit.*, 2.2.2), two free abelian groups $Y = \mathbb{Z}[I]$ and $X = \mathrm{Hom}_{\mathbb{Z}}(Y, \mathbb{Z})$ together with

(a) the canonical bilinear pairing $\langle , \rangle : Y \times X \longrightarrow \mathbb{Z}$;

(b) an obvious embedding $I \hookrightarrow Y$ $(i \mapsto i)$ and an embedding $I \hookrightarrow X$ $(i \mapsto i')$, such that $\langle i, j' \rangle = i \cdot j$ for any $i, j \in I$.

We will call X *the lattice of weights*, and Y *the lattice of coroots*. An element of X will be typically denoted by $\lambda, \mu, \nu, \ldots$; and an element of Y will be typically denoted by $\alpha, \beta, \gamma, \ldots$.

11.3. We consider the finite dimensional algebra U over the field k defined as in the section 1.3 of [AJS]. We also consider the category $\mathcal{C}$ of finite dimensional X-graded U-modules defined as in the section 2.3 of [AJS].

11.3.1. The algebra U is given by generators $E_i, F_i, K_i^{\pm 1}$, $i \in I$, subject to relations

(z) $K_i \cdot K_i^{-1} = 1$; $K_i K_j = K_j K_i$;

(a) $K_j E_i = \zeta^{j \cdot i} E_i K_j$;

(b) $K_j F_i = \zeta^{-j \cdot i} F_i K_j$;

(c) $E_i F_j - F_j E_i = \delta_{ij} \frac{K_i - K_i^{-1}}{\zeta - \zeta^{-1}}$;

(d) $E_i^l = F_j^l = 0$;

(e) $E_i E_j - E_j E_i = 0$ if $i \cdot j = 0$; $E_i^2 E_j - (\zeta + \zeta^{-1}) E_i E_j E_i + E_i E_j^2 = 0$ if $i \cdot j = -1$;

(f) $F_i F_j - F_j F_i = 0$ if $i \cdot j = 0$; $F_i^2 F_j - (\zeta + \zeta^{-1}) F_i F_j F_i + F_i F_j^2 = 0$ if $i \cdot j = -1$.

The algebra U has a unique k-algebra X-grading $U = \oplus U_\mu$ for which $|E_i| = i'$, $|F_i| = -i'$, $|K_i| = 0$.

We define a comultiplication

$$\Delta : U \longrightarrow U \otimes U \tag{182}$$

as a unique k-algebra mapping such that

$$\Delta(K^{\pm 1}) = K^{\pm 1} \otimes K^{\pm 1};$$

$$\Delta(E_i) = E_i \otimes K_i + 1 \otimes E_i;$$

$$\Delta(F_i) = F_i \otimes 1 + K_i^{-1} \otimes F_i.$$

This makes U a Hopf algebra (with obvious unit and counit).

11.3.2. The category $\mathcal{C}$ is by definition a category of finite dimensional X-graded k-vector spaces $V = \oplus_{\mu \in X} V_\mu$, equipped with a left action of U such that the U-action is compatible with the X-grading, and

$$K_i x = \zeta^{\langle i,\mu \rangle} x$$

for $x \in V_\mu$, $i \in I$.

Since U is a Hopf algebra, $\mathcal{C}$ has a canonical structure of a *tensor category*.

11.4. We define an algebra $\mathfrak{u}$ having generators $\theta_i, \epsilon_i, K_i^{\pm 1}$, $i \in I$, subject to relations

(z) $K_i \cdot K_i^{-1} = 1$; $K_i K_j = K_j K_i$;

(a) $K_j \epsilon_i = \zeta^{j \cdot i} \epsilon_i K_j$;

(b) $K_j \theta_i = \zeta^{-j \cdot i} \theta_i K_j$;

(c) $\epsilon_i \theta_j - \zeta^{i \cdot j} \theta_j \epsilon_i = \delta_{ij}(1 - K_i^{-2})$

(d) if $f \in \mathrm{Ker}(S) \subset {}'\mathfrak{f}$ (see (65)) then $f = 0$;

(e) the same as (d) for the free algebra $\mathfrak{E}$ on the generators ϵ_i.

11.4.1. Let us define the comultiplication

$$\Delta : \mathfrak{u} \longrightarrow \mathfrak{u} \otimes \mathfrak{u} \tag{183}$$

by the formulas

$$\Delta(K_i^{\pm 1}) = K_i^{\pm 1} \otimes K_i^{\pm 1};$$

$$\Delta(\theta_i) = \theta_i \otimes 1 + K_i^{-1} \otimes \theta_i;$$

$$\Delta(\epsilon_i) = \epsilon_i \otimes 1 + K_i^{-1} \otimes \epsilon_i$$

and the condition that Δ is a morphism of k-algebras.

This makes $\mathfrak{u}$ a Hopf algebra (with obvious unit and counit).

$\mathfrak{u}$ is an X-graded k-algebra, with an X-grading defined uniquely by the conditions $|K_i^{\pm 1}| = 0$; $|\theta_i| = -i'$; $|\epsilon_i| = i'$.

11.5. We define $\tilde{\mathcal{C}}$ as a category of finite dimensional X-graded vector spaces $V = \oplus V_\lambda$, equipped with a structure of a left $\mathfrak{u}$-module compatible with X-gradings and such that

$$K_i x = \zeta^{\langle i,\lambda \rangle} x$$

for $x \in V_\lambda$, $i \in I$.

Since $\mathfrak{u}$ is a Hopf algebra, $\tilde{\mathcal{C}}$ is a tensor category.

11.6. Recall that for any $\Lambda \in X$ we have defined in 5.2 the X-graded $\mathfrak{f}$-module $L(\Lambda)$. It is a quotient-module of the Verma module $V(\Lambda)$, and it inherits its X-grading from the one of $V(\Lambda)$ (see 2.15). Thus $L(\Lambda) = \oplus L(\Lambda)_\lambda$, and we define the action of generators K_i on $L(\Lambda)_\lambda$ as multiplication by $\zeta^{\langle i,\lambda\rangle}$. Finally, we define the action of generators ϵ_i on $V(\Lambda)$ as in 2.16. These operators on $V(\Lambda)$ satisfy the relations (a) — (c) above.

We check immediately that this action descends to the quotient $L(\Lambda)$. Moreover, it follows from Theorem 2.23 that these operators acting on $L(\Lambda)$ satisfy the relations (a) — (e) above. So we have constructed the action of $\mathfrak{u}$ on $L(\Lambda)$, therefore we can regard it as an object of $\tilde{\mathcal{C}}$.

11.6.1. **Lemma.** *$L(\Lambda)$ is an irreducible object in $\tilde{\mathcal{C}}$.*

Proof. Let $I(\Lambda)$ be the maximal proper homogeneous (with respect to X-grading) submodule of $V(\Lambda)$ (the sum of all homogeneous submodules not containing v_Λ). Then $V(\Lambda)/I(\Lambda)$ is irreducible, so it suffices to prove that $I(\Lambda) = \ker(S_\Lambda)$. The inclusion $\ker(S_\Lambda) \subset I(\Lambda)$ is obvious. Let us prove the opposite inclusion. Let $y \in I(\Lambda)$. It is enough to check that $S_\Lambda(y,x) = 0$ for any $x \in V(\Lambda)$ of the form $\theta_{i_1}\ldots\theta_{i_n}v_\Lambda$. By (22) we have $S_\Lambda(y,\theta_{i_1}\ldots\theta_{i_n}v_\Lambda) = S_\Lambda(\epsilon_{i_n}\ldots\epsilon_{i_1}y, v_\Lambda) = 0$ since $\epsilon_{i_n}\ldots\epsilon_{i_1}y \in I(\Lambda)$. □

11.7. Let us consider elements $E_i, F_i \in \mathfrak{u}$ given by the following formulas:

$$E_i = \frac{\zeta^2}{\zeta - \zeta^{-1}}\epsilon_i K_i;\ F_i = \theta_i \tag{184}$$

It is immediate to check that these elements satisfy the relations 11.3.1 (a) — (c). Moreover, one checks without difficulty that

$$\theta_i\theta_j - \theta_j\theta_i \in \mathrm{Ker}(S) \text{ if } i \cdot j = 0,$$

and

$$\theta_i^2\theta_j - (\zeta + \zeta^{-1})\theta_i\theta_j\theta_i + \theta_j\theta_i^2 \in \mathrm{Ker}(S) \text{ if } i \cdot j = -1$$

(cf. [SV2], 1.16). Also, it is immediate that

$$S(\theta_i^a, \theta_i^a) = \prod_{p=1}^{a} \frac{1-\zeta^{2p}}{1-\zeta^2}.$$

It follows that $\theta_i^l \in \mathrm{Ker}(S)$ for all i.

It follows easily that the formulas (184) define a surjective morphism of algebras

$$R : U \longrightarrow \mathfrak{u} \tag{185}$$

Moreover, one checks at once that R is a map of Hopf algebras.

Therefore, R induces a tensor functor

$$Q : \tilde{\mathcal{C}} \longrightarrow \mathcal{C} \tag{186}$$

which is an embedding of a full subcategory.

11.8. **Theorem.** *Q is an equivalence.*

Proof. It is enough to check that $\tilde{\mathcal{C}}$ contains enough projectives for $\mathcal{C}$ (see e.g. Lemma A.15. of [KL] IV).

First of all, we know from [AJS], section 4.1, that the simple $\mathfrak{u}$-modules $L(\Lambda), \Lambda \in X$ exhaust the list of simple objects of $\mathcal{C}$. Second, we know from [APW2], Theorem 4.6 and Remark 4.7, that the module $L(-\rho)$ is projective where $-\rho \in X$ is characterized by the property $\langle i, -\rho\rangle = -1$ for any $i \in I$. Finally we know, say, from [APW2], Remark 4.7, and [APW1], Lemma 9.11, that the set of modules $\{L(\Lambda) \otimes L(-\rho), \Lambda \in X\}$ is an ample system of projectives for $\mathcal{C}$. □

11.9. Denote by $\mathfrak{u}^0$ (resp., U^0) the subalgebra of $\mathfrak{u}$ (resp., of U) generated by $K_i^{\pm 1}$, $i \in I$. Obviously, both algebras are isomorphic to the ring of Laurent polynomials in K_i, and the map R induces an identity isomorphism between them.

Denote by $\mathfrak{u}^{\leq 0}$ (resp., by $\mathfrak{u}^-$) the subalgebra of $\mathfrak{u}$ generated by $\theta_i, K_i^{\pm 1}$ (resp., by θ_i), $i \in I$. The last algebra may be identified with $\mathfrak{f}$. As a vector space, $\mathfrak{u}^{\leq 0}$ is isomorphic to $\mathfrak{f} \otimes \mathfrak{u}^0$.

Denote by $U^- \subset U$ the subalgebra generated by $F_i, i \in I$, and by $U^{\leq 0} \subset U$ the subalgebra generated by F_i, $K_i^{\pm 1}, i \in I$. As a vector space it is isomorphic to $U^- \otimes U^0$.

11.10. **Theorem.** *(a) R is an isomorphism;*

(b) R induces an isomorphism $U^- \xrightarrow{\sim} \mathfrak{f}$.

Proof. Evidently it is enough to prove b). We know that R is surjective, and that U^- is finite dimensional. So it suffices to prove that $\dim U^- \leq \dim \mathfrak{f}$. We know by [L1] 36.1.5. that $\dim U^- = \dim L(-\rho)$. On the other hand, the map $\mathfrak{f} \longrightarrow L(-\rho), f \mapsto f(v_{-\rho})$ is surjective by construction. □

12. Non-simply laced case

In the non-simply laced case all main results of this part hold true as well. However, the definitions need some minor modifications. In this section we will describe them.

12.1. We will use terminology and notations from [L1], especially from Chapters 1-3. Let us fix a Cartan datum $(I, \cdot)$ of finite type, not necessarily simply laced, cf. *loc. cit.*, 2.1.3. Let $(Y = \mathbb{Z}[I], X = \mathrm{Hom}(Y, \mathbb{Z}), \langle , \rangle, I \hookrightarrow X, I \hookrightarrow Y)$ be the simply connected root datum associated with $(I, \cdot)$, *loc.cit.*, 2.2.2.

We set $d_i := \frac{i \cdot i}{2}$, $i \in I$; these numbers are positive integers. We set $\zeta_i := \zeta^{d_i}$. The embedding $I \hookrightarrow X$ sends $i \in I$ to i' such that $\langle d_j j, i'\rangle = j \cdot i$ for all $i, j \in I$.

12.2. The category $\mathcal{C}$ is defined in the same way as in the simply laced case, where the definition of the Hopf algebra U should be modified as follows (cf. [AJS], 1.3).

By definition, U has generators $E_i, F_i, K_i^{\pm 1}$, $i \in I$, subject to relations

(z) $K_i \cdot K_i^{-1} = 1;\ K_i K_j = K_j K_i;$

(a) $K_j E_i = \zeta^{\langle j, i'\rangle} E_i K_j;$

(b) $K_j F_i = \zeta^{-\langle j, i'\rangle} F_i K_j;$

(c) $E_iF_j - F_jE_i = \delta_{ij}\frac{\tilde{K}_i - \tilde{K}_i^{-1}}{\zeta_i - \zeta_i^{-1}}$;

(d) $E_i^l = F_j^l = 0$;

(e) $\sum_{p=0}^{1-\langle i,j'\rangle}(-1)^p E_i^{(p)} E_j E_i^{(1-\langle i,j'\rangle - p)} = 0$ for $i \neq j$;

(f) $\sum_{p=0}^{1-\langle i,j'\rangle}(-1)^p F_i^{(p)} F_j F_i^{(1-\langle i,j'\rangle - p)} = 0$ for $i \neq j$,

where we have used the notations: $\tilde{K}_i := K_i^{d_i}$; $G_i^{(p)} := G_i^p/[p]_i^!$, $G = E$ or F,

$$[p]_i^! := \prod_{a=1}^{p} \frac{\zeta_i^a - \zeta_i^{-a}}{\zeta_i - \zeta_i^{-1}}.$$

The X-grading on U is defined in the same way as in the simply laced case.

The comultiplication is defined as

$$\Delta(K_i) = K_i \otimes K_i;$$
$$\Delta(E_i) = E_i \otimes \tilde{K}_i + 1 \otimes E_i;$$
$$\Delta(F_i) = F_i \otimes 1 + \tilde{K}_i^{-1} \otimes F_i.$$

12.2.1. *Remark.* This algebra is very close to (and presumably coincides with) the algebra **U** from [L1], 3.1, specialized to $v = \zeta$. We use the opposite comultiplication, though.

12.3. The definition of the Hopf algebra $\mathfrak{u}$ should be modified as follows. It has generators $\theta_i, \epsilon_i, K_i^{\pm 1}$, $i \in I$, subject to relations

(z) $K_i \cdot K_i^{-1} = 1$; $K_iK_j = K_jK_i$;

(a) $K_j\epsilon_i = \zeta^{\langle j,i'\rangle}\epsilon_i K_j$;

(b) $K_j\theta_i = \zeta^{-\langle j,i'\rangle}\theta_i K_j$;

(c) $\epsilon_i\theta_j - \zeta^{i\cdot j}\theta_j\epsilon_i = \delta_{ij}(1 - \tilde{K}_i^{-2})$

(d) if $f \in \mathrm{Ker}(S) \subset {}'\mathfrak{f}$ (see (65)) then $f = 0$;

(e) the same as (d) for the free algebra $\mathfrak{E}$ on the generators ϵ_i.

The comultiplication is defined as

$$\Delta(K_i^{\pm 1}) = K_i^{\pm 1} \otimes K_i^{\pm 1};$$
$$\Delta(\theta_i) = \theta_i \otimes 1 + \tilde{K}_i^{-1} \otimes \theta_i;$$
$$\Delta(\epsilon_i) = \epsilon_i \otimes 1 + \tilde{K}_i^{-1} \otimes \epsilon_i$$

The category $\tilde{\mathcal{C}}$ is defined in the same way as in the simply laced case.

12.4. We define an X-grading on the free ${}'\mathfrak{f}$-module $V(\Lambda)$ with generator v_Λ by setting $V(\Lambda)_\Lambda = \mathsf{k} \cdot v_\Lambda$ and assuming that operators θ_i decrease the grading by i'.

The definition of the form S_Λ on $V(\Lambda)$ should be modified as follows. It is a unique bilinear from such that $S_\Lambda(v_\Lambda, v_\Lambda) = 1$ and $S(\theta_i x, y) = S(x, \epsilon_i y)$ where the operators $\epsilon_i : V(\Lambda) \longrightarrow V(\Lambda)$ are defined by the requirements $\epsilon_i(v_\Lambda) = 0$,

$$\epsilon_i(\theta_j x) = \zeta^{i\cdot j}\theta_j\epsilon_i(x) + \delta_{ij}[\langle i, \lambda\rangle]_{\zeta_i} x$$

for $x \in V(\Lambda)_\lambda$.

We define $L(\Lambda)$ as a quotient $V(\Lambda)/\mathrm{Ker}(S)$. As in the simply laced case, $L(\Lambda)$ is naturally an object of $\mathcal{C}$, and the same argument proves that it is irreducible.

12.5. We define the morphism

$$R: U \longrightarrow \mathrm{u} \tag{187}$$

by the formulas

$$R(E_i) = \frac{\zeta_i^2}{\zeta_i - \zeta_i^{-1}} \epsilon_i \tilde{K}_i;\ R(F_i) = \theta_i;\ R(K_i) = K_i \tag{188}$$

Using [L1], 1.4.3, one sees immediately that it is correctly defined morphism of algebras. It follows at once from the definitions that R is a morphism of Hopf algebras. Hence, we get a tensor functor

$$Q: \tilde{\mathcal{C}} \longrightarrow \mathcal{C} \tag{189}$$

and the same proof as in 11.8 shows that Q is an *equivalence of categories*.

It is a result of primary importance for us. It implies in particular that all irreducibles in $\mathcal{C}$ (as well as their tensor products), come from $\tilde{\mathcal{C}}$.

12.6. Suppose we are given $\Lambda_0, \ldots, \Lambda_{-n} \in X$ and $\nu \in \mathbb{N}[I]$. Let $\pi: J \longrightarrow I$, $\pi: {}^{\pi}\mathbb{A} \longrightarrow A_\nu$ denote the same as in 6.12. We will use the notations for spaces and functors from Section 10.

The definition of the local system $\mathcal{I}(\Lambda_0, \ldots, \Lambda_{-n}; \nu)$ from *loc. cit.* should be modified: it should have half-monodromies $\zeta_j^{-(\pi(j),\Lambda_i)}$ for $i \in (n)$, $j \in J$, the other formulas stay without change.

After that, the standard sheaves are defined as in *loc. cit.* Now we are arriving at the main results of this part. The proof is the same as the proof of theorems 10.6 and 10.8, taking into account the previous algebraic remarks.

12.7. **Theorem.** *Let $L(\Lambda_0), \ldots, L(\Lambda_{-n})$ be irreducibles of $\mathcal{C}$, $\lambda \in X$, $\lambda = \sum_{m=0}^n \Lambda_{-m} - \sum_i \nu_i i'$ for some $\nu_i \in \mathbb{N}$. Set $\nu = \sum \nu_i i \in \mathbb{N}[I]$.*

Given a bijection $\eta: J \xrightarrow{\sim} [N]$, we have natural isomorphisms

$$\phi_{!*}^{(\eta)}: \Phi_\nu({}^{\psi}\mathcal{I}_\nu(\Lambda_0, \ldots, \Lambda_{-n})_{!*}) \xrightarrow{\sim} (L(\Lambda_0) \otimes \ldots \otimes L(\Lambda_{-n}))_\lambda \tag{190}$$

A change of η multiplies these isomorphisms by the sign of the corresponding permutation of $[N]$. □

12.8. **Theorem.** *In the notations of the previous theorem we have natural isomorphisms*

$$\phi_{!*,0}^{(\eta)}: {}^{\psi}\mathcal{I}_\nu(\Lambda_0, \ldots, \Lambda_{-n})_{!*0} \xrightarrow{\sim} C_{\mathfrak{f}}^{\bullet}(L(\Lambda_0) \otimes \ldots \otimes L(\Lambda_{-n}))_\lambda \tag{191}$$

where we used the notation $C^\bullet(\ldots)_\lambda$ for ${}_\nu C^\bullet(\ldots)$. A change of η multiplies these isomorphisms by the sign of the corresponding permutation of $[N]$. □

Part III. TENSOR CATEGORIES ARISING FROM CONFIGURATION SPACES

1. Introduction

1.1. In Chapter 1 we associate with every Cartan matrix of finite type and a non-zero complex number ζ an abelian artinian category $\mathcal{FS}$. We call its objects *finite factorizable sheaves.* They are certain infinite collections of perverse sheaves on configuration spaces, subject to a compatibility ("factorization") and finiteness conditions.

In Chapter 2 the tensor structure on $\mathcal{FS}$ is defined using functors of nearby cycles. It makes $\mathcal{FS}$ a braided tensor category.

In Chapter 3 we define, using vanishing cycles functors, an exact tensor functor

$$\Phi : \mathcal{FS} \longrightarrow \mathcal{C}$$

to the category $\mathcal{C}$ connected with the corresponding quantum group, cf. [AJS], 1.3 and [FS] II, 11.3, 12.2.

In Chapter 4 we prove

1.2. **Theorem.** *Φ is an equivalence of categories.*

One has to distinguish two cases.

(i) ζ is not root of unity. In this case it is wellknown that $\mathcal{C}$ is semisimple. This case is easier to treat; 1.2 is Theorem 18.4.

(ii) ζ is a root of unity. This is of course the most interesting case; 1.2 is Theorem 17.1.

Φ may be regarded as a way of localizing u-modules from category $\mathcal{C}$ to the origin of the affine line $\mathbb{A}^1$. More generally, in order to construct tensor structure on $\mathcal{FS}$, we define for each finite set K certain categories ${}^K\mathcal{FS}$ along with the functor g_K : $\mathcal{FS}^K \longrightarrow {}^K\mathcal{FS}$ (see the Sections 9, 10) which may be regarded as a way of localizing K-tuples of u-modules to K-tuples of points of the affine line. In the parts IV, V we will show how to localize u-modules to the points of an arbitrary smooth curve. For example, the case of a projective line is already quite interesting, and is connected with "semiinfinite" cohomology of quantum groups.

We must warn the reader that the proofs of some technical topological facts are only sketched in this part. The full details will appear later on.

1.3. The construction of the space $\mathcal{A}$ in Section 2 is inspired by the idea of "semi-infinite space of divisors on a curve" one of us learnt from A.Beilinson back in 1990. The construction of the braiding local system $\mathcal{I}$ in Section 3 is very close in spirit to P.Deligne's letter [D1]. In terms of this letter, all the local systems $\mathcal{I}^\alpha_\mu$ arise from the semisimple braided tensor category freely generated by the irreducibles $i \in I$ with the square of R-matrix: $i \otimes j \longrightarrow j \otimes i \longrightarrow i \otimes j$ given by the scalar matrix $\zeta^{i \cdot j}$.

We are very grateful to B.Feigin for the numerous inspiring discussions, and to P.Deligne and L.Positselsky for the useful comments concerning the definition of morphisms in ${}^K\mathcal{FS}$ in 9.6.

1.4. **Notations.** We will use all the notations of parts I and II. References to *loc. cit.* will look like Z.1.1 where Z=I or II.

During the whole part we fix a Cartan datum $(I,\cdot)$ of finite type and denote by $(Y=\mathbb{Z}[I]; X=\mathrm{Hom}(Y,\mathbb{Z}); I\hookrightarrow Y, i\mapsto i; I\hookrightarrow X, i\mapsto i')$ the simply connected root datum associated with $(I,\cdot)$, [L1], 2.2.2. Given $\alpha=\sum a_i i\in Y$, we will denote by α' the element $\sum a_i i'\in X$. This defines an embedding

$$Y\hookrightarrow X \tag{192}$$

We will use the notation $d_i:=\frac{i\cdot i}{2}$. We have $\langle j',d_i i\rangle=i\cdot j$. We will denote by A the $I\times I$-matrix $(\langle i,j'\rangle)$. We will denote by $\lambda,\mu\mapsto\lambda\cdot\mu$ a unique $\mathbb{Z}[\frac{1}{\det\, A}]$- valued scalar product on X such that (192) respects scalar products. We have

$$\lambda\cdot i'=\langle\lambda,d_i i\rangle \tag{193}$$

for each $\lambda\in X,\ i\in I$.

We fix a non-zero complex number ζ' and suppose that our ground field k contains ζ'. We set $\zeta:=(\zeta')^{\det\, A}$; for $a=\frac{c}{\det\, A}$, $c\in\mathbb{Z}$, we will use the notation $\zeta^a:=(\zeta')^c$.

We will use the following partial orders on X and Y. For $\alpha=\sum a_i i,\ \beta=\sum b_i i\in Y$ we write $\alpha\leq\beta$ if $a_i\leq b_i$ for all i. For $\lambda,\mu\in X$ we write $\lambda\geq\mu$ if $\lambda-\mu=\alpha'$ for some $\alpha\in Y,\ \alpha\geq 0$.

1.5. If X_1,X_2 are topological spaces, $\mathcal{K}_i\in\mathcal{D}(X_i)$, $i=1,2$, we will use the notation $\mathcal{K}_1\boxtimes\mathcal{K}_2$ for $p_1^*\mathcal{K}_1\otimes p_2^*\mathcal{K}_2$ (where $p_i: X_1\times X_2\longrightarrow X_i$ are projections). If J is a finite set, $|J|$ will denote its cardinality.

For a constructible complex $\mathcal{K}$, $SS(\mathcal{K})$ will denote the singular support of $\mathcal{K}$ (micro-support in the terminology of [KS], cf. *loc. cit.*, ch. V).

2. Space $\mathcal{A}$

2.1. We will denote by $\mathbb{A}^1$ the complex affine line with a fixed coordinate t. Given real c, c', $0 \leq c < c'$, we will use the notations $D(c) = \{t \in \mathbb{A}^1 |\ |t| < c\}$; $\bar{D}(c) = \{t \in \mathbb{A}^1 |\ |t| \leq c\}$; $\mathbb{A}^1_{>c} := \{t \in \mathbb{A}^1 |\ |t| > c\}$; $D(c, c') = \mathbb{A}^1_{>c} \cap D(c')$.

Recall that we have introduced in II.6.12 configuration spaces $\mathcal{A}_\alpha$ for $\alpha \in \mathbb{N}[I]$. If $\alpha = \sum a_i i$, the space $\mathcal{A}_\alpha$ parametrizes configurations of I-colored points $\mathbf{t} = (t_j)$ on $\mathbb{A}^1$, such that there are precisely a_i points of color i.

2.2. Let us introduce some open subspaces of $\mathcal{A}_\alpha$. Given a sequence

$$\vec{\alpha} = (\alpha_1, \ldots, \alpha_p) \in \mathbb{N}[I]^p \tag{194}$$

and a sequence of real numbers

$$\vec{d} = (d_1 \ldots, d_{p-1}) \tag{195}$$

such that $0 < d_{p-1} < d_{p-2} \ldots < d_1$, $p \geq 2$, we define an open subspace

$$\mathcal{A}^{\vec{\alpha}}(\vec{d}) \subset \mathcal{A}_\alpha \tag{196}$$

which parametrizes configurations $\mathbf{t}$ such that α_p of points t_j lie inside the disk $D(d_{p-1})$, for $2 \leq i \leq p-1$, α_i of points lie inside the annulus $D(d_{i-1}, d_i)$, and α_1 of points lie inside the ring $\mathcal{A}^1_{>d_1}$.

For $p = 1$, we set $\mathcal{A}^\alpha(\emptyset) := \mathcal{A}_\alpha$.

By definition, a configuration space of empty collections of points consists of one point. For example, so is $\mathcal{A}^0(\emptyset)$.

2.2.1. *Cutting.* Given $i \in [p-1]$ define subsequences

$$\vec{d}_{\leq i} = (d_1, \ldots, d_i);\ \vec{d}_{\geq i} = (d_i, \ldots, d_{p-1}) \tag{197}$$

and

$$\vec{\alpha}_{\leq i} = (\alpha_1, \ldots, \alpha_i, 0);\ \vec{\alpha}_{\geq i} = (0, \alpha_{i+1}, \ldots, \alpha_p) \tag{198}$$

We have obvious *cutting isomorphisms*

$$c_i : \mathcal{A}^{\vec{\alpha}}(\vec{d}) \xrightarrow{\sim} \mathcal{A}^{\vec{\alpha}_{\leq i}}(\vec{d}_{\leq i}) \times \mathcal{A}^{\vec{\alpha}_{\geq i}}(\vec{d}_{\geq i}) \tag{199}$$

satisfying the following compatibility:

for $i < j$ the square

$$\begin{array}{ccccc} & & \mathcal{A}^{\vec{\alpha}}(\vec{d}) & & \\ & \swarrow^{c_j} & & \searrow^{c_i} & \\ \mathcal{A}^{\vec{\alpha}_{\leq j}}(\vec{d}_{\leq j}) \times \mathcal{A}^{\vec{\alpha}_{\geq j}}(\vec{d}_{\geq j}) & & & & \mathcal{A}^{\vec{\alpha}_{\leq i}}(\vec{d}_{\leq i}) \times \mathcal{A}^{\vec{\alpha}_{\geq i}}(\vec{d}_{\geq i}) \\ & \searrow^{c_i \times \mathrm{id}} & & \swarrow^{\mathrm{id} \times c_j} & \\ & & \mathcal{A}^{\vec{\alpha}_{\leq i}}(\vec{d}_{\leq i}) \times \mathcal{A}^{\vec{\alpha}_{\geq i;\leq j}}(\vec{d}_{\geq i;\leq j}) \times \mathcal{A}^{\vec{\alpha}_{\geq j}}(\vec{d}_{\geq j}) & & \end{array}$$

commutes.

2.2.2. *Dropping.* For i as above, let $\partial_i \vec{d}$ denote the subsequence of $\vec{d}$ obtained by dropping d_i, and set

$$\partial_i \vec{\alpha} = (\alpha_1, \ldots, \alpha_{i-1}, \alpha_i + \alpha_{i+1}, \alpha_{i+2}, \ldots, \alpha_p) \tag{200}$$

We have obvious open embeddings

$$\mathcal{A}^{\partial_i \vec{\alpha}}(\partial_i \vec{d}) \hookrightarrow \mathcal{A}^{\vec{\alpha}}(\vec{d}) \tag{201}$$

2.3. Let us define $\mathcal{A}^{\vec{\alpha}}_{\mu}(\vec{d})$ as the space $\mathcal{A}^{\vec{\alpha}}(\vec{d})$ equipped with an additional index $\mu \in X$. One should understand μ as a weight assigned to the origin in $\mathbb{A}^1$. We will abbreviate the notation $\mathcal{A}^{\alpha}_{\mu}(\emptyset)$ to $\mathcal{A}^{\alpha}_{\mu}$.

Given a triple $(\vec{\alpha}, \vec{d}, \mu)$ as above, let us define its i-cutting — two triples $(\vec{\alpha}_{\leq i}, \vec{d}_{\leq i}, \mu_{\leq i})$ and $(\vec{\alpha}_{\geq i}, \vec{d}_{\geq i}, \mu)$, where

$$\vec{\mu}_{\leq i} = \vec{\mu} - (\sum_{j=i+1}^{p} \alpha_j)'.$$

We will also consider triples $(\partial_i \vec{\alpha}, \partial_i \vec{d}, \partial_i \mu)$ where $\partial_i \mu = \mu$ if $i < p-1$, and $\partial_{p-1}\mu = \mu - \alpha'_p$.

The cutting isomorphisms (199) induce isomorphisms

$$\mathcal{A}^{\vec{\alpha}}_{\mu}(\vec{d}) \xrightarrow{\sim} \mathcal{A}^{\vec{\alpha}_{\leq i}}_{\mu_{\leq i}}(\vec{d}_{\leq i}) \times \mathcal{A}^{\vec{\alpha}_{\geq i}}_{\mu}(\vec{d}_{\geq i}) \tag{202}$$

2.4. For each $\mu \in X$, $\alpha = \sum a_i i, \beta = \sum b_i i \in \mathbb{N}[I]$,

$$\sigma = \sigma^{\alpha,\beta}_{\mu} : \mathcal{A}^{\alpha}_{\mu} \hookrightarrow \mathcal{A}^{\alpha+\beta}_{\mu+\beta'} \tag{203}$$

will denote a closed embedding which adds b_i points of color i equal to 0.

For $d > 0$, $\mathcal{A}^{(\alpha,\beta)}_{\mu+\beta'}(d)$ is an open neighbourhood of $\sigma(\mathcal{A}^{\alpha}_{\mu})$ in $\mathcal{A}^{\beta+\alpha}_{\mu+\beta'}$.

2.5. By definition, $\mathcal{A}$ is a collection of all spaces $\mathcal{A}^{\vec{\alpha}}_{\mu}(\vec{d})$ as above, together with the cutting isomorphisms (202) and the closed embeddings (203).

2.6. Given a coset $c \in X/Y$ (where we regard Y as embedded into X by means of a map $i \mapsto i'$), we define $\mathcal{A}_c$ as a subset of $\mathcal{A}$ consisting of $\mathcal{A}^{\alpha}_{\mu}$ such that $\mu \in c$. Note that the closed embeddings σ, as well as cutting isomorphisms act inside $\mathcal{A}_c$. This subset will be called *a connected component* of $\mathcal{A}$. The set of connected components will be denoted $\pi_0(\mathcal{A})$. Thus, we have canonically $\pi_0(\mathcal{A}) \cong X/Y$.

2.7. We will be interested in two stratifications of spaces $\mathcal{A}_\mu^\alpha$. We will denote by $\overset{\bullet}{\mathcal{A}}{}_\mu^\alpha \subset \mathcal{A}_\mu^\alpha$ the complement

$$A_\mu^\alpha - \bigcup_{\beta<\alpha} \sigma(\mathcal{A}^\beta_{\mu-\beta'+\alpha'}).$$

We define a *toric stratification* of $\mathcal{A}_\mu^\alpha$ as

$$\mathcal{A}_\mu^\alpha = \coprod \sigma(\overset{\bullet}{\mathcal{A}}{}^\beta_{\mu-\beta'+\alpha'}).$$

Another stratification of $\mathcal{A}_\mu^\alpha$ is *the principal stratification* defined in II.7.14. Its open stratum will be denoted by $\overset{\circ}{\mathcal{A}}{}_\mu^\alpha \subset \mathcal{A}_\mu^\alpha$. Unless specified otherwise, we will denote the prinicipal stratification on spaces $\mathcal{A}_\mu^\alpha$, as well as the induced stratifications on its subspaces, by $\mathcal{S}$.

The sign $\circ$ (resp., $\bullet$) over a subspace of $\mathcal{A}_\mu^\alpha$ will denote the intersection of this subspace with $\overset{\circ}{\mathcal{A}}{}_\mu^\alpha$ (resp., with $\overset{\bullet}{\mathcal{A}}{}_\mu^\alpha$).

3. Braiding local system $\mathcal{I}$

3.1. **Local systems $\mathcal{I}_\mu^\alpha$.** Let us recall some definitions from II. Let $\alpha = \sum_i a_i i \in \mathbb{N}[I]$ be given. Following II.6.12, let us choose an *unfolding* of α, i.e. a set J together with a map $\pi : J \longrightarrow I$ such that $|\pi^{-1}(i)| = a_i$ for all i. We define the group $\Sigma_\pi := \{\sigma \in \mathrm{Aut}(J) |\ \sigma\circ\pi = \pi\}$.

We define ${}^\pi\mathbb{A}$ as an affine space with coordinates t_j, $j \in J$; it is equipped with the principal stratification defined by hyperplanes $t_j = 0$ and $t_i = t_j$, cf. II.7.1. The group Σ_π acts on ${}^\pi\mathbb{A}$ by permutations of coordinates, respecting the stratification. By definition, $\mathcal{A}_\alpha = {}^\pi\mathbb{A}/\Sigma_\pi$. We will denote by the same letter π the canonical projection ${}^\pi\mathbb{A} \longrightarrow \mathcal{A}_\alpha$.

If ${}^\pi\overset{\circ}{\mathbb{A}} \subset {}^\pi\mathbb{A}$ denotes the open stratum of the principal stratification, $\pi({}^\pi\overset{\circ}{\mathbb{A}}) = \overset{\circ}{\mathcal{A}}_\pi$, and the restriction of π to ${}^\pi\overset{\circ}{\mathbb{A}}$ is unramified covering.

Suppose a weight $\mu \in X$ is given. Let us define a one dimensional local system ${}^\pi\mathcal{I}_\mu$ over ${}^\pi\overset{\circ}{\mathbb{A}}$ by the procedure II.8.1. Its fiber over each positive chamber $C \in \pi_0({}^\pi\overset{\circ}{\mathbb{A}}_\mathbb{R})$ is identified with k; and monodromies along standard paths shown on II, Fig. 5 (a), (b) are given by the formulas

$$^C T_{ij} = \zeta^{-\pi(i)\cdot\pi(j)},\ ^C T_{i0} = \zeta^{2\mu\cdot\pi(i)'} \tag{204}$$

respectively (cf. (193)). (Note that, by technical reasons, this definition differs by the sign from that of II.8.2 and II.12.6).

We have a canonical Σ_π-equivariant structure on ${}^\pi\mathcal{I}_\mu$, i.e. a compatible system of isomorphisms

$$i_\sigma : {}^\pi\mathcal{I}_\mu \xrightarrow{\sim} \sigma^*\,{}^\pi\mathcal{I}_\mu,\ \sigma \in \Sigma_\pi, \tag{205}$$

defined uniquely by the condition that

$$(i_\sigma)_C = \mathrm{id}_\mathsf{k} :\ ({}^\pi\mathcal{I}_\mu)_C = \mathsf{k} \xrightarrow{\sim} (\sigma^*\,{}^\pi\mathcal{I}_\mu)_{\sigma C} = \mathsf{k}$$

for all (or for some) chamber C. As a consequence, the group Σ_π acts on the local system $\pi_*\,{}^\pi\mathcal{I}_\mu$.

Let $\mathrm{sgn} : \Sigma_\pi \longrightarrow \{\pm 1\}$ denote the sign character. We define a one-dimensional local system $\mathcal{I}^\alpha_\mu$ over $\mathring{\mathcal{A}}^\alpha_\mu = \mathring{\mathcal{A}}_\alpha$ as follows:

$$\mathcal{I}^\alpha_\mu := (\pi_* \, {}^\pi\mathcal{I}_\mu)^{\mathrm{sgn}} \tag{206}$$

where the superscript $(\bullet)^{\mathrm{sgn}}$ denotes the subsheaf of sections x such that $\sigma x = \mathrm{sgn}(\sigma)x$ for all $\sigma \in \Sigma_\pi$. Cf. II.8.16.

Alternatively, we can define this local system as follows. By the descent, there exists a unique local system $\mathcal{I}'_\mu$ over $\mathcal{A}_\alpha$ such that $\pi^*\mathcal{I}'_\mu$ is equal to ${}^\pi\mathcal{I}_\mu$ with the equivariant structure described above. In fact,

$$\mathcal{I}'_\mu = (\pi_* \, {}^\pi\mathcal{I}_\mu)^{\Sigma_\pi}$$

where the superscript $(\bullet)^{\Sigma_\pi}$ denotes invariants. We have

$$\mathcal{I}^\alpha_\mu = \mathcal{I}'_\mu \otimes \mathcal{S}ign \tag{207}$$

where $\mathcal{S}ign$ denotes the one-dimensional local system over $\mathcal{A}_\alpha$ associated with the sign representation $\pi_1(\mathcal{A}_\alpha) \longrightarrow \Sigma_\pi \xrightarrow{\mathrm{sgn}} \{\pm 1\}$.

This definition does not depend (up to a canonical isomorphism) upon the choice of an unfolding.

3.2. For each triple $(\vec{\alpha}, \vec{d}, \mu)$ as in the previous section, let us denote by $\mathcal{I}^{\vec{\alpha}}_\mu(\vec{d})$ the restriction of $\mathcal{I}^\alpha_\mu$ to the subspace $\mathring{\mathcal{A}}^{\vec{\alpha}}_\mu(\vec{d}) \subset \mathring{\mathcal{A}}^\alpha(\mu)$ where $\alpha \in \mathbb{N}[I]$ is the sum of components of $\vec{\alpha}$.

Let us define *factorization isomorphisms*

$$\phi_i = \phi^{\vec{\alpha}}_{\mu;i}(\vec{d}) : \mathcal{I}^{\vec{\alpha}}_\mu(\vec{d}) \xrightarrow{\sim} \mathcal{I}^{\vec{\alpha}_{\le i}}_{\mu_{\le i}}(\vec{d}_{\le i}) \boxtimes \mathcal{I}^{\vec{\alpha}_{\ge i}}_\mu(\vec{d}_{\ge i}) \tag{208}$$

(we are using identifications (202)). By definition, we have canonical identifications of the stalks of all three local systems over a point with real coordinates, with k. We define (208) as a unique isomorphism acting as identity when restricted to such a stalk. We will omit irrelevant indices from the notation for ϕ if there is no risk of confusion.

3.3. **Associativity.** These isomorphisms have the following *associativity property.* For all $i < j$, diagrams

$$\begin{array}{ccccc}
 & & \mathcal{I}(\vec{d}) & & \\
 & \swarrow^{\phi_j} & & \searrow^{\phi_i} & \\
\mathcal{I}(\vec{d}_{\le j}) \boxtimes \mathcal{I}(\vec{d}_{\ge j}) & & & & \mathcal{I}(\vec{d}_{\le i}) \boxtimes \mathcal{I}(\vec{d}_{\ge i}) \\
 & \searrow_{\phi_i \boxtimes \mathrm{id}} & & \swarrow_{\mathrm{id} \boxtimes \phi_j} & \\
 & & \mathcal{I}(\vec{d}_{\le i}) \boxtimes \mathcal{I}(\vec{d}_{\ge i; \le j}) \boxtimes \mathcal{I}(\vec{d}_{\ge j}) & &
\end{array}$$

are commutative.

In fact, it is enough to check the commutativity restricted to some fiber $(\bullet)_C$, where it is obvious.

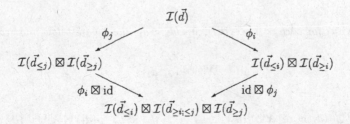

3.4. The collection of local systems $\mathcal{I} = \{\mathcal{I}^\alpha_\mu\}$ together with factorization isomorphisms (208) will be called *the braiding local system* (over $\overset{\circ}{\mathcal{A}}$).

The couple $(\mathcal{A}, \mathcal{I})$ will be called *the semi-infinite configuration space associated with the Cartan datum* $(I, \cdot)$ *and parameter* ζ.

3.5. Let $j : \overset{\circ}{\mathcal{A}}{}^{\vec\alpha}_\mu(\vec d) \hookrightarrow \overset{\bullet}{\mathcal{A}}{}^{\vec\alpha}_\mu(\vec d)$ denote an embedding; let us define a preverse sheaf

$$\overset{\bullet}{\mathcal{I}}{}^{\vec\alpha}_\mu(\vec d) := j_{!*}\mathcal{I}^{\vec\alpha}_\mu(\vec d)[\dim\ A^{\vec\alpha}_\mu] \in \mathcal{M}(\overset{\bullet}{\mathcal{A}}{}^{\vec\alpha}_\mu(\vec d); \mathcal{S})$$

By functoriality, factorization isomorphisms (208) induce analogous isomorphisms (denoted by the same letter)

$$\phi_i = \phi^{\vec\alpha}_{\mu;i}(\vec d) : \overset{\bullet}{\mathcal{I}}{}^{\vec\alpha}_\mu(\vec d) \xrightarrow{\sim} \overset{\bullet}{\mathcal{I}}{}^{\vec\alpha_{\leq i}}_{\mu_{\leq i}}(\vec d_{\leq i}) \boxtimes \overset{\bullet}{\mathcal{I}}{}^{\vec\alpha_{\geq i}}_{\mu}(\vec d_{\geq i}) \tag{209}$$

satisfying the associativity property completely analogous to 3.3.

4. Factorizable sheaves

4.1. The aim of this section is to define certain k-linear category $\widetilde{\mathcal{FS}}$. Its objects will be called *factorizable sheaves* (over $(\mathcal{A}, \mathcal{I})$). By definition, $\widetilde{\mathcal{FS}}$ is a direct product of k-categories $\widetilde{\mathcal{FS}}_c$, where c runs through $\pi_0(\mathcal{A})$ (see 2.6). Objects of $\widetilde{\mathcal{FS}}_c$ will be called *factorizable sheaves supported at* $\mathcal{A}_c$.

In what follows we pick c, and denote by $X_c \subset X$ the corresponding coset modulo Y.

4.2. **Definition.** A factorizable sheaf $\mathcal{X}$ over $(\mathcal{A}, \mathcal{I})$ supported at $\mathcal{A}_c$ *is the following collection of data:*

(a) a weight $\lambda \in X_c$*; it will be denoted by* $\lambda(\mathcal{X})$*;*

(b) for each $\alpha \in \mathbb{N}[I]$*, a sheaf* $\mathcal{X}^\alpha \in \mathcal{M}(\mathcal{A}^\alpha_\lambda; \mathcal{S})$*;*

we will denote by $\mathcal{X}^{\vec\alpha}(\vec d)$ *perverse sheaves over* $\mathcal{A}^{\vec\alpha}_\lambda(\vec d)$ *obtained by taking the restrictions with respect to the embeddings* $\mathcal{A}^{\vec\alpha}_\lambda(\vec d) \hookrightarrow \mathcal{A}^\alpha_\lambda$*;*

(c) for each $\alpha, \beta \in \mathbb{N}[I]$, $d > 0$, *a* factorization isomorphism

$$\psi^{\alpha,\beta}(d) : \mathcal{X}^{(\alpha,\beta)}(d) \xrightarrow{\sim} \overset{\bullet}{\mathcal{I}}{}^{(\alpha,0)}_{\lambda-\beta'}(d) \boxtimes \mathcal{X}^{(0,\beta)}(d) \tag{210}$$

such that

(associativity) *for each* $\alpha, \beta, \gamma \in \mathbb{N}[I]$, $0 < d_2 < d_1$, *the square below must commute:*

$$\begin{array}{ccc}
 & \mathcal{X}^{(\alpha,\beta,\gamma)}(d_1, d_2) & \\
\swarrow{\scriptstyle\psi} & & \searrow{\scriptstyle\psi} \\
\overset{\bullet}{\mathcal{I}}{}^{(\alpha,\beta,0)}_{\lambda-\gamma'}(d_1, d_2) \boxtimes \mathcal{X}^{(0,\gamma)}(d_2) & & \overset{\bullet}{\mathcal{I}}{}^{(\alpha,0)}_{\lambda-\beta'-\gamma'}(d_1) \boxtimes \mathcal{X}^{(0,\beta,\gamma)}(d_1, d_2) \\
{\scriptstyle\phi\boxtimes\mathrm{id}}\searrow & & \swarrow{\scriptstyle\mathrm{id}\boxtimes\psi} \\
 & \overset{\bullet}{\mathcal{I}}{}^{(\alpha,0)}_{\lambda-\beta'-\gamma'}(d_1) \boxtimes \overset{\bullet}{\mathcal{I}}{}^{(0,\beta,0)}_{\lambda-\gamma'}(d_1, d_2) \boxtimes \mathcal{X}^{(0,\gamma)}(d_2) &
\end{array}$$

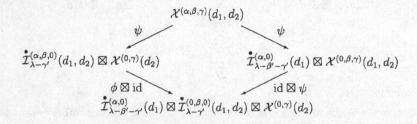

4.2.1. Remark that with these definitions, the braiding local system $\mathcal{I}$ resembles a "coalgebra", and a factorizable sheaf — a "comodule" over it.

4.3. **Remark.** Note an immediate corollary of the factorization axiom. We have isomorphisms

$$\mathcal{X}^{(\alpha,0)}(d) \cong \mathcal{X}^0 \otimes \overset{\bullet}{\mathcal{I}}{}_{\lambda}^{(\alpha,0)}(d) \tag{211}$$

(where $\mathcal{X}^0$ is simply a vector space).

Our next aim is to define morphisms between factorizable sheaves.

4.4. Let $\mathcal{X}$ be as above. For each $\mu \geq \lambda$, $\mu = \lambda + \beta'$, and $\alpha \in \mathbb{N}[I]$, let us define a sheaf $\mathcal{X}_\mu^\alpha \in \mathcal{M}(\mathcal{A}_\mu^\alpha; \mathcal{S})$ as $\sigma_* \mathcal{X}^{\alpha-\beta}$. For example, $\mathcal{X}_\lambda^\alpha = \mathcal{X}^\alpha$. By taking restriction, the sheaves $\mathcal{X}_\mu^{\vec{\alpha}}(\vec{d}) \in \mathcal{M}(\mathcal{A}_\mu^{\vec{\alpha}}(\vec{d}); \mathcal{S})$ are defined.

Suppose $\mathcal{X}, \mathcal{Y}$ are two factorizable sheaves supported at $\mathcal{A}_c$, $\lambda = \lambda(\mathcal{X})$, $\nu = \lambda(\mathcal{Y})$. Let $\mu \in X$, $\mu \geq \lambda$, $\mu \geq \nu$, $\alpha, \beta \in \mathbb{N}[I]$. By definition we have canonical isomorphisms

$$\theta = \theta_\mu^{\beta,\alpha}: \ \mathrm{Hom}_{\mathcal{A}_\mu^\alpha}(\mathcal{X}_\mu^\alpha, \mathcal{Y}_\mu^\alpha) \xrightarrow{\sim} \mathrm{Hom}_{\mathcal{A}_{\mu+\beta'}^{\alpha+\beta}}(\mathcal{X}_{\mu+\beta'}^{\alpha+\beta}, \mathcal{Y}_{\mu+\beta'}^{\alpha+\beta}) \tag{212}$$

The maps (210) induce analogous isomorphisms

$$\psi_\mu^{\alpha,\beta}(d): \mathcal{X}_\mu^{(\alpha,\beta)}(d) \xrightarrow{\sim} \overset{\bullet}{\mathcal{I}}{}_{\mu-\beta'}^{(\alpha,0)}(d) \boxtimes \mathcal{X}_\mu^{(0,\beta)}(d) \tag{213}$$

which satisfy the same associativity property as in 4.2.

For $\alpha \geq \beta$ let us define maps

$$\tau_\mu^{\alpha\beta}: \mathrm{Hom}_{\mathcal{A}_\mu^\alpha}(\mathcal{X}_\mu^\alpha, \mathcal{Y}_\mu^\alpha) \longrightarrow \mathrm{Hom}_{\mathcal{A}_\mu^\beta}(\mathcal{X}_\mu^\beta, \mathcal{Y}_\mu^\beta) \tag{214}$$

as compositions

$$\mathrm{Hom}_{\mathcal{A}_\mu^\alpha}(\mathcal{X}_\mu^\alpha, \mathcal{Y}_\mu^\alpha) \xrightarrow{res} \mathrm{Hom}_{\mathcal{A}_\mu^{(\alpha-\beta,\beta)}(d)}(\mathcal{X}_\mu^{(\alpha-\beta,\beta)}(d), \mathcal{Y}_\mu^{(\alpha-\beta,\beta)}(d)) \overset{\psi}{\xrightarrow{\sim}} \tag{215}$$

$$\overset{\psi}{\xrightarrow{\sim}} \mathrm{Hom}_{\mathcal{A}_{\mu-\beta'}^{(\alpha-\beta,0)}(d)}(\overset{\bullet}{\mathcal{I}}{}_{\mu-\beta'}^{(\alpha-\beta,0)}(d), \overset{\bullet}{\mathcal{I}}{}_{\mu-\beta'}^{(\alpha-\beta,0)}(d)) \otimes \mathrm{Hom}_{\mathcal{A}_\mu^{(0,\beta)}(d)}(\mathcal{X}_\mu^{(0,\beta)}(d), \mathcal{Y}_\mu^{(0,\beta)}(d)) =$$

$$= \mathsf{k} \otimes_{\mathsf{k}} \mathrm{Hom}_{\mathcal{A}_\mu^{(0,\beta)}(d)}(\mathcal{X}_\mu^{(0,\beta)}(d), \mathcal{Y}_\mu^{(0,\beta)}(d)) = \mathrm{Hom}_{\mathcal{A}_\mu^{(0,\beta)}(d)}(\mathcal{X}_\mu^{(0,\beta)}(d), \mathcal{Y}_\mu^{(0,\beta)}(d)) \xrightarrow{\sim}$$

$$\xrightarrow{\sim} \mathrm{Hom}_{\mathcal{A}_\mu^\beta}(\mathcal{X}_\mu^\beta, \mathcal{Y}_\mu^\beta)$$

where we have chosen some $d > 0$, the first map is the restriction, the second one is induced by the factorization isomorphism, the last one is inverse to the restriction. This definition does not depend on the choice of d.

The associativity axiom implies that these maps satisfy an obvious transitivity property. They are also compatible in the obvious way with the isomorphisms θ.

We define the space $\mathrm{Hom}_{\widetilde{\mathcal{FS}}_c}(\mathcal{X}, \mathcal{Y})$ as the following inductive-projective limit

$$\mathrm{Hom}_{\widetilde{\mathcal{FS}}_c}(\mathcal{X}, \mathcal{Y}) := \lim_{\substack{\rightarrow \\ \mu}} \lim_{\substack{\leftarrow \\ \beta}} \mathrm{Hom}(\mathcal{X}_\mu^\beta, \mathcal{Y}_\mu^\beta) \tag{216}$$

where the inverse limit is over $\beta \in \mathbb{N}[I]$, the transition maps being $\tau_\mu^{\alpha\beta}$, μ being fixed, and the direct limit over $\mu \in X$ such that $\mu \geq \lambda$, $\mu \geq \nu$, the transition maps being induced by (212).

With these spaces of homomorphisms, factorizable sheaves supported at $\mathcal{A}_c$ form a k-linear category to be denoted by $\widetilde{\mathcal{FS}}_c$ (the composition of morphisms is obvious).

As we have already mentioned, the category $\widetilde{\mathcal{FS}}$ is by definition a product $\prod_{c\in\pi_0(\mathcal{A})}\widetilde{\mathcal{FS}}_C$. Thus, an object $\mathcal{X}$ of $\widetilde{\mathcal{FS}}$ is a direct sum $\oplus_{c\in\pi_0(\mathcal{A})}\mathcal{X}_c$, where $\mathcal{X}_c\in\widetilde{\mathcal{FS}}_c$. If $\mathcal{X}\in\widetilde{\mathcal{FS}}_c$, $\mathcal{Y}\in\widetilde{\mathcal{FS}}_{c'}$, then

$$\mathrm{Hom}_{\widetilde{\mathcal{FS}}}(\mathcal{X},\mathcal{Y})=\mathrm{Hom}_{\widetilde{\mathcal{FS}}_c}(\mathcal{X},\mathcal{Y})$$

if $c=c'$, and 0 otherwise.

4.5. Let $\mathcal{V}ect_f$ denote the category of finite dimensional k-vector spaces. Recall that in II.7.14 the functors of "vanishing cycles at the origin"

$$\Phi_\alpha:\mathcal{M}(\mathcal{A}^\alpha_\mu;\mathcal{S})\longrightarrow\mathcal{V}ect_f$$

have been defined.

Given $\mathcal{X}\in\widetilde{\mathcal{FS}}_c$, let us define for each $\lambda\in X_c$ a vector space

$$\Phi_\lambda(\mathcal{X}):=\Phi_\alpha(\mathcal{X}^\alpha_{\lambda+\alpha'})\tag{217}$$

where $\alpha\in\mathbb{N}[I]$ is such that $\lambda+\alpha'\geq\lambda(\mathcal{X})$. If $\lambda\in X-X_c$, we set $\Phi_\lambda(\mathcal{X})=0$.

Due to the definition of the sheaves $\mathcal{X}^\alpha_\mu$, 4.4, this vector space does not depend on a choice of α, up to a unique isomorphism.

This way we get an exact functor

$$\Phi:\widetilde{\mathcal{FS}}_c\longrightarrow\mathcal{V}ect^X_f\tag{218}$$

to the category of X-graded vector spaces with finite dimensional components which induces an exact functor

$$\Phi:\widetilde{\mathcal{FS}}\longrightarrow\mathcal{V}ect^X_f\tag{219}$$

4.6. **Lemma.** *If $\Phi(\mathcal{X})=0$ then $\mathcal{X}=0$.*

Proof. We may suppose that $\mathcal{X}\in\widetilde{\mathcal{FS}}_c$ for some c. Let $\lambda=\lambda(\mathcal{X})$. Let us prove that for every $\alpha=\sum a_i i\in\mathbb{N}[I]$, $\mathcal{X}^\alpha=0$. Let us do it by induction on $|\alpha|:=\sum a_i$. We have $\mathcal{X}^0=\Phi_\lambda(\mathcal{X})=0$ by assumption.

Given an arbitrary α, it is easy to see from the factorizability and induction hypothesis that $\mathcal{X}^\alpha$ is supported at the origin of $\mathcal{A}_\alpha$. Since $\Phi_{\lambda-\alpha}(\mathcal{X})=0$, we conclude that $\mathcal{X}^\alpha=0$. □

5. Finite sheaves

5.1. **Definition.** *A factorizable sheaf $\mathcal{X}$ is called* finite *if $\Phi(X)$ is finite dimensional.*

This is equivalent to saying that there exists only finite number of $\alpha\in\mathbb{N}[I]$ such that $\Phi_\alpha(\mathcal{X}^\alpha)\neq0$ (or $SS(\mathcal{X}^\alpha)$ contains the conormal bundle to the origin $0\in\mathcal{A}^\alpha_\lambda$, where $\lambda:=\lambda(\mathcal{X})$).

5.2. **Definition.** *The category of finite factorizable sheaves (FFS for short) is a full subcategory $\mathcal{FS} \subset \widetilde{\mathcal{FS}}$ whose objects are finite factorizable sheaves.*

We set $\mathcal{FS}_c := \mathcal{FS} \cap \widetilde{\mathcal{FS}}_c$ for $c \in \pi_0(\mathcal{A})$.

This category is our main character. It is clear that $\mathcal{FS}$ is a strictly full subcategory of $\widetilde{\mathcal{FS}}$ closed with respect to taking subobjects and quotients.

The next stabilization lemma is important.

5.3. **Lemma.** *Let $\mathcal{X}, \mathcal{Y}$ be two FFS's supported at the same connected component of $\mathcal{A}$. For a fixed $\mu \geq \lambda(\mathcal{X}), \lambda(\mathcal{Y})$ there exists $\alpha \in \mathbb{N}[I]$ such that for any $\beta \geq \alpha$ the transition map*

$$\tau_\mu^{\beta\alpha} : \mathrm{Hom}_{\mathcal{A}_\mu^\beta}(\mathcal{X}_\mu^\beta, \mathcal{Y}_\mu^\beta) \longrightarrow \mathrm{Hom}_{\mathcal{A}_\mu^\alpha}(\mathcal{X}_\mu^\alpha, \mathcal{Y}_\mu^\alpha)$$

is an isomorphism.

Proof. Let us introduce a finite set

$$N_\mu(\mathcal{Y}) := \{\alpha \in \mathbb{N}[I] |\ \Phi_\alpha(\mathcal{Y}_\mu^\alpha) \neq 0\}.$$

Let us pick $\beta \in \mathbb{N}[I]$. Consider a non-zero map $f : \mathcal{X}_\mu^\beta \longrightarrow \mathcal{Y}_\mu^\beta$. For each $\alpha \leq \beta$ we have a map $f^\alpha := \tau_\mu^{\beta\alpha}(f) : \mathcal{X}_\mu^\alpha \longrightarrow \mathcal{Y}_\mu^\alpha$. Let us consider subsheaves $\mathcal{Z}^\alpha := \mathrm{Im}(f^\alpha) \subset \mathcal{Y}_\mu^\alpha$. These subsheaves satisfy an obvious factorization property.

Let us consider the toric stratification of $\mathcal{A}_\mu^\beta$. For each $\alpha \leq \beta$ set $\mathcal{A}^\alpha := \sigma(\mathcal{A}_{\mu+\alpha'-\beta'}^\alpha) \subset \mathcal{A}_\mu^\beta$; $\overset{\bullet}{\mathcal{A}}{}^\alpha := \sigma(\overset{\bullet}{\mathcal{A}}{}_{\mu+\alpha'-\beta'}^\alpha)$. Thus, the subspaces $\overset{\bullet}{\mathcal{A}}{}^\alpha$ are strata of the toric stratification. We have $\alpha_1 \leq \alpha_2$ iff $\mathcal{A}^{\alpha_1} \subset \mathcal{A}^{\alpha_2}$.

Let γ denote a maximal element in the set $\{\alpha |\ \mathcal{Z}^\beta|_{\overset{\bullet}{\mathcal{A}}{}^\alpha} \neq 0\}$. Then it is easy to see that $\mathcal{Z}^{\beta-\gamma}$ is a non-zero scyscraper on $\mathcal{A}^{\beta-\gamma}$ supported at the origin. Therefore, $\Phi_{\beta-\gamma}(\mathcal{Z}^{\beta-\gamma}) \neq 0$, whence $\Phi_{\beta-\gamma}(\mathcal{Y}^{\beta-\gamma}) \neq 0$, i.e. $\beta - \gamma \in N_\mu(\mathcal{Y})$.

Suppose that for some $\alpha \leq \beta$, $\tau_\mu^{\beta\alpha}(f) = 0$. Then

$$\mathcal{Z}^\beta|_{\mathcal{A}(\beta-\alpha,\alpha)(d)} = 0$$

for every $d > 0$. It follows that if $\mathcal{Z}^\beta|_{\overset{\bullet}{\mathcal{A}}{}^\delta} \neq 0$ then $\delta \not\geq \beta - \alpha$.

Let us apply this remark to δ equal to γ above. Suppose that $\gamma = \sum c_i i$, $\beta = \sum b_i i$, $\alpha = \sum a_i i$. There exists i such that $c_i < b_i - a_i$. Recall that $\beta - \gamma \in N_\mu(\mathcal{Y})$. Consequently, we have

5.3.1. **Corollary.** *Suppose that $\alpha \geq \delta$ for all $\delta \in N_\mu(\mathcal{Y})$. Then all the maps*

$$\tau_\mu^{\beta\alpha} : \mathrm{Hom}(\mathcal{X}_\mu^\beta, \mathcal{Y}_\mu^\beta) \longrightarrow \mathrm{Hom}(\mathcal{X}_\mu^\alpha, \mathcal{Y}_\mu^\alpha),$$

$\beta \geq \alpha$, are injective. □

Since all the spaces $\mathrm{Hom}(\mathcal{X}_\mu^\alpha, \mathcal{Y}_\mu^\alpha)$ are finite dimensional due to the constructibility of our sheaves, there exists an α such that all $\tau_\mu^{\beta\alpha}$ are isomorphisms. Lemma is proven. □

5.4. For $\lambda \in X_c$ let us denote by $\mathcal{FS}_{c;\leq\lambda} \subset \mathcal{FS}_c$ the full subcategory whose objects are FFS's $\mathcal{X}$ such that $\lambda(\mathcal{X}_c) \leq \lambda$. Obviously $\mathcal{FS}_c$ is a filtered union of these subcategories.

We have obvious functors

$$p_\lambda^\beta : \mathcal{FS}_{c;\leq\lambda} \longrightarrow \mathcal{M}(\mathcal{A}_\lambda^\beta; \mathcal{S}),\ \mathcal{X} \mapsto \mathcal{X}_\lambda^\beta \tag{220}$$

The previous lemma claims that for every $\mathcal{X}, \mathcal{Y} \in \mathcal{FS}_{c;\leq\lambda}$ there exists $\alpha \in \mathbb{N}[I]$ such that for every $\beta \geq \alpha$ the map

$$p_\lambda^\beta : \mathrm{Hom}_{\mathcal{FS}}(\mathcal{X}, \mathcal{Y}) \longrightarrow \mathrm{Hom}_{\mathcal{A}_\lambda^\beta}(\mathcal{X}_\lambda^\beta, \mathcal{Y}_\lambda^\beta)$$

is an isomorphism. (Obviously, a similar claim holds true for any finite number of FFS's.)

5.5. **Lemma.** *$\mathcal{FS}$ is an abelian artinian category.*

Proof. $\mathcal{FS}$ is abelian by Stabilization lemma. Each object has finite length by Lemma 4.6. □

6. Standard sheaves

6.1. For $\Lambda \in X$, let us define factorizable sheaves $\mathcal{M}(\Lambda), D\mathcal{M}(\Lambda)_{\zeta^{-1}}$ and $\mathcal{L}(\Lambda)$ as follows. (The notation $D\mathcal{M}(\Lambda)_{\zeta^{-1}}$ will be explained in 13.3 below).

Set

$$\lambda(\mathcal{M}(\Lambda)) = \lambda(D\mathcal{M}(\Lambda)_{\zeta^{-1}}) = \lambda(\mathcal{L}(\Lambda)) = \Lambda.$$

For $\alpha \in \mathbb{N}[I]$ let j denote the embedding $\overset{\bullet}{\mathcal{A}}_\alpha \hookrightarrow \mathcal{A}_\alpha$. We define

$$\mathcal{M}(\Lambda)^\alpha = j_!\overset{\bullet}{\mathcal{I}}{}^\alpha_\Lambda;\ D\mathcal{M}(\Lambda)^\alpha_{\zeta^{-1}} = j_*\overset{\bullet}{\mathcal{I}}{}^\alpha_\Lambda;\ \mathcal{L}(\Lambda)^\alpha = j_{!*}\overset{\bullet}{\mathcal{I}}{}^\alpha_\Lambda.$$

The factorization isomorphisms are defined by functoriality from these isomorphisms for $\overset{\bullet}{\mathcal{I}}$.

Thus, the collections $\{\mathcal{M}(\Lambda)^\alpha\}_\alpha$, etc. form factorizable sheaves to be denoted by $\mathcal{M}(\Lambda), D\mathcal{M}(\Lambda)_{\zeta^{-1}}$ and $\mathcal{L}(\Lambda)$ respectively. Obviously, we have a canonical morphism

$$m : \mathcal{M}(\Lambda) \longrightarrow D\mathcal{M}(\Lambda)_{\zeta^{-1}} \tag{221}$$

and $\mathcal{L}(\Lambda)$ is equal to its image.

6.2. **Theorem.** (i) *The factorizable sheaves $\mathcal{L}(\Lambda)$ are finite.*

(ii) *They are irreducible objects of $\mathcal{FS}$, non-isomorphic for different Λ, and they exhaust all irreducibles in $\mathcal{FS}$, up to isomorphism.*

Proof. (i) follows from II.8.18.

(ii) Since the sheaves $\mathcal{L}(\Lambda)^\alpha$ are irreducible as objects of $\mathcal{M}(\mathcal{A}_\alpha; \mathcal{S})$, the irreducibility of $\mathcal{L}(\Lambda)$ follows easily. It is clear that they are non-isomorphic (consider the highest component).

Suppose $\mathcal{X}$ is an irreducible FFS, $\lambda = \lambda(\mathcal{X})$. Let $\alpha \in \mathbb{N}[X]$ be a minimal among β such that $\Phi_{\lambda-\beta}(\mathcal{X}) \neq 0$; set $\Lambda = \lambda - \alpha$. By factorizability and the universal property of !-extension, there exists a morphism if FS's $f : \mathcal{M}(\Lambda) \longrightarrow \mathcal{X}$ such that $\Phi_\Lambda(f) \neq 0$ (hence is a monomorphism). It follows from irreducibility of $\mathcal{L}(\Lambda)$ that the composition $\mathrm{Ker}(f) \longrightarrow \mathcal{M}(\Lambda) \longrightarrow \mathcal{L}(\Lambda)$ is equal to zero, hence f factors through a non-zero morphism $\mathcal{L}(\Lambda) \longrightarrow \mathcal{X}$ which must be an isomorphism. □

6.3. Let us look more attentively at the sheaf $\mathcal{L}(0)$.

Let $\tilde{\mathcal{A}}^{\alpha} \subset \mathcal{A}^{\alpha}$ denote the open stratum of the *diagonal* stratification, i.e. the complement to the diagonals. Thus, $\overset{\circ}{\mathcal{A}}{}^{\alpha} \subset \tilde{\mathcal{A}}^{\alpha}$. Let $\tilde{\mathcal{I}}^{\alpha}$ denote the local system over $\tilde{\mathcal{A}}^{\alpha}$ defined in the same way as local systems $\mathcal{I}^{\alpha}_{\mu}$, but using only "diagonal" monodromies, cf. II.6.3.

One sees immediately that $\mathcal{L}(0)^{\alpha}$ is equal to the middle extenstion of $\tilde{\mathcal{I}}^{\alpha}$.

Chapter 2. Tensor structure

7. Marked disk operad

7.1. Let K be a finite set. If T is any set, we will denote by T^K the set of all mappings $K \longrightarrow T$; elements of T^K will be denoted typically by $\vec{x} = (x_k)_{k\in K}$.

We will use the following partial orders on $X^K, \mathbb{N}[I]^K$. For $\vec{\lambda} = (\lambda_k),\ \vec{\mu} = (\mu_k) \in X^K$, we write $\vec{\lambda} \geq \vec{\mu}$ iff $\lambda_k \geq \mu_k$ for all k. An order on $\mathbb{N}[I]^K$ is defined in the same manner.

For $\vec{\alpha} = (\alpha_k) \in \mathbb{N}[I]^K$ we will use the notation α for the sum of its components $\sum_{k\in K} \alpha_k$; the same agreement will apply to X^K.

$\mathbb{A}^K$ will denote the complex affine space with fixed coordinates u_k, $k \in K$; $\mathring{\mathbb{A}}^K \subset \mathbb{A}^K$ will denote the open stratum of the diagonal stratification.

7.2. **Trees.** We will call *a tree* a couple

$$\tau = (\sigma, \vec{d}) \tag{222}$$

where σ, to be called *the shape* of τ,

$$\sigma = (K_p \xrightarrow{\rho_{p-1}} K_{p-1} \xrightarrow{\rho_{p-2}} \ldots \xrightarrow{\rho_1} K_1 \xrightarrow{\rho_0} K_0)$$

is a sequence of epimorhisms of finite sets, such that $\mathrm{card}(K_0) = 1$, $\vec{d} = (d_0, d_1, \ldots, d_p)$, to be called *the thickness* of τ — a tuple of real numbers such that $d_0 = 1 > d_1 > \ldots > d_p \geq 0$.

We will use a notation ρ_{ab} for composition $K_a \longrightarrow K_{a-1} \longrightarrow \ldots \longrightarrow K_b$, $a > b$.

A number $p \geq 0$ will be called *the height* of τ and denoted $\mathrm{ht}(\tau)$. Elements $k \in K_i$ will be called *branches of height* i; d_i will be called *the thickness of* k. A unique branch of height 0 will be called *bole* and denoted by $*(\tau)$.

The set K_p will be called *the base* of τ and denoted K_τ; we will also say that τ is K_p*-based*; we will denote d_p by d_τ. We will use notation $K(\tau)$ for the set $\coprod_{i=0}^{p} K_i$ and $K'(\tau)$ for $\coprod_{i=0}^{p-1} K_i$.

A tree of height one will be called *elementary*. A tree τ whose branches of height $\mathrm{ht}(\tau)$ have thickness 0, will be called *grown up*; otherwise it will be called *young*. We will assign to every tree τ a grown up tree $\tilde{\tau}$ by changing the thickness of the thinnest branches to zero.

Thus, an elementary tree is essentially a finite set and a real $0 \leq d < 1$; a grown up elementary tree is essentally a finite set.

7.2.1. *Cutting.* Suppose we have a tree τ as above, and an integer i, $0 < i < p$. We define the operation of *cutting* τ at level i. It produces from τ new trees $\tau_{\leq i}$ and $\tau_{\geq k}$, $k \in K_i$. Namely,

$$\tau_{\leq i} = (\sigma_{\leq i}, \vec{d}_{\leq i}),$$

where $\sigma_{\leq i} = (K_i \longrightarrow K_{i-1} \longrightarrow \ldots \longrightarrow K_0)$ and $\vec{d}_{\leq i} = (d_0, d_1, \ldots, d_i)$.

Second, for $k \in K_i$

$$\tau_{\geq k} = (\sigma_{\geq k}, \vec{d}_{>i})$$

where

$$\sigma_{\geq k} = (\rho_{pi}^{-1}(k) \longrightarrow \rho_{p-1,i}^{-1}(k) \longrightarrow \ldots \longrightarrow \rho_i^{-1}(k) \longrightarrow \{k\}),$$
$$d_{>i} = (1, d_{i+1}d_i^{-1}, \ldots, d_p d_i^{-1}).$$

7.2.2. For $0 < i \leq p$ we will denote by $\partial_i\tau$ a tree $(\partial_i\sigma, \partial_i\vec{d})$ where

$$\partial_i\sigma = (K_p \longrightarrow \ldots \longrightarrow \hat{K}_i \longrightarrow \ldots \longrightarrow K_0),$$

and $\partial_i\vec{d}$ is obtained from $\vec{d}$ by omitting d_i.

7.3. **Operad of disks.** For $r \in \mathbb{R}_{>0} \cup \{\infty\}$, $z \in \mathbb{C}$, we define an open disk $D(z;r) := \{u \in \mathbb{A}^1 |\ |u-z| < r\}$, and a closed disk $\bar{D}(z;r) := \{u \in \mathbb{A}^1 |\ |u-z| \leq r\}$.

For a tree (222) we define a space

$$\mathcal{O}(\tau) = \mathcal{O}(\sigma;\vec{d})$$

parametrizing all collections $\vec{D} = (\bar{D}_k)_{k\in K_\tau}$ of closed disks, such that $D_{*(\tau)} = D(0;1)$, for $k \in K_i$ the disk $\bar{D}_k$ has radius d_i, for fixed $i \in [p]$ the disks $\bar{D}_k$, $k \in K_i$, do not intersect, and for each $i \in [0, p-1]$ and each $k \in K_{i+1}$ we have $\bar{D}_k \subset D_{\rho_i(k)}$.

Sometimes we will call such a collection *a configuration of disks shaped by a tree* τ.

7.3.1. Given such a configuration, we will use the notation

$$\mathring{D}_k(\tau) = D_k - \bigcup_{l\in\rho_i^{-1}(k)} (\bar{D}_l) \tag{223}$$

if $k \in K_i$ and $i < p$, and we set $\mathring{D}_k(\tau) = D_k$ if $i = p$.

If $\tau = (K \longrightarrow \{*\}; d)$ is an elementary tree, we will use the notation $\mathcal{O}(K;d)$ for $\mathcal{O}(\tau)$; if $d = 0$, we will abrreviate the notation to $\mathcal{O}(K)$.

We have obvious embeddings

$$\mathcal{O}(K;d) \hookrightarrow \mathcal{O}(K) \tag{224}$$

and

$$\mathcal{O}(K) \hookrightarrow \mathring{\mathbb{A}}^K \tag{225}$$

this one is a homotopy equivalence.

We have open embeddings

$$\mathcal{O}(\tau) \hookrightarrow \mathcal{O}(\tilde{\tau}) \tag{226}$$

obtained by changing the radius of smallest discs to zero.

7.3.2. *Substitution.* For each tree τ and $0 < i < \mathrm{ht}(\tau)$ we have the following *substitution isomorphisms*

$$\mathcal{O}(\tau) \cong \mathcal{O}(\tau_{\leq i}) \times \prod_{k \in K_i} \mathcal{O}(\tau_{\geq k}) \tag{227}$$

In fact, a configuration of disks shaped by a tree τ is the same as a configuration shaped by $\tau_{\leq i}$, and for each $k \in K_i$ a configuration shaped by $\tau_{>k}$ inside D_k (playing the role of D_0; here we have to make a dilation by d_i^{-1}).

These isomorphisms satisfy obvious quadratic relations connected with pairs $0 < i < j < \mathrm{ht}(\tau)$. We leave their formulation to the reader.

7.4. **Enhanced trees.** We will call an *enhanced tree* a couple $(\tau, \vec{\alpha})$ where τ is a tree and $\vec{\alpha} \in \mathbb{N}[I]^{K'(\tau)}$. Vector $\vec{\alpha}$ will be called *enhancement* of τ.

Let us define cutting for enhanced trees. Given τ and i as in 7.2.1, let us note that $K'(\tau_{\leq i})$ and $K'(\tau_{\geq k})$ are subsets of $K'(\tau)$. We define

$$\vec{\alpha}_{\leq i} \in \mathbb{N}[I]^{K'(\tau_{\leq i})},\ \vec{\alpha}_{\geq k} \in \mathbb{N}[I]^{K'(\tau_{\geq k})}$$

as the corresponding subsequences of $\vec{\alpha}$.

Let us define operations ∂_i for enhanced trees. Namely, in the setup of 7.2.2, we define $\partial_i\vec{\alpha} = (\alpha'_k) \in \mathbb{N}[I]^{K'(\partial_i\tau)}$ as follows. If $i = p$ then $K'(\partial_p\tau) \subset K'(\tau)$, and we define $\partial_p\vec{\alpha}$ as a corresponding subsequence. If $i < p$, we set $\alpha'_k = \alpha_k$ if $k \in K_j,\ j > i$ or $j < i-1$. If $j = i-1$, we set

$$\alpha'_k = \alpha_k + \sum_{l \in \rho_{i-1}^{-1}(k)} \alpha_l.$$

7.5. **Enhanced disk operad.** Given an enhanced tree $(\tau, \vec{\alpha})$, let us define a configuration space $\mathcal{A}^{\vec{\alpha}}(\tau)$ as follows. Its points are couples $(\vec{D}, \mathbf{t})$, where $\vec{D} \in \mathcal{O}(\tau)$ and $\mathbf{t} = (t_j)$ is an α-colored configuration in $\mathbb{A}^1$ (see II.6.12) such that

for each $k \in K'(\tau)$ exactly α_k points lie inside $\overset{\circ}{D}_k(\tau)$ if $k \notin K_\tau$ (resp., inside $D_k(\tau)$ if $k \in K_\tau$) (see 7.3.1).

In particular, all points lie inside $D_{*(\tau)} = D(0;1)$ and outside $\bigcup_{k \in K_\tau} \bar{D}_k$ if τ is young. This space is an open subspace of the product $\mathcal{O}(\tau) \times \mathcal{A}_\alpha$.

We will also use a notation

$$\mathcal{A}^\alpha(L; d) := \mathcal{A}^\alpha(L \longrightarrow \{*\}; d)$$

for elementary trees and $\mathcal{A}^\alpha(L)$ for $\mathcal{A}^\alpha(L; 0)$.

The isomorphisms (227) induce isomorphisms

$$\mathcal{A}^{\vec{\alpha}}(\tau) \cong \mathcal{A}^{\vec{\alpha}_{\leq i}}(\tau_{\leq i}) \times \prod_{k \in K_i} \mathcal{A}^{\vec{\alpha}_{\geq k}}(\tau_{\geq k}) \tag{228}$$

We have embeddings

$$d_i : \mathcal{A}^{\vec{\alpha}}(\tau) \longrightarrow \mathcal{A}^{\partial_i\vec{\alpha}}(\partial_i\tau),\ 0 < i \leq p, \tag{229}$$

— dropping all disks $D_k,\ k \in K_i$.

We have obvious open embeddings

$$\mathcal{A}^{\vec{\alpha}}(\tau) \hookrightarrow \mathcal{A}^\alpha(K_\tau; d_\tau) \hookrightarrow \mathcal{A}^\alpha(K_\tau) \tag{230}$$

7.6. **Marked trees.** We will call *a marked tree* a triple $(\tau, \vec{\alpha}, \vec{\mu})$ where $(\tau, \vec{\alpha})$ is an enhanced tree, and $\vec{\mu} \in X^{K_\tau}$. We will call $\mathrm{ht}(\tau)$ *the height* of this marked tree.

Let us define operations ∂_i, $0 < i \leq p = \mathrm{ht}(\tau)$ for marked trees. Namely, for $i < p$ we set $\partial_i \vec{\mu} = \vec{\mu}$. For $i = p$ we define $\partial_p \vec{\mu}$ as $(\mu'_k)_{k \in K_{p-1}}$, where

$$\mu'_k = \sum_{l \in \rho_{p-1}^{-1}(k)} \mu_l - \alpha'_l.$$

Let us define cutting for marked trees. Namely, for $1 \leq i < p$ we define $\vec{\mu}_{\leq i}$ as $\partial_{i+1} \ldots \partial_{p-1} \partial_p \vec{\mu}$.

Next, for $k \in K_i$ we have $K_{\tau_{\geq k}} \subset K_\tau$, and we define $\vec{\mu}_{\geq k}$ as a corresponding subsequence of $\vec{\mu}$.

7.7. **Marked disk operad.** Now we can introduce our main objects. For each marked tree $(\tau, \vec{\alpha}, \vec{\mu})$ we define $\mathcal{A}^{\vec{\alpha}}_{\vec{\mu}}(\tau)$ as a topological space $\mathcal{A}^{\vec{\alpha}}(\tau)$ defined above, together with a marking $\vec{\mu}$ of the tree τ considered as an additional index assigned to this space.

We will regard $\mathcal{A}^{\vec{\alpha}}_{\vec{\mu}}(\tau)$ as a space whose points are configurations $(\vec{D}, \mathrm{t}) \in \mathcal{A}^{\vec{\alpha}}(\tau)$, together with a marking of smallest disks D_k, $k \in K_\tau$, by weights μ_k.

As above, we will use abbreviations $\mathcal{A}^{\alpha}_{\vec{\mu}}(L; d)$ for $\mathcal{A}^{\alpha}_{\vec{\mu}}(L \longrightarrow \{*\}; d)$ (where $\vec{\mu} \in X^L$) and $\mathcal{A}^{\alpha}_{\vec{\mu}}(L)$ for $\mathcal{A}^{\alpha}_{\vec{\mu}}(L; 0)$.

We have natural open embeddings

$$d_i : \mathcal{A}^{\vec{\alpha}}_{\vec{\mu}} \longrightarrow \mathcal{A}^{\partial_i \vec{\alpha}}_{\partial_i \vec{\mu}}(\partial_i \tau),\ 0 < i \leq p, \tag{231}$$

and

$$\mathcal{A}^{\vec{\alpha}}(\tau) \hookrightarrow \mathcal{A}^{\alpha}(K_\tau; d_\tau) \hookrightarrow \mathcal{A}^{\alpha}(K_\tau) \tag{232}$$

induced by the corresponding maps without marking.

The substitution isomorphisms (228) induce isomorphisms

$$\mathcal{A}^{\vec{\alpha}}_{\vec{\mu}}(\tau) \cong \mathcal{A}^{\vec{\alpha}_{\leq i}}_{\vec{\mu}_{\leq i}}(\tau_{\leq i}) \times \prod_{k \in K_i} \mathcal{A}^{\vec{\alpha}_{\geq k}}_{\vec{\mu}_{\geq k}}(\tau_{\geq k}) \tag{233}$$

7.8. We define closed embeddings

$$\sigma = \sigma^{\alpha}_{\vec{\mu};\vec{\beta}} : \mathcal{A}^{\alpha}_{\vec{\mu}}(K) \longrightarrow \mathcal{A}^{\alpha+\beta}_{\vec{\mu}+\vec{\beta}'}(K) \tag{234}$$

where $\vec{\beta} = (\beta_k)_{k \in K}$, $\beta_k = \sum_i b^i_k \cdot i$ and $\beta = \sum_k \beta_k$. By definition, σ leaves points u_k intact (changing their markings) and adds b^i_k copies of points of color i equal to u_k.

7.9. **Stratifications.** We set

$$\overset{\bullet}{\mathcal{A}}{}^{\alpha}_{\vec{\mu}}(K) := \mathcal{A}^{\alpha}_{\vec{\mu}}(K) - \bigcup_{\vec{\gamma} > 0} \sigma(\mathcal{A}^{\alpha-\gamma}_{\vec{\mu}-\vec{\gamma}}(K))$$

We define a *toric stratification* of $\mathcal{A}^{\alpha}_{\vec{\mu}}(K)$ as

$$\mathcal{A}^{\alpha}_{\vec{\mu}}(K) = \coprod_{\vec{\beta} < \vec{\alpha}} \sigma(\overset{\bullet}{\mathcal{A}}{}^{\beta}_{\vec{\mu}-\vec{\alpha}'+\vec{\beta}'}(K))$$

A *principal stratification* on $\mathcal{A}^{\alpha}_{\vec{\mu}}(K)$ is defined as follows. The space $\mathcal{A}^{\alpha}_{\vec{\mu}}(K)$ is a quotient of $\overset{\circ}{\mathbb{A}}{}^K \times \mathbb{A}^J$ where $\pi : J \longrightarrow I$ is an unfolding of α (cf. II.6.12). We define the principal stratification as the image of the diagonal stratification on $\overset{\circ}{\mathbb{A}}{}^K \times \mathbb{A}^J$ under the canonical projection $\overset{\circ}{\mathbb{A}}{}^K \times \mathbb{A}^J \longrightarrow \mathcal{A}^{\alpha}_{\vec{\mu}}(K)$. We will denote by $\overset{\circ}{\mathcal{A}}{}^{\alpha}_{\vec{\mu}}(K)$ the open stratum of the principal stratification.

8. Cohesive local systems ${}^K\mathcal{I}$

8.1. Let us fix a non-empty finite set K. Suppose we are given $\vec{\mu} \in X^K$ and $\alpha \in \mathbb{N}[I]$. Let us pick an unfolding of α, $\pi : J \longrightarrow I$. Let

$$\pi^{\alpha}_{\vec{\mu}} : (D(0;1)^K \times D(0;1)^J)^{\circ} \longrightarrow \overset{\circ}{\mathcal{A}}{}^{\alpha}_{\vec{\mu}}(K) \tag{235}$$

denote the canonical projection (here $(D(0;1)^K \times D(0;1)^J)^{\circ}$ denotes the open stratum of the diagonal stratification).

Let us define a one dimensional local system ${}^{\pi}\mathcal{I}_{\vec{\mu}}$ by the same procedure as in 3.1. Its fiber over each positive chamber $C \in \pi_0((D(0;1)^K \times D(0;1)^J)^{\circ}_{\mathbb{R}})$ is identified with k. Monodromies along the standard paths are given by the formulas

$${}^{C}T_{ij} = \zeta^{-\pi(i)\cdot\pi(j)},\ {}^{C}T_{ik} = \zeta^{2\mu_k\cdot\pi(i)'},\ {}^{C}T_{km} = \zeta^{-\mu_k\cdot\mu_m}, \tag{236}$$

$i, j \in J,\ i \neq j;\ k, m \in K,\ k \neq m$. Here ${}^{C}T_{ij}$ and ${}^{C}T_{km}$ are half-circles, and ${}^{C}T_{ik}$ are full circles. This definition essentially coincides with II.12.6, except for an overall sign.

We define a one-dimensional local system $\mathcal{I}^{\alpha}_{\vec{\mu}}(K)$ over $\overset{\circ}{\mathcal{A}}{}^{\alpha}_{\vec{\mu}}(K)$ as

$$\mathcal{I}^{\alpha}_{\vec{\mu}} := (\pi_*\, {}^{\pi}\mathcal{I}_{\vec{\mu}})^{\mathrm{sgn}} \tag{237}$$

where the superscript $(\bullet)^{\mathrm{sgn}}$ has the same meaning as in 3.1.

For each non-empty subset $L \subset K$ we can take a part of weights $\vec{\mu}_L = (\mu_k)_{k\in L}$ and get a local system $\mathcal{I}^{\alpha}_{\vec{\mu}}(L)$ over $\overset{\circ}{\mathcal{A}}{}^{\alpha}_{\vec{\mu}}(L)$.

For each marked tree $(\tau, \vec{\alpha}, \vec{\mu})$ with $K_\tau \subset K$, we define the local system $\mathcal{I}^{\vec{\alpha}}_{\vec{\mu}}(\tau)$ as the restriction of $\mathcal{I}^{\alpha}_{\vec{\mu}}(K_\tau)$ with respect to embedding (232).

8.2. **Factorization.** . The same construction as in 3.2 defines *factorization isomorphisms*

$$\phi_i = \phi^{\vec{\alpha}}_{i;\vec{\mu}}(\tau) : \mathcal{I}^{\vec{\alpha}}_{\vec{\mu}}(\tau) \cong \mathcal{I}^{\vec{\alpha}_{\leq i}}_{\vec{\mu}_{\leq i}}(\tau_{\leq i}) \boxtimes \boxed{\times}_{k\in K_i} \mathcal{I}^{\vec{\alpha}_{\geq k}}_{\vec{\mu}_{\geq k}}(\tau_{\geq k}) \tag{238}$$

They satisfy the property of

8.3. **Associativity.** For all $0 < i < j < p$ squares

$$\begin{array}{ccc} & \mathcal{I}(\tau) & \\ {\scriptstyle \phi_j}\swarrow & & \searrow{\scriptstyle \phi_i} \\ \mathcal{I}(\tau_{\leq j}) \boxtimes \boxed{\times}_{l\in K_j} \mathcal{I}(\tau_{\geq l}) & & \mathcal{I}(\tau_{\leq i}) \boxtimes \boxed{\times}_{k\in K_i} \mathcal{I}(\tau_{\geq k}) \\ {\scriptstyle \phi_i \boxtimes \mathrm{id}}\searrow & & \swarrow{\scriptstyle \mathrm{id} \boxtimes \phi_j} \\ & \mathcal{I}(\tau_{\leq i}) \boxtimes \boxed{\times}_{k\in K_i} \mathcal{I}(\tau_{\geq k;\leq j}) \boxtimes \boxed{\times}_{l\in K_j} \mathcal{I}(\tau_{\geq l}) & \end{array}$$

commute. (To unburden the notation we have omitted irrelevant indices — they are restored uniquely.)

8.4. The collection of local systems ${}^K\mathcal{I} = \{\mathcal{I}^{\alpha}_{\vec{\mu}}(L),\ L \subset K\}$, together with the factorization isomorphisms defined above, will be called *the cohesive local system over* ${}^K\overset{\circ}{\mathcal{A}}$.

8.5. Let us define perverse sheaves

$$\overset{\bullet}{\mathcal{I}}{}^{\vec{\alpha}}_{\vec{\mu}}(\tau) := j_{!*}\mathcal{I}^{\vec{\alpha}}_{\vec{\mu}}(\tau)[\dim\ \mathcal{A}^{\vec{\alpha}}_{\vec{\mu}}(\tau)] \in \mathcal{M}(\overset{\bullet}{\mathcal{A}}{}^{\vec{\alpha}}_{\vec{\mu}}(\tau);\mathcal{S})$$

where $j : \overset{\circ}{\mathcal{A}}{}^{\vec{\alpha}}_{\vec{\mu}}(\tau) \hookrightarrow \overset{\bullet}{\mathcal{A}}{}^{\vec{\alpha}}_{\vec{\mu}}(\tau)$ denotes the embedding. By functoriality, the factorization isomorphisms (238) induce isomorphisms

$$\phi_i = \phi^{\vec{\alpha}}_{i;\vec{\mu}}(\tau) : \overset{\bullet}{\mathcal{I}}{}^{\vec{\alpha}}_{\vec{\mu}}(\tau) \cong \overset{\bullet}{\mathcal{I}}{}^{\vec{\alpha}_{\leq i}}_{\vec{\mu}_{\leq i}}(\tau_{\leq i}) \boxtimes \boxed{\times}_{k\in K_i} \overset{\bullet}{\mathcal{I}}{}^{\vec{\alpha}_{\geq k}}_{\vec{\mu}_{\geq k}}(\tau_{\geq k}) \qquad (239)$$

These isomorphisms satisfy an associativity property completely analogous to 8.3; one should only replace $\mathcal{I}$ by $\overset{\bullet}{\mathcal{I}}$ in the diagrams.

9. Factorizable sheaves over ${}^K\mathcal{A}$

We keep the assumptions of the previous section.

9.1. The first goal of this section is to define a k-linear category ${}^K\widetilde{\mathcal{FS}}$ whose objects will be called *factorizable sheaves (over* $({}^K\mathcal{A},\ {}^K\mathcal{I})$*)*. Similarly to $\widetilde{\mathcal{FS}}$, this category is by definition a product of k-categories

$${}^K\widetilde{\mathcal{FS}} = \prod_{\vec{c}\in\pi_0(\mathcal{A})^K} {}^K\widetilde{\mathcal{FS}}_{\vec{c}}. \qquad (240)$$

Objects of ${}^K\widetilde{\mathcal{FS}}_{\vec{c}}$ will be called *factorizable sheaves supported at* $\vec{c}$.

9.2. **Definition.** *A factorizable sheaf* $\mathcal{X}$ *over* $({}^K\mathcal{A},\ {}^K\mathcal{I})$ *supported at* $\vec{c} = (c_k) \in \pi_0(\mathcal{A})^K$ *is the following collection of data:*

(a) a K*-tuple of weights* $\vec{\lambda} = (\lambda_k) \in X^K$ *such that* $\lambda_k \in X_{c_k}$*, to be denoted by* $\vec{\lambda}(\mathcal{X})$*;*

(b) for each $\alpha \in \mathbb{N}[I]$ *a sheaf* $\mathcal{X}^{\alpha}(K) \in \mathcal{M}(\mathcal{A}^{\alpha}_{\vec{\lambda}}(K);\mathcal{S})$.

Taking restrictions, as in 8.1, we get for each K*-based enhanced tree* $(\tau,\vec{\alpha})$ *sheaves* $\mathcal{X}^{\vec{\alpha}}(\tau) \in \mathcal{M}(\mathcal{A}^{\vec{\alpha}}_{\vec{\lambda}}(\tau);\mathcal{S})$.

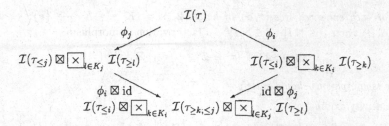

(c) For each enhanced tree $(\tau, \vec{\alpha})$ *of height* 2, $\tau = (K \xrightarrow{\mathrm{id}} K \longrightarrow \{*\};\ (1, d, 0))$, $\vec{\alpha} = (\alpha, \vec{\beta})$ *where* $\alpha \in \mathbb{N}[I]$; $\vec{\beta} \in \mathbb{N}[I]^K$, *a* factorization isomorphism

$$\psi(\tau) : \mathcal{X}^{(\alpha,\vec{\beta})}(\tau) \cong \overset{\bullet}{\mathcal{I}}{}^{\alpha}_{\vec{\lambda}(\tau)_{\leq 1}}(\tau_{\leq 1}) \boxtimes \mathcal{X}^{(0,\vec{\beta})}(\tau) \tag{241}$$

These isomorphisms should satisfy

Associativity axiom.

For all enhanced trees $(\tau, \vec{\alpha})$ *of height* 3, $\tau = (K \xrightarrow{\mathrm{id}} K \xrightarrow{\mathrm{id}} K \longrightarrow \{*\};\ (1, d_1, d_2, 0))$, $\vec{\alpha} = (\alpha, \vec{\beta}, \vec{\gamma})$ *where* $\alpha \in \mathbb{N}[I]$; $\vec{\beta}, \vec{\gamma} \in \mathbb{N}[I]^K$, *the square*

$$\begin{array}{ccc}
 & \mathcal{X}^{(\alpha,\vec{\beta},\vec{\gamma})}(\tau) & \\
 \swarrow^{\psi_2} & & \searrow^{\psi_1} \\
\overset{\bullet}{\mathcal{I}}{}^{(\alpha,\vec{\beta})}_{\vec{\lambda}(\tau)_{\leq 2}}(\tau_{\leq 2}) \boxtimes \mathcal{X}^{(0,\vec{0},\vec{\gamma})}(\tau) & & \overset{\bullet}{\mathcal{I}}{}^{\alpha}_{\vec{\lambda}(\tau)_{\leq 1}}(\tau_{\leq 1}) \boxtimes \mathcal{X}^{(0,\vec{\beta},\vec{\gamma})}(\tau) \\
 \searrow^{\phi_1 \boxtimes \mathrm{id}} & & \swarrow^{\mathrm{id} \boxtimes \psi_2} \\
 & \overset{\bullet}{\mathcal{I}}{}^{\alpha}_{\vec{\lambda}(\tau)_{\leq 1}}(\tau_{\leq 1}) \boxtimes \boxed{\times}_{k\in K}\, \overset{\bullet}{\mathcal{I}}{}^{\beta_k}_{\vec{\lambda}(\tau)_{\geq k; \leq 2}}(\tau_{\geq k; \leq 2}) \boxtimes \mathcal{X}^{(0,\vec{0},\vec{\gamma})}(\tau) &
\end{array}$$

commutes.

9.3. Let $\mathcal{X}$ be as above. For each $\vec{\mu} \in X^K$, $\vec{\mu} \geq \vec{\lambda}$, so that $\vec{\mu} = \vec{\lambda} + \vec{\beta}'$ for some $\vec{\beta} \in \mathbb{N}[I]^K$, and $\alpha \in \mathbb{N}[I]$, let us define a sheaf $\mathcal{X}^{\alpha}_{\vec{\mu}}(K) \in \mathcal{M}(\mathcal{A}^{\alpha}_{\vec{\mu}}(K); \mathcal{S})$ as $\sigma_* \mathcal{X}^{\alpha-\beta}(K)$. For example, $\mathcal{X}^{\alpha}_{\vec{\lambda}}(K) = \mathcal{X}^{\alpha}(K)$.

Taking restrictions, the sheaves $\mathcal{X}^{\vec{\alpha}}_{\vec{\mu}}(\tau) \in \mathcal{M}(\mathcal{A}^{\vec{\alpha}}_{\vec{\mu}}(\tau); \mathcal{S})$ for all K-based trees τ are defined.

9.4. Suppose $\mathcal{X}, \mathcal{Y}$ are two factorizable sheaves supported at $\vec{c}$, $\vec{\lambda} = \vec{\lambda}(\mathcal{X})$, $\vec{\nu} = \vec{\lambda}(\mathcal{Y})$. Let $\vec{\mu} \in X^K$, $\vec{\mu} \geq \vec{\lambda}$, $\vec{\mu} \geq \vec{\nu}$. By definition, we have canonical isomorphisms

$$\theta = \theta^{\alpha}_{\vec{\mu};\vec{\beta}} :\ \mathrm{Hom}_{\mathcal{A}^{\alpha}_{\vec{\mu}}(K)}(\mathcal{X}^{\alpha}_{\vec{\mu}}(K), \mathcal{Y}^{\alpha}_{\vec{\mu}}(K)) \xrightarrow{\sim} \mathrm{Hom}_{\mathcal{A}^{\alpha+\beta}_{\vec{\mu}+\vec{\beta}'}(K)}(\mathcal{X}^{\alpha+\beta}_{\vec{\mu}+\vec{\beta}'}(K), \mathcal{Y}^{\alpha+\beta}_{\vec{\mu}+\vec{\beta}'}(K)) \tag{242}$$

for each $\alpha \in \mathbb{N}[I]$, $\vec{\beta} \in \mathbb{N}[I]^K$.

9.4.1. Suppose we are given $\vec{\beta} = (\beta_k) \in \mathbb{N}[I]^K$. Let $\beta = \sum_k \beta_k$ as usually. Choose a real d, $0 < d < 1$.

Consider a marked tree $(\tau_d,\ (0, \vec{\beta}),\ \vec{\mu})$ where

$$\tau_d = (K \xrightarrow{\mathrm{id}} K \longrightarrow \{*\};\ (0, d, 1)).$$

We have the restriction homomorphism

$$\xi_{\vec{\mu};\vec{\beta};d} : \mathrm{Hom}_{\mathcal{A}^{\beta}_{\vec{\mu}}(K)}(\mathcal{X}^{\beta}_{\vec{\mu}}(K), \mathcal{Y}^{\beta}_{\vec{\mu}}(K)) \longrightarrow \mathrm{Hom}_{\mathcal{A}^{(0,\vec{\beta})}_{\vec{\mu}}(\tau_d)}(\mathcal{X}^{(0,\vec{\beta})}_{\vec{\mu}}(\tau_d), \mathcal{Y}^{(0,\vec{\beta})}_{\vec{\mu}}(\tau_d)) \tag{243}$$

Suppose we are given $'\vec{\beta} = ('\beta_k) \in \mathbb{N}[I]^K$ such that $'\vec{\beta} \leq \vec{\beta}$. Let $'\beta = \sum_k {'\beta_k}$ as usually. Choose a real ε, $0 < \varepsilon < d$.

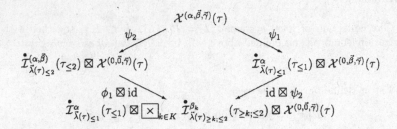

The restriction and the factorization isomorphisms ψ induce the map

$$\eta^{\vec{\beta};d}_{\vec{\mu};\,'\vec{\beta};\varepsilon}: \mathrm{Hom}_{\mathcal{A}^{(0,\vec{\beta})}_{\vec{\mu}}(\tau_d)}(\mathcal{X}^{(0,\vec{\beta})}_{\vec{\mu}}(\tau_d), \mathcal{Y}^{(0,\vec{\beta})}_{\vec{\mu}}(\tau_d)) \longrightarrow \mathrm{Hom}_{\mathcal{A}^{(0,\,'\vec{\beta})}_{\vec{\mu}}(\tau_\varepsilon)}(\mathcal{X}^{(0,\,'\vec{\beta})}_{\vec{\mu}}(\tau_\varepsilon), \mathcal{Y}^{(0,\,'\vec{\beta})}_{\vec{\mu}}(\tau_\varepsilon)) \tag{244}$$

The associativity axiom implies that these maps satisfy an obvious transitivity property.

We define the space $\mathrm{Hom}_{K\widetilde{\mathcal{FS}}}(\mathcal{X},\mathcal{Y})$ as the following inductive-projective limit

$$\mathrm{Hom}_{K\widetilde{\mathcal{FS}}}(\mathcal{X},\mathcal{Y}) := \underset{\rightarrow}{\lim}\ \underset{\leftarrow}{\lim}\ \mathrm{Hom}_{\mathcal{A}^{\alpha}_{\vec{\mu}}(K)}(\mathcal{X}^{\alpha}_{\vec{\mu}}(K), \mathcal{Y}^{\alpha}_{\vec{\mu}}(K)) \tag{245}$$

where the inverse limit is understood as follows. Its elements are collections of maps

$$\{f^{\alpha}_K : \mathcal{X}^{\alpha}_{\vec{\mu}}(K) \longrightarrow \mathcal{Y}^{\alpha}_{\vec{\mu}}(K)\}$$

given for all $\alpha \in \mathbb{N}[I]$, $\vec{\beta} \in \mathbb{N}[I]^K$, such that for every α, $'\vec{\beta} \leq \vec{\beta}$, $0 < \varepsilon < d < 1$ as above, we have

$$\eta^{\vec{\beta};d}_{\vec{\mu};\,'\vec{\beta};\varepsilon}\xi_{\vec{\mu};\vec{\beta};d}(f^{\beta}_K) = \xi_{\vec{\mu};\,'\vec{\beta};\varepsilon}(f^{'\beta}_K)$$

$\vec{\mu}$ being fixed. The direct limit is taken over $\vec{\mu} \in X^K$ such that $\vec{\mu} \geq \vec{\lambda}$, $\vec{\mu} \geq \vec{\nu}$, the transition maps being induced by (242).

With these spaces of homomorphisms, factorizable sheaves supported at $\vec{c}$ form a k-linear category to be denoted by ${}^K\widetilde{\mathcal{FS}}_{\vec{c}}$. As we have already mentioned, the category of factorizable sheaves ${}^K\widetilde{\mathcal{FS}}$ is by definition the product (240).

FINITE SHEAVES

9.5. **Definition.** *A sheaf $\mathcal{X} \in {}^K\widetilde{\mathcal{FS}}_{\vec{c}}$ is called* finite *if there exists only finitely many $\vec{\beta} \in \mathbb{N}[I]^K$ such that the singular support of $\mathcal{X}^{\alpha}_{\vec{\lambda}}(K)$ contains the conormal bundle to $\sigma^{\alpha-\beta}_{\vec{\lambda}-\vec{\beta}';\vec{\beta}}(\mathcal{A}^{\alpha-\beta}_{\vec{\lambda}-\vec{\beta}'})$ (see (234)) for $\alpha \geq \beta = \sum_k \beta_k$.*

A sheaf $\mathcal{X} = \oplus_{\vec{c}}\mathcal{X}_{\vec{c}} \in {}^K\widetilde{\mathcal{FS}}$, $\mathcal{X}_{\vec{c}} \in {}^K\widetilde{\mathcal{FS}}_{\vec{c}}$ is called finite if all $\mathcal{X}_{\vec{c}}$ are finite.

9.6. Suppose we are given finite sheaves $\mathcal{X},\mathcal{Y} \in {}^K\mathcal{FS}_{\vec{c}}$; and $\vec{\mu} \geq \vec{\lambda}(\mathcal{X}), \vec{\lambda}(\mathcal{Y})$. As in the proof of the Lemma 5.3, one can see that there exists $'\vec{\beta} \in \mathbb{N}[I]^K$ such that for any $\vec{\beta} \geq {}'\vec{\beta}$ the map

$$\eta^{'\vec{\beta};d}_{\vec{\mu};\vec{\beta};\varepsilon}: \mathrm{Hom}_{\mathcal{A}^{(0,\vec{\beta})}_{\vec{\mu}}(\tau_d)}(\mathcal{X}^{(0,\vec{\beta})}_{\vec{\mu}}(\tau_d), \mathcal{Y}^{(0,\vec{\beta})}_{\vec{\mu}}(\tau_d)) \longrightarrow \mathrm{Hom}_{\mathcal{A}^{(0,\,'\vec{\beta})}_{\vec{\mu}}(\tau_\varepsilon)}(\mathcal{X}^{(0,\,'\vec{\beta})}_{\vec{\mu}}(\tau_\varepsilon), \mathcal{Y}^{(0,\,'\vec{\beta})}_{\vec{\mu}}(\tau_\varepsilon)) \tag{246}$$

is an isomorphism. We will identify all the spaces $\mathrm{Hom}_{\mathcal{A}^{(0,\vec{\beta})}_{\vec{\mu}}(\tau_d)}(\mathcal{X}^{(0,\vec{\beta})}_{\vec{\mu}}(\tau_d), \mathcal{Y}^{(0,\vec{\beta})}_{\vec{\mu}}(\tau_d))$ with the help of the above isomorphisms, and we will denote this stabilized space by $\overline{\mathrm{Hom}}_{K\mathcal{FS}}(\mathcal{X},\mathcal{Y})$. Evidently, it does not depend on a choice of $'\vec{\beta}$.

Quite similarly to the *loc.cit* one can see that for any $\vec{\beta} \geq {}'\vec{\beta}$ the map

$$\xi_{\vec{\mu};\vec{\beta};d}: \mathrm{Hom}_{\mathcal{A}^{\beta}_{\vec{\mu}}(K)}(\mathcal{X}^{\beta}_{\vec{\mu}}(K), \mathcal{Y}^{\beta}_{\vec{\mu}}(K)) \longrightarrow \mathrm{Hom}_{\mathcal{A}^{(0,\vec{\beta})}_{\vec{\mu}}(\tau_d)}(\mathcal{X}^{(0,\vec{\beta})}_{\vec{\mu}}(\tau_d), \mathcal{Y}^{(0,\vec{\beta})}_{\vec{\mu}}(\tau_d)) \tag{247}$$

is an injection.

Thus we may view $\mathrm{Hom}_{\mathcal{A}^{\beta}_{\vec{\mu}}(K)}(\mathcal{X}^{\beta}_{\vec{\mu}}(K), \mathcal{Y}^{\beta}_{\vec{\mu}}(K))$ as the subspace of $\overline{\mathrm{Hom}}_{{}^K\mathcal{FS}}(\mathcal{X}, \mathcal{Y})$.

We define $\mathrm{Hom}_{{}^K\mathcal{FS}}(\mathcal{X}, \mathcal{Y}) \subset \overline{\mathrm{Hom}}_{{}^K\mathcal{FS}}(\mathcal{X}, \mathcal{Y})$ as the projective limit of the system of subspaces $\mathrm{Hom}_{\mathcal{A}^{\beta}_{\vec{\mu}}(K)}(\mathcal{X}^{\beta}_{\vec{\mu}}(K), \mathcal{Y}^{\beta}_{\vec{\mu}}(K))$, $\vec{\beta} \geq {}'\vec{\beta}$.

With such definition of morphisms finite factorizable sheaves supported at $\vec{c}$ form an abelian category to be denoted by ${}^K\mathcal{FS}_{\vec{c}}$. We set by definition

$$ {}^K\mathcal{FS} = \prod_{\vec{c} \in \pi_0(\mathcal{A})^K} {}^K\mathcal{FS}_{\vec{c}} \tag{248} $$

10. Gluing

10.1. Let

$$ \mathcal{A}^{\alpha}_{\mu;1} \subset \mathcal{A}^{\alpha}_{\mu} $$

denote an open configuration subspace parametrizing configurations lying entirely inside the unit disk $D(0;1)$. Due to monodromicity, the restriction functors

$$ \mathcal{M}(\mathcal{A}^{\alpha}_{\mu}; \mathcal{S}) \longrightarrow \mathcal{M}(\mathcal{A}^{\alpha}_{\mu;1}; \mathcal{S}) $$

are equivalences.

Let $\{*\}$ denote a one-element set. We have closed embeddings

$$ i : \mathcal{A}^{\alpha}_{\mu;1} \hookrightarrow \mathcal{A}^{\alpha}_{\mu}(\{*\}), $$

which identify the first space with the subspace of the second one consisting of configurations with the small disk centered at 0. The inverse image functors

$$ i^*[-1] : \mathcal{M}(\mathcal{A}^{\alpha}_{\mu}(\{*\}); \mathcal{S}) \longrightarrow \mathcal{M}(\mathcal{A}^{\alpha}_{\mu}; \mathcal{S}) \tag{249} $$

are equivalences, again due to monodromicity. Thus, we get equivalences

$$ \mathcal{M}(\mathcal{A}^{\alpha}_{\mu}; \mathcal{S}) \xrightarrow{\sim} \mathcal{M}(\mathcal{A}^{\alpha}_{\mu}(\{*\}; \mathcal{S}) $$

which induce canonical equivalences

$$ \widetilde{\mathcal{FS}} \xrightarrow{\sim} \widetilde{\mathcal{FS}}^{\{*\}} \tag{250} $$

and

$$ \mathcal{FS} \xrightarrow{\sim} \mathcal{FS}^{\{*\}} \tag{251} $$

Using these equivalences, we will sometimes identify these categories.

10.2. **Tensor product of categories.** Let $\mathcal{B}_1, \mathcal{B}_2$ be k-linear abelian categories. Their tensor product category $\mathcal{B}_1 \otimes \mathcal{B}_2$ is defined in §5 of [D2]. It comes together with a canonical right biexact functor $\mathcal{B}_1 \times \mathcal{B}_2 \longrightarrow \mathcal{B}_1 \otimes \mathcal{B}_2$, and it is the initial object among such categories.

10.2.1. *Basic Example.* Let M_i, $i = 1, 2$, be complex algebraic varieties equipped with algebraic Whitney stratifications $\mathcal{S}_i$. Let $\mathcal{B}_i = \mathcal{M}(M_i; \mathcal{S}_i)$. Then

$$ \mathcal{B}_1 \otimes \mathcal{B}_2 = \mathcal{M}(M_1 \times M_2; \mathcal{S}_1 \times \mathcal{S}_2). $$

The canonical functor $\mathcal{B}_1 \times \mathcal{B}_2 \longrightarrow \mathcal{B}_1 \otimes \mathcal{B}_2$ sends $(\mathcal{X}_1, \mathcal{X}_2)$ to $\mathcal{X}_1 \boxtimes \mathcal{X}_2$.

10.2.2. Recall the notations of 9.4.1. Let us consider the following category $\mathcal{FS}^{\otimes K}$. Its objects are the collections of perverse sheaves $\mathcal{X}_{\vec{\mu}}^{(0,\vec{\beta})}(\tau_d)$ on the spaces $\mathcal{A}_{\vec{\mu}}^{(0,\vec{\beta})}(\tau_d)$ for sufficiently small d, satisfying the usual factorization and finiteness conditions. The morphisms are defined via the inductive-projective system with connecting maps $\eta^{\vec{\beta};d}_{\vec{\mu};\ '\vec{\beta};\varepsilon}$. Using the above Basic Example, one can see easily that the category $\mathcal{FS}^{\otimes K}$ is canonically equivalent to $\mathcal{FS}\otimes\ldots\otimes\mathcal{FS}$ (K times) which justifies its name.

By definition, the category ${}^K\mathcal{FS}$ comes together with the functor $p_K:\ {}^K\mathcal{FS}\longrightarrow \mathcal{FS}^{\otimes K}$ injective on morphisms. In effect,

$$\mathrm{Hom}_{{}^K\mathcal{FS}}(\mathcal{X},\mathcal{Y})\hookrightarrow \overline{\mathrm{Hom}}_{{}^K\mathcal{FS}}(\mathcal{X},\mathcal{Y})=\mathrm{Hom}_{\mathcal{FS}^{\otimes K}}(p_K(\mathcal{X}),p_K(\mathcal{Y})).$$

Let us construct a functor in the opposite direction.

10.3. **Gluing of factorizable sheaves.** For each $0<d<1$ let us consider a tree

$$\tau_d=(K\overset{\mathrm{id}}{\longrightarrow}K\longrightarrow\{*\};(1,d,0)).$$

Suppose we are given $\alpha\in\mathbb{N}[I]$. Let $\mathcal{V}(\alpha)$ denote the set of all enhancements $\vec{\alpha}=(\alpha_*;(\alpha_k)_{k\in K})$ of τ such that $\alpha_*+\sum_{k\in K}\alpha_k=\alpha$. Obviously, the open subspaces $\mathcal{A}^{\vec{\alpha}}(\tau_d)\subset\mathcal{A}^\alpha(K)$, for varying d and $\vec{\alpha}\in\mathcal{V}(\alpha)$, form an open covering of $\mathcal{A}^\alpha(K)$.

Suppose we are given a collection of factorizable sheaves $\mathcal{X}_k\in\mathcal{FS}_{c_k}$, $k\in K$. Set $\vec{\lambda}=(\lambda(\mathcal{X}_k))\in X^K$. For each $d,\vec{\alpha}$ as above consider a sheaf

$$\mathcal{X}^{\vec{\alpha}}(\tau_d):=\overset{\bullet}{\mathcal{I}}{}^{\alpha_*}_{\vec{\lambda}_{\le 1}}(\tau_{d;\le 1})\boxtimes\boxed{\times}_{k\in K}\mathcal{X}_k^{\alpha_k}$$

over $\mathcal{A}^{\vec{\alpha}}_{\vec{\lambda}}(\tau_d)$.

Non-trivial pairwise intersections of the above open subspaces look as follows. For $0<d_2<d_1<1$, consider a tree of height 3

$$\varsigma=\varsigma_{d_1,d_2}=(K\overset{\mathrm{id}}{\longrightarrow}K\overset{\mathrm{id}}{\longrightarrow}K\longrightarrow\{*\};(1,d_1,d_2,0)).$$

We have $\partial_1\varsigma=\tau_{d_2}$, $\partial_2\varsigma=\tau_{d_1}$. Let $\vec{\beta}=(\beta_*,(\beta_{1;k})_{k\in K},(\beta_{2;k})_{k\in K})$ be an enhancement of ς. Set $\vec{\alpha}_1=\partial_2\vec{\beta}$, $\vec{\alpha}_2=\partial_1\vec{\beta}$. Note that $\vec{\beta}$ is defined uniquely by $\vec{\alpha}_1,\vec{\alpha}_2$. We have

$$\mathcal{A}^{\vec{\beta}}_{\vec{\lambda}}(\varsigma)=\mathcal{A}^{\vec{\alpha}_1}_{\vec{\lambda}}(\tau_{d_1})\cap\mathcal{A}^{\vec{\alpha}_2}_{\vec{\lambda}}(\tau_{d_2}).$$

Due to the factorization property for sheaves $\overset{\bullet}{\mathcal{I}}$ and $\mathcal{X}_k$ we have isomorphisms

$$\mathcal{X}^{\vec{\alpha}_1}(\tau_{d_1})|_{\mathcal{A}^{\vec{\beta}}_{\vec{\lambda}}(\varsigma)}\cong\overset{\bullet}{\mathcal{I}}{}^{\partial_3\vec{\beta}}_{\partial_3\vec{\lambda}}(\partial_3\varsigma)\boxtimes\boxed{\times}_{k\in K}(\mathcal{X}_k)^{\beta_{2;k}}_{\lambda_k},$$

and

$$\mathcal{X}^{\vec{\alpha}_2}(\tau_{d_2})|_{\mathcal{A}^{\vec{\beta}}_{\vec{\lambda}}(\varsigma)}\cong\overset{\bullet}{\mathcal{I}}{}^{\partial_3\vec{\beta}}_{\partial_3\vec{\lambda}}(\partial_3\varsigma)\boxtimes\boxed{\times}_{k\in K}(\mathcal{X}_k)^{\beta_{2;k}}_{\lambda_k}$$

Taking composition, we get isomorphisms

$$\phi^{\vec{\alpha}_1,\vec{\alpha}_2}_{d_1,d_2}:\mathcal{X}^{\vec{\alpha}_1}(\tau_{d_1})|_{\mathcal{A}^{\vec{\alpha}_1}_{\vec{\lambda}}(\tau_{d_1})\cap\mathcal{A}^{\vec{\alpha}_2}_{\vec{\lambda}}(\tau_{d_2})}\overset{\sim}{\longrightarrow}\mathcal{X}^{\vec{\alpha}_2}(\tau_{d_2})|_{\mathcal{A}^{\vec{\alpha}_1}_{\vec{\lambda}}(\tau_{d_1})\cap\mathcal{A}^{\vec{\alpha}_2}_{\vec{\lambda}}(\tau_{d_2})}\qquad(252)$$

From the associativity of the factorization for the sheaves $\overset{\bullet}{\mathcal{I}}$ and $\mathcal{X}_k$ it follows that the isomorphisms (252) satisfy the cocycle condition; hence they define a sheaf $\mathcal{X}^\alpha(K)$ over $\mathcal{A}^\alpha_{\vec{\lambda}}(K)$.

Thus, we have defined a collection of sheaves $\{\mathcal{X}^\alpha(K)\}$. Using the corresponding data for the sheaves $\mathcal{X}_k$, one defines easily factorization isomorphisms 9.2 (d) and check that they satisfy the associativity property. One also sees immediately that the collection of sheaves $\{\mathcal{X}^\alpha(K)\}$ is finite. We leave this verification to the reader.

This way we get maps

$$\prod_k \mathrm{Ob}(\mathcal{FS}_{c_k}) \longrightarrow \mathrm{Ob}({}^K\mathcal{FS}_{\vec{c}}),\ \vec{c} = (c_k)$$

which extend by additivity to the map

$$g_K : \mathrm{Ob}(\mathcal{FS}^K) \longrightarrow \mathrm{Ob}({}^K\mathcal{FS}) \tag{253}$$

To construct the functor

$$g_K : \mathcal{FS}^K \longrightarrow {}^K\mathcal{FS} \tag{254}$$

it remains to define g_K on morphisms.

Given two collections of finite factorizable sheaves $\mathcal{X}_k, \mathcal{Y}_k \in \mathcal{FS}_{c_k}$, $k \in K$, let us choose $\vec{\lambda} = (\lambda_k)_{k\in K}$ such that $\lambda_k \geq \lambda(\mathcal{X}_k), \lambda(\mathcal{Y}_k)$ for all $k \in K$. Suppose we have a collection of morphisms $f_k : \mathcal{X}_k \longrightarrow \mathcal{Y}_k$, $k \in K$; that is the maps $f_k^{\alpha_k} : (\mathcal{X}_k)^{\alpha_k}_{\lambda_k} \longrightarrow (\mathcal{Y}_k)^{\alpha_k}_{\lambda_k}$ given for any $\alpha_k \in \mathbb{N}[I]$ compatible with factorizations.

Given $\alpha \in \mathbb{N}[I]$ and an enhancement $\vec{\alpha} \in \mathcal{V}(\alpha)$ as above we define the morphism $f^{\vec{\alpha}}(\tau_d) : \mathcal{X}^{\vec{\alpha}}(\tau_d) \longrightarrow \mathcal{Y}^{\vec{\alpha}}(\tau_d)$ over $\mathcal{A}^{\vec{\alpha}}_{\vec{\lambda}}(\tau_d)$ as follows: it is the tensor product of the identity on $\overset{\bullet}{\mathcal{I}}{}^{\alpha_\bullet}_{\vec{\lambda}_{\leq 1}}(\tau_{d;\leq 1})$ with the morphisms $f_k^{\alpha_k} : (\mathcal{X}_k)^{\alpha_k}_{\lambda_k} \longrightarrow (\mathcal{Y}_k)^{\alpha_k}_{\lambda_k}$.

One sees easily as above that the morphisms $f^{\vec{\alpha}}(\tau_d)$ glue together to give a morphism $f^\alpha(K) : \mathcal{X}^\alpha(K) \longrightarrow \mathcal{Y}^\alpha(K)$; as α varies they provide a morphism $f(K) : g_K((\mathcal{X}_k)) \longrightarrow g_K((\mathcal{Y}_k))$. Thus we have defined the desired functor g_K. Obviously it is K-exact, so by universal property it defines the same named functor

$$g_K : \mathcal{FS}^{\otimes K} \longrightarrow {}^K\mathcal{FS} \tag{255}$$

By the construction, the composition $p_K \circ g_K : \mathcal{FS}^{\otimes K} \longrightarrow \mathcal{FS}^{\otimes K}$ is isomorphic to the identity functor. Recalling that p_K is injective on morphisms we see that g_K and p_K are quasiinverse. Thus we get

10.4. **Theorem.** *The functors p_K and g_K establish a canonical equivalence*

$$ {}^K\mathcal{FS} \xrightarrow{\sim} \mathcal{FS}^{\otimes K} \quad \square$$

11. Fusion

BRAIDED TENSOR CATEGORIES

In this part we review the definition of a braided tensor category following Deligne, [D1].

11.1. Let $\mathcal{C}$ be a category, Y a locally connected locally simply connected topological space. By a *local system* over Y with values in $\mathcal{C}$ we will mean a locally constant sheaf over Y with values in $\mathcal{C}$. They form a category to be denoted by $\mathcal{L}ocsys(Y;\mathcal{C})$.

11.1.1. We will use the following basic example. If X is a complex algebraic variety with a Whitney stratification $\mathcal{S}$ then the category $\mathcal{M}(X\times Y;\mathcal{S}\times\mathcal{S}_{Y;tr})$ is equivalent to $\mathcal{L}ocsys(Y;\mathcal{M}(X;\mathcal{S}))$. Here $\mathcal{S}_{Y;tr}$ denotes the trivial stratification of Y, i.e. the first category consists of sheaves smooth along Y.

11.2. Let $\pi: K\longrightarrow L$ be an epimorphism of non-empty finite sets. We will use the notations of 7.3. For real ϵ,δ such that $1>\epsilon>\delta>0$, consider a tree

$$\tau_{\pi;\epsilon,\delta}=(K\overset{\pi}{\longrightarrow}L\longrightarrow\{*\};\ (1,\epsilon,\delta))$$

We have an isomorphism which is a particular case of (227):

$$\mathcal{O}(\tau_{\pi;\epsilon,\delta})\cong\mathcal{O}(L;\epsilon)\times\prod_{l\in L}\mathcal{O}(K_l;\delta\epsilon^{-1})$$

where $K_l:=\pi^{-1}(l)$.

11.2.1. **Lemma** *There exists essentially unique functor*

$$r_\pi:\mathcal{L}ocsys(\mathcal{O}(K);\mathcal{C})\longrightarrow\mathcal{L}ocsys(\mathcal{O}(L)\times\prod_l\mathcal{O}(K_l);\mathcal{C})$$

such that for each ϵ,δ as above the square

$$\begin{array}{ccc}\mathcal{L}ocsys(\mathcal{O}(K);\mathcal{C}) & \xrightarrow{r_\pi} & \mathcal{L}ocsys(\mathcal{O}(L)\times\prod_l\mathcal{O}(K_l);\mathcal{C})\\ \downarrow & & \downarrow\\ \mathcal{L}ocsys(\tau_{\pi;\epsilon,\delta};\mathcal{C}) & \xrightarrow{\sim} & \mathcal{L}ocsys(\mathcal{O}(L;\epsilon)\times\prod_l\mathcal{O}(K_l;\delta\epsilon^{-1});\mathcal{C})\end{array}$$

commutes.

Proof follows from the remark that $\mathcal{O}(L)$ is a union of its open subspaces

$$\mathcal{O}(L)=\bigcup_{\epsilon>0}\mathcal{O}(L;\epsilon).\ \square$$

11.3. Let $\mathcal{C}$ be a category. A *braided tensor structure* on $\mathcal{C}$ is the following collection of data.

(i) For each non-empty finite set K a functor

$$\otimes_K : \mathcal{C}^K \longrightarrow \mathcal{L}ocsys(\mathcal{O}(K); \mathcal{C}),\ \{X_k\} \mapsto \otimes_K X_k$$

from the K-th power of $\mathcal{C}$ to the category of local systems (locally constant sheaves) over the space $\mathcal{O}(K)$ with values in $\mathcal{C}$ (we are using the notations of 7.3).

We suppose that $\otimes_{\{*\}} X$ is the constant local system with the fiber X.

(ii) For each $\pi : K \longrightarrow L$ as above a natural isomorphism

$$\phi_\pi : (\otimes_K X_k)|_{\mathcal{O}(L)\times\prod\mathcal{O}(K_l)} \xrightarrow{\sim} \otimes_L (\otimes_{K_l} X_k).$$

To simplify the notation, we will write this isomorphism in the form

$$\phi_\pi : \otimes_K X_k \xrightarrow{\sim} \otimes_L (\otimes_{K_l} X_k),$$

implying that in the left hand side we must take restriction.

These isomorphisms must satisfy the following

Associativity axiom. For each pair of epimorphisms $K \xrightarrow{\pi} L \xrightarrow{\rho} M$ the square

$$\begin{array}{ccccc} & & \otimes_K X_k & & \\ & \swarrow^{\phi_{\rho\pi}} & & \searrow^{\phi_\pi} & \\ \otimes_M(\otimes_{K_m} X_k) & & & & \otimes_L(\otimes_{K_l} X_k) \\ & \searrow_{\otimes_M \phi_{\pi|K_m}} & & \swarrow_{\phi_\rho} & \\ & & \otimes_M(\otimes_{L_m}(\otimes_{K_l} X_k)) & & \end{array}$$

where $K_m := (\rho\pi)^{-1}(m),\ L_m := \rho^{-1}(m)$, commutes.

11.4. The connection with the conventional definition is as follows. Given two objects $X_1, X_2 \in \mathrm{Ob}\,\mathcal{C}$, define an object $X_1 \dot{\otimes} X_2$ as the fiber of $\otimes_{\{1,2\}} X_k$ at the point $(1/3, 1/2)$. We have natural isomorphisms

$$A_{X_1,X_2,X_3} : X_1\dot{\otimes}(X_2\dot{\otimes}X_3) \xrightarrow{\sim} (X_1\dot{\otimes}X_2)\dot{\otimes}X_3$$

coming from isomorphisms ϕ associated with two possible order preserving epimorphic maps $\{1,2,3\} \longrightarrow \{1,2\}$, and

$$R_{X_1,X_2} : X_1\dot{\otimes}X_2 \xrightarrow{\sim} X_2\dot{\otimes}X_1$$

coming from the standard half-circle monodromy. Associativity axiom for ϕ is equivalent to the the usual compatibilities for these maps.

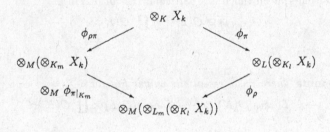

11.5. Now suppose that the data 11.3 is given for *all* (possibly empty) tuples and all (not necessarily epimorphic) maps. The space $\mathcal{O}(\emptyset)$ is by definition one point, and a local system $\otimes_\emptyset$ over it is simply an object of $\mathcal{C}$; let us denote it $\mathbf{1}$ and call a *unit* of our tensor structure. In this case we will say that $\mathcal{C}$ is a braided tensor category with unit.

In the conventional language, we have natural isomorphisms

$$\mathbf{1}\overset{\bullet}{\otimes}X \overset{\sim}{\longrightarrow} X$$

(they correspond to $\{2\} \hookrightarrow \{1,2\}$) satisfying the usual compatibilities with A and R.

FUSION FUNCTORS

11.6. Let

$$\mathcal{A}_{\alpha;1} \subset \mathcal{A}_\alpha$$

denote the open subspace parametrizing configurations lying inside the unit disk $D(0;1)$.

Let K be a non-empty finite set. Obviously, $\mathcal{A}^\alpha(K) = \mathcal{A}_{\alpha;1} \times \mathcal{O}(K)$, and we have the projection

$$\mathcal{A}^\alpha(K) \longrightarrow \mathcal{O}(K).$$

Note also that we have an evident open embedding $\mathcal{O}(K) \hookrightarrow D(0;1)^K$.

Our aim in this part is to define certain *fusion functors*

$$\Psi_K : \mathcal{D}(\mathcal{A}^\alpha(K)) \longrightarrow \mathcal{D}^{mon}(\mathcal{A}^\alpha(\{*\}) \times \mathcal{O}(K))$$

where $(\bullet)^{mon}$ denotes the full subcategory of complexes smooth along $\mathcal{O}(K)$. The construction follows the classical definition of nearby cycles functor, [D3].

11.7. **Poincaré groupoid.** We start with a topological notation. Let X be a connected locally simply connected topological space. Let us denote by $\widetilde{X \times X}$ the space whose points are triples (x, y, p), where $x, y \in X$; p is a homotopy class of paths in X connecting x with y. Let

$$c_X : \widetilde{X \times X} \longrightarrow X \times X \tag{256}$$

be the evident projection. Note that for a fixed $x \in X$, the restriction of c_X to $c_X^{-1}(X \times \{x\})$ is a universal covering of X with a group $\pi_1(X; x)$.

11.8. Consider the diagram with cartesian squares

$$\begin{array}{ccccccc}
\mathcal{A}^\alpha(\{*\}) \times \mathcal{O}(K) & \xrightarrow{\tilde{\Delta}} & \widehat{\mathcal{A}_\alpha(K)} \times \mathcal{O}(K) & \xleftarrow{\tilde{j}} & \mathcal{A}^\alpha(K) \times \mathcal{O}(K) & \xleftarrow{\tilde{c}} & \mathcal{A}^\alpha\widetilde{(K) \times \mathcal{O}}(K) \\
\downarrow & & \downarrow & & \downarrow & & \downarrow \\
D(0,1) \times \mathcal{O}(K) & \xrightarrow{\Delta} & D(0;1)^K \times \mathcal{O}(K) & \xleftarrow{j} & \mathcal{O}(K) \times \mathcal{O}(K) & \xleftarrow{c} & \mathcal{O}\widetilde{(K) \times \mathcal{O}}(K)
\end{array}$$

where we have denoted $\widehat{\mathcal{A}^\alpha(K)} := \mathcal{A}_{\alpha;1} \times D(0;1)^K$. Here Δ is induced by the diagonal embedding $D(0;1) \hookrightarrow D(0;1)^K$, j — by the open embedding $\mathcal{O}(K) \hookrightarrow D(0;1)^K$, c is the map (256). The upper horizontal arrows are defined by pull-back.

We define Ψ_K as a composition

$$\Psi_K = \tilde{\Delta}^* \tilde{j}_* \tilde{c}_* \tilde{c}^* p^*[1]$$

where $p : \mathcal{A}^\alpha(K) \times \mathcal{O}(K) \longrightarrow \mathcal{A}^\alpha(K)$ is the projection.

This functor is t-exact and induces an exact functor

$$\Psi_K : \mathcal{M}(\mathcal{A}^\alpha(K); \mathcal{S}) \longrightarrow \mathcal{M}(\mathcal{A}^\alpha(\{*\}) \times \mathcal{O}(K); \mathcal{S} \times \mathcal{S}_{tr}) \tag{257}$$

where $\mathcal{S}_{tr}$ denotes the trivial stratification of $\mathcal{O}(K)$.

11.9. Set

$$\mathcal{A}^\alpha(K)_d := \mathcal{A}_{\alpha;1} \times \mathcal{O}(K; d)$$

The category $\mathcal{M}(\mathcal{A}^\alpha(K); \mathcal{S})$ is equivalent to the "inverse limit" "$\varprojlim$"$\mathcal{M}(\mathcal{A}^\alpha(K)_d; \mathcal{S})$.

Let $\pi : K \longrightarrow L$ be an epimorphism. Consider a configuration space

$$\mathcal{A}^\alpha(\tau_{\pi;d}) := \mathcal{A}_{\alpha;1} \times \mathcal{O}(\tau_{\pi;d})$$

where $\tau_{\pi;d} := \tau_{\pi;d,0}$. An easy generalization of the definition of Ψ_K yields a functor

$$\Psi_{\pi;d} : \mathcal{M}(\mathcal{A}^\alpha(\tau_{\pi;d})) \longrightarrow \mathcal{M}(\mathcal{A}^\alpha(L)_d \times \prod_{l \in L} \mathcal{O}(K_l))$$

(In what follows we will omit for brevity stratifications from the notations of abelian categories $\mathcal{M}(\bullet)$, implying that we use the principal stratification on all configuration spaces $\mathcal{A}^\alpha(\bullet)$ and the trivial stratification on spaces $\mathcal{O}(\bullet)$, i.e. our sheaves are smooth along these spaces.) Passing to the limit over $d > 0$ we conclude that there exists essentially unique functor

$$\Psi_{K \longrightarrow L} : \mathcal{M}(\mathcal{A}^\alpha(K)) \longrightarrow \mathcal{M}(\mathcal{A}^\alpha(L) \times \prod_l \mathcal{O}(K_l)) \tag{258}$$

such that all squares

$$\begin{array}{ccc} \mathcal{M}(\mathcal{A}^\alpha(K)) & \xrightarrow{\Psi_{K\longrightarrow L}} & \mathcal{M}(\mathcal{A}^\alpha(L)\times\prod_l \mathcal{O}(K_l)) \\ \downarrow & & \downarrow \\ \mathcal{M}(\mathcal{A}^\alpha(\tau_{\pi;d})) & \xrightarrow{\Psi_{\pi;d}} & \mathcal{M}(\mathcal{A}^\alpha(L)_d\times\prod_l \mathcal{O}(K_l)) \end{array}$$

commute (the vertical arrows being restrictions). If $L = \{*\}$, we return to Ψ_K.

11.10. **Lemma.** *All squares*

$$\begin{array}{ccc} \mathcal{M}(\mathcal{A}^\alpha(K)) & \xrightarrow{\Psi_K} & \mathcal{M}(\mathcal{A}^\alpha(\{*\})\times\mathcal{O}(K)) \\ {\scriptstyle \Psi_{K\longrightarrow L}}\downarrow & & \downarrow{\scriptstyle r_\pi} \\ \mathcal{M}(\mathcal{A}^\alpha(L)\times\prod_l \mathcal{O}(K_l)) & \xrightarrow{\Psi_L} & \mathcal{M}(\mathcal{A}^\alpha(\{*\})\times\mathcal{O}(L)\times\prod_l \mathcal{O}(K_l)) \end{array}$$

2-commute. More precisely, there exist natural isomorphisms

$$\phi_{K\longrightarrow L} : r_\pi\circ\Psi_K \xrightarrow{\sim} \Psi_L\circ\Psi_{K\longrightarrow L}.$$

These isomorphisms satisfy a natural cocycle condition (associated with pairs of epimorphisms $K\longrightarrow L\longrightarrow M$).

11.11. Applying the functors Ψ_K componentwise, we get functors

$$\Psi_K : {}^K\mathcal{FS}\longrightarrow \mathcal{L}ocsys(\mathcal{O}(K);\mathcal{FS});$$

taking composition with the gluing functor g_K, (254), we get functors

$$\otimes_K : \mathcal{FS}^K\longrightarrow \mathcal{L}ocsys(\mathcal{O}(K);\mathcal{FS}) \qquad (259)$$

It follows from the previous lemma that these functors define a braided tensor structure on $\mathcal{FS}$.

11.12. Let us define a unit object in $\mathcal{FS}$ as $\mathbf{1}_{\mathcal{FS}} = \mathcal{L}(0)$ (cf. 6.3). One can show that it is a unit for the braided tensor structure defined above.

Chapter 3. Functor Φ

12. FUNCTOR Φ

12.1. Recall the category $\mathcal{C}$ defined in II.11.3.2 and II.12.2.

Our main goal in this section will be the construction of a tensor functor $\Phi: \mathcal{FS} \longrightarrow \mathcal{C}$.

12.2. Recall that we have already defined in 4.5 a functor

$$\Phi : \mathcal{FS} \longrightarrow \mathcal{V}ect_f^X.$$

Now we will construct natural transformations

$$\epsilon_i : \Phi_\lambda(\mathcal{X}) \longrightarrow \Phi_{\lambda+i'}(\mathcal{X})$$

and

$$\theta_i : \Phi_{\lambda+i'}(\mathcal{X}) \longrightarrow \Phi_\lambda(\mathcal{X}).$$

We may, and will, assume that $\mathcal{X} \in \mathcal{FS}_c$ for some c. If $\lambda \notin X_c$ then there is nothing to do.

Suppose that $\lambda \in X_c$; pick $\alpha \in \mathbb{N}[I]$ such that $\lambda + \alpha' \geq \lambda(\mathcal{X})$. By definition.

$$\Phi_\lambda(\mathcal{X}) = \Phi_\alpha(\mathcal{X}^\alpha_{\lambda+\alpha'})$$

where Φ_α is defined in II.7.14 (the definition will be recalled below).

12.3. Pick an unfolding $\pi: J \longrightarrow I$ of α, II.6.12; we will use the same notation for the canonical projection

$$\pi : {}^\pi\mathbb{A} \longrightarrow \mathcal{A}^\alpha_{\lambda+\alpha'} = \mathcal{A}_\alpha.$$

Let N be the dimension of $\mathcal{A}_\alpha$.

12.4. Recall some notations from II.8.4. For each $r \in [0, N]$ we have denoted by $\mathcal{P}_r(J;1)$ the set of all maps

$$\varrho : J \longrightarrow [0, r]$$

such that $\varrho(J)$ contains $[r]$. Let us assign to such ϱ the real point $w_\varrho = (\varrho(j))_{j\in J} \in {}^\pi\mathbb{A}$.

There exists a unique positive facet of $\mathcal{S}_\mathbb{R}$, F_ϱ containing w_ϱ. This establishes a bijection between $\mathcal{P}_r(J;1)$ and the set $\mathcal{F}ac_r$ of r-dimensional positive facets. At the same time we have fixed on each F_ϱ a point w_ϱ. This defines cells $D^+_\varrho := D^+_{F_\varrho}$, $S^+_\varrho := S^+_{F_\varrho}$, cf. II.7.2.

Note that this "marking" of positive facets is Σ_π-invariant. In particular, the group Σ_π permutes the above mentioned cells.

We will denote by $\{0\}$ the unique zero-dimensional facet.

12.5. Given a complex $\mathcal{K}$ from the bounded derived category $\mathcal{D}^b(\mathcal{A}_\alpha)$, its inverse image $\pi^*\mathcal{K}$ is correctly defined as an element of the *equivariant* derived category $\mathcal{D}^b({}^\pi\mathbb{A}, \Sigma_\pi)$ obtained by localizing the category of Σ_π-equivariant complexes on ${}^\pi\mathbb{A}$. The direct image π_* acts between equivariant derived categories

$$\pi_* : \mathcal{D}^b({}^\pi\mathbb{A}, \Sigma_\pi) \longrightarrow \mathcal{D}^b(\mathcal{A}_\alpha, \Sigma_\pi)$$

(the action of Σ_π on $\mathcal{A}_\pi$ being trivial).

We have the functor of Σ_π-invariants

$$(\bullet)^{\Sigma_\pi} : \mathcal{D}^b(\mathcal{A}_\alpha, \Sigma_\pi) \longrightarrow \mathcal{D}^b(\mathcal{A}_\alpha) \tag{260}$$

12.5.1. **Lemma.** *For every $\mathcal{K} \in \mathcal{D}^b(\mathcal{A}_\alpha)$ the canonical morphism*

$$\mathcal{K} \longrightarrow (\pi_*\pi^*\mathcal{K})^{\Sigma_\pi}$$

is an isomorphism.

Proof. The claim may be checked fiberwise. Taking of a fiber commutes with taking Σ_π-invariants since our group Σ_π is finite and we are living over the field of characteristic zero, hence $(\bullet)^{\Sigma_\pi}$ is exact. After that, the claim follows from definitions. □

12.5.2. **Corollary.** *For every $\mathcal{K} \in \mathcal{D}^b(\mathcal{A}_\alpha)$ we have canonical isomorphism*

$$R\Gamma(\mathcal{A}_\alpha; \mathcal{K}) \xrightarrow{\sim} R\Gamma({}^\pi\mathbb{A}; \pi^*\mathcal{K})^{\Sigma_\pi} \ \square \tag{261}$$

12.6. Following II.7.13, consider the sum of coordinates function

$$\sum t_j : {}^\pi\mathbb{A} \longrightarrow \mathbb{A}^1;$$

and for $\mathcal{L} \in \mathcal{D}({}^\pi\mathbb{A}; \mathcal{S})$ let $\Phi_{\sum t_j}(\mathcal{L})$ denote the fiber at the origin of the corresponding vanishing cycles functor. If $H \subset {}^\pi\mathbb{A}$ denotes the inverse image $(\sum t_j)^{-1}(\{1\})$ then we have canonical isomorphisms

$$\Phi_{\sum t_j}(\mathcal{L}) \cong R\Gamma({}^\pi\mathbb{A}, H; \mathcal{L}) \cong \Phi^+_{\{0\}}(\mathcal{L}) \tag{262}$$

The first one follows from the definition of vanishing cycles and the second one from homotopy argument.

Note that the if $\mathcal{L} = \pi^*\mathcal{K}$ for some $\mathcal{K} \in \mathcal{D}(\mathcal{A}_\alpha; \mathcal{S})$ then the group Σ_π operates canonically on all terms of (262), and the isomorphisms are Σ_π-equivariant.

Let us use the same notation

$$\sum t_j : \mathcal{A}_\alpha \longrightarrow \mathbb{A}^1$$

for the descended function, and for $\mathcal{K} \in \mathcal{D}(\mathcal{A}_\alpha; \mathcal{S})$ let $\Phi_{\sum t_j}(\mathcal{K})$ denote the fiber at the origin of the corresponding vanishing cycles functor. If $\mathcal{K}$ belongs to $\mathcal{M}(\mathcal{A}_\alpha; \mathcal{S})$ then $\Phi_{\sum t_j}(\mathcal{K})$ reduces to a single vector space and this is what we call $\Phi_\alpha(\mathcal{K})$.

If $\bar{H}$ denotes $\pi(H) = (\sum t_j)^{-1}(\{1\}) \subset \mathcal{A}_\alpha$ then we have canonical isomorphism

$$\Phi_{\sum t_j}(\mathcal{K}) \cong R\Gamma(\mathcal{A}_\alpha, \bar{H}; \mathcal{K})$$

12.7. **Corollary.** (i) *For every $\mathcal{K} \in \mathcal{D}(\mathcal{A}_\alpha; \mathcal{S})$ we have a canonical isomorphism*

$$i_\pi : \Phi_{\sum t_j}(\mathcal{K}) \xrightarrow{\sim} \Phi^+_{\{0\}}(\pi^*\mathcal{K})^{\Sigma_\pi}. \tag{263}$$

(ii) This isomorphism does not depend on the choice of an unfolding $\pi : J \longrightarrow I$.

Let us explain what (ii) means. Suppose $\pi' : J' \longrightarrow I$ be another unfolding of α. There exists (a non unique) isomorphism

$$\rho : J \xrightarrow{\sim} J'$$

such that $\pi' \circ \rho = \pi$. It induces isomorphisms

$$\rho^* : \Sigma_{\pi'} \xrightarrow{\sim} \Sigma_\pi$$

(conjugation by ρ), and

$$\rho^* : \Phi^+_{\{0\}}((\pi')^*\mathcal{K}) \xrightarrow{\sim} \Phi^+_{\{0\}}(\pi^*\mathcal{K})$$

such that

$$\rho^*(\sigma x) = \rho^*(\sigma)\rho^*(x),$$

$\sigma \in \Sigma_{\pi'}$, $x \in \Phi^+_{\{0\}}((\pi')^*\mathcal{K})$. Passing to invariants, we get an isomorphism

$$\rho^* : \Phi^+_{\{0\}}((\pi')^*\mathcal{K})^{\Sigma_{\pi'}} \xrightarrow{\sim} \Phi^+_{\{0\}}(\pi^*\mathcal{K})^{\Sigma_\pi}.$$

Now (ii) means that $i_\pi \circ \rho^* = i_{\pi'}$. As a consequence, the last isomorphism does not depend on the choice of ρ.

Proof. Part (i) follows from the preceding discussion and 12.5.2.

As for (ii), it suffices to prove that any automorphism $\rho : J \xrightarrow{\sim} J$ respecting π induces the identity automorphism of the space of invariants $\Phi^+_{\{0\}}(\pi^*\mathcal{K})^{\Sigma_\pi}$. But the action of ρ on the space $\Phi^+_{\{0\}}(\pi^*\mathcal{K})$ comes from the action of Σ_π on this space, and our claim is obvious. □

In computations the right hand side of (263) will be used as a *de facto* definition of Φ_α.

12.8. Suppose that $\alpha = \sum a_i i$; pick an i such that $a_i > 0$.

Let us introduce the following notation. For a partition $J = J_1 \coprod J_2$ and a positive d let $\mathbb{A}^{J_1,J_2}(d)$ denote an open suspace of ${}^\pi\mathbb{A}$ consisting of all points $\mathbf{t} = (t_j)$ such that $|t_j| > d$ (resp., $|t_j| < d$) if $j \in J_1$ (resp., $j \in J_2$).

We have obviously

$$\pi^{-1}(\mathcal{A}^{i,\alpha-i}(d)) = \coprod_{j\in\pi^{-1}(i)} \mathbb{A}^{\{j\},J-\{j\}}(d) \tag{264}$$

12.9. For $j \in \pi^{-1}(i)$ let $\pi_j : J - \{j\} \longrightarrow I$ denote the restriction of π; it is an unfolding of $\alpha - i$. The group Σ_{π_j} may be identified with the subgroup of Σ_π consisting of automorphisms stabilizing j. For $j', j'' \in J$ let $(j'j'')$ denotes their transposition. We have

$$\Sigma_{\pi_{j''}} = (j'j'')\Sigma_{\pi_{j'}}(j'j'')^{-1} \tag{265}$$

For a fixed $j \in \pi^{-1}(i)$ we have a partition into cosets

$$\Sigma_\pi = \coprod_{j'\in\pi^{-1}(i)} \Sigma_{\pi_j}(jj') \tag{266}$$

12.10. For $j \in J$ let F_j denote a one-dimensional facet corresponding to the map $\varrho_j : J \longrightarrow [0,1]$ sending all elements to 0 except for j being sent to 1 (cf. 12.4).

For $\mathcal{K} \in \mathcal{D}(\mathcal{A}_\alpha; \mathcal{S})$ we have canonical and variation maps

$$v_j : \Phi^+_{\{0\}}(\pi^*\mathcal{K}) \rightleftarrows \Phi^+_{F_j}(\pi^*\mathcal{K}) : u_j$$

defined in II, (89), (90). Taking their sum, we get maps

$$v_i : \Phi^+_{\{0\}}(\pi^*\mathcal{K}) \rightleftarrows \oplus_{j\in\pi^{-1}(i)} \Phi^+_{F_j}(\pi^*\mathcal{K}) : u_i \qquad (267)$$

Note that the group Σ_π operates naturally on both spaces and both maps v_i and u_i respect this action.

Let us consider more attentively the action of Σ_π on $\oplus_{j\in\pi^{-1}(i)} \Phi^+_{F_j}(\pi^*\mathcal{K})$. A subgroup Σ_{π_j} respects the subspace $\Phi^+_{F_j}(\pi^*\mathcal{K})$. A transposition $(j'j'')$ maps $\Phi^+_{F_{j'}}(\pi^*\mathcal{K})$ isomorphically onto $\Phi^+_{F_{j''}}(\pi^*\mathcal{K})$.

Let us consider the space of invariants

$$(\oplus_{j\in\pi^{-1}(i)} \Phi^+_{F_j}(\pi^*\mathcal{K}))^{\Sigma_\pi}$$

For every $k \in \pi^{-1}(i)$ the obvious projection induces isomorphism

$$(\oplus_{j\in\pi^{-1}(i)} \Phi^+_{F_j}(\pi^*\mathcal{K}))^{\Sigma_\pi} \xrightarrow{\sim} (\Phi^+_{F_k}(\pi^*\mathcal{K}))^{\Sigma_{\pi_k}};$$

therefore for two different $k, k' \in \pi^{-1}(i)$ we get an isomorphism

$$i_{kk'} : (\Phi^+_{F_k}(\pi^*\mathcal{K}))^{\Sigma_{\pi_k}} \xrightarrow{\sim} (\Phi^+_{F_{k'}}(\pi^*\mathcal{K}))^{\Sigma_{\pi_{k'}}} \qquad (268)$$

Obviously, it is induced by transposition (kk').

12.11. Let us return to the situation 12.2 and apply the preceding discussion to $\mathcal{K} = \mathcal{X}^\alpha_{\lambda+\alpha'}$. We have by definition

$$\Phi_\lambda(\mathcal{X}) = \Phi_{\sum t_j}(\mathcal{X}^\alpha_{\lambda+\alpha'}) \cong \Phi^+_{\{0\}}(\pi^*\mathcal{X}^\alpha_{\lambda+\alpha'})^{\Sigma_\pi} \qquad (269)$$

On the other hand, let us pick some $k \in \pi^{-1}(i)$ and a real d, $0 < d < 1$. The subspace

$$F_k^\perp(d) \subset \mathbb{A}^{\{j\},J-\{j\}}(d) \qquad (270)$$

consisting of points (t_j) with $t_k = 1$, is a transversal slice to the face F_k. Consequently, the factorization isomorphism for $\mathcal{X}^\alpha_{\lambda+\alpha'}$ lifted to $\mathbb{A}^{\{j\},J-\{j\}}(d)$ induces isomorphism

$$\Phi^+_{F_k}(\pi^*\mathcal{X}^\alpha_{\lambda+\alpha'}) \cong \Phi^+_{\{0\}}(\pi_k^*\mathcal{X}^{\alpha-i}_{\lambda+\alpha'}) \otimes (\mathcal{I}^i_{\lambda+i'})_{\{1\}} = \Phi^+_{\{0\}}(\pi_k^*\mathcal{X}^{\alpha-i}_{\lambda+\alpha'})$$

Therefore we get isomorphisms

$$\Phi_{\lambda+i'}(\mathcal{X}) = \Phi^+_{\{0\}}(\pi_k^*\mathcal{X}^{\alpha-i}_{\lambda+\alpha'})^{\Sigma_{\pi_k}} \cong \Phi^+_{F_k}(\pi^*\mathcal{X}^\alpha_{\lambda+\alpha'})^{\Sigma_{\pi_k}} \cong \qquad (271)$$
$$\cong (\oplus_{j\in\pi^{-1}(i)} \Phi^+_{F_j}(\pi^*\mathcal{X}^\alpha_{\lambda+\alpha'}))^{\Sigma_\pi}$$

It follows from the previous discussion that this isomorphism does not depend on the intermediate choice of $k \in \pi^{-1}(i)$.

Now we are able to define the operators θ_i, ϵ_i:

$$\epsilon_i : \Phi_\lambda(\mathcal{X}) \rightleftarrows \Phi_{\lambda+i'}(\mathcal{X}) : \theta_i$$

By definition, they are induced by the maps u_i, v_i (cf. (267)) respectively (for $\mathcal{K} = \mathcal{X}^\alpha_{\lambda+\alpha'}$) after passing to Σ_π-invariants and taking into account isomorphisms (269) and (271).

12.12. **Theorem.** *The operators* ϵ_i *and* θ_i *satisfy the relations* II.12.3, *i.e. the previous construction defines functor*

$$\tilde{\Phi}: \mathcal{FS} \longrightarrow \tilde{\mathcal{C}}$$

where the category $\tilde{\mathcal{C}}$ *is defined as in* loc. cit.

Proof will occupy the rest of the section.

12.13. Let $\mathfrak{u}^+$ (resp., $\mathfrak{u}^-$) denote the subalgebra of $\mathfrak{u}$ generated by all ϵ_i (resp., θ_i). For $\beta \in \mathbb{N}[I]$ let $\mathfrak{u}^\pm_\beta \subset \mathfrak{u}^\pm$ denote the corresponding homogeneous component.

The proof will go as follows. First, relations II.12.3 (z), (a), (b) are obvious. We will do the rest in three steps.

Step 1. Check of (d). This is equivalent to showing that the action of operators θ_i correctly defines maps

$$\mathfrak{u}^-_\beta \otimes \Phi_{\lambda+\beta'}(\mathcal{X}) \longrightarrow \Phi_\lambda(\mathcal{X})$$

for all $\beta \in \mathbb{N}[I]$.

Step 2. Check of (e). This is equivalent to showing that the action of operators ϵ_i correctly defines maps

$$\mathfrak{u}^+_\beta \otimes \Phi_\lambda(\mathcal{X}) \longrightarrow \Phi_{\lambda+\beta'}(\mathcal{X})$$

for all $\beta \in \mathbb{N}[I]$.

Step 3. Check of (c).

12.14. Let us pick an arbitrary $\beta = \sum b_i i \in \mathbb{N}[I]$ and $\alpha \in \mathbb{N}[I]$ such that $\lambda + \alpha' \geq \lambda(\mathcal{X})$ and $\alpha \geq \beta$. We pick the data from 12.3. In what follows we will generalize the considerations of 12.8 — 12.11.

Let $U(\beta)$ denote the set of all subsets $J' \subset J$ such that $|J' \cap \pi^{-1}(i)| = b_i$ for all i. Thus, for such J', $\pi_{J'} := \pi|_{J'} : J' \longrightarrow I$ is an unfolding of β and $\pi_{J-J'}$ — an unfolding of $\alpha - \beta$. We have a disjoint sum decomposition

$$\pi^{-1}(\mathcal{A}^{\beta,\alpha-\beta}(d)) = \coprod_{J' \in U(\beta)} \mathbb{A}^{J',J-J'}(d) \tag{272}$$

(cf. (264)).

12.15. For $J' \in U(\beta)$ let $F_{J'}$ denote a one-dimensional facet corresponding to the map $\varrho_{J'} : J \longrightarrow [0,1]$ sending $j \notin J'$ to 0 $j \in J'$ — to 1 (cf. 12.4).

For $\mathcal{K} \in \mathcal{D}(\mathcal{A}_\alpha; \mathcal{S})$ we have canonical and variation maps

$$v_{J'} : \Phi^+_{\{0\}}(\pi^*\mathcal{K}) \rightleftarrows \Phi^+_{F_{J'}}(\pi^*\mathcal{K}) : u_{J'}$$

Taking their sum, we get maps

$$v_\beta : \Phi^+_{\{0\}}(\pi^*\mathcal{K}) \rightleftarrows \oplus_{J' \in U(\beta)} \Phi^+_{F_{J'}}(\pi^*\mathcal{K}) : u_\beta \tag{273}$$

The group Σ_π operates naturally on both spaces and both maps v_β and u_β respect this action.

A subgroup $\Sigma_{J'}$ respects the subspace $\Phi^+_{F_{J'}}(\pi^*\mathcal{K})$. The projection induces an isomorphism

$$(\oplus_{J' \in U(\beta)} \Phi^+_{F_{J'}}(\pi^*\mathcal{K}))^{\Sigma_\pi} \xrightarrow{\sim} \Phi^+_{F_{J'}}(\pi^*\mathcal{K})^{\Sigma_{J'}}.$$

We have the crucial

12.16. **Lemma.** *Factorization isomorphism for $\mathcal{X}$ induces canonical isomorphism*

$$\mathfrak{u}_\beta^- \otimes \Phi_{\lambda+\beta'}(\mathcal{X}) \cong (\oplus_{J'\in U(\beta)} \Phi^+_{F_{J'}}(\pi^*\mathcal{X}^\alpha_{\lambda+\alpha'}))^{\Sigma_\pi} \tag{274}$$

Proof. The argument is the same as in 12.11, using II, Thm. 6.16. □

12.17. As a consequence, passing to Σ_π-invariants in (273) (for $\mathcal{K} = \mathcal{X}^\alpha_{\lambda+\alpha'}$) we get the maps

$$\epsilon_\beta : \Phi_\lambda(\mathcal{X}) \rightleftarrows \mathfrak{u}_\beta^- \otimes \Phi_{\lambda+\beta'}(\mathcal{X}) : \theta_\beta$$

12.18. **Lemma.** *The maps θ_β provide $\Phi(\mathcal{X})$ with a structure of a left module over the negative subalgebra $\mathfrak{u}^-$.*

Proof. We must prove the associativity. It follows from the associativity of factorization isomorphisms. □

This lemma proves relations II.12.3 (d) for operators θ_i, completing Step 1 of the proof of our theorem.

12.19. Now let us consider operators

$$\epsilon_\beta : \Phi_\lambda(\mathcal{X}) \longrightarrow \mathfrak{u}_\beta^- \otimes \Phi_{\lambda+\beta'}(\mathcal{X}).$$

By adjointness, they induce operators

$$\mathfrak{u}_\beta^{-*} \otimes \Phi_\lambda(\mathcal{X}) \longrightarrow \Phi_{\lambda+\beta'}(\mathcal{X})$$

The bilinear form S, II.2.10, induces isomorhisms

$$S : \mathfrak{u}_\beta^{-*} \xrightarrow{\sim} \mathfrak{u}_\beta^-;$$

let us take their composition with the isomorphism of algebras

$$\mathfrak{u}^- \xrightarrow{\sim} \mathfrak{u}^+$$

sending θ_i to ϵ_i. We get isomorphisms

$$S' : \mathfrak{u}_\beta^{-*} \xrightarrow{\sim} \mathfrak{u}_\beta^+$$

Using S', we get from ϵ_β operators

$$\mathfrak{u}_\beta^+ \otimes \Phi_\lambda(\mathcal{X}) \longrightarrow \Phi_{\lambda+\beta'}(\mathcal{X})$$

12.19.1. **Lemma.** *The above operators provide $\Phi(\mathcal{X})$ with a structure of a left $\mathfrak{u}^+$-module.*

For $\beta = i$ they coincide with operators ϵ_i defined above.

This lemma completes Step 2, proving relations II.12.3 (e) for generators ϵ_i.

12.20. Now we will perform the last Step 3 of the proof, i.e. prove the relations II.12.3 (c) between operators ϵ_i, θ_j. Consider a square

$$\begin{array}{ccccc} & & \Phi_{\lambda+i'}(\mathcal{X}) & & \\ & \overset{\epsilon_i}{\nearrow} & & \overset{\theta_j}{\searrow} & \\ \Phi_\lambda(\mathcal{X}) & & & & \Phi_{\lambda-j'+i'}(\mathcal{X}) \\ & \underset{\theta_j}{\searrow} & & \underset{\epsilon_i}{\nearrow} & \\ & & \Phi_{\lambda-j'}(\mathcal{X}) & & \end{array}$$

We have to prove that

$$\epsilon_i\theta_j - \zeta^{i\cdot j}\theta_j\epsilon_i = \delta_{ij}(1 - \zeta^{-2\lambda\cdot i'}) \tag{275}$$

(cf. (193)).

As before, we may and will suppose that $\mathcal{X} \in \mathcal{FS}_c$ and $\lambda \in X_c$ for some c. Choose $\alpha \in \mathbb{N}[I]$ such that $\alpha \geq i$, $\alpha \geq j$ and $\lambda - j' + \alpha' \geq \lambda(\mathcal{X})$. The above square may be identified with the square

$$\begin{array}{ccccc} & & \Phi_{\alpha-i-j}(\mathcal{X}^{\alpha-i-j}_{\lambda-j'+\alpha'}) & & \\ & \overset{\epsilon_i}{\nearrow} & & \overset{\theta_j}{\searrow} & \\ \Phi_{\alpha-j}(\mathcal{X}^{\alpha-j}_{\lambda-j'+\alpha'}) & & & & \Phi_{\alpha-i}(\mathcal{X}^{\alpha-i}_{\lambda-j'+\alpha'}) \\ & \underset{\theta_j}{\searrow} & & \underset{\epsilon_i}{\nearrow} & \\ & & \Phi_{\alpha}(\mathcal{X}^{\alpha}_{\lambda-j'+\alpha'}) & & \end{array}$$

12.21. Choose an unfolding $\pi : J \longrightarrow I$ of α; let $\Sigma = \Sigma_\pi$ be its automorphism group, and $\pi : \mathbb{A} = {}^\pi\mathbb{A} \longrightarrow \mathcal{A}_\alpha$ denote the corresponding covering. We will reduce our proof to certain statements about (vanishing cycles of) sheaves on $\mathbb{A}$.

Let us introduce a vector space

$$V = H^0\Phi^+_{\{0\}}(\pi^*\mathcal{X}^\alpha_{\lambda-j'+\alpha'});$$

the group Σ operates on it, and we have

$$\Phi_\alpha(\mathcal{X}^\alpha_{\lambda-j'+\alpha'}) \cong V^\Sigma \tag{276}$$

For each $k \in J$ we have a positive one-dimensional facet $F_k \subset \mathbb{A}_\mathbb{R}$ defined as in 12.10. Denote

$$V_k = H^0\Phi^+_{F_k}(\pi^*\mathcal{X}^\alpha_{\lambda-j'+\alpha'});$$

we have canonically

$$\Phi_{\alpha-p}(\mathcal{X}^{\alpha-p}_{\lambda-j'+\alpha'}) \cong (\oplus_{k\in\pi^{-1}(p)}V_k)^\Sigma \tag{277}$$

for each $p \in I$, $p \leq \alpha$, cf. 12.11.

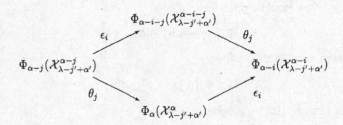

12.21.1. We have to extend considerations of 12.11 to two-dimensional facets. For each pair of different $k, l \in J$ such that $\pi(k) = i, \pi(l) = j$, let F_{kl} denote a two-dimensional positive facet corresponding to the map $\varrho : J \longrightarrow [0,2]$ sending k to 1, l — to 2 and all other elements — to zero (cf. 12.4). Set

$$V_{kl} = H^0\Phi^+_{F_{kl}}(\pi^*\mathcal{X}^\alpha_{\lambda-j'+\alpha'}).$$

Again, due to equivariance of our sheaf, the group Σ operates on $\oplus V_{kl}$ in such a way that

$$\sigma(V_{kl}) = V_{\sigma(k)\sigma(l)}.$$

Let

$$\pi_{kl} : J - \{k,l\} \longrightarrow I$$

be the restriction of π. It is an unfolding of $\alpha - i - j$; let Σ_{kl} denote its automorphism group. Pick d_1, d_2 such that $0 < d_2 < 1 < d_1 < 2$. The subspace

$$F^\perp_{kl} \subset \mathbb{A}^{\{l\},\{k\},J-\{k,l\}}(d_1,d_2) \qquad (278)$$

consisting of all points (t_j) such that $t_k = 1$ and $t_l = 2$ is a transversal slice to f_{kl}. Consequently, factorization axiom implies canonical isomorphism

$$\Phi^+_{F_{kl}}(\pi^*\mathcal{X}^\alpha_{\lambda-j'+\alpha'}) \cong \Phi^+_{\{0\}}(\pi^*_{kl}\mathcal{X}^{\alpha-i-j}_{\lambda-j'+\alpha'}) \otimes (\mathcal{I}^i_{\lambda+i'})_{\{1\}} \otimes (\mathcal{I}^j_\lambda)_{\{2\}} = \Phi^+_{\{0\}}(\pi^*_{kl}\mathcal{X}^{\alpha-i-j}_{\lambda-j'+\alpha'})$$

Symmetry. Interchanging k and l, we get isomorphisms

$$t : V_{kl} \xrightarrow{\sim} V_{lk} \qquad (279)$$

Passing to Σ-invariants, we get isomorphisms

$$\Phi_{\alpha-i-j}(\mathcal{X}^{\alpha-i-j}_{\lambda-j'+\alpha'}) = \Phi^+_{\{0\}}(\pi^*_{kl}\mathcal{X}^{\alpha-i-j}_{\lambda-j'+\alpha'})^{\Sigma_{kl}} \cong \qquad (280)$$
$$\cong \Phi^+_{F_{kl}}(\pi^*\mathcal{X}^\alpha_{\lambda-j'+\alpha'})^{\Sigma_{kl}} \cong (\oplus_{k\in\pi^{-1}(i),l\in\pi^{-1}(j),k\neq l} V_{kl})^\Sigma$$

(cf. (271)).

12.22. The canonical and variation maps induce linear operators

$$v_k : V \rightleftarrows V_k : u^k$$

and

$$v^k_{lk} : V_k \rightleftarrows V_{lk} : u^{lk}_k$$

which are Σ-equivariant in the obvious sense. Taking their sum, we get operators

$$V \rightleftarrows \oplus_{k\in\pi^{-1}(p)} V_k \rightleftarrows \oplus_{l\in\pi^{-1}(q),k\in\pi^{-1}(p)} V_{lk}$$

which induce, after passing to Σ-invariants, operators ϵ_p, ϵ_q and θ_p, θ_q.

Our square takes a form

$$\begin{array}{ccc}
(\oplus_{k\in\pi^{-1}(j),l\in\pi^{-1}(i),k\neq l} V_{lk})^\Sigma & \overset{t}{\xrightarrow{\sim}} & (\oplus_{k\in\pi^{-1}(j),l\in\pi^{-1}(i),k\neq l} V_{kl})^\Sigma \\
\Big\uparrow \sum v^k_{lk} & & \Big\downarrow \sum u^{kl}_l \\
(\oplus_{k\in\pi^{-1}(j)} V_k)^\Sigma & & (\oplus_{l\in\pi^{-1}(i)} V_l)^\Sigma \\
\sum u^k \searrow & & \nearrow \sum v_l \\
 & V^\Sigma &
\end{array}$$

Now we will formulate two relations between u and v which imply the necessary relations between ϵ and θ.

12.23. **Lemma.** *Suppose that $i = j$ and $\pi(k) = i$. Consider operators*

$$u^k : V_k \rightleftarrows V : v_k.$$

The composition $v_k u^k$ is equal to the multiplication by $1 - \zeta^{-2\lambda \cdot i'}$.

Proof. Consider the transversal slice $F_k^\perp(d)$ to the face F_k as in 12.11. It follows from the definition of the canonical and variation maps, II.7.10, that composition $v_k u^k$ is equal to $1 - T^{-1}$ where T is the monodromy acquired by of $\Phi^+_{\{0\}}(\pi_k^* \mathcal{X}^{\alpha-i}_{\lambda-i'+\alpha'})$ when the point t_k moves around the disc of radius d where all other points are living. By factorization, $T = \zeta^{2\lambda \cdot i'}$. $\square$

12.24. **Lemma.** *For $k \in \pi^{-1}(j)$, $l \in \pi^{-1}(i)$, $k \neq l$, consider the pentagon*

$$\begin{array}{ccccc}
V_{lk} & & \overset{t}{\underset{\sim}{\longrightarrow}} & & V_{kl} \\
v^k_{lk} \uparrow & & & & \downarrow u^{kl}_l \\
V_k & & & & V_l \\
& u^k \searrow & & \nearrow v_l & \\
& & V & &
\end{array}$$

We have

$$v_l u^k = \zeta^{i \cdot j} u^{kl}_k \circ t \circ v^k_{lk}.$$

This lemma is a consequence of the following more general statement.

12.25. Let $\mathcal{K} \in \mathcal{D}(\mathcal{A}; \mathcal{S})$ be arbitrary. We have naturally

$$\Phi^+_{F_{kl}}(\mathcal{K}) \cong \Phi^+_{\{0\}}(\mathcal{K}|_{F^\perp_{kl}(d_1,d_2)}[-2])$$

Let

$$t : \Phi^+_{F_{kl}}(\mathcal{K}) \xrightarrow{\sim} \Phi^+_{F_{lk}}(\mathcal{K})$$

be the monodromy isomorphism induced by the travel of the point t_l in the upper half plane to the position to the left of t_k (outside the disk of radius d_1).

12.25.1. **Lemma.** *The composition*

$$v^{\{0\}}_{F_l} \circ u^{F_k}_{\{0\}} : \Phi^+_{F_k}(\mathcal{K}) \longrightarrow \Phi^+_{F_l}(\mathcal{K})$$

is equal to $u^{F_{kl}}_{F_l} \circ t \circ v^{F_k}_{F_{lk}}$.

12.26. It remains to note that due to 12.22 the desired relation (275) is a formal consequence of lemmas 12.23 and 12.24. This completes the proof of Theorem 12.12. $\square$

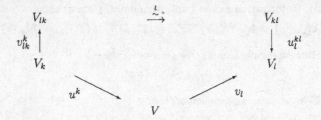

12.27. Taking composition of $\tilde{\Phi}$ with an inverse to the equivalence Q, II (143), we get a functor

$$\Phi : \mathcal{FS} \longrightarrow \mathcal{C} \tag{281}$$

13. Main properties of Φ

TENSOR PRODUCTS

13.1. **Theorem.** *Φ is a tensor functor, i.e. we have natural isomorphisms*

$$\Phi(\mathcal{X}\dot{\otimes}\mathcal{Y}) \xrightarrow{\sim} \Phi(\mathcal{X}) \otimes \Phi(\mathcal{Y})$$

satisfying all necessary compatibilities.

Proof follows from the Additivity theorem, II.9.3. □

DUALITY

13.2. Let $\tilde{\mathcal{C}}_{\zeta^{-1}}$ denote the category $\tilde{\mathcal{C}}$ with the value of parameter ζ changed to ζ^{-1}. The notations $\mathcal{FS}_{\zeta^{-1}}$, etc. will have the similar meaning.

Let us define a functor

$$D : \tilde{\mathcal{C}}^{opp} \longrightarrow \tilde{\mathcal{C}}_{\zeta^{-1}} \tag{282}$$

as follows. By definition, for $V = \oplus V_\lambda \in Ob\ \tilde{\mathcal{C}}^{opp} = Ob\ \tilde{\mathcal{C}}$,

$$(DV)_\lambda = V_\lambda^*,$$

and operators

$$\theta_{i,DV} : (DV)_\lambda \rightleftarrows (DV)_{\lambda - i'} : \epsilon_{i,DV}$$

are defined as

$$\epsilon_{i,DV} = \theta^*_{i,V};\ \theta_{i,DV} = -\zeta^{2\lambda \cdot i'}\epsilon^*_{i,V}.$$

On morphisms D is defined in the obvious way. One checks directly that D is well defined, respects tensor structures, and is an equivalence.

13.3. Let us define a functor

$$D : \mathcal{FS}^{opp} \longrightarrow \mathcal{FS}_{\zeta^{-1}} \tag{283}$$

as follows. For $\mathcal{X} \in Ob\ \mathcal{FS}^{opp} = Ob\ \mathcal{FS}$ we set $\lambda(D\mathcal{X}) = \lambda(\mathcal{X})$; $(D\mathcal{X})^\alpha = D(\mathcal{X}^\alpha)$ where D in the right hand side is Verdier duality. Factorization isomorphisms for $D\mathcal{X}$ are induced in the obvious way from factorization isomorphisms for $\mathcal{X}$. The value of D on morphisms is defined in the obvious way.

D is a tensor equivalence.

13.4. **Theorem.** *Functor $\tilde{\Phi}$ commutes with D.*

Proof. Our claim is a consequence of the following topological remarks.

13.5. Consider a standard affine space $\mathbb{A} = \mathbb{A}^J$ with a principal stratification $\mathcal{S}$ as in II.7. Let $\mathcal{K} \in \mathcal{D}(\mathbb{A}; \mathcal{S})$, let F_j be the one-dimensional facet corresponding to an element $j \in J$ as in 12.10. Consider a transversal slice $F_j^{\perp}(d)$ as in 12.11. We have canonically

$$\Phi^+_{F_j}(\mathcal{K}) \cong \Phi_{\{0\}}(\mathcal{K}|_{F_j^{\perp}(d)}[-1]);$$

when the point t_j moves counterclockwise around the disk of radius d, $\Phi^+_j(\mathcal{K})$ acquires monodromy

$$T_j : \Phi^+_j(\mathcal{K}) \xrightarrow{\sim} \Phi^+_j(\mathcal{K}).$$

13.5.1. **Lemma.** *Consider canonical and variation maps*

$$v_{D\mathcal{K}} : \Phi^+_{\{0\}}(D\mathcal{K}) \rightleftarrows \Phi^+_{F_j}(D\mathcal{K}) : u_{D\mathcal{K}}$$

Let us identify $\Phi^+_{\{0\}}(D\mathcal{K}), \Phi^+_{F_j}(D\mathcal{K})$ with $\Phi^+_{\{0\}}(\mathcal{K})^*$ and $\Phi^+_{F_j}(\mathcal{K})^*$ respectively. Then the maps become

$$v_{D\mathcal{K}} = u^*_{\mathcal{K}};\ u_{D\mathcal{K}} = -v^*_{\mathcal{K}} \circ T^*_j$$

The theorem follows from this lemma. □

STANDARD OBJECTS

13.6. **Theorem.** *We have naturally*

$$\Phi(\mathcal{L}(\Lambda)) \cong L(\Lambda)$$

for all $\Lambda \in X$.

Combining this with Theorem 6.2, we get

13.7. **Theorem.** *Φ induces bijection between sets of isomorphism classes of irreducibles.* □

13.8. **Verma modules.** Let $\mathfrak{u}^{\geq 0} \subset \mathfrak{u}$ denote the subalgebra generated by $\epsilon_i, K_i, K^{-1}_i$, $i \in I$. For $\Lambda \in X$, let χ_Λ denote a one-dimensional representation of $\mathfrak{u}^{\geq 0}$ generated by a vector v_Λ, with the action

$$\epsilon_i v_\Lambda = 0,\ K_i v_\Lambda = \zeta^{\langle \Lambda, i\rangle} v_\Lambda.$$

Let $M(\Lambda)$ denote the induced $\mathfrak{u}$-module

$$M(\Lambda) = \mathfrak{u} \otimes_{\mathfrak{u}^{\geq 0}} \chi_\Lambda.$$

Equipped with an obvious X-grading, $M(\Lambda)$ becomes an object of $\tilde{\mathcal{C}}$. We will also use the same notation for the corresponding object of $\mathcal{C}$.

13.9. **Theorem.** *The factorizable sheaves $\mathcal{M}(\Lambda)$ are finite. We have naturally*

$$\Phi(\mathcal{M}(\Lambda)) \cong M(\Lambda).$$

Proof is given in the next two subsections.

13.10. Let us consider the space $\mathbb{A}$ as in 13.5. Suppose that $\mathcal{K} \in \mathcal{D}(\mathbb{A}; \mathcal{S})$ has the form $\mathcal{K} = j_! j^* \mathcal{K}$ where

$$j : \mathbb{A} - \bigcup_{j \in J} \{t_j = 0\} \hookrightarrow \mathbb{A}.$$

Let F_Δ be the positive facet whose closure is the main diagonal.

13.10.1. **Lemma.** *The canonical map*

$$u : \Phi^+_{F_\Delta}(\mathcal{K}) \longrightarrow \Phi^+_{\{0\}}(\mathcal{K})$$

is an isomorphism.

Proof. Pick $j_0 \in J$, and consider a subspace $Y = \{t_{j_0} = 0\} \cup \{t_{j_0} = 1\} \subset \mathbb{A}$. Set $\mathcal{K}' = k_! k^* \mathcal{K}$ where

$$k : \mathbb{A} - \bigcup_{j \in J} \{t_j = 0\} - \bigcup_{j \in J} \{t_j = 1\} \hookrightarrow \mathbb{A}.$$

We have

$$\Phi^+_{\{0\}}(\mathcal{K}) = R\Gamma(\mathbb{A}; \mathcal{K}')$$

On the other hand by homotopy we have

$$\Phi^+_{F_\Delta}(\mathcal{K}) \cong R\Gamma(\{t_{j_0} = c\}; \mathcal{K}')[-1]$$

where c is any real between 0 and 1. Let us compute $R\Gamma(\mathbb{A}; \mathcal{K}')$ using the Leray spectral sequence of a projection

$$p : \mathbb{A} \longrightarrow \mathbb{A}^1, \ (t_j) \mapsto t_{j_0}.$$

The complex $p_* \mathcal{K}'$ is equal to zero at the points $\{0\}$ and $\{1\}$, and is constant with the fiber $R\Gamma(\{t_{j_0} = c\}; \mathcal{K}')$ over c. It follows that

$$R\Gamma(\mathbb{A}; \mathcal{K}') \cong R\Gamma(\mathbb{A}^1; p_* \mathcal{K}') \cong R\Gamma(\{t_{j_0} = c\}; \mathcal{K}')[-1],$$

and the inverse to this isomorphism may be identified with u. □

13.11. Suppose we have $\alpha \in \mathbb{N}[I]$, let $\pi : J \longrightarrow I$; $\pi : \mathbb{A} \longrightarrow \mathcal{A}_\alpha$ be the corresponding unfolding. Let us apply the previous lemma to $\mathcal{K} = \pi^* \mathcal{M}(\Lambda)^\alpha$. Note that after passing to Σ_π-invariants, the map u becomes

$$u_\alpha^- \longrightarrow \tilde{\Phi}_\alpha(\mathcal{M}(\Lambda))$$

by Theorem II.6.16. This identifies homogeneous components of $\tilde{\Phi}_\alpha(\mathcal{M}(\Lambda))$ with the components of the Verma module. After that, the action of ϵ_i and θ_i is identified with the action of u on it. The theorem is proven. □

Chapter 4. Equivalence

14. Truncation functors

14.1. Recall the notations of 5.4. We fix a coset $X_c \subset X$, and we denote the subcategory $\mathcal{FS}_{\leq\lambda;c} \subset \mathcal{FS}$ by $\mathcal{FS}_{\leq\lambda}$ for simplicity until further notice.

Given $\lambda \in X$, we will denote by $\mathcal{C}_{\leq\lambda} \subset \mathcal{C}$ the full subcategory of all u-modules V such that $V_\mu \neq 0$ implies $\mu \leq \lambda$. We denote by q_λ the embedding functor $\mathcal{C}_\lambda \hookrightarrow \mathcal{C}$. Obviously, $\Phi(\mathcal{FS}_{\leq\lambda}) \subset \mathcal{C}_\lambda$.

In this section we will construct functors

$$\sigma_\lambda^!, \sigma_\lambda^* : \mathcal{FS} \longrightarrow \mathcal{FS}_{\leq\lambda}$$

and

$$q_\lambda^!, q_\lambda^* : \mathcal{C} \longrightarrow \mathcal{C}_{\leq\lambda},$$

such that $\sigma_\lambda^!$ (resp., σ_λ^*) is right (resp., left) adjoint to σ_λ and $q_\lambda^!$ (resp., q_λ^*) is right (resp., left) adjoint to q_λ.

14.2. First we describe $\sigma_\lambda^*, \sigma_\lambda^!$. Given a factorizable sheaf $\mathcal{X} = \{\mathcal{X}^\alpha\}$ with $\lambda(\mathcal{X}) = \mu \geq \lambda$ we define FS's $\mathcal{Y} := \sigma_\lambda^*\mathcal{X}$ and $\mathcal{Z} := \sigma_\lambda^!\mathcal{X}$ as follows.

We set $\lambda(\mathcal{Y}) = \lambda(\mathcal{Z}) = \lambda$. For $\alpha \in \mathbb{N}[I]$ we set

$$\mathcal{Y}^\alpha = L^0\sigma^*\mathcal{X}^{\alpha+\mu-\lambda}$$

if $\alpha + \mu - \lambda \in \mathbb{N}[I]$ and 0 otherwise, and

$$\mathcal{Z}^\alpha = R^0\sigma^!\mathcal{X}^{\alpha+\mu-\lambda}$$

if $\alpha + \mu - \lambda \in \mathbb{N}[I]$ and 0 otherwise. Here σ denotes the canonical closed embedding (cf. 2.4)

$$\sigma : \mathcal{A}_\lambda^\alpha \hookrightarrow \mathcal{A}_\mu^{\alpha+\mu-\lambda},$$

and $L^0\sigma^*$ (resp., $R^0\sigma^!$) denotes the zeroth perverse cohomology of σ^* (resp., of $\sigma^!$). The factorization isomorphisms for $\mathcal{Y}$ and $\mathcal{Z}$ are induced from those for $\mathcal{X}$; associativity is obvious.

14.3. **Lemma.** *Let $\mathcal{M} \in \mathcal{FS}_{\leq\lambda}$, $\mathcal{X} \in \mathcal{FS}$. Then*

$$\mathrm{Hom}_{\mathcal{FS}}(\mathcal{X}, \mathcal{M}) = \mathrm{Hom}_{\mathcal{FS}_{\leq\lambda}}(\sigma_\lambda^*\mathcal{X}, \mathcal{M})$$

and

$$\mathrm{Hom}_{\mathcal{FS}}(\mathcal{M}, \mathcal{X}) = \mathrm{Hom}_{\mathcal{FS}_{\leq\lambda}}(\mathcal{M}, \sigma_\lambda^!\mathcal{X})$$

Proof. Let $\mu = \lambda(\mathcal{X})$. We have

$$\mathrm{Hom}_{\mathcal{A}_\mu^\alpha}(\mathcal{X}_\mu^\alpha, \mathcal{M}_\mu^\alpha) = \mathrm{Hom}_{\mathcal{A}_\mu^{\alpha-\mu+\lambda}}(\sigma_\lambda^*\mathcal{X}_\lambda^{\alpha-\mu+\lambda}, \mathcal{M}_\lambda^{\alpha-\mu+\lambda})$$

by the usual adjointness. Passing to projective limit in α, we get the desired result for σ^*. The proof for $\sigma^!$ is similar. □

14.4. Given $\lambda \leq \mu \in X_c$, we denote by $\sigma_{\lambda\leq\mu}$ the embedding of the full subcategory

$$\sigma_{\lambda\leq\mu} : \mathcal{FS}_{\leq\lambda} \hookrightarrow \mathcal{FS}_{\leq\mu}.$$

Obviously, the functor

$$\sigma^*_{\lambda\leq\mu} := \sigma^*_\lambda \circ \sigma_\mu : \mathcal{FS}_{\leq\mu} \longrightarrow \mathcal{FS}_{\leq\lambda}$$

is left adjoint to $\sigma_{\lambda\leq\mu}$. Similarly, $\sigma^!_{\lambda\leq\mu} := \sigma^!_\lambda \circ \sigma_\mu$ is the right adjoint to $\sigma_{\lambda\leq\mu}$.

For $\lambda \leq \mu \leq \nu$ we have obvious transitivities

$$\sigma^*_{\lambda\leq\mu}\sigma^*_{\mu\leq\nu} = \sigma^*_{\lambda\leq\nu};\ \sigma^!_{\lambda\leq\mu}\sigma^!_{\mu\leq\nu} = \sigma^!_{\lambda\leq\nu}.$$

14.5. For each $\alpha \in \mathbb{N}[I]$ and $i \in I$ such that $\alpha \geq i$ let $j^\alpha_{\nu-i'\leq\nu}$ denote the open embedding

$$j^\alpha_{\nu-i'\leq\nu} : \mathcal{A}^\alpha_\nu - \sigma(\mathcal{A}^{\alpha-i}_{\nu-i'}) \hookrightarrow \mathcal{A}^\alpha_\nu.$$

Note that the complement of this open subspace is a divisor, so the corresponding extension by zero and by $*$ functors are t-exact, cf. [BBD], 4.1.10 (i). Let us define functors

$$j_{\nu-i'\leq\nu!}, j_{\nu-i'\leq\nu*} : \mathcal{FS}_{\leq\nu} \longrightarrow \mathcal{FS}_{\leq\nu}$$

as follows. For a factorizable sheaf $\mathcal{X} = \{\mathcal{X}^\alpha_\nu\} \in \mathcal{FS}_{\leq\nu}$ we set

$$(j_{\nu-i'\leq\nu!}\mathcal{X})^\alpha_\nu = j^\alpha_{\nu-i'\leq\nu!}j^{\alpha*}_{\nu-i'\leq\nu}\mathcal{X}^\alpha_\nu$$

and

$$(j_{\nu-i'\leq\nu*}\mathcal{X})^\alpha_\nu = j^\alpha_{\nu-i'\leq\nu*}j^{\alpha*}_{\nu-i'\leq\nu}\mathcal{X}^\alpha_\nu,$$

the factorization isomorphisms being induced from those for $\mathcal{X}$.

14.6. **Lemma.** *We have natural in $\mathcal{X} \in \mathcal{FS}_{\leq\nu}$ exact sequences*

$$j_{\nu-i',\nu!}\mathcal{X} \longrightarrow \mathcal{X} \xrightarrow{a} \sigma^*_{\nu-i',\nu}\mathcal{X} \longrightarrow 0$$

and

$$0 \longrightarrow \sigma^!_{\nu-i',\nu}\mathcal{X} \xrightarrow{a'} \mathcal{X} \longrightarrow j_{\nu-i',\nu*}\mathcal{X}$$

where the maps a and a' are the adjunction morphisms.

Proof. Evidently follows from the same claim at each finite level, which is [BBD], 4.1.10 (ii). □

14.7. Recall (see 13.3) that we have the Verdier duality functor

$$D : \mathcal{FS}^{opp} \longrightarrow \mathcal{FS}_{\zeta^{-1}}.$$

By definition, $D(\mathcal{FS}^{opp}_{\leq\lambda}) \subset \mathcal{FS}_{\zeta^{-1};\leq\lambda}$ for all λ.

We have functorial isomorphisms

$$D \circ \sigma^*_\lambda \cong \sigma^!_\lambda \circ D;\ D \circ \sigma^*_{\nu-i',\nu} \cong \sigma^!_{\nu-i',\nu} \circ D$$

and

$$D \circ j_{\nu-i',\nu*} \cong j_{\nu-i',\nu!} \circ D$$

After applying D, one of the exact sequences in 14.6 becomes another one.

14.8. Let us turn to the category $\mathcal{C}$. Below we will identify $\mathcal{C}$ with $\tilde{\mathcal{C}}$ using the equivalence Q, cf. II.12.5. In other words, we will regard objects of $\mathcal{C}$ as $\mathfrak{u}$-modules.

For $\lambda \in X_c$ functors $q^!_\lambda$ and q^*_λ are defined as follows. For $V \in \mathcal{C}$, $q^!_\lambda V$ (resp., $q^*_\lambda V$) is the maximal subobject (resp., quotient) of V belonging to the subcategory $\mathcal{C}_\lambda$.

For $\lambda \leq \mu \in X_c$ let $q_{\lambda\leq\mu}$ denotes an embedding of a full subcategory

$$q_{\lambda\leq\mu} : \mathcal{C}_{\leq\lambda} \hookrightarrow \mathcal{C}_{\leq\mu}$$

Define $q^!_{\lambda\leq\mu} := q^!_\lambda \circ q_\mu$; $q^*_{\lambda\leq\mu} := q^!_\lambda \circ q_\mu$. Obviously, the first functor is right adjoint, and the second one is left adjoint to $q_{\lambda\leq\mu}$. They have an obvious transitivity property.

14.9. Recall that in 13.2 the weight preserving duality equivalence

$$D : \mathcal{C}^{opp} \longrightarrow \mathcal{C}_{\zeta^{-1}}$$

is defined. By definition, $D(\mathcal{C}^{opp}_{\leq\lambda}) \subset \mathcal{C}_{\zeta^{-1};\leq\lambda}$ for all λ.

We have functorial isomorphisms

$$D \circ q^*_\lambda \cong q^!_\lambda \circ D;\ D \circ q^*_{\nu-i',\nu} \cong q^!_{\nu-i',\nu} \circ D.$$

14.10. Given $i \in I$, let us introduce a "Levi" subalgebra $\mathfrak{l}_i \subset \mathfrak{u}$ generated by $\theta_j,\ \epsilon_j,\ j \neq i$, and $K_i^{\pm}$. Let $\mathfrak{p}_i \subset \mathfrak{u}$ denote the "parabolic" subalgebra generated by $\mathfrak{l}_i$ and ϵ_i.

The subalgebra $\mathfrak{l}_i$ projects isomorphically to $\mathfrak{p}_i/(\epsilon_i)$ where (ϵ_i) is a two-sided ideal generated by ϵ_i. Given an $\mathfrak{l}_i$-module V, we can consider it as a $\mathfrak{p}_i$-module by restriction of scalars for the projection $\mathfrak{p}_i \longrightarrow \mathfrak{p}_i/(\epsilon_i) \cong \mathfrak{l}_i$, and form the induced $\mathfrak{u}$-module $\mathrm{Ind}^{\mathfrak{u}}_{\mathfrak{p}_i} V$ — "generalized Verma".

14.11. Given an $\mathfrak{u}$-module $V \in \mathcal{C}_{\leq\nu}$, let us consider a subspace

$${}_iV = \oplus_{\alpha\in\mathbb{N}[I-\{i\}]} V_{\nu-\alpha'} \subset V.$$

It is an X-graded $\mathfrak{p}_i$-submodule of V. Consequently, we have a canonical element

$$\pi \in \mathrm{Hom}_{\mathcal{C}}(\mathrm{Ind}^{\mathfrak{u}}_{\pi_i}\ {}_iV, V) = \mathrm{Hom}_{\mathfrak{p}_i}({}_iV, V)$$

corresponding to the embedding ${}_iV \hookrightarrow V$.

We will also consider the dual functor

$$V \mapsto D^{-1}\mathrm{Ind}^{\mathfrak{u}}_{\mathfrak{p}_i}\ {}_i(DV).$$

By duality, we have a natural morphism in $\mathcal{C}$, $V \longrightarrow D^{-1}\mathrm{Ind}^{\mathfrak{u}}_{\mathfrak{p}_i}\ {}_i(DV)$.

14.12. **Lemma.** *We have natural in $V \in \mathcal{C}_{\leq\nu}$ exact sequences*

$$\mathrm{Ind}^{\mathfrak{u}}_{\mathfrak{p}_i}\ {}_iV \xrightarrow{\pi} V \longrightarrow q^*_{\nu-i'\leq\nu}V \longrightarrow 0$$

and

$$0 \longrightarrow q^!_{\nu-i'\leq\nu}V \longrightarrow V \longrightarrow D^{-1}\mathrm{Ind}^{\mathfrak{u}}_{\mathfrak{p}_i}\ {}_i(DV).$$

where the arrows $V \longrightarrow q^*_{\nu-i'\leq\nu}V$ and $q^!_{\nu-i'\leq\nu}V \longrightarrow V$ are adjunction morphisms.

Proof. Let us show the exactness of the first sequence. By definition, $q^*_{\nu-i'\leq\nu}V$ is the maximal quotient of V lying in the subcategory $\mathcal{C}_{\nu-i'\leq\nu} \subset \mathcal{C}$. Obviously, $\mathrm{Coker}\,\pi \in \mathcal{C}_{\nu-i',\nu}$. It remains to show that for any morphism $h : V \longrightarrow W$ with $W \in \mathcal{C}_{\nu-i'}$, the composition $h\circ\pi : \mathrm{Ind}^{\mathfrak{u}}_{\mathfrak{p}_i}\ {}_iV \longrightarrow W$ is zero. But $\mathrm{Hom}_{\mathcal{C}}(\mathrm{Ind}^{\mathfrak{u}}_{\mathfrak{p}_i}\ {}_iV, W) = \mathrm{Hom}_{\mathfrak{p}_i}({}_iV, W) = 0$ by weight reasons.

The second exact sequence is the dual to the first one. □

14.13. **Lemma.** *We have natural in $\mathcal{X} \in \mathcal{FS}_\nu$ isomorphisms*

$$\Phi j_{\nu-i'\leq\nu !}\mathcal{X} \xrightarrow{\sim} \mathrm{Ind}^{\mathfrak{u}}_{\mathfrak{p}_i\ i}(\Phi\mathcal{X})$$

and

$$\Phi j_{\nu-i'\leq\nu *}\mathcal{X} \xrightarrow{\sim} D^{-1}\mathrm{Ind}^{\mathfrak{u}}_{\mathfrak{p}_i\ i}(D\Phi\mathcal{X})$$

such that the diagram

$$\begin{array}{ccccc}
\Phi j_{\nu-i',\nu !}\mathcal{X} & \longrightarrow & \Phi\mathcal{X} & \longrightarrow & \Phi j_{\nu-i',\nu *}\mathcal{X} \\
\downarrow & & \Vert & & \downarrow \\
\mathrm{Ind}^{\mathfrak{u}}_{\mathfrak{p}_i\ i}(\Phi\mathcal{X}) & \longrightarrow & \Phi\mathcal{X} & \longrightarrow & D^{-1}\mathrm{Ind}^{\mathfrak{u}}_{\mathfrak{p}_i\ i}(D\Phi\mathcal{X})
\end{array}$$

commutes.

14.14. **Lemma.** *Let $\lambda \in X_c$. We have natural in $\mathcal{X} \in \mathcal{FS}$ isomorphisms*

$$\Phi\sigma^*_\lambda\mathcal{X} \cong q^*_\lambda\Phi\mathcal{X}$$

and

$$\Phi\sigma^!_\lambda\mathcal{X} \cong q^!_\lambda\Phi\mathcal{X}$$

Proof follows at once from lemmas 14.13, 14.6 and 14.12. □

15. Rigidity

15.1. **Lemma.** *Let $\mathcal{X} \in \mathcal{FS}_{\leq 0}$. Then the natural map*

$$a : \mathrm{Hom}_{\mathcal{FS}}(\mathcal{L}(0), \mathcal{X}) \longrightarrow \mathrm{Hom}_{\mathcal{C}}(L(0), \Phi(\mathcal{X}))$$

is an isomorphism.

Proof. We know already that a is injective, so we have to prove its surjectivity. Let $\mathcal{K}(0)$ (resp., $K(0)$) denote the kernel of the projection $\mathcal{M}(0) \longrightarrow \mathcal{L}(0)$ (resp., $M(0) \longrightarrow L(0)$). Consider a diagram with exact rows:

$$\begin{array}{ccccccc} 0 & \longrightarrow & \mathrm{Hom}(\mathcal{L}(0),\mathcal{X}) & \longrightarrow & \mathrm{Hom}(\mathcal{M}(0),\mathcal{X}) & \longrightarrow & \mathrm{Hom}(\mathcal{K}(0),\mathcal{X}) \\ & & \downarrow a & & \downarrow b & & \downarrow \\ 0 & \longrightarrow & \mathrm{Hom}(L(0),\Phi(\mathcal{X})) & \longrightarrow & \mathrm{Hom}(M(0),\Phi(\mathcal{X})) & \longrightarrow & \mathrm{Hom}(K(0),\Phi(\mathcal{X})) \end{array}$$

All vertical rows are injective. On the other hand, $\mathrm{Hom}(M(0),\Phi(\mathcal{X})) = \Phi(\mathcal{X})_0$. The last space is isomorphic to a generic stalk of $\mathcal{X}_0^\alpha$ for each $\alpha \in \mathbb{N}[I]$, which in turn is isomorphic to $\mathrm{Hom}_{\mathcal{FS}}(\mathcal{M}(0),\mathcal{X})$ by the universal property of the shriek extension. Consequently, b is isomorphism by the equality of dimensions. By diagram chase, we conclude that a is isomorphism. □

15.2. **Lemma.** *For every $\mathcal{X} \in \mathcal{FS}$ the natural maps*

$$\mathrm{Hom}_{\mathcal{FS}}(\mathcal{L}(0),\mathcal{X}) \longrightarrow \mathrm{Hom}_{\mathcal{C}}(L(0),\Phi(\mathcal{X}))$$

and

$$\mathrm{Hom}_{\mathcal{FS}}(\mathcal{X},\mathcal{L}(0)) \longrightarrow \mathrm{Hom}_{\mathcal{C}}(\Phi(\mathcal{X}),L(0))$$

are isomorphisms.

Proof. We have

$$\mathrm{Hom}_{\mathcal{FS}}(\mathcal{L}(0),\mathcal{X}) = \mathrm{Hom}_{\mathcal{FS}}(\mathcal{L}(0),\sigma_0^!\mathcal{X})$$

(by lemma 14.3)

$$= \mathrm{Hom}_{\mathcal{C}}(L(0),\Phi(\sigma_0^!\mathcal{X}))$$

(by the previous lemma)

$$= \mathrm{Hom}_{\mathcal{C}}(L(0),q_0^!\Phi(\mathcal{X}))$$

(by lemma 14.14)

$$= \mathrm{Hom}_{\mathcal{C}}(L(0),\Phi(\mathcal{X})).$$

This proves the first isomorphism. The second one follows by duality. □

15.3. Recall (see [KL]IV, Def. A.5) that an object X of a tensor category is called *rigid* if there exists another object X^* together with morphisms

$$i_X : \mathbf{1} \longrightarrow X \otimes X^*$$

and

$$e_X : X^* \otimes X \longrightarrow \mathbf{1}$$

such that the compositions

$$X = \mathbf{1} \otimes X \overset{i_X\otimes\mathrm{id}}{\longrightarrow} X \otimes X^* \otimes X \overset{\mathrm{id}\otimes e_X}{\longrightarrow} X$$

and

$$X^* = X^* \otimes \mathbf{1} \overset{\mathrm{id}\otimes i_X}{\longrightarrow} X^* \otimes X \otimes X^* \overset{e_X\otimes\mathrm{id}}{\longrightarrow} X^*$$

are equal to id_X and id_{X^*} respectively. A tensor category is called rigid if all its objects are rigid.

15.4. **Theorem.** *All irreducible objects $\mathcal{L}(\lambda)$, $\lambda \in X$, are rigid in $\mathcal{FS}$.*

Proof. We know (cf. [AJS], 7.3) that $\mathcal{C}$ is rigid. Moreover, there exists an involution $\lambda \mapsto \bar{\lambda}$ on X such that $L(\lambda)^* = L(\bar{\lambda})$. Let us define $\mathcal{L}(\lambda)^* := \mathcal{L}(\bar{\lambda})$; $i_{\mathcal{L}(\lambda)}$ corresponds to $i_{L(\lambda)}$ under identification

$$\mathrm{Hom}_{\mathcal{FS}}(\mathcal{L}(0), \mathcal{L}(\lambda) \otimes \mathcal{L}(\bar{\lambda})) = \mathrm{Hom}_{\mathcal{C}}(L(0), L(\lambda) \otimes L(\bar{\lambda}))$$

and $e_{\mathcal{L}(\lambda)}$ corresponds to $e_{L(\lambda)}$ under identification

$$\mathrm{Hom}_{\mathcal{FS}}(\mathcal{L}(\bar{\lambda}) \otimes \mathcal{L}(\lambda), \mathcal{L}(0)) = \mathrm{Hom}_{\mathcal{C}}(L(\bar{\lambda}) \otimes L(\lambda), L(0)),$$

cf. 15.2. □

16. Steinberg sheaf

In this section we assume that l is a positive number prime to 2, 3 and that ζ' is a primitive $(l \cdot \det A)$-th root of 1 (recall that $\zeta = (\zeta')^{\det A}$).

We fix a weight $\lambda_0 \in X$ such that $\langle i, \lambda_0 \rangle = -1 (\mathrm{mod}\ l)$ for any $i \in I$. Our goal in this section is the proof of the following

16.1. **Theorem.** *The FFS $\mathcal{L}(\lambda_0)$ is a projective object of the category $\mathcal{FS}$.*

Proof. We have to check that $\mathrm{Ext}^1(\mathcal{L}(\lambda_0), \mathcal{X}) = 0$ for any FFS $\mathcal{X}$. By induction on the length of $\mathcal{X}$ it is enough to prove that $\mathrm{Ext}^1(\mathcal{L}(\lambda_0), \mathcal{L}) = 0$ for any simple FFS $\mathcal{L}$.

16.2. To prove vanishing of Ext^1 in $\mathcal{FS}$ we will use the following principle. Suppose $\mathrm{Ext}^1(\mathcal{X}, \mathcal{Y}) \neq 0$, and let

$$0 \longrightarrow \mathcal{Y} \longrightarrow \mathcal{Z} \longrightarrow \mathcal{X} \longrightarrow 0$$

be the corresponding nonsplit extension. Let us choose a weight λ which is bigger than of $\lambda(\mathcal{X}), \lambda(\mathcal{Y}), \lambda(\mathcal{Z})$. Then for any $\alpha \in \mathbb{N}[I]$ the sequence

$$0 \longrightarrow \mathcal{Y}^\alpha_\lambda \longrightarrow \mathcal{Z}^\alpha_\lambda \longrightarrow \mathcal{X}^\alpha_\lambda \longrightarrow 0$$

is also exact, and for $\alpha \gg 0$ it is also nonsplit (see lemma 5.3). That is, for $\alpha \gg 0$ we have $\mathrm{Ext}^1(\mathcal{X}^\alpha_\lambda, \mathcal{Y}^\alpha_\lambda) \neq 0$ *in the category of all perverse sheaves on the space $\mathcal{A}^\alpha_\lambda$*. This latter Ext can be calculated purely topologically. So its vanishing gives a criterion of Ext^1-vanishing in the category $\mathcal{FS}$.

16.3. In calculating $\mathrm{Ext}^1(\mathcal{L}(\lambda_0), \mathcal{L}(\mu))$ we will distinguish between the following three cases:

a) $\mu \not\geq \lambda_0$;

b) $\mu = \lambda_0$;

c) $\mu > \lambda_0$.

16.4. Let us treat the case a).

16.4.1. **Lemma.** *For any $\alpha \in \mathbb{N}[I]$ the sheaf $\mathcal{L}(\lambda_0)^\alpha_{\lambda_0}$ is the shriek-extension from the open stratum of toric stratification of $\mathcal{A}^\alpha_{\lambda_0}$.*

Proof. Due to the factorization property it is enough to check that the stalk of $\mathcal{L}(\lambda_0)^\alpha_{\lambda_0}$ at the point $\{0\} \in \mathcal{A}^\alpha_{\lambda_0}$ vanishes for any $\alpha \in \mathbb{N}[I], \alpha \neq 0$. By the Theorem II.8.23, we have $(\mathcal{L}(\lambda_0)^\alpha_{\lambda_0})_0 = {}_\alpha C^\bullet_{\mathfrak{f}}(L(\lambda_0)) \simeq 0$ since $L(\lambda_0)$ is a free $\mathfrak{f}$-module by [L1] 36.1.5 and Theorem II.11.10(b) and, consequently, $C^\bullet_{\mathfrak{f}}(L(\lambda_0)) \simeq H^0_{\mathfrak{f}}(L(\lambda_0)) = \mathsf{k}$ and has weight zero. □

Returning to the case a), let us choose $\nu \in X$, $\nu \geq \lambda_0, \nu \geq \mu$. For any α, the sheaf $\mathcal{L}' := \mathcal{L}(\mu)^\alpha_\nu$ is supported on the subspace

$$\mathcal{A}' := \sigma(\mathcal{A}^{\alpha+\mu-\nu}_\mu) \subset \mathcal{A}^\alpha_\nu$$

and $\mathcal{L}'' := \mathcal{L}(\lambda_0)^\alpha_\nu$ — on the subspace

$$\mathcal{A}'' := \sigma(\mathcal{A}^{\alpha+\lambda_0-\nu}_{\lambda_0}) \subset \mathcal{A}^\alpha_\nu.$$

Let i denote a closed embedding

$$i : \mathcal{A}'' \hookrightarrow \mathcal{A}^\alpha_\nu$$

and j an open embedding

$$j : \overset{\circ}{\mathcal{A}}{}'' := \mathcal{A}'' - \mathcal{A}'' \cap \mathcal{A}' \hookrightarrow \mathcal{A}''.$$

We have by adjointness

$$R\mathrm{Hom}_{\mathcal{A}^\alpha_\nu}(\mathcal{L}'', \mathcal{L}') = R\mathrm{Hom}_{\mathcal{A}''}(\mathcal{L}'', i^!\mathcal{L}') =$$

(by the previous lemma)

$$= R\mathrm{Hom}_{\overset{\circ}{\mathcal{A}}{}''}(j^*\mathcal{L}'', j^*i^!\mathcal{L}') = 0$$

since obviously $j^*i^!\mathcal{L}' = 0$. This proves the vanishing in the case a).

16.5. In case (b), suppose

$$0 \longrightarrow \mathcal{L}(\lambda_0) \longrightarrow \mathcal{X} \longrightarrow \mathcal{L}(\lambda_0) \longrightarrow 0$$

is a nonsplit extension in $\mathcal{FS}$. Then for $\alpha \gg 0$ the restriction of $\mathcal{X}^\alpha_{\lambda_0}$ to the open toric stratum of $\mathcal{A}^\alpha_{\lambda_0}$ is a nonsplit extension

$$0 \longrightarrow \dot{\mathcal{I}}^\alpha_{\lambda_0} \longrightarrow \dot{\mathcal{X}}^\alpha_{\lambda_0} \longrightarrow \dot{\mathcal{I}}^\alpha_{\lambda_0} \longrightarrow 0$$

(in the category of all perverse sheaves on $\dot{\mathcal{A}}^\alpha_{\lambda_0}$) (we can restrict to the open toric stratum because of lemma 16.4.1). This contradicts to the factorization property of FFS $\mathcal{X}$. This contradiction completes the consideration of case (b).

16.6. In case (c), suppose $\mathrm{Ext}^1(\mathcal{L}(\lambda_0), \mathcal{L}(\mu)) \neq 0$ whence $\mathrm{Ext}^1(\mathcal{L}(\lambda_0)^\alpha_\mu, \mathcal{L}(\mu)^\alpha_\mu) \neq 0$ for some $\alpha \in \mathbb{N}[I]$ by the principle 16.2. Here the latter Ext is taken in the category of all perverse sheaves on $\mathcal{A}^\alpha_\mu$. We have $\mu - \lambda_0 = \beta'$ for some $\beta \in \mathbb{N}[I], \beta \neq 0$.

Let us consider the closed embedding

$$\sigma : \mathcal{A}' := \mathcal{A}^{\alpha-\beta}_{\lambda_0} \hookrightarrow \mathcal{A}^\alpha_\mu;$$

let us denote by j an embedding of the open toric stratum

$$j : \dot{\mathcal{A}}' := \dot{\mathcal{A}}^{\alpha-\beta}_{\lambda_0} \hookrightarrow \mathcal{A}'.$$

As in the previous case, we have

$$R\mathrm{Hom}_{\mathcal{A}^\alpha_\mu}(\mathcal{L}(\lambda_0)^\alpha_\mu, \mathcal{L}(\mu)^\alpha_\mu) = R\mathrm{Hom}_{\mathcal{A}'}(\mathcal{L}(\lambda_0)^\alpha_\mu, \sigma^! \mathcal{L}(\mu)^\alpha_\mu) = R\mathrm{Hom}_{\overset{\bullet}{\mathcal{A}'}}(j^*\mathcal{L}(\lambda_0)^\alpha_\mu, j^*\sigma^!\mathcal{L}(\mu)^\alpha_\mu).$$

We claim that $j^*\sigma^!\mathcal{L}(\mu)^\alpha_\mu = 0$. Since the sheaf $\mathcal{L}(\mu)^\alpha_\mu$ is Verdier auto-dual up to replacing ζ by ζ^{-1}, it suffices to check that $j^*\sigma^*\mathcal{L}(\mu)^\alpha_\mu = 0$.

To prove this vanishing, by factorization property of $\mathcal{L}(\mu)$, it is enough to check that the stalk of the sheaf $\mathcal{L}(\mu)^\beta_\mu$ at the point $\{0\} \in \mathcal{A}^\beta_\mu$ vanishes.

By the Theorem II.8.23, we have $(\mathcal{L}(\mu)^\beta_\mu)_0 = {}_\beta C^\bullet_{\mathfrak{f}}(L(\mu))$. By the Theorem II.11.10 and Shapiro Lemma, we have ${}_\beta C^\bullet_{\mathfrak{f}}(L(\mu)) \simeq C^\bullet_{\dot{U}}(M(\lambda_0) \otimes L(\mu))$.

By the Theorem 36.1.5. of [L1], the canonical projection $M(\lambda_0) \longrightarrow L(\lambda_0)$ is an isomorphism. By the autoduality of $L(\lambda_0)$ we have $C^\bullet_{\dot{U}}(L(\lambda_0) \otimes L(\mu)) \simeq R\mathrm{Hom}^\bullet_{\dot{U}}(L(\lambda_0), L(\mu)) \simeq 0$ since $L(\lambda_0)$ is a projective $\dot{U}$-module, and $\mu \neq \lambda_0$.

This completes the case c) and the proof of the theorem. □

17. Equivalence

We keep the assumptions of the previous section.

17.1. Theorem. *Functor $\Phi : \mathcal{FS} \longrightarrow \mathcal{C}$ is an equivalence.*

17.2. Lemma. *For any $\lambda \in X$ the FFS $\mathcal{L}(\lambda_0)\overset{\bullet}{\otimes}\mathcal{L}(\lambda)$ is projective.*

As λ ranges through X, these sheaves form an ample system of projectives in $\mathcal{FS}$.

Proof. We have

$$\mathrm{Hom}(\mathcal{L}(\lambda_0)\overset{\bullet}{\otimes}\mathcal{L}(\lambda), ?) = \mathrm{Hom}(\mathcal{L}(\lambda_0), \mathcal{L}(\lambda)^* \overset{\bullet}{\otimes} ?)$$

by the rigidity, and the last functor is exact since $\overset{\bullet}{\otimes}$ is a biexact functor in $\mathcal{FS}$, and $\mathcal{L}(\lambda_0)$ is projective. Therefore, $\mathcal{L}(\lambda_0)\overset{\bullet}{\otimes}\mathcal{L}(\lambda)$ is projective.

To prove that these sheaves form an ample system of projectives, it is enough to show that for each $\mu \in X$ there exists λ such that $\mathrm{Hom}(\mathcal{L}(\lambda_0)\overset{\bullet}{\otimes}\mathcal{L}(\lambda), \mathcal{L}(\mu)) \neq 0$. We have

$$\mathrm{Hom}(\mathcal{L}(\lambda_0)\overset{\bullet}{\otimes}\mathcal{L}(\lambda), \mathcal{L}(\mu)) = \mathrm{Hom}(\mathcal{L}(\lambda), \mathcal{L}(\lambda_0)^*\overset{\bullet}{\otimes}\mathcal{L}(\mu)).$$

Since the sheaves $\mathcal{L}(\lambda)$ exhaust irreducibles in $\mathcal{FS}$, there exists λ such that $\mathcal{L}(\lambda)$ embeds into $\mathcal{L}(\lambda_0)^*\overset{\bullet}{\otimes}\mathcal{L}(\mu)$, hence the last group is non-zero. □

17.3. Proof of 17.1. As λ ranges through X, the modules $\Phi(\mathcal{L}(\lambda_0)\overset{\bullet}{\otimes}\mathcal{L}(\lambda)) = L(\lambda_0) \otimes L(\lambda)$ form an ample system of projectives in $\mathcal{C}$. By the Lemma A.15 of [KL]IV we only have to show that

$$\Phi : \mathrm{Hom}_{\mathcal{FS}}(\mathcal{L}(\lambda_0)\overset{\bullet}{\otimes}\mathcal{L}(\lambda), \mathcal{L}(\lambda_0)\overset{\bullet}{\otimes}\mathcal{L}(\mu)) \longrightarrow \mathrm{Hom}_{\mathcal{C}}(L(\lambda_0) \otimes L(\lambda), L(\lambda_0) \otimes L(\mu))$$

is an isomorphism for any $\lambda, \mu \in X$. We already know that it is an injection. Therefore, it remains to compare the dimensions of the spaces in question. We have

$$\dim \mathrm{Hom}_{\mathcal{FS}}(\mathcal{L}(\lambda_0)\overset{\bullet}{\otimes}\mathcal{L}(\lambda), \mathcal{L}(\lambda_0)\overset{\bullet}{\otimes}\mathcal{L}(\mu)) = \dim \mathrm{Hom}_{\mathcal{FS}}(\mathcal{L}(\lambda_0), \mathcal{L}(\lambda_0)\overset{\bullet}{\otimes}\mathcal{L}(\mu)\overset{\bullet}{\otimes}\mathcal{L}(\lambda)^*)$$

by rigidity,

$$= [\mathcal{L}(\lambda_0)\dot{\otimes}\mathcal{L}(\mu)\dot{\otimes}\mathcal{L}(\lambda)^* : \mathcal{L}(\lambda_0)]$$

because $\mathcal{L}(\lambda_0)$ is its own indecomposable projective cover in $\mathcal{FS}$,

$$= [L(\lambda_0) \otimes L(\mu) \otimes L(\lambda)^* : L(\lambda_0)]$$

since Φ induces an isomorphism of K-rings of the categories $\mathcal{FS}$ and $\mathcal{C}$,

$$= \dim \mathrm{Hom}_{\mathcal{C}}(L(\lambda_0) \otimes L(\lambda), L(\lambda_0) \otimes L(\mu))$$

by the same argument applied to $\mathcal{C}$. The theorem is proven. □

18. The case of generic ζ

In this section we suppose that ζ is not a root of unity.

18.1. Recall the notations of II.11,12. We have the algebra U defined in II.12.2, the algebra $\mathfrak{u}$ defined in II.12.3, and the homomorphism $R: U \longrightarrow \mathfrak{u}$ defined in II.12.5.

18.1.1. **Lemma.** *The map $R: U \longrightarrow \mathfrak{u}$ is an isomorphism.*

Proof follows from [R], no. 3, Corollaire. □

18.2. We call $\Lambda \in X$ dominant if $\langle i, \Lambda\rangle \geq 0$ for any $i \in I$. An irreducible U-module $L(\Lambda)$ is finite dimensional if only if Λ is dominant. Therefore we will need a larger category $\mathcal{O}$ containing all irreducibles $L(\Lambda)$.

Define $\mathcal{O}$ as a category consisting of all X-graded U-modules $V = \oplus_{\mu\in X} V_\mu$ such that

a) V_μ is finite dimensional for any $\mu \in X$;

b) there exists $\lambda = \lambda(V)$ such that $V_\mu = 0$ if $\mu \not\geq \lambda(V)$.

18.2.1. **Lemma** *The category $\mathcal{O}$ is equivalent to the usual category $\mathcal{O}_{\mathfrak{g}}$ over the classical finite dimensional Lie algebra $\mathfrak{g}$.*

Proof. See [F1]. □

18.3. Let W denote the Weyl group of our root datum. For $w \in W$, $\lambda \in X$ let $w \cdot \lambda$ denote the usual action of W on X centered at $-\rho$.

Finally, for $\Lambda \in X$ let $M(\Lambda) \in \mathcal{O}$ denote the U^--free Verma module with highest weight Λ.

18.3.1. **Corollary.** *Let $\mu, \nu \in X$ be such that $W \cdot \mu \neq W \cdot \nu$. Then* $\mathrm{Ext}^\bullet(M(\nu), L(\mu)) = 0$.

Proof. $M(\nu)$ and $L(\mu)$ have different central characters. □

18.4. **Theorem.** *Functor $\Phi : \mathcal{FS} \longrightarrow \mathcal{C}$ is an equivalence.*

Proof. We know that $\Phi(\mathcal{L}(\Lambda)) \simeq L(\Lambda)$ for any $\Lambda \in X$. So $\Phi(\mathcal{X})$ is finite dimensional iff all the irreducible subquotients of $\mathcal{X}$ are of the form $\mathcal{L}(\lambda)$, λ dominant. By virtue of Lemma 18.2.1 above the category $\mathcal{C}$ is semisimple: it is equivalent to the category of finite dimensional $\mathfrak{g}$-modules. It consists of finite direct sums of modules $L(\lambda)$, λ dominant. So to prove the Theorem it suffices to check semisimplicity of $\mathcal{FS}$. Thus the Theorem follows from

18.5. **Lemma.** *Let $\mu, \nu \in X$ be the dominant weights. Then* $\mathrm{Ext}^1(\mathcal{L}(\mu), \mathcal{L}(\nu)) = 0$.

Proof. We will distinguish between the following two cases:

(a) $\mu = \nu$;

(b) $\mu \neq \nu$.

In calculating Ext^1 we will use the principle 16.2. The argument in case (a) is absolutely similar to the one in section 16.5, and we leave it to the reader.

In case (b) either $\mu - \nu \notin Y \subset X$ — and then the sheaves $\mathcal{L}(\nu)$ and $\mathcal{L}(\mu)$ are supported on the different connected components of $\mathcal{A}$, whence Ext^1 obviously vanishes, — or there exists $\lambda \in X$ such that $\lambda \geq \mu, \nu$. Let us fix such λ. Suppose $\mathrm{Ext}^1(\mathcal{L}(\mu), \mathcal{L}(\nu)) \neq 0$. Then according to the principle 16.2 there exists $\alpha \in \mathbb{N}[I]$ such that $\mathrm{Ext}^1(\mathcal{L}(\mu)^\alpha_\lambda, \mathcal{L}(\nu)^\alpha_\lambda) \neq 0$. The latter Ext is taken in the category of all perverse sheaves on $\mathcal{A}^\alpha_\lambda$.

We have $\mathrm{Ext}^1(\mathcal{L}(\mu)^\alpha_\lambda, \mathcal{L}(\nu)^\alpha_\lambda) = R^1\Gamma(\mathcal{A}^\alpha_\lambda, D(\mathcal{L}(\mu)^\alpha_\lambda \otimes D\mathcal{L}(\nu)^\alpha_\lambda))$ where D stands for Verdier duality, and $\otimes$ denotes the usual tensor product of constructible comlexes.

We will prove that

$$\mathcal{L}(\mu)^\alpha_\lambda \otimes D\mathcal{L}(\nu)^\alpha_\lambda = 0 \tag{284}$$

and hence we will arrive at the contradiction. Equality (284) is an immediate corollary of the lemma we presently formulate.

For $\beta \leq \alpha$ let us consider the canonical embedding

$$\sigma : \overset{\bullet}{\mathcal{A}}{}^{\alpha-\beta} \hookrightarrow \mathcal{A}^\alpha$$

and denote its image by $\overset{\bullet}{\mathcal{A}}{}'$ (we omit the lower case indices).

18.5.1. **Lemma.** *(i) If $\sigma^*\mathcal{L}(\mu)^\alpha_\lambda \neq 0$ then $\lambda - \beta \in W \cdot \mu$.*

(ii) If $\sigma^!\mathcal{L}(\mu)^\alpha_\lambda \neq 0$ then $\lambda - \beta \in W \cdot \mu$.

To deduce Lemma 18.5 from this lemma we notice first that the sheaf $\mathcal{L}(\mu)^\alpha_\lambda$ is Verdier autodual up to replacing ζ by ζ^{-1}. Second, since the W-orbits of ν and μ are disjoint, we see that over any toric stratum $\overset{\bullet}{\mathcal{A}}{}' \subset \mathcal{A}^\alpha$ at least one of the factors of (284) vanishes.

It remains to prove Lemma 18.5.1. We will prove (i), while (ii) is just dual. Let us denote $\beta + \mu - \lambda$ by γ. If $\gamma \notin \mathbb{N}[I]$ then (i) is evident. Otherwise, by the factorizability condition it is enough to check that the stalk of $\mathcal{L}(\mu)^\gamma_\mu$ at the origin in $\mathcal{A}^\gamma_\mu$ vanishes if $\mu - \gamma \notin W \cdot \mu$. Let us denote $\mu - \gamma$ by χ.

By the Theorem II.8.23, we have $(\mathcal{L}(\mu)^{\gamma}_{\mu})_0 = {}_{\gamma}C^{\bullet}_{U^-}(L(\mu)) \simeq C^{\bullet}_{U}(M(\chi) \otimes L(\mu))$ which is dual to $\mathrm{Ext}^{\bullet}_{U}(M(\chi), L(\mu))$. But the latter Ext vanishes by the Corollary 18.3.1 since $W \cdot \chi \neq W \cdot \mu$. □

This completes the proof of Lemma 18.5 together with Theorem 18.4.

Part IV. LOCALIZATION ON $\mathbb{P}^1$

1. Introduction

1.1. Given a collection of m finite factorizable sheaves $\{\mathcal{X}_k\}$, we construct here some perverse sheaves over configuration spaces of points on a projective line $\mathbb{P}^1$ with m additional marked points.

We announce here (with sketch proof) the computation of the cohomology spaces of these sheaves. They turn out to coincide with certain "semiinfinite" Tor spaces of the corresponding $\mathfrak{u}$-modules introduced by S.Arkhipov. For a precise formulation see Theorem 8.11.

This result is strikingly similar to the following hoped-for picture of affine Lie algebra representation theory, explained to us by A.Beilinson. Let M_1, M_2 be two modules over an affine Lie algebra $\hat{\mathfrak{g}}$ on the critical level. One hopes that there is a localization functor which associates to these modules perverse sheaves $\Delta(M_1), \Delta(M_2)$ over the semiinfinite flag space $\hat{G}/\hat{B}^0$. Suppose that $\Delta(M_1)$ and $\Delta(M_2)$ are equivariant with respect to the opposite Borel subgroups of $\hat{G}$. Then the intersection S of their supports is finite dimensional, and one hopes that

$$R^\bullet\Gamma(S, \Delta(M_1) \otimes \Delta(M_2)) = \mathrm{Tor}^{\hat{\mathfrak{g}}}_{\frac{\infty}{2}-\bullet}(M_1, M_2)$$

where in the right hand side we have the Feigin (Lie algebra) semiinfinite homology.

As a corollary of Theorem 8.11 we get a description of local systems of conformal blocks in WZW models in genus zero (cf. [MS]) as natural subquotients of some semisimple local systems of geometric origin. In particular, these local systems are semisimple themselves.

1.2. We are grateful to G.Lusztig for the permission to use his unpublished results. Namely, Theorem 6.2.1 about braiding in the category $\mathcal{C}$ is due to G.Lusztig. Chapter 2 (semiinfinite homological algebra in $\mathcal{C}$) is an exposition of the results due to S.Arkhipov (see [Ark]).

We are also grateful to A.Kirillov, Jr. who explained to us how to handle the conformal blocks of non simply laced Lie algebras.

1.3. Unless specified otherwise, we will keep the notations of parts I,II,III. For $\alpha = \sum_i a_i i \in \mathbb{N}[I]$ we will use the notation $|\alpha| := \sum_i a_i$.

References to *loc.cit.* will look like Z.1.1 where Z=I, II or III.

We will keep assumptions of III.1.4 and III.16. In particular, a "quantization" parameter ζ will be a primitive l-th root of unity where l is a fixed positive number prime to $2, 3$.

Chapter 1. Gluing over $\mathbb{P}^1$

2. Cohesive local system

2.1. Notations. Let $\alpha \in \mathbb{N}[X]$, $\alpha = \sum a_\mu \mu$. We denote by suppα the subset of X consisting of all μ such that $a_\mu \neq 0$. Let $\pi: J \longrightarrow X$ be an *unfolding* of α, that is a map of sets such that $|\pi^{-1}(\mu)| = a_\mu$ for any $\mu \in X$. As always, Σ_π denotes the group of automorphisms of J preserving the fibers of π.

$\mathbb{P}^1$ will denote a complex projective line. The J-th cartesian power $\mathbb{P}^{1J}$ will be denoted by $\mathcal{P}^J$. The group Σ_π acts naturally on $\mathcal{P}^J$, and the quotient space $\mathcal{P}^J/\Sigma_\pi$ will be denoted by $\mathcal{P}^\alpha$.

$\mathcal{P}^{oJ}$ (resp., $\mathcal{P}^{o\alpha}$) stands for the complement to diagonals in $\mathcal{P}^J$ (resp., in $\mathcal{P}^\alpha$).

$T\mathcal{P}^J$ stands for the total space of the tangent bundle to $\mathcal{P}^J$; its points are couples $((x_j),(\tau_j))$ where $(x_j) \in \mathcal{P}^J$ and τ_j is a tangent vector at x_j. An open subspace

$$\dot{T}\mathcal{P}^J \subset T\mathcal{P}^J$$

consists of couples with $\tau_j \neq 0$ for all j. So, $\dot{T}\mathcal{P}^J \longrightarrow \mathcal{P}^J$ is a $(\mathbb{C}^*)^J$-torsor. We denote by $T\mathcal{P}^{oJ}$ its restriction to $\mathcal{P}^{oJ}$. The group Σ_π acts freely on $T\mathcal{P}^{oJ}$, and we denote the quotient $T\mathcal{P}^{oJ}/\Sigma_\pi$ by $T\mathcal{P}^{o\alpha}$.

The natural projection

$$T\mathcal{P}^{oJ} \longrightarrow T\mathcal{P}^{o\alpha}$$

will be denoted by π, or sometimes by π_J.

2.2. Let $\mathbb{P}^1_{st}$ (st for "standard") denote "the" projective line with fixed coordinate z; $D_\epsilon \subset \mathbb{P}^1_{st}$ — the open disk of radius ϵ centered at $z = 0$; $D := D_1$. We will also use the notation $D_{(\epsilon,1)}$ for the open annulus $D - \overline{D}_\epsilon$ (bar means the closure).

The definitions of $D^J, D^\alpha, D^{oJ}, D^{o\alpha}, \dot{T}D^J, TD^{oJ}, TD^{o\alpha}$, etc., copy the above definitions, with D replacing $\mathbb{P}^1$.

2.3. Given a finite set K, let $\tilde{\mathcal{P}}^K$ denote the space of K-tuples $(u_k)_{k\in K}$ of algebraic isomorphisms $\mathbb{P}^1_{st} \xrightarrow{\sim} \mathbb{P}^1$ such that the images $u_k(D)$ do not intersect.

Given a K-tuple $\vec{\alpha} = (\alpha_k) \in \mathbb{N}[X]^K$ such that $\alpha = \sum_k \alpha_k$, define a space

$$\mathcal{P}^{\vec{\alpha}} := \tilde{\mathcal{P}}^K \times \prod_{k\in K} \dot{T}D^{\alpha_k}$$

and an open subspace

$$\mathcal{P}^{o\vec{\alpha}} := \tilde{\mathcal{P}}^K \times \prod_{k\in K} TD^{o\alpha_k}.$$

We have an evident "substitution" map

$$q_{\vec{\alpha}} : \mathcal{P}^{\vec{\alpha}} \longrightarrow \dot{T}\mathcal{P}^\alpha$$

which restricts to $q_{\vec{\alpha}} : \mathcal{P}^{o\vec{\alpha}} \longrightarrow T\mathcal{P}^{o\alpha}$.

2.3.1. In the same way we define the spaces $TD^{\vec{\alpha}}, TD^{o\vec{\alpha}}$.

2.3.2. Suppose that we have an epimorphism $\xi : L \longrightarrow K$, denote $L_k := \xi^{-1}(k)$. Assume that each α_k is in turn decomposed as

$$\alpha_k = \sum_{l \in L_k} \alpha_l, \ \alpha_l \in \mathbb{N}[X];$$

set $\vec{\alpha}_k = (\alpha_l) \in \mathbb{N}[X]^{L_k}$. Set $\vec{\alpha}_L = (\alpha_l) \in \mathbb{N}[X]^L$.

Let us define spaces

$$\mathcal{P}^{\vec{\alpha}_L;\xi} = \tilde{\mathcal{P}}^K \times \prod_{k \in K} \overset{\bullet}{T}D^{\vec{\alpha}_k}$$

and

$$\mathcal{P}^{o\vec{\alpha}_L;\xi} = \tilde{\mathcal{P}}^K \times \prod_{k \in K} TD^{o\vec{\alpha}_k}.$$

We have canonical substitution maps

$$q^1_{\vec{\alpha}_L;\xi} : \mathcal{P}^{\vec{\alpha}_L;\xi} \longrightarrow \mathcal{P}^{\vec{\alpha}}$$

and

$$q^2_{\vec{\alpha}_L;\xi} : \mathcal{P}^{\vec{\alpha}_L;\xi} \longrightarrow \mathcal{P}^{\vec{\alpha}_L}.$$

Obviously,

$$q_{\vec{\alpha}} \circ q^1_{\vec{\alpha}_L;\xi} = q_{\vec{\alpha}_L} \circ q^2_{\vec{\alpha}_L;\xi}.$$

2.4. **Balance function.** Consider a function $n : X \longrightarrow \mathbb{Z}[\frac{1}{2 \det A}]$ such that

$$n(\mu + \nu) = n(\mu) + n(\nu) + \mu \cdot \nu$$

It is easy to see that n can be written in the following form:

$$n(\mu) = \frac{1}{2}\mu \cdot \mu + \mu \cdot \nu_0$$

for some $\nu_0 \in X$. From now on we fix such a function n and hence the corresponding ν_0.

2.5. For an arbitrary $\alpha \in \mathbb{N}[X]$, let us define a one-dimensional local system $\mathcal{I}^\alpha_D$ on $TD^{o\alpha}$. We will proceed in the same way as in III.3.1.

Pick an unfolding of α, $\pi : J \longrightarrow X$. Define a local system $\mathcal{I}^J_D$ on TD^{oJ} as follows: its stalk at each point $((\tau_j), (x_j))$ where all x_j are real, and all the tangent vectors τ_j are real and directed to the right, is k. Monodromies are:

— x_i moves counterclockwise around x_j: monodromy is $\zeta^{-2\pi(i)\cdot\pi(j)}$;

— τ_j makes a counterclockwise circle: monodromy is $\zeta^{-2n(\pi(j))}$.

This local system has an evident Σ_π-equivariant structure, and we define a local system $\mathcal{I}^\alpha_D$ as

$$\mathcal{I}^\alpha_D = (\pi_* \mathcal{I}^J_D)^{\mathrm{sgn}}$$

where $\pi : TD^{oJ} \longrightarrow TD^{o\alpha}$ is the canonical projection, and $(\bullet)^{\mathrm{sgn}}$ denotes the subsheaf of skew Σ_π-invariants.

2.6. We will denote the unique homomorphism

$$\mathbb{N}[X] \longrightarrow X$$

identical on X, as $\alpha \mapsto \alpha^\sim$.

2.6.1. **Definition.** *An element* $\alpha \in \mathbb{N}[X]$ *is called* admissible *if* $\alpha^\sim \equiv -2\nu_0 \bmod lY$.

2.7. We have a canonical "1-jet at 0" map

$$p_K : \tilde{\mathcal{P}}^K \longrightarrow T\mathcal{P}^{oK}$$

2.8. **Definition.** *A cohesive local system (CLS) (over $\mathbb{P}^1$) is the following collection of data:*

(i) for each admissible $\alpha \in \mathbb{N}[X]$ a one-dimensional local system $\mathcal{I}^\alpha$ over $T\mathcal{P}^{o\alpha}$;

(ii) for each decomposition $\alpha = \sum_{k\in K} \alpha_k$, $\alpha_k \in \mathbb{N}[X]$, a factorization isomorphism

$$\phi_{\vec{\alpha}} : q_{\vec{\alpha}}^* \mathcal{I}^\alpha \xrightarrow{\sim} p_K^* \pi_K^* \mathcal{I}^{\alpha_K} \boxtimes \boxed{\times}_k \mathcal{I}_D^{\widetilde{\alpha_k}}$$

Here $\alpha_K := \sum_k \widetilde{\alpha_k} \in \mathbb{N}[X]$ (note that α_K is obviously admissible); $\pi_K : T\mathcal{P}^{oK} \longrightarrow T\mathcal{P}^{o\alpha}$ is the symmetrization map.

These isomorphisms must satisfy the following

Associativity axiom. *In the assumptions of 2.3.2 the equality*

$$\phi_{\vec{\alpha}_L;\xi} \circ q^{1*}_{\vec{\alpha}_L;\xi}(\phi_{\vec{\alpha}}) = q^{2*}_{\vec{\alpha}_L;\xi}(\phi_{\vec{\alpha}_L})$$

should hold. Here $\phi_{\vec{\alpha}_L;\xi}$ is induced by the evident factorization isomorphisms for local systems on the disk $\mathcal{I}_D^{\vec{\alpha}_k}$.

Morphisms between CLS's are defined in the obvious way.

2.9. **Theorem.** *Cohesive local systems over $\mathbb{P}^1$ exist. Every two CLS's are isomorphic. The group of automorphisms of a CLS is k^*.*

This theorem is a particular case of a more general theorem, valid for curves of arbitrary genus, to be proved in Part V. We leave the proof in the case of $\mathbb{P}^1$ to the interested reader.

3. Gluing

3.1. Let us define an element $\rho \in X$ by the condition $\langle \rho, i \rangle = 1$ for all $i \in I$. From now on we choose a balance function n, cf. 2.4, in the form

$$n(\mu) = \frac{1}{2}\mu \cdot \mu + \mu \cdot \rho.$$

It has the property that $n(-i') = 0$ for all $i \in I$. Thus, in the notations of *loc. cit.* we set

$$\nu_0 = \rho.$$

We pick a corresponding CLS $\mathcal{I} = \{\mathcal{I}^\beta,\ \beta \in \mathbb{N}[X]\}$.

Given $\alpha = \sum a_i i \in \mathbb{N}[I]$ and $\vec{\mu} = (\mu_k) \in X^K$, we define an element

$$\alpha_{\vec{\mu}} = \sum a_i \cdot (-i') + \sum_k \mu_k \in \mathbb{N}[X]$$

where the sum in the right hand side is a formal one. We say that a pair $(\vec{\mu}, \alpha)$ is *admissible* if $\alpha_{\vec{\mu}}$ is admissible in the sense of the previous section, i.e.

$$\sum_k \mu_k - \alpha \equiv -2\rho \bmod lY.$$

Note that given $\vec{\mu}$, there exists $\alpha \in \mathbb{N}[I]$ such that $(\vec{\mu}, \alpha)$ is admissible if and only if $\sum_k \mu_k \in Y$; if this holds true, such elements α form an obvious countable set.

We will denote by

$$e : \mathbb{N}[I] \longrightarrow \mathbb{N}[X]$$

a unique homomorphism sending $i \in I$ to $-i' \in X$.

3.2. Let us consider the space $\dot{T\mathcal{P}}^K \times \dot{T\mathcal{P}}^{e(\alpha)}$; its points are quadruples $((z_k), (\tau_k), (x_j), (\omega_j))$ where $(z_k) \in \mathcal{P}^K$, τ_k — a non-zero tangent vector to $\mathbb{P}^1$ at z_k, $(x_j) \in \mathcal{P}^{e(\alpha)}$, ω_j — a non-zero tangent vector at x_j. To a point z_k is assigned a weight μ_k, and to x_j — a weight $-\pi(j)'$. Here $\pi : J \longrightarrow I$ is an unfolding of α (implicit in the notation $(x_j) = (x_j)_{j\in J}$).

We will be interested in some open subspaces:

$$\dot{T\mathcal{P}}^{\alpha}_{\vec{\mu}} := T\mathcal{P}^{oK} \times \dot{T\mathcal{P}}^{e(\alpha)} \subset \dot{T\mathcal{P}}^{\alpha_{\vec{\mu}}}$$

and

$$T\mathcal{P}^{o\alpha}_{\vec{\mu}} \subset \dot{T\mathcal{P}}^{\alpha}_{\vec{\mu}}$$

whose points are quadruples $((z_k), (\tau_k), (x_j), (\omega_j)) \in \dot{T\mathcal{P}}^{\alpha}_{\vec{\mu}}$ with all $z_k \neq x_j$. We have an obvious symmetrization projection

$$p^{\alpha}_{\vec{\mu}} : T\mathcal{P}^{o\alpha}_{\vec{\mu}} \longrightarrow T\mathcal{P}^{o\alpha_{\vec{\mu}}}.$$

Define a space

$$\mathcal{P}^{\alpha}_{\vec{\mu}} = T\mathcal{P}^{oK} \times \mathcal{P}^{e(\alpha)};$$

its points are triples $((z_k), (\tau_k), (x_j))$ where (z_k), (τ_k) and (x_j) are as above; and to z_k and x_j the weights as above are assigned. We have the canonical projection

$$\dot{T\mathcal{P}}^{\alpha}_{\vec{\mu}} \longrightarrow \mathcal{P}^{\alpha}_{\vec{\mu}}.$$

We define the open subspaces

$$\mathcal{P}^{o\alpha}_{\vec{\mu}} \subset \mathcal{P}^{\bullet\alpha}_{\vec{\mu}} \subset \mathcal{P}^{\alpha}_{\vec{\mu}}.$$

Here the $\bullet$-subspace (resp., o-subspace) consists of all $((z_k), (\tau_k), (x_j))$ with $z_k \neq x_j$ for all k, j (resp., with all z_k and x_j distinct).

We define the *principal stratification* $\mathcal{S}$ of $\mathcal{P}^{\alpha}_{\vec{\mu}}$ as the stratification generated by subspaces $z_k = x_j$ and $x_j = x_{j'}$ with $\pi(j) \neq \pi(j')$. Thus, $\mathcal{P}^{o\alpha}_{\vec{\mu}}$ is the open stratum of $\mathcal{S}$. As usually, we will denote by the same letter the induced stratifications on subspaces.

The above projection restricts to

$$T\mathcal{P}^{o\alpha}_{\vec{\mu}} \longrightarrow \mathcal{P}^{o\alpha}_{\vec{\mu}}.$$

3.3. Factorization structure.

3.3.1. Suppose we are given $\vec{\alpha} \in \mathbb{N}[I]^K$, $\beta \in \mathbb{N}[I]$; set $\alpha := \sum_k \alpha_k$. Define a space

$$\mathcal{P}^{\vec{\alpha},\beta}_{\vec{\mu}} \subset \tilde{\mathcal{P}}^K \times \prod_k D^{\alpha_k} \times \mathcal{P}^{e(\beta)}$$

consisting of all collections $((u_k), ((x^{(k)}_j)_k), (y_j))$ where $(u_k) \in \tilde{\mathcal{P}}^K$, $(x^{(k)}_j)_k \in D^{\alpha_k}$, $(y_j) \in \mathcal{P}^{e(\beta)}$, such that

$$y_j \in \mathbb{P}^1 - \bigcup_{k\in K} \overline{u_k(D)}$$

for all j (the bar means closure).

We have canonical maps

$$q_{\vec{\alpha},\beta} : \mathcal{P}_{\vec{\mu}}^{\vec{\alpha},\beta} \longrightarrow \mathcal{P}_{\vec{\mu}}^{\alpha+\beta},$$

assigning to $((u_k),((x_j^{(k)})_k),(y_j))$ a configuration $(u_k(0)),(\overset{\bullet}{u}_k(\tau)),(u_k(x_j^{(k)})),(y_j)),$ where τ is the unit tangent vector to D at 0, and

$$p_{\vec{\alpha},\beta} : \mathcal{P}_{\vec{\mu}}^{\vec{\alpha},\beta} \longrightarrow \prod_k D^{\alpha_k} \times \mathcal{P}_{\vec{\mu}-\vec{\alpha}}^{\bullet\beta}$$

sending $((u_k),((x_j^{(k)})_k),(y_j))$ to $((u_k(0)),(\overset{\bullet}{u}_k(\tau)),(y_j))$.

3.3.2. Suppose we are given $\vec{\alpha},\vec{\beta} \in \mathbb{N}[I]^K$, $\gamma \in \mathbb{N}[I]$; set $\alpha := \sum_k \alpha_k$, $\beta := \sum_k \beta_k$. Define a space $D^{\alpha,\beta}$ consisting of couples $(D_\epsilon,(x_j))$ where $D_\epsilon \subset D$ is some smaller disk $(0<\epsilon<1)$, and $(x_j) \in D^{\alpha+\beta}$ is a configuration such that α points dwell inside D_ϵ, and β points — outside $\overline{D_\epsilon}$. We have an evident map

$$q_{\alpha,\beta} : D^{\alpha,\beta} \longrightarrow D^{\alpha+\beta}.$$

Let us define a space

$$\mathcal{P}_{\vec{\mu}}^{\vec{\alpha},\vec{\beta},\gamma} \subset \tilde{\mathcal{P}}^K \times \prod_{k\in K} D^{\alpha_k,\beta_k} \times \mathcal{P}^{e(\gamma)}$$

consisting of all triples $((u_k),\mathbf{x},(y_j))$ where $(u_k) \in \tilde{\mathcal{P}}^K$, $\mathbf{x} \in \prod_k D^{\alpha_k,\beta_k}$, $(y_j) \in \mathcal{P}^{e(\gamma)}$ such that

$$y_j \in \mathbb{P}^1 - \bigcup_k \overline{u_k(D)}.$$

We have obvious projections

$$q^1_{\vec{\alpha},\vec{\beta},\gamma} : \mathcal{P}_{\vec{\mu}}^{\vec{\alpha},\vec{\beta},\gamma} \longrightarrow \mathcal{P}_{\vec{\mu}}^{\vec{\alpha}+\vec{\beta},\gamma}$$

and

$$q^2_{\vec{\alpha},\vec{\beta},\gamma} : \mathcal{P}_{\vec{\mu}}^{\vec{\alpha},\vec{\beta},\gamma} \longrightarrow \mathcal{P}_{\vec{\mu}}^{\vec{\alpha},\beta+\gamma}$$

such that

$$q_{\vec{\alpha}+\vec{\beta},\gamma} \circ q^1_{\vec{\alpha},\vec{\beta},\gamma} = q_{\vec{\alpha},\beta+\gamma} \circ q^2_{\vec{\alpha},\vec{\beta},\gamma}.$$

We will denote the last composition by $q_{\vec{\alpha},\vec{\beta},\gamma}$.

We have a natural projection

$$p_{\vec{\alpha},\vec{\beta},\gamma} : \mathcal{P}_{\vec{\mu}}^{\vec{\alpha},\vec{\beta},\gamma} \longrightarrow \prod_k D_{\frac{1}{2}}^{\alpha_k} \times \prod_k D_{(\frac{1}{2},1)}^{\beta_k} \times \mathcal{P}_{\vec{\mu}-\vec{\alpha}-\vec{\beta}}^{\bullet\gamma}.$$

3.4. Let us consider a local system $p_{\vec{\mu}.}^{\alpha *}\mathcal{I}^{\alpha_{\vec{\mu}}}$ over $T\mathcal{P}_{\vec{\mu}}^{o\alpha}$. By our choice of the balance function n, its monodromies with respect to the rotating of tangent vectors ω_j at points x_j corresponding to negative simple roots, are trivial. Therefore it descends to a unique local system over $\mathcal{P}_{\vec{\mu}}^{o\alpha}$, to be denoted by $\mathcal{I}_{\mu}^{\alpha}$.

We define a perverse sheaf

$$\mathcal{I}_{\vec{\mu}}^{\bullet\alpha} := j_{!*}\mathcal{I}_{\vec{\mu}}^{\alpha}[\dim \mathcal{P}_{\vec{\mu}}^{\alpha}] \in \mathcal{M}(\mathcal{P}_{\vec{\mu}}^{\bullet\alpha};\mathcal{S}).$$

3.5. **Factorizable sheaves over $\mathbb{P}^1$.** Suppose we are given a K-tuple of FFS's $\{\mathcal{X}_k\}$, $\mathcal{X}_k \in \mathcal{FS}_{c_k}$, $k \in K$, $c_k \in X/Y$, where $\sum_k c_k = 0$. Let us pick $\vec{\mu} = (\vec{\mu}_k) \geq (\lambda(\mathcal{X}_k))$.

Let us call a *factorizable sheaf over* $\mathbb{P}^1$ *obtained by gluing the sheaves* $\mathcal{X}_k$ the following collection of data which we will denote by $g(\{\mathcal{X}_k\})$.

(i) For each $\alpha \in \mathbb{N}[I]$ such that $(\vec{\mu}, \alpha)$ is admissible, a sheaf $\mathcal{X}^{\alpha}_{\vec{\mu}} \in \mathcal{M}(\mathcal{P}^{\alpha}_{\vec{\mu}}; \mathcal{S})$.

(ii) For each $\vec{\alpha} = (\alpha_k) \in \mathbb{N}[I]^K$, $\beta \in \mathbb{N}[I]$ such that $(\mu, \alpha + \beta)$ is admissible (where $\alpha = \sum \alpha_k$), a *factorization isomorphism*

$$\phi_{\vec{\alpha},\beta} : q^*_{\vec{\alpha},\beta}\mathcal{X}^{\alpha+\beta}_{\vec{\mu}} \xrightarrow{\sim} p^*_{\vec{\alpha},\beta}((\boxed{\times}_{k\in K} \mathcal{X}^{\alpha_k}_{\mu_k}) \boxtimes \mathcal{I}^{\bullet\beta}_{\vec{\mu}-\vec{\alpha}}).$$

These isomorphisms should satisfy

Associativity property. The following two isomorphisms

$$q^*_{\vec{\alpha},\vec{\beta},\gamma}\mathcal{X}^{\alpha+\beta+\gamma}_{\vec{\mu}} \xrightarrow{\sim} p^*_{\vec{\alpha},\vec{\beta},\gamma}((\boxed{\times}_k \mathcal{X}^{\alpha_k}_{\mu_k}) \boxtimes (\boxed{\times}_k \mathcal{I}^{\bullet\beta_k}_{\mu_k-\alpha_k}) \boxtimes \mathcal{I}^{\bullet\gamma}_{\vec{\mu}-\vec{\alpha}-\vec{\beta}})$$

are equal:

$$\psi_{\vec{\beta},\gamma} \circ q^{2*}_{\vec{\alpha},\vec{\beta},\gamma}(\phi_{\vec{\alpha},\beta+\gamma}) = \phi_{\vec{\alpha},\vec{\beta}} \circ q^{1*}_{\vec{\alpha},\vec{\beta},\gamma}(\phi_{\vec{\alpha}+\vec{\beta},\gamma}).$$

Here $\psi_{\vec{\beta},\gamma}$ is the factorization isomorphism for $\mathcal{I}^\bullet$, and $\phi_{\vec{\alpha},\vec{\beta}}$ is the tensor product of factorization isomorphisms for the sheaves $\mathcal{X}_k$.

3.6. **Theorem.** *There exists a unique up to a canonical isomorphism factorizable sheaf over* $\mathbb{P}^1$ *obtained by gluing the sheaves* $\{\mathcal{X}_k\}$.

Proof is similar to III.10.3. □

Chapter 2. Semiinfinite cohomology

In this chapter we discuss, following essentially [Ark], the "Semiinfinite homological algebra" in the category $\mathcal{C}$.

4. Semiinfinite functors Ext and Tor in $\mathcal{C}$

4.1. Let us call an u-module $\mathfrak{u}^-$*-induced* (resp., $\mathfrak{u}^+$*-induced*) if it is induced from some $\mathfrak{u}^{\geq 0}$ (resp., $\mathfrak{u}^{\leq 0}$)-module.

4.1.1. **Lemma.** *If M is a $\mathfrak{u}^-$-induced, and N is $\mathfrak{u}^+$-induced then $M \otimes_{\mathsf{k}} N$ is u-projective.*

Proof. An induced module has a filtration whose factors are corresponding Verma modules. For Verma modules the claim is easy. □

4.2. **Definition.** Let $M^\bullet = \oplus_{\lambda \in X} M^\bullet_\lambda$ be a complex (possibly unbounded) in $\mathcal{C}$. We say that $M^\bullet$ is *concave* (resp. *convex*) if it satisfies the properties (a) and (b) below.

(a) There exists $\lambda_0 \in X$ such that for any $\lambda \in X$, if $M^\bullet_\lambda \neq 0$ then $\lambda \geq \lambda_0$ (resp. $\lambda \leq \lambda_0$).

(b) For any $\mu \in X$ the subcomplex $\oplus_{\lambda \leq \mu} M^\bullet_\lambda$ (resp. $\oplus_{\lambda \geq \mu} M^\bullet_\lambda$) is finite.

We will denote the category of concave (resp., convex) complexes by $\mathcal{C}^\uparrow$ (resp., $\mathcal{C}^\downarrow$).

4.3. Let $V \in \mathcal{C}$. We will say that a surjection $\phi : P \longrightarrow V$ is *good* if it satisfies the following properties:

(a) P is $\mathfrak{u}^-$-induced;

(b) Let $\mu \in \operatorname{supp} P$ be an extremal point, that is, there is no $\lambda \in \operatorname{supp} P$ such that $\lambda > \mu$. Then $\mu \notin \operatorname{supp}(\ker \phi)$.

For any V there exists a good surjection as above. Indeed, denote by p the projection $p : M(0) \longrightarrow L(0)$, and take for ϕ the map $p \otimes \mathrm{id}_V$.

4.4. Iterating, we can construct a $\mathfrak{u}^-$-induced convex left resolution of $\mathsf{k} = L(0)$. Let us pick such a resolution and denote it by $P^\bullet_\swarrow$:

$$\cdots \longrightarrow P^{-1}_\swarrow \longrightarrow P^0_\swarrow \longrightarrow L(0) \longrightarrow 0$$

We will denote by

$$^* : \mathcal{C} \longrightarrow \mathcal{C}^{opp}$$

the rigidity in $\mathcal{C}$ (see e.g. [AJS], 7.3). We denote by $P^\bullet_\nearrow$ the complex $(P^\bullet_\swarrow)^*$. It is a $\mathfrak{u}^-$-induced concave right resolution of k. The fact that $P^\bullet_\nearrow$ is $\mathfrak{u}^-$-induced follows since $\mathfrak{u}^-$ is Frobenius (see e.g. [PW]).

4.5. In a similar manner, we can construct a $\mathfrak{u}^+$-induced concave left resolution of k. Let us pick such a resolution and denote it $P^\bullet_\nwarrow$:

$$\cdots \longrightarrow P^{-1}_\nwarrow \longrightarrow P^0_\nwarrow \longrightarrow L(0) \longrightarrow 0$$

We denote by $P^\bullet_\searrow$ the complex $(P^\bullet_\nwarrow)^*$. It is a $\mathfrak{u}^+$-induced convex right resolution of k.

4.5.1. For $M \in \mathcal{C}$ we denote by $P^\bullet_\swarrow(M)$ (resp., $P^\bullet_\nearrow(M)$, $P^\bullet_\nwarrow(M)$, $P^\bullet_\searrow(M)$) the resolution $P^\bullet_\swarrow \otimes_k M$ (resp., $P^\bullet_\nearrow \otimes_k M$, $P^\bullet_\nwarrow \otimes_k M$, $P^\bullet_\searrow \otimes_k M$) of M.

4.6. We denote by $\mathcal{C}_r$ the category of X-graded right $\mathfrak{u}$-modules $V = \oplus_{\lambda \in X} V_\lambda$ such that

$$K_i|_{V_\lambda} = \zeta^{-\langle i,\lambda\rangle}$$

(note the change of a sign!), the operators E_i, F_i acting as $E_i : V_\lambda \longrightarrow V_{\lambda+i'}$, $F_i : V_\lambda \longrightarrow V_{\lambda-i'}$.

4.6.1. Given $M \in \mathcal{C}$, we define $M^\vee \in \mathcal{C}_r$ as follows: $(M^\vee)_\lambda = (M_{-\lambda})^*$, $E_i : (M^\vee)_\lambda \longrightarrow (M^\vee)_{\lambda+i'}$ is the transpose of $E_i : M_{-\lambda-i'} \longrightarrow M_{-\lambda}$, similarly, F_i on $M^\vee$ is the transpose of F_i on M.

This way we get an equivalence

$$^\vee : \mathcal{C}^{opp} \xrightarrow{\sim} \mathcal{C}_r.$$

Similarly, one defines an equivalence $^\vee : \mathcal{C}_r^{opp} \longrightarrow \mathcal{C}$, and we have an obvious isomorphism $^\vee \circ {}^\vee \cong \mathrm{Id}$.

4.6.2. Given $M \in \mathcal{C}$, we define $sM \in \mathcal{C}_r$ as follows: $(sM)_\lambda = M_\lambda$; $xg = (sg)x$ for $x \in M, g \in \mathfrak{u}$ where

$$s : \mathfrak{u} \longrightarrow \mathfrak{u}^{opp}$$

is the antipode defined in [AJS], 7.2. This way we get an equivalence

$$s : \mathcal{C} \xrightarrow{\sim} \mathcal{C}_r$$

One defines an equivalence $s : \mathcal{C}_r \xrightarrow{\sim} \mathcal{C}$ in a similar manner. The isomorphism of functors $s \circ s \cong \mathrm{Id}$ is constructed in *loc. cit.*, 7.3.

Note that the rigidity $*$ is just the composition

$$* = s \circ {}^\vee.$$

4.6.3. We define the categories $\mathcal{C}_r^\uparrow$ and $\mathcal{C}_r^\downarrow$ in the same way as in 4.2.

For $V \in \mathcal{C}_r$ we define $P^\bullet_\swarrow(V)$ as

$$P^\bullet_\swarrow(V) = sP^\bullet_\swarrow(sV);$$

and $P^\bullet_\nearrow(V), P^\bullet_\searrow(V), P^\bullet_\nwarrow(V)$ in a similar way.

4.7. **Definition.** (i) Let $M, N \in \mathcal{C}$. We define

$$\mathrm{Ext}_{\mathcal{C}}^{\frac{\infty}{2}+\bullet}(M,N) := H^\bullet(\mathrm{Hom}_{\mathcal{C}}(P^\bullet_\searrow(M), P^\bullet_\nearrow(N))).$$

(ii) Let $V \in \mathcal{C}_r$, $N \in \mathcal{C}$. We define

$$\mathrm{Tor}^{\mathcal{C}}_{\frac{\infty}{2}+\bullet}(V,N) := H^{-\bullet}(P^\bullet_\swarrow(V) \otimes_{\mathcal{C}} P^\bullet_\searrow(N)). \ \square$$

Here we understand $\mathrm{Hom}_{\mathcal{C}}(P^\bullet_\searrow(M), P^\bullet_\nearrow(N)$ and $P^\bullet_\swarrow(V) \otimes_{\mathcal{C}} P^\bullet_\searrow(N)$ as simple complexes associated with the corresponding double complexes. Note that due to our boundedness properties of weights of our resolutions, these double complexes are bounded. Therefore all $\mathrm{Ext}^{\frac{\infty}{2}+i}$ and $\mathrm{Tor}_{\frac{\infty}{2}+i}$ spaces are finite dimensional, and are non-zero only for finite number of $i \in \mathbb{Z}$.

4.8. **Lemma.** *For $M, N \in \mathcal{C}$ there exist canonical nondegenerate pairings*

$$\mathrm{Ext}_{\mathcal{C}}^{\frac{\infty}{2}+n}(M,N) \otimes \mathrm{Tor}^{\mathcal{C}}_{\frac{\infty}{2}+n}(N^{\vee}, M) \longrightarrow \mathsf{k}.$$

Proof. There is an evident non-degenenerate pairing

$$\mathrm{Hom}_{\mathcal{C}}(M,N) \otimes (N^{\vee} \otimes_{\mathcal{C}} M) \longrightarrow \mathsf{k}.$$

It follows that the complexes computing Ext and Tor are also canonically dual. □

4.9. **Theorem.** (i) *Let $M, N \in \mathcal{C}$. Let $R^{\bullet}_{\searrow}(M)$ be a $\mathfrak{u}^+$-induced convex right resolution of M, and $R^{\bullet}_{\nearrow}(N)$ — a $\mathfrak{u}^-$-induced concave right resolution of N. Then there is a canonical isomorphism*

$$\mathrm{Ext}_{\mathcal{C}}^{\frac{\infty}{2}+\bullet}(M,N) \cong H^{\bullet}(\mathrm{Hom}_{\mathcal{C}}(R^{\bullet}_{\searrow}(M), R^{\bullet}_{\nearrow}(N))).$$

(ii) *Let $V \in \mathcal{C}_r, N \in \mathcal{C}$. Let $R^{\bullet}_{\swarrow}(V)$ be a $\mathfrak{u}^-$-induced convex left resolution of V, and $R^{\bullet}_{\searrow}(N)$ — a $\mathfrak{u}^+$-induced convex right resolution of N lying in $\mathcal{C}^{\downarrow}$. Then there is a canonical isomorphism*

$$\mathrm{Tor}^{\mathcal{C}}_{\frac{\infty}{2}+\bullet}(V,N) \cong H^{-\bullet}(R^{\bullet}_{\swarrow}(V) \otimes_{\mathcal{C}} R^{\bullet}_{\searrow}(N)).$$

Proof will occupy the rest of the section.

4.10. **Lemma.** *Let $V \in \mathcal{C}_r$; let $R^{\bullet}_i$, $i = 1, 2$, be two $\mathfrak{u}^-$-induced convex left resolutions of V. There exists a third $\mathfrak{u}^-$-induced convex left resolution $R^{\bullet}$ of V, together with two termwise surjective maps*

$$R^{\bullet} \longrightarrow R^{\bullet}_i,\ i = 1, 2,$$

inducing identity on V.

Proof. We will construct $R^{\bullet}$ inductively, from right to left. Let

$$R^{\bullet}_i:\ \ldots \xrightarrow{d_i^{-2}} R_i^{-1} \xrightarrow{d_i^{-1}} R_i^0 \xrightarrow{\epsilon_i} V.$$

First, define $L_0 := R_1^0 \times_V R_2^0$. We denote by δ the canonical map $L^0 \longrightarrow V$, and by q_i^0 the projections $L^0 \longrightarrow R_i^0$. Choose a good surjection $\phi_0 : R^0 \longrightarrow L^0$ and define $p_i^0 := q_i^0 \circ \phi_0 : R^0 \longrightarrow R_i^0$; $\epsilon := \delta \circ \phi_0 : R^0 \longrightarrow V$.

Set $K_i^{-1} := \ker \epsilon_i$; $K := \ker \epsilon$. The projections p_i^0 induce surjections $p_i^0 : K^{-1} \longrightarrow K_i^{-1}$. Let us define

$$L^{-1} := \ker((d_1^{-1} - p_1^0, d_2^{-1} - p_2^0) : R_1^{-1} \oplus K^{-1} \oplus R_2^{-1} \longrightarrow K_1^{-1} \oplus K_2^{-1}).$$

We have canonical projections $q_i^{-1} : L^{-1} \longrightarrow R_i^{-1}$, $\delta^{-1} : L^{-1} \longrightarrow K^{-1}$. Choose a good surjection $\phi_{-1} : R^{-1} \longrightarrow L^{-1}$ and write $d^{-1} : R^{-1} \longrightarrow R^0$ for $\delta^{-1} \circ \phi_{-1}$ composed with the inclusion $K^{-1} \hookrightarrow R^0$. We define $p_i^{-1} := q_i^{-1} \circ \phi_{-1}$.

We have just described an induction step, and we can proceed in the same manner. One sees directly that the left $\mathfrak{u}^-$-induced resolution $R^{\bullet}$ obtained this way actually lies in $\mathcal{C}_r^{\downarrow}$. □

4.11. Let $N \in \mathcal{C}$, and let $R^\bullet_\searrow$ be a $\mathfrak{u}^+$-induced convex right resolution of N. For $n \geq 0$ let $b_{\geq n}(R^\bullet_\searrow)$ denote the stupid truncation:

$$0 \longrightarrow R^0_\searrow \longrightarrow \ldots \longrightarrow R^n_\searrow \longrightarrow 0 \longrightarrow \ldots$$

For $m \geq n$ we have evident truncation maps $b_{\geq m}(R^\bullet_\searrow) \longrightarrow b_{\geq n}(R^\bullet_\searrow)$.

4.11.1. **Lemma.** *Let $R^\bullet_\swarrow$ be a $\mathfrak{u}^-$-induced left resolution of a module $V \in \mathcal{C}_r$. We have*

$$H^\bullet(R^\bullet_\swarrow \otimes_{\mathcal{C}} R^\bullet_\searrow) = \lim_{\leftarrow n} H^\bullet(R^\bullet_\swarrow \otimes_{\mathcal{C}} b_{\leq n} R^\bullet_\searrow).$$

For every $i \in \mathbb{Z}$ the inverse system

$$\{H^i(R^\bullet_\swarrow \otimes_{\mathcal{C}} b_{\leq n} R^\bullet_\searrow)\}$$

stabilizes.

Proof. All spaces $H^i(R^\bullet_\swarrow \otimes_{\mathcal{C}} b_{\leq n} R^\bullet_\searrow)$ and only finitely many weight components of $R^\bullet_\swarrow$ and $R^\bullet_\searrow$ contribute to H^i. □

4.12. **Proof of Theorem 4.9.** Let us consider case (ii), and prove that $H^\bullet(R^\bullet_\swarrow(V) \otimes_{\mathcal{C}} R^\bullet_\searrow(N))$ does not depend, up to a canonical isomorphism, on the choice of a resolution $R^\bullet_\swarrow(V)$. The other independences are proved exactly in the same way.

Let $R^\bullet_i$, $i = 1, 2$, be two left $\mathfrak{u}^-$-induced left convex resolutions of V. According to Lemma 4.10, there exists a third one, $R^\bullet$, projecting onto $R^\bullet_i$. Let us prove that the projections induce isomorphisms

$$H^\bullet(R^\bullet \otimes_{\mathcal{C}} R^\bullet_\searrow(N)) \xrightarrow{\sim} H^\bullet(R^\bullet_i \otimes_{\mathcal{C}} R^\bullet_\searrow(N)).$$

By Lemma 4.11.1, it suffices to prove that

$$H^\bullet(R^\bullet \otimes_{\mathcal{C}} b_{\leq n} R^\bullet_\searrow(N)) \xrightarrow{\sim} H^\bullet(R^\bullet_i \otimes_{\mathcal{C}} b_{\leq n} R^\bullet_\searrow(N)).$$

for all n. Let $Q^\bullet_i$ be a cone of $R^\bullet \longrightarrow R^\bullet_i$. It is an exact $\mathfrak{u}^-$-induced convex complex bounded from the right. It is enough to check that $H^\bullet(Q^\bullet_i \otimes_{\mathcal{C}} b_{\leq n} R^\bullet_\searrow(N)) = 0$.

Note that for $W \in \mathcal{C}_r, M \in \mathcal{C}$ we have canonically

$$W \otimes_{\mathcal{C}} M = (W \otimes sM) \otimes_{\mathcal{C}} \mathsf{k}.$$

Thus

$$H^\bullet(Q^\bullet_i \otimes_{\mathcal{C}} b_{\leq n} R^\bullet_\searrow(N)) = H^\bullet((Q^\bullet_i \otimes s b_{\leq n} R^\bullet_\searrow(N)) \otimes_{\mathcal{C}} \mathsf{k}) = 0,$$

since $(Q^\bullet_i \otimes s b_{\leq n} R^\bullet_\searrow(N)$ is an exact bounded from the right complex, consisting of modules which are tensor products of $\mathfrak{u}^+$-induced and $\mathfrak{u}^-$-induced, hence $\mathfrak{u}$-projective modules (see Lemma 4.1.1).

4.12.1. It remains to show that if p' and p'' are two maps between $\mathfrak{u}^-$-induced convex resolutions of V, $R^\bullet_1 \longrightarrow R^\bullet_2$, inducing identity on V, then the isomorphisms

$$H^\bullet(R^\bullet_1 \otimes_{\mathcal{C}} R^\bullet_\searrow(N)) \xrightarrow{\sim} H^\bullet(R^\bullet_2 \otimes_{\mathcal{C}} R^\bullet_\searrow(N))$$

induced by p' and p'', coincide. Arguing as above, we see that it is enough to prove this with $R^\bullet_\searrow(N)$ replaced by $b_{\leq n} R^\bullet_\searrow(N)$. This in turn is equivalent to showing that two isomorphisms

$$H^\bullet((R^\bullet_1 \otimes s b_{\leq n} R^\bullet_\searrow(N)) \otimes_{\mathcal{C}} \mathsf{k}) \xrightarrow{\sim} H^\bullet((R^\bullet_2 \otimes s b_{\leq n} R^\bullet_\searrow(N)) \otimes_{\mathcal{C}} \mathsf{k})$$

coincide. But $R_i^\bullet \otimes sb_{\leq n} R_\searrow^\bullet(N)$ are complexes of projective u-modules, and the morphisms $p' \otimes \mathrm{id}$ and $p'' \otimes \mathrm{id}$ induce the same map on cohomology, hence they are homotopic; therefore they induce homotopic maps after tensor multiplication by k.

This completes the proof of the theorem. □

5. Some calculations

We will give a recipe for calculation of $\mathrm{Tor}^{\mathcal{C}}_{\frac{\infty}{2}+\bullet}$ which will prove useful for the next chapter.

5.1. Recall that in III.13.2 the duality functor

$$D : \mathcal{C}_{\zeta^{-1}} \longrightarrow \mathcal{C}^{opp}$$

has been defined (we identify $\mathcal{C}$ with $\tilde{\mathcal{C}}$ as usually). We will denote objects of $\mathcal{C}_{\zeta^{-1}}$ by letters with the subscript $(\bullet)_{\zeta^{-1}}$.

Note that $DL(0)_{\zeta^{-1}} = L(0)$.

Let us describe duals to Verma modules. For $\lambda \in X$ let us denote by $M^+(\lambda)$ the Verma module with respect to the subalgebra $\mathfrak{u}^+$ with the lowest weight λ, that is

$$M^+(\lambda) := \mathrm{Ind}^{\mathfrak{u}}_{\mathfrak{u}^{\leq 0}} \chi_\lambda$$

where χ_λ is an evident one-dimensional representaion of $\mathfrak{u}^{\leq 0}$ corresponding to the character λ.

5.1.1. **Lemma.** *We have*

$$DM(\lambda)_{\zeta^{-1}} = M^+(\lambda - 2(l-1)\rho).$$

Proof follows from [AJS], Lemma 4.10. □

5.2. Let us denote by $K^\bullet$ a two term complex in $\mathcal{C}$

$$L(0) \longrightarrow DM(0)_{\zeta^{-1}}$$

concentrated in degrees 0 and 1, the morphism being dual to the canonical projection $M(0)_{\zeta^{-1}} \longrightarrow L(0)_{\zeta^{-1}}$.

For $n \geq 1$ define a complex

$$K_n^\bullet := b_{\geq 0}(K^{\bullet \otimes n}[1]);$$

it is concentrated in degrees from 0 to $n-1$. For example, $K_1^\bullet = DM(0)_{\zeta^{-1}}$.

For $n \geq 1$ we will denote by

$$\xi : K_n^\bullet \longrightarrow K_{n+1}^\bullet$$

the map induced by the embedding $L(0) \hookrightarrow DM(0)_{\zeta^{-1}}$.

We will need the following evident properties of the system $\{K_n^\bullet, \xi_n\}$:

(a) $K_n^\bullet$ is $\mathfrak{u}^+$-induced;

(b) $K_n^\bullet$ is exact off degrees 0 and $n-1$; $H^0(K_n^\bullet) = \mathsf{k}$. ξ_n induces identity map between $H^0(K_n^\bullet)$ and $H^0(K_{n+1}^\bullet)$.

(c) For a fixed $\mu \in X$ there exists $m \in \mathbb{N}$ such that for any n we have $(b_{\geq m} K_n^\bullet)_{\geq \mu} = 0$. Here for $V = \oplus_{\lambda \in X} \in \mathcal{C}$ we set

$$V_{\geq \mu} := \oplus_{\lambda \geq \mu} V_\lambda.$$

5.3. Let $V \in \mathcal{C}_r$; let $R_\swarrow^\bullet(V)$ be a $\mathfrak{u}^-$-induced convex left resolution of V. Let $N \in \mathcal{C}$.

5.3.1. **Lemma.** (i) *For a fixed $k \in \mathbb{Z}$ the direct system $\{H^k(R_\swarrow^\bullet(V) \otimes_{\mathcal{C}} (K_n^\bullet \otimes N)), \xi_n\}$ stabilizes.*

(ii) *We have a canonical isomorphism*

$$\mathrm{Tor}^{\mathcal{C}}_{\frac{\infty}{2}+\bullet}(V, N) \cong \lim_{\to n} H^{-\bullet}(R_\swarrow^\bullet(V) \otimes_{\mathcal{C}} (K_n^\bullet \otimes N)).$$

Proof. (i) is similar to Lemma 4.11.1. (ii) By Theorem 4.9 we can use any $\mathfrak{u}^+$-induced right convex resolution of N to compute $\mathrm{Tor}^{\mathcal{C}}_{\frac{\infty}{2}+\bullet}(V, N)$. Now extend $K_n^\bullet \otimes N$ to a $\mathfrak{u}^+$-induced convex resolution of N and argue like in the proof of Lemma 4.11.1 again. □

5.4. Recall the notations of 4.4 and take for $R_\swarrow^\bullet(V)$ the resolution $P_\swarrow^\bullet(V) = P_\swarrow^\bullet \otimes V$. Then

$$H^\bullet(P_\swarrow^\bullet(V) \otimes_{\mathcal{C}} (K_n^\bullet \otimes N)) = H^\bullet(V \otimes_{\mathcal{C}} (P_\swarrow^\bullet \otimes K_n^\bullet \otimes N)).$$

Note that $P_\swarrow^\bullet \otimes K_n^\bullet \otimes N$ is a right bounded complex quasi-isomorphic to $K_n^\bullet \otimes N$. The terms of $P_\swarrow^\bullet \otimes K_n^\bullet$ are $\mathfrak{u}$-projective by Lemma 4.1.1, hence the terms of $P_\swarrow^\bullet \otimes K_n^\bullet \otimes N$ are projective by rigidity of $\mathcal{C}$. Therefore,

$$H^{-\bullet}(V \otimes_{\mathcal{C}} (P_\swarrow^\bullet \otimes K_n^\bullet \otimes N)) = \mathrm{Tor}^{\mathcal{C}}_\bullet(V, K_n^\bullet \otimes N).$$

Here $\mathrm{Tor}^{\mathcal{C}}_\bullet(*, *)$ stands for the zeroth weight component of $\mathrm{Tor}^{\mathfrak{u}}_\bullet(*, *)$.

Putting all the above together, we get

5.5. **Corollary.** *For a fixed $k \in \mathbb{Z}$ the direct system $\{\mathrm{Tor}^{\mathcal{C}}_k(V, K_n^\bullet \otimes N)\}$ stabilizes. We have*

$$\mathrm{Tor}^{\mathcal{C}}_{\frac{\infty}{2}+\bullet}(V, N) = \lim_{\to n} \mathrm{Tor}^{\mathcal{C}}_\bullet(V, K_n^\bullet \otimes N). \ \square$$

5.6. Dually, consider complexes $DK^\bullet_{n,\zeta^{-1}}$. They form a projective system

$$\{\ldots \longrightarrow DK^\bullet_{n+1,\zeta^{-1}} \longrightarrow DK^\bullet_{n,\zeta^{-1}} \longrightarrow \ldots\}$$

These complexes enjoy properties dual to (a) — (c) above.

5.7. **Theorem.** *For every $k \in \mathbb{Z}$ we have canonical isomorphisms*

$$\mathrm{Tor}^{\mathcal{C}}_{\frac{\infty}{2}+k}(V, N) \cong \lim_{\leftarrow m} \lim_{\to n} H^{-k}((V \otimes sDK^\bullet_{m,\zeta^{-1}}) \otimes_{\mathcal{C}} (K_n^\bullet \otimes N)).$$

Both the inverse and the direct systems actually stabilize.

Proof follows from Lemma 5.3.1. We leave details to the reader. □

Here is an example of calculation of $\mathrm{Tor}^{\mathcal{C}}_{\frac{\infty}{2}+\bullet}$.

5.8. **Lemma.** $\mathrm{Tor}^{\mathcal{C}}_{\frac{\infty}{2}+\bullet}(\mathsf{k}, L(2(l-1)\rho)) = \mathsf{k}$ *in degree* 0.

Proof. According to Lemma 4.8 it suffices to prove that $\mathrm{Ext}_{\mathcal{C}}^{\frac{\infty}{2}+\bullet}(L(2(l-1)\rho), L(0)) = \mathsf{k}$. Choose a $\mathfrak{u}^+$-induced right convex resolution

$$L(2(l-1)\rho) \xrightarrow{\epsilon} R^\bullet_\searrow$$

such that

$$R^0_\searrow = DM(2(l-1)\rho)_{\zeta^{-1}} = M^+(0),$$

and all the weights in $R^{\geq 1}_\searrow$ are $< 2(l-1)\rho$.

Similarly, choose a $\mathfrak{u}^-$-free right concave resolution

$$L(0) \xrightarrow{\epsilon} R^\bullet_\nearrow$$

such that

$$R^0_\nearrow = M(2(l-1)\rho) = DM^+(0)_{\zeta^{-1}},$$

and all the weights of $R^{\geq 1}_\nearrow$ are > 0. By Theorem 4.9 we have

$$\mathrm{Ext}_{\mathcal{C}}^{\frac{\infty}{2}+\bullet}(L(2(l-1)\rho), L(0)) = H^\bullet(\mathrm{Hom}^\bullet_{\mathcal{C}}(R^\bullet_\searrow, R^\bullet_\nearrow)).$$

Therefore it is enough to prove that

(a) $\mathrm{Hom}_{\mathcal{C}}(R^0_\searrow, R^0_\nearrow) = \mathsf{k}$;

(b) $\mathrm{Hom}_{\mathcal{C}}(R^m_\searrow, R^n_\nearrow) = 0$ for $(m,n) \neq (0,0)$.

(a) is evident. Let us prove (b) for, say, $n > 0$. $R^m_\searrow$ has a filtration with successive quotients of type $M^+(\lambda)$, $\lambda \leq 0$; similarly, $R^n_\searrow$ has a filtration with successive quotients of type $DM^+(\mu)_{\zeta^{-1}}$, $\mu > 0$. We have $\mathrm{Hom}_{\mathcal{C}}(M^+(\lambda), DM^+(\mu)_{\zeta^{-1}}) = 0$, therefore $\mathrm{Hom}_{\mathcal{C}}(R^m_\searrow, R^n_\nearrow) = 0$. The proof for $m > 0$ is similar. Lemma is proven. □

CONFORMAL BLOCKS AND $\mathit{Tor}^{\mathcal{C}}_{\frac{\infty}{2}+\bullet}$

5.9. Let $M \in \mathcal{C}$. We have a canonical embedding

$$\mathrm{Hom}_{\mathcal{C}}(\mathsf{k}, M) \hookrightarrow M$$

which identifies $\mathrm{Hom}_{\mathcal{C}}(\mathsf{k}, M)$ with the maximal trivial subobject of M. Dually, we have a canonical epimorhism

$$M \longrightarrow \mathrm{Hom}_{\mathcal{C}}(M, \mathsf{k})^*$$

which identifies $\mathrm{Hom}_{\mathcal{C}}(M, \mathsf{k})^*$ with the maximal trivial quotient of M. Let us denote by $\langle M \rangle$ the image of the composition

$$\mathrm{Hom}_{\mathcal{C}}(\mathsf{k}, M) \longrightarrow M \longrightarrow \mathrm{Hom}_{\mathcal{C}}(M, \mathsf{k})^*$$

Thus, $\langle M \rangle$ is canonically a subquotient of M.

One sees easily that if $N \subset M$ is a trivial direct summand of M which is maximal, i.e. not contained in greater direct summand, then we have a canonical isomorphism $\langle M \rangle \xrightarrow{\sim} N$. By this reason, we will call $\langle M \rangle$ *the maximal trivial direct summand* of M.

5.10. Let

$$\Delta_l = \{\lambda \in X | \langle i, \lambda + \rho \rangle > 0, \text{ for all } i \in I; \ \langle \gamma, \lambda + \rho \rangle < l\}$$

denote the first alcove. Here $\gamma \in \mathcal{R} \subset Y$ is the highest coroot.

For $\lambda_1, \dots, \lambda_n \in \Delta_l$, *the space of conformal blocks* is defined as

$$\langle L(\lambda_1), \dots, L(\lambda_n) \rangle := \langle L(\lambda_1) \otimes \dots \otimes L(\lambda_n) \rangle$$

(see e.g. [A] and Lemma 9.3 below).

5.11. **Corollary.** *The space of conformal blocks $\langle L(\lambda_1), \dots, L(\lambda_n) \rangle$ is canonically a subquotient of* $\mathrm{Tor}^{\mathcal{C}}_{\frac{\infty}{2}+0}(\mathsf{k}, L(\lambda_1) \otimes \dots \otimes L(\lambda_n) \otimes L(2(l-1)\rho))$.

Proof follows easily from the definition of $\langle \bullet \rangle$ and Lemma 5.8. □

5.12. Let us consider an example showing that $\langle L(\lambda_1), \dots, L(\lambda_n) \rangle$ is in general a *proper* subquotient of $\mathrm{Tor}^{\mathcal{C}}_{\frac{\infty}{2}+0}(\mathsf{k}, L(\lambda_1) \otimes \dots \otimes L(\lambda_n) \otimes L(2(l-1)\rho))$.

We leave the following to the reader.

5.12.1. **Exercise.** *Let $P(0)$ be the indecomposable projective cover of $L(0)$. We have* $\mathrm{Tor}^{\mathcal{C}}_{\frac{\infty}{2}+0}(\mathsf{k}, P(0)) = \mathsf{k}$. □

We will construct an example featuring $P(0)$ as a direct summand of $L(\lambda_1) \otimes \dots \otimes L(\lambda_n)$.

Let us take a root datum of type $sl(2)$; take $l = 5$, $n = 4$, $\lambda_1 = \lambda_2 = 2$, $\lambda_3 = \lambda_4 = 3$ (we have identified X with $\mathbb{Z}$).

In our case $\rho = 1$, so $2(l-1)\rho = 8$. Note that $P(0)$ has highest weight 8, and it is a unique indecomposable projective with the highest weight 8. So, if we are able to find a projective summand of highest weight 0 in $V = L(2) \otimes L(2) \otimes L(3) \otimes L(3)$ then $V \otimes L(8)$ will contain a projective summand of highest weight 8, i.e. $P(0)$.

Let U_{k} denote the quantum group with divided powers over k (see [L2], 8.1). The algebra $\mathfrak{u}$ lies inside U_{k}. It is wellknown that all irreducibles $L(\lambda)$, $\lambda \in \Delta_l$, lift to simple U_{k}-modules $\widetilde{L(\lambda)}$ and for $\lambda_1, \dots, \lambda_n \in \Delta$ the U_{k}-module $\widetilde{L(\lambda_1)} \otimes \dots \otimes \widetilde{L(\lambda_n)}$ is a direct sum of irreducibles $\widetilde{L(\lambda)}$, $\lambda \in \Delta_l$, and indecomposable projectives $\widetilde{P(\lambda)}$, $\lambda \geq 0$ (see, e.g. [A]).

Thus $\widetilde{L(2)} \otimes \widetilde{L(2)} \otimes \widetilde{L(3)} \otimes \widetilde{L(3)}$ contains an indecomposable projective summand with the highest weight 10, i.e. $\widetilde{P(8)}$. One can check easily that when restricted to $\mathfrak{u}$, $\widetilde{P(8)}$ remains projective and contains a summand $P(-2)$. But the highest weight of $P(-2)$ is zero.

We conclude that $L(2) \otimes L(2) \otimes L(3) \otimes L(3) \otimes L(8)$ contains a projective summand $P(0)$, whence

$$\langle L(2), L(2), L(3), L(3) \rangle \neq \mathrm{Tor}^{\mathcal{C}}_{\frac{\infty}{2}+0}(\mathsf{k}, L(2) \otimes L(2) \otimes L(3) \otimes L(3) \otimes L(8)).$$

Chapter 3. Global sections

6. Braiding and balance in $\mathcal{C}$ and $\mathcal{FS}$

6.1. Let U_k be the quantum group with divided powers, cf [L2], 8.1. Let ${}_R\mathcal{C}$ be the category of finite dimensional integrable U_k-modules defined in [KL]IV, §37. It is a rigid braided tensor category. The braiding, i.e. family of isomorphisms

$$\tilde{R}_{V,W} : V \otimes W \xrightarrow{\sim} W \otimes V, \ V, W \in {}_R\mathcal{C},$$

satisfying the usual constraints, has been defined in [L1], Ch. 32.

6.2. As u is a subalgebra of U_k, we have the restriction functor preserving X-grading

$$\Upsilon : {}_R\mathcal{C} \longrightarrow \mathcal{C}.$$

The following theorem is due to G.Lusztig (private communication).

6.2.1. **Theorem.** (a) *There is a unique braided structure* $(R_{V,W}, \theta_V)$ *on* $\mathcal{C}$ *such that the restriction functor* Υ *commutes with braiding.*

(b) *Let* $V = L(\lambda)$, *and let* μ *be the highest weight of* $W \in \mathcal{C}$, *i.e.* $W_\mu \neq 0$ *and* $W_\nu \neq 0$ *implies* $\nu \leq \mu$. *Let* $x \in V,\ y \in W_\mu$. *Then*

$$R_{V,W}(x \otimes y) = \zeta^{\lambda \cdot \mu} y \otimes x;$$

(c) *Any braided structure on* $\mathcal{C}$ *enjoying the property* (b) *above coincides with that defined in* (a). □

6.3. Recall that an automorphism $\tilde{\theta} = \{\tilde{\theta}_V : V \xrightarrow{\sim} V\}$ of the identity functor of ${}_R\mathcal{C}$ is called *balance* if for any $V, W \in {}_R\mathcal{C}$ we have

$$\tilde{R}_{W,V} \circ \tilde{R}_{V,W} = \tilde{\theta}_{V \otimes W} \circ (\tilde{\theta}_V \otimes \tilde{\theta}_W)^{-1}.$$

The following proposition is an easy application of the results of [L1], Chapter 32.

6.4. **Proposition.** *The category* ${}_R\mathcal{C}$ *admits a unique balance* $\tilde{\theta}$ *such that*

— *if* $\tilde{L}(\lambda)$ *is an irreducible in* ${}_R\mathcal{C}$ *with the highest weight* λ, *then* $\tilde{\theta}$ *acts on* $\tilde{L}(\lambda)$ *as multiplication by* $\zeta^{n(\lambda)}$. □

Here $n(\lambda)$ denotes the function introduced in 3.1.

Similarly to 6.2.1, one can prove

6.5. **Theorem.** (a) *There is a unique balance* θ *on* $\mathcal{C}$ *such that* Υ *commutes with balance;*

(b) $\theta_{L(\lambda)} = \zeta^{n(\lambda)}$;

(c) *If* θ' *is a balance in* $\mathcal{C}$ *having property* (b), *then* $\theta' = \theta$. □

6.6. According to Deligne's ideology, [D1], the gluing construction of 3.6 provides the category $\mathcal{FS}$ with the balance $\theta^{\mathcal{FS}}$. Recall that the braiding $R^{\mathcal{FS}}$ has been defined in III.11.11 (see also 11.4). It follows easily from the definitions that $(\Phi(R^{\mathcal{FS}}), \Phi(\theta^{\mathcal{FS}}))$ satisfy the properties (b)(i) and (ii) above. Therefore, we have $(\Phi(R^{\mathcal{FS}}), \Phi(\theta^{\mathcal{FS}})) = (R, \theta)$, i.e. Φ is an equivalence of braided balanced categories.

7. Global sections over $\mathcal{A}(K)$

7.1. Let K be a finite non-empty set, $|K| = n$, and let $\{\mathcal{X}_k\}$ be a K-tuple of finite gactorizable sheaves. Let $\lambda_k := \lambda(\mathcal{X}_k)$ and $\lambda = \sum_k \lambda_k$; let $\alpha \in \mathbb{N}[I]$. Consider the sheaf $\mathcal{X}^\alpha(K)$ over $\mathcal{A}^\alpha_\lambda(K)$ obtained by gluing $\{\mathcal{X}_k\}$, cf. III.10.3. Thus

$$\mathcal{X}^\alpha(K) = g_K(\{\mathcal{X}_k\})^\alpha_\lambda$$

in the notations of *loc. cit.*

We will denote by η, or sometimes by η^α, or η^α_K the projection $\mathcal{A}^\alpha(K) \longrightarrow \mathcal{O}(K)$. We are going to describe $R\eta_*\mathcal{X}^\alpha(K)[-n]$. Note that it is an element of $\mathcal{D}(\mathcal{O}(K))$ which is smooth, i.e. its cohomology sheaves are local systems.

7.2. Let $V_1, \ldots, V_n \in \mathcal{C}$. Recall (see II.3) that $C^\bullet_{\mathfrak{u}^-}(V_1 \otimes \ldots \otimes V_n)$ denotes the Hochschild complex of the $\mathfrak{u}^-$-module $V_1 \otimes \ldots \otimes V_n$. It is naturally X-graded, and its λ-component is denoted by the subscript $(\bullet)_\lambda$ as usually.

Let us consider a homotopy point $\mathbf{z} = (z_1, \ldots, z_n) \in \mathcal{O}(K)$ where all z_i are real, $z_1 < \ldots < z_n$. Choose a bijection $K \xrightarrow{\sim} [n]$. We want to describe a stalk $R\eta_*\mathcal{X}^\alpha(K)_{\mathbf{z}}[-n]$. The following theorem generalizes Theorem II.8.23. The proof is similar to *loc. cit*, cf. III.12.16, and will appear later.

7.3. **Theorem.** *There is a canonical isomorphism, natural in $\mathcal{X}_i$,*

$$R\eta_*\mathcal{X}^\alpha(K)_{\mathbf{z}}[-n] \cong C^\bullet_{\mathfrak{u}^-}(\Phi(\mathcal{X}_1) \otimes \ldots \otimes \Phi(\mathcal{X}_n))_{\lambda-\alpha}. \ \square$$

7.4. The group $\pi_1(\mathcal{O}(K); \mathbf{z})$ is generated by counterclockwise loops of z_{k+1} around z_k, σ_i, $k = 1, \ldots, n-1$. Let σ_k act on $\Phi(\mathcal{X}_1) \otimes \ldots \otimes \Phi(\mathcal{X}_n)$ as

$$\mathrm{id} \otimes \ldots \otimes R_{\Phi(\mathcal{X}_{k+1}),\Phi(\mathcal{X}_k)} \circ R_{\Phi(\mathcal{X}_k),\Phi(\mathcal{X}_{k+1})} \otimes \ldots \otimes \mathrm{id}.$$

This defines an action of $\pi_1(\mathcal{O}(K); \mathbf{z})$ on $\Phi(\mathcal{X}_1) \otimes \ldots \otimes \Phi(\mathcal{X}_n)$, whence we get an action of this group on $C^\bullet_{\mathfrak{u}^-}(\Phi(\mathcal{X}_1) \otimes \ldots \otimes \Phi(\mathcal{X}_n))$ respecting the X-grading. Therefore we get a complex of local systems over $\mathcal{O}(K)$; let us denote it $C^\bullet_{\mathfrak{u}^-}(\Phi(\mathcal{X}_1) \otimes \ldots \otimes \Phi(\mathcal{X}_n))^\heartsuit$.

7.5. **Theorem.** *There is a canonical isomorphism in $\mathcal{D}(\mathcal{O}(K))$*

$$R\eta_*\mathcal{X}^\alpha(K)[-n] \xrightarrow{\sim} C^\bullet_{\mathfrak{u}^-}(\Phi(\mathcal{X}_1) \otimes \ldots \otimes \Phi(\mathcal{X}_n))^\heartsuit_{\lambda-\alpha}.$$

Proof follows from 6.6 and Theorem 7.3. $\square$

7.6. **Corollary.** *Set $\lambda_\infty := \alpha + 2(l-1)\rho - \lambda$. There is a canonical isomorphism in $\mathcal{D}(\mathcal{O}(K))$*

$$R\eta_*\mathcal{X}^\alpha(K)[-n] \xrightarrow{\sim} C^\bullet_{\mathfrak{u}}(\Phi(\mathcal{X}_1)\otimes\ldots\otimes\Phi(\mathcal{X}_n)\otimes DM(\lambda_\infty)_{\zeta^{-1}})^\heartsuit_0.$$

Proof. By Shapiro's lemma, we have a canonical morphism of complexes which is a quasiisomorphism

$$C^\bullet_{\mathfrak{u}^-}(\Phi(\mathcal{X}_1)\otimes\ldots\otimes\Phi(\mathcal{X}_n))_{\lambda-\alpha} \longrightarrow C^\bullet_{\mathfrak{u}}(\Phi(\mathcal{X}_1)\otimes\ldots\otimes\Phi(\mathcal{X}_n)\otimes M^+(\alpha-\lambda))_0.$$

By Lemma 5.1.1, $M^+(\alpha-\lambda) = DM(\lambda_\infty)_{\zeta^{-1}}$. □

8. Global sections over $\mathcal{P}$

8.1. Let J be a finite set, $|J| = m$, and $\{\mathcal{X}_j\}$ a J-tuple of finite factorizable sheaves. Set $\mu_j := \lambda(\mathcal{X}_j)$, $\vec{\mu} = (\mu_j) \in X^J$. Let $\alpha \in \mathbb{N}[I]$ be such that $(\vec{\mu},\alpha)$ is admissible, cf. 3.1. Let $\mathcal{X}^\alpha_{\vec{\mu}}$ be the preverse sheaf on $\mathcal{P}^\alpha_{\vec{\mu}}$ obtained by gluing the sheaves $\mathcal{X}_j$, cf. 3.5 and 3.6.

Note that the group $\mathrm{PGL}_2(\mathbb{C}) = \mathrm{Aut}(\mathbb{P}^1)$ operates naturally on $\mathcal{P}^\alpha_{\vec{\mu}}$ and the sheaf $\mathcal{X}^\alpha_{\vec{\mu}}$ is equivariant with respect to this action.

Let

$$\bar{\eta} : \mathcal{P}^\alpha_{\vec{\mu}} \longrightarrow T\mathcal{P}^{oJ}$$

denote the natural projection; we will denote this map also by $\bar{\eta}_J$ or $\bar{\eta}^\alpha_J$. Note that $\bar{\eta}$ commutes with the natural action of $\mathrm{PGL}_2(\mathbb{C})$ on these spaces. Therefore $R\bar{\eta}_*\mathcal{X}^\alpha_{\vec{\mu}}$ is a smooth $\mathrm{PGL}_2(\mathbb{C})$-equivariant complex on $T\mathcal{P}^{oJ}$. Our aim in this section will be to compute this complex algebraically.

Note that $R\bar{\eta}_*\mathcal{X}^\alpha_{\vec{\mu}}$ descends uniquely to the quotient

$$\underline{T\mathcal{P}}^{oJ} := T\mathcal{P}^{oJ}/\mathrm{PGL}_2(\mathbb{C})$$

8.2. Let us pick a bijection $J \xrightarrow{\sim} [m]$. Let $\underline{Z}$ be a contractible real submanifold of $\underline{T\mathcal{P}}^{oJ}$ defined in [KL]II, 13.1 (under the name $\underline{\mathcal{V}}_0$). Its points are configurations $(z_1,\tau_1,\ldots,z_m,\tau_m)$ such that $z_j \in \mathbb{P}^1(\mathbb{R}) = S \subset \mathbb{P}^1(\mathbb{C})$; the points z_j lie on S in this cyclic order; they orient S in the same way as $(0,1,\infty)$ does; the tangent vectors τ_j are real and compatible with this orientation.

8.3. **Definition.** An m-tuple of weights $\vec{\mu} \in X^m$ is called *positive* if

$$\sum_{j=1}^m \mu_j + (1-l)2\rho \in \mathbb{N}[I] \subset X.$$

If this is so, we will denote

$$\alpha(\vec{\mu}) := \sum_{j=1}^m \mu_j + (1-l)2\rho \; \square$$

8.4. **Theorem.** *Let $\mathcal{X}_1, \ldots, \mathcal{X}_m \in \mathcal{FS}$. Let $\vec{\mu}$ be a positive m-tuple of weights, $\mu_j \geq \lambda(\mathcal{X}_j)$, and let $\alpha = \alpha(\vec{\mu})$. Let $\mathcal{X}^\alpha_{\vec{\mu}}$ be the sheaf on $\mathcal{P}^\alpha_{\vec{\mu}}$ obtained by gluing the sheaves $\mathcal{X}_j$. There is a canonical isomorphism*

$$R^\bullet \bar{\eta}_* \mathcal{X}^\alpha_{\vec{\mu}}[-2m]_Z \cong \mathrm{Tor}^{\mathcal{C}}_{\frac{\infty}{2}-\bullet}(\mathsf{k}, \Phi(\mathcal{X}_1) \otimes \ldots \otimes \Phi(\mathcal{X}_m)).$$

Proof is sketched in the next few subsections.

8.5. **Two-sided Čech resolutions.** The idea of the construction below is inspired by [B1], p. 40.

Let P be a topological space, $\mathcal{U} = \{U_i |\ i = 1, \ldots, N\}$ an open covering of P. Let $j_{i_0 i_1 \ldots i_a}$ denote the embedding

$$U_{i_0} \cap \ldots \cap U_{i_a} \hookrightarrow P.$$

Given a sheaf $\mathcal{F}$ on P, we have a canonical morphism

$$\mathcal{F} \longrightarrow \check{C}^\bullet(\mathcal{U}; \mathcal{F}) \tag{285}$$

where

$$\check{C}^a(\mathcal{U}; \mathcal{F}) = \oplus_{i_0 < i_1 < \ldots < i_a} j_{i_0 i_1 \ldots i_a *} j^*_{i_0 i_1 \ldots i_a} \mathcal{F},$$

the differential being the usual Čech one.

Dually, we define a morphism

$$\check{C}_\bullet(\mathcal{U}; \mathcal{F}) \longrightarrow \mathcal{F} \tag{286}$$

where

$$\check{C}_\bullet(\mathcal{U}; \mathcal{F}) : \ 0 \longrightarrow \check{C}_N(\mathcal{U}; \mathcal{F}) \longrightarrow \check{C}_{N-1}(\mathcal{U}; \mathcal{F}) \longrightarrow \ldots \longrightarrow \check{C}_0(\mathcal{U}; \mathcal{F}) \longrightarrow 0$$

where

$$\check{C}_a(\mathcal{U}; \mathcal{F}) = \oplus_{i_0 < i_1 < \ldots < i_a} j_{i_0 i_1 \ldots i_a !} j^*_{i_0 i_1 \ldots i_a} \mathcal{F}.$$

If $\mathcal{F}$ is injective then the arrows (285) and (286) are quasiisomorphisms.

Suppose we have a second open covering of P, $\mathcal{V} = \{V_j |\ j = 1, \ldots, N\}$. Let us define sheaves

$$\check{C}^a_b(\mathcal{U}, \mathcal{V}; \mathcal{F}) := \check{C}_b(\mathcal{V}; \check{C}^a(\mathcal{U}; \mathcal{F}));$$

they form a bicomplex. Let us consider the associated simple complex $\check{C}^\bullet(\mathcal{U}, \mathcal{V}; \mathcal{F})$, i.e.

$$\check{C}^i(\mathcal{U}, \mathcal{V}; \mathcal{F}) = \oplus_{a-b=i}\ \check{C}^a_b(\mathcal{U}, \mathcal{V}; \mathcal{F}).$$

It is a complex concentrated in degrees from $-N$ to N. We have canonical morphisms

$$\mathcal{F} \longrightarrow \check{C}^\bullet(\mathcal{U}; \mathcal{F}) \longleftarrow \check{C}^\bullet(\mathcal{U}, \mathcal{V}; \mathcal{F})$$

If $\mathcal{F}$ is injective then both arrows are quasiisomorphisms, and the above functors are exact on injective sheaves. Therefore, they pass to derived categories, and we get a functor $\mathcal{K} \mapsto \check{C}^\bullet(\mathcal{U}, \mathcal{V}; \mathcal{K})$ from the bounded derived category $\mathcal{D}^b(P)$ to the bounded filtered derived category $\mathcal{DF}(P)$. This implies

8.6. **Lemma.** *Suppose that $\mathcal{K} \in \mathcal{D}^b(P)$ is such that $R^i\Gamma(P; \check{C}^a_b(\mathcal{U}, \mathcal{V}; \mathcal{K})) = 0$ for all a, b and all $i \neq 0$. Then we have a canonical isomorphism in $\mathcal{D}^b(P)$,*

$$R\Gamma(P; \mathcal{K}) \xrightarrow{\sim} R^0\Gamma(P; \check{C}^\bullet(\mathcal{U}, \mathcal{V}; \mathcal{K})). \ \square$$

8.7. Returning to the assumptions of theorem 8.4, let us pick a point $\mathbf{z} = (z_1, \tau_1, \ldots, z_m, \tau_m) \in T\mathcal{P}^{oJ}$ such that z_j are real numbers $z_1 < \ldots < z_m$ and tangent vectors are directed to the right.

By definition, we have canonically

$$R\bar{\eta}_*(\mathcal{P}^{\alpha}_{\bar{\mu}}; \mathcal{X}^{\alpha}_{\bar{\mu}})_{\underline{\mathbf{z}}} = R\Gamma(\mathcal{P}^{\alpha}; \mathcal{K})$$

where

$$\mathcal{K} := \mathcal{X}^{\alpha}_{\bar{\mu}}|_{\bar{\eta}^{-1}(\mathbf{z})}[-2m].$$

Let us pick $N \geq |\alpha|$ and reals $p_1, \ldots, p_N, q_1, \ldots, q_N$ such that

$$p_1 < \ldots < p_N < z_1 < \ldots < z_m < q_N < \ldots < q_1.$$

Let us define two open coverings $\mathcal{U} = \{U_i |\ i = 1, \ldots, N\}$ and $\mathcal{V} = \{V_i |\ i = 1, \ldots, N\}$ of the space $\mathcal{P}^{\alpha}$ where

$$U_i = \mathcal{P}^{\alpha} - \bigcup_k \{t_k = p_i\};\ V_i = \mathcal{P}^{\alpha} - \bigcup_k \{t_k = q_i\},$$

where t_k denote the standard coordinates.

8.8. **Lemma.** (i) *We have*

$$R^i\Gamma(\mathcal{P}^{\alpha}; \check{C}^a_b(\mathcal{U}, \mathcal{V}; \mathcal{K})) = 0$$

for all a, b *and all* $i \neq 0$.

(ii) *We have canonical isomorphism*

$$R^0\Gamma(\mathcal{P}^{\alpha}; \check{C}^{\bullet}(\mathcal{U}, \mathcal{V}; \mathcal{K})) \cong sDK^{\bullet}_{N,\zeta^{-1}} \otimes_{\mathcal{C}} (K^{\bullet}_N \otimes \Phi(\mathcal{X}_1) \otimes \ldots \otimes \Phi(\mathcal{X}_m)),$$

in the notations of 5.7.

Proof (sketch). We should regard the computation of $R\Gamma(\mathcal{P}^{\alpha}; \check{C}^a_b(\mathcal{U}, \mathcal{V}; \mathcal{K}))$ as the computation of global sections over $\mathcal{P}^{\alpha}$ of a sheaf obtained by gluing $\mathcal{X}_j$ into points z_j, the Verma sheaves $\mathcal{M}(0)$ or irreducibles $\mathcal{L}(0)$ into the points p_j, and dual sheaves $D\mathcal{M}(0)_{\zeta^{-1}}$ or $D\mathcal{L}(0)_{\zeta^{-1}}$ into the points q_j.

Using $\mathrm{PGL}_2(\mathbb{R})$-invariance, we can move one of the points p_j to infinity. Then, the desired global sections are reduced to global sections over an affine space $\mathcal{A}^{\alpha}$, which are calculated by means of Theorem 7.3.

Note that in our situation all the sheaves $\check{C}^a_b(\mathcal{U}, \mathcal{V}; \mathcal{K})$ actually belong to the abelian category $\mathcal{M}(\mathcal{P}^{\alpha})$ of perverse sheaves. So $\check{C}^{\bullet}(\mathcal{U}, \mathcal{V}; \mathcal{K})$ is a resolution of $\mathcal{K}$ in $\mathcal{M}(\mathcal{P}^{\alpha})$. □

8.9. The conclusion of 8.4 follow from the previous lemma and Theorem 5.7. □

8.10. The group $\pi_1(\underline{T\mathcal{P}}^{om}, \mathcal{Z})$ operates on the spaces $\mathrm{Tor}^{\mathcal{C}}_{\frac{\infty}{2}+\bullet}(\mathsf{k}, \Phi(\mathcal{X}_1) \otimes \ldots \otimes \Phi(\mathcal{X}_m))$ via its action on the object $\Phi(\mathcal{X}_1) \otimes \ldots \otimes \Phi(\mathcal{X}_m)$ induced by the braiding and balance in $\mathcal{C}$. Let us denote by

$$\mathrm{Tor}^{\mathcal{C}}_{\frac{\infty}{2}+\bullet}(\mathsf{k}, \Phi(\mathcal{X}_1) \otimes \ldots \otimes \Phi(\mathcal{X}_m))^{\heartsuit}$$

the corresponding local system on $\underline{T\mathcal{P}}^{om}$.

8.11. **Theorem.** *There is a canonical isomorphism of local systems on $\underline{\mathcal{TP}}^{om}$:*

$$R^{\bullet-2m}\bar{\eta}_*\mathcal{X}^{\alpha}_{\vec{\mu}} \cong \mathrm{Tor}^{\mathcal{C}}_{\frac{\infty}{2}-\bullet}(\mathsf{k}, \Phi(\mathcal{X}_1) \otimes \ldots \otimes \Phi(\mathcal{X}_m))^{\heartsuit}$$

Proof follows immediately from 6.6 and Theorem 8.4. □

9. Application to conformal blocks

9.1. In applications to conformal blocks we will encounter the roots of unity ζ of not necessarily odd degree l. So we have to generalize all the above considerations to the case of arbitrary l.

The definitions of the categories $\mathcal{C}$ and $\mathcal{FS}$ do not change (for the category $\mathcal{C}$ the reader may consult [AP], §3). The construction of the functor $\Phi : \mathcal{FS} \longrightarrow \mathcal{C}$ and the proof that Φ is an equivalence repeats the one in III word for word.

Here we list the only minor changes (say, in the definition of the Steinberg module) following [L1] and [AP].

9.1.1. So suppose ζ is a primitive root of unity of an *even* degree l.

We define $\ell := \frac{l}{2}$. For the sake of unification of notations, in case l is *odd* we define $\ell := l$. For $i \in I$ we define $\ell_i := \frac{\ell}{(\ell, d_i)}$ where (ℓ, d_i) stands for the greatest common divisor of ℓ and d_i.

For a coroot $\alpha \in \mathcal{R} \in Y$ we can find an element w of the Weyl group W and a simple coroot $i \in Y$ such that $w(i) = \alpha$ (notations of [L1], 2.3). We define $\ell_\alpha := \frac{\ell}{(\ell, d_i)}$, and the result does not depend on a choice of i and w.

We define $\gamma_0 \in \mathcal{R}$ to be the highest coroot, and $\beta_0 \in \mathcal{R}$ to be the coroot dual to the highest root. Note that $\gamma_0 = \beta_0$ iff our root datum is simply laced.

9.1.2. We define

$$Y_\ell := \{\lambda \in X | \lambda \cdot \mu \in \ \ell\mathbb{Z} \text{ for any } \mu \in X\}$$

One should replace the congruence modulo lY in the Definition 2.6.1 and in 3.1 by the congruence modulo Y_ℓ.

We define $\rho_\ell \in X$ as the unique element such that $\langle i, \rho_\ell \rangle = \ell_i - 1$ for any $i \in I$.

Then the *Steinberg module* $L(\rho_\ell)$ is irreducible projective in $\mathcal{C}$ (see [AP] 3.14).

Note also that ρ_ℓ is the highest weight of $\mathfrak{u}^+$.

One has to replace all the occurences of $(l-1)2\rho$ in the above sections by $2\rho_\ell$.

In particular, the new formulations of the Definition 8.3 and the Theorem 8.4 force us to make the following changes in 3.1 and 3.4.

In 3.1 we choose a balance function n in the form

$$n(\mu) = \frac{1}{2}\mu \cdot \mu - \mu \cdot \rho_\ell$$

In other words, we set $\nu_0 = -\rho_\ell$. *This balance function does not necessarily have the property that* $n(-i') \equiv 0 \bmod l$. *It is only true that* $n(-i') \equiv 0 \bmod \ell$.

We say that a pair $(\vec{\mu}, \alpha)$ is admissible if $\sum_k \mu_k - \alpha \equiv 2\rho_\ell \bmod Y_\ell$.

9.1.3. The last change concerns the definition of the first alcove in 5.10. The corrected definition reads as follows:

if $\ell_i = \ell$ for any $i \in I$, then

$$\Delta_l = \{\lambda \in X| \; \langle i, \lambda + \rho \rangle > 0, \text{ for all } i \in I; \; \langle \gamma_0, \lambda + \rho \rangle < \ell\};$$

if not, then

$$\Delta_l = \{\lambda \in X| \; \langle i, \lambda + \rho \rangle > 0, \text{ for all } i \in I; \; \langle \beta_0, \lambda + \rho \rangle < \ell_{\beta_0}\}$$

9.2. Let $\hat{\mathfrak{g}}$ denote the affine Lie algebra associated with $\mathfrak{g}$:

$$0 \longrightarrow \mathbb{C} \longrightarrow \hat{\mathfrak{g}} \longrightarrow \mathfrak{g}((\epsilon)) \longrightarrow 0.$$

Let $\tilde{\mathcal{O}}_\kappa$ be the category of integrable $\hat{\mathfrak{g}}$-modules with the central charge $\kappa - h$ where h stands for the dual Coxeter number of $\mathfrak{g}$. It is a semisimple balanced braided rigid tensor category (see e.g. [MS] or [F2]).

Let $\mathcal{O}_{-\kappa}$ be the category of $\mathfrak{g}$-integrable $\hat{\mathfrak{g}}$-modules of finite length with the central charge $-\kappa - h$. It is a balanced braided rigid tensor (bbrt) category (see [KL]). Let $\tilde{\mathcal{O}}_{-\kappa}$ be the semisimple subcategory of $\mathcal{O}_{-\kappa}$ formed by direct sums of simple $\hat{\mathfrak{g}}$-modules with highest weights in the alcove ∇_κ:

$$\nabla_\kappa := \{\lambda \in X| \; \langle i, \lambda + \rho \rangle > 0, \text{ for all } i \in I; \; \langle \beta_0, \lambda + \rho \rangle < \kappa\}$$

The bbrt structure on $\mathcal{O}_{-\kappa}$ induces the one on $\tilde{\mathcal{O}}_{-\kappa}$, and one can construct an equivalence

$$\tilde{\mathcal{O}}_\kappa \xrightarrow{\sim} \tilde{\mathcal{O}}_{-\kappa}$$

respecting bbrt structure (see [F2]). D.Kazhdan and G.Lusztig have constructed an equivalence

$$\mathcal{O}_{-\kappa} \xrightarrow{\sim} {}_R\mathcal{C}_\zeta$$

(notations of 6.1) respecting bbrt structure (see [KL] and [L3]). Here $\zeta = \exp(\frac{\pi\sqrt{-1}}{d\kappa})$ where $d = \max_{i \in I} d_i$. Thus $l = 2d\kappa$, and $\ell = d\kappa$.

Note that the alcoves ∇_κ and Δ_l (see 9.1.3) coincide.

The Kazhdan-Lusztig equivalence induces an equivalence

$$\tilde{\mathcal{O}}_{-\kappa} \xrightarrow{\sim} \tilde{\mathcal{O}}_\zeta$$

where $\tilde{\mathcal{O}}_\zeta$ is the semisimple subcategory of ${}_R\mathcal{C}_\zeta$ formed by direct sums of simple U_k-modules $\tilde{L}(\lambda)$ with $\lambda \in \Delta$ (see [A] and [AP]). The bbrt structure on ${}_R\mathcal{C}_\zeta$ induces the one on $\tilde{\mathcal{O}}_\zeta$, and the last equivalence respects bbrt structure. We denote the composition of the above equivalences by

$$\phi : \tilde{\mathcal{O}}_\kappa \xrightarrow{\sim} \tilde{\mathcal{O}}_\zeta.$$

Given any bbrt category $\mathcal{B}$ and objects $L_1, \ldots, L_m \in \mathcal{B}$ we obtain a local system $\mathrm{Hom}_{\mathcal{B}}(\mathbf{1}, L_1 \otimes \ldots \otimes L_m)^\heartsuit$ on $T\mathcal{P}^{om}$ with monodromies induced by the action of braiding and balance on $L_1 \otimes \ldots \otimes L_m$.

Here and below we write a superscript $X^\heartsuit$ to denote a local system over $T\mathcal{P}^{om}$ with the fiber at a standard real point $z_1 < \ldots < z_m$ with tangent vectors looking to the right, equal to X.

Thus, given $L_1, \ldots, L_m \in \tilde{\mathcal{O}}_\kappa$, the local system

$$\mathrm{Hom}_{\tilde{\mathcal{O}}_\kappa}(\mathbf{1}, L_1 \tilde{\otimes} \ldots \tilde{\otimes} L_m)^\heartsuit$$

called *local system of conformal blocks* is isomorphic to the local system $\mathrm{Hom}_{\tilde{\mathcal{O}}_\zeta}(\mathbf{1}, \phi(L_1) \tilde{\otimes} \ldots \tilde{\otimes} \phi(L_m))^\heartsuit$. Here $\tilde{\otimes}$ will denote the tensor product in "tilded" categories.

To unburden the notations we leave out the subscript ζ in ${}_R\mathcal{C}_\zeta$ from now on.

For an object $X \in {}_R\mathcal{C}$ let us define a vector space $\langle X \rangle_{{}_R\mathcal{C}}$ in the same manner as in 5.9, i.e. as an image of the canonical map from the maximal trivial subobject of X to the maximal trivial quotient of X. Given $X_1, \ldots, X_m \in {}_R\mathcal{C}$, we denote

$$\langle X_1, \ldots X_m \rangle := \langle X_1 \otimes \ldots \otimes X_m \rangle_{{}_R\mathcal{C}}.$$

9.2.1. **Lemma.** *We have an isomorphism of local systems*

$$\mathrm{Hom}_{\tilde{\mathcal{O}}_\zeta}(\mathbf{1}, \phi(L_1) \tilde{\otimes} \ldots \tilde{\otimes} \phi(L_m))^\heartsuit \cong \langle \phi(L_1), \ldots, \phi(L_m) \rangle^\heartsuit_{{}_R\mathcal{C}}$$

Proof. Follows from [A]. □

9.3. **Lemma.** *The restriction functor $\Upsilon : {}_R\mathcal{C} \longrightarrow \mathcal{C}$ (cf. 6.2) induces isomorphism*

$$\langle \phi(L_1), \ldots, \phi(L_m) \rangle_{{}_R\mathcal{C}} \xrightarrow{\sim} \langle \Upsilon\phi(L_1), \ldots, \Upsilon\phi(L_m) \rangle_{\mathcal{C}}.$$

Proof. We must prove that if $\lambda_1, \ldots, \lambda_m \in \Delta$, $\widetilde{L(\lambda_1)}, \ldots, \widetilde{L(\lambda_m)}$ are corresponding simples in ${}_R\mathcal{C}$, and $L(\lambda_i) = \Upsilon\widetilde{L(\lambda_i)}$ — the corresponding simples in $\mathcal{C}$, then the maximal trivial direct summand of $\widetilde{L(\lambda_1)} \otimes \ldots \otimes \widetilde{L(\lambda_m)}$ in ${}_R\mathcal{C}$ maps isomorphically to the maximal trivial direct summand of $L(\lambda_1) \otimes \ldots \otimes L(\lambda_m)$ in $\mathcal{C}$.

According to [A], [AP], $\widetilde{L(\lambda_1)} \otimes \ldots \otimes \widetilde{L(\lambda_m)}$ is a direct sum of a module $\widetilde{L(\lambda_1)} \tilde{\otimes} \ldots \tilde{\otimes} \widetilde{L(\lambda_m)} \in \mathcal{O}_\zeta$ and a negligible module $N \in {}_R\mathcal{C}$. Here *negligible* means that any endomorphism of N has quantum trace zero (see *loc. cit.*). Moreover, it is proven in *loc cit.* that N is a direct summand of $W \otimes M$ for some $M \in {}_R\mathcal{C}$ where $W = \oplus_{\omega \in \Omega} \widetilde{L(\omega)}$,

$$\Omega = \{\omega \in X \mid \langle i, \omega + \rho \rangle > 0 \text{ for all } i \in I;\ \langle \beta_0, \omega + \rho \rangle = \kappa\}$$

being the affine wall of the first alcove. By *loc. cit.*, W is negligible. Since $\Upsilon\widetilde{L(\omega)} = L(\omega)$, $\omega \in \Omega$ and since Υ commutes with braiding, balance and rigidity, we see that the modules $L(\omega)$ are negligible in $\mathcal{C}$. Hence ΥW is negligible, and $\Upsilon W \otimes \Upsilon M$ is negligible, and finally ΥN is negligible. This implies that ΥN cannot have trivial summands (since $L(0)$) is not negligible).

We conclude that

$$\langle \Upsilon(\widetilde{L(\lambda_1)} \otimes \ldots \otimes \widetilde{L(\lambda_m)}) \rangle_{\mathcal{C}} = \langle \Upsilon\widetilde{L(\lambda_1)} \tilde{\otimes} \ldots \tilde{\otimes} \Upsilon\widetilde{L(\lambda_m)} \rangle_{\mathcal{C}} =$$
$$\langle \widetilde{L(\lambda_1)} \tilde{\otimes} \ldots \tilde{\otimes} \widetilde{L(\lambda_m)} \rangle_{{}_R\mathcal{C}} = \langle \widetilde{L(\lambda_1)} \otimes \ldots \otimes \widetilde{L(\lambda_m)} \rangle_{{}_R\mathcal{C}} \ \square$$

9.4. Corollary 5.11 implies that the local system

$$\langle \Upsilon\phi(L_1), \dots, \Upsilon\phi(L_m)\rangle_{\mathcal{C}}^{\heartsuit}$$

is canonically a subquotient of the local system

$$\mathrm{Tor}^{\mathcal{C}}_{\frac{\infty}{2}+0}(\mathsf{k}, \Upsilon\phi(L_1) \otimes \dots \otimes \Upsilon\phi(L_m) \otimes L(2\rho_\ell)^{\heartsuit}$$

(the action of monodromy being induced by braiding and balance on the first m factors).

9.5. Let us fix a point $\infty \in \mathbb{P}^1$ and a nonzero tangent vector $v \in T_\infty \mathbb{P}^1$. This defines an open subset

$$T\mathcal{A}^{om} \subset T\mathcal{P}^{om}$$

and the locally closed embedding

$$\xi : T\mathcal{A}^{om} \hookrightarrow T\mathcal{P}^{om+1}.$$

Given $\lambda_1, \dots, \lambda_m \in \Delta$, we consider the integrable $\hat{\mathfrak{g}}$-modules $\hat{L}(\lambda_1), \dots, \hat{L}(\lambda_m)$ of central charge $\kappa - h$.

Suppose that

$$\lambda_1 + \dots + \lambda_m = \alpha \in \mathbb{N}[I] \subset X.$$

We define $\lambda_\infty := 2\rho_\ell$, and $\vec{\lambda} := (\lambda_1, \dots, \lambda_m, \lambda_\infty)$. Note that $\vec{\lambda}$ is positive and $\alpha = \alpha(\vec{\lambda})$, in the notations of 8.3.

Denote by $\mathcal{X}^{\alpha}_{\vec{\lambda}}$ the sheaf on $\mathcal{P}^{\alpha}_{\vec{\lambda}}$ obtained by gluing $\mathcal{L}(\lambda_1), \dots, \mathcal{L}(\lambda_m), \mathcal{L}(\lambda_\infty)$. Note that

$$\mathcal{X}^{\alpha}_{\vec{\lambda}} = j_{!*}\mathcal{I}^{\alpha}_{\vec{\lambda}}$$

where $j : \mathcal{P}^{o\alpha}_{\vec{\lambda}} \hookrightarrow \mathcal{P}^{\alpha}_{\vec{\lambda}}$.

Consider the local system of conformal blocks

$$\mathrm{Hom}_{\tilde{\mathcal{O}}_\kappa}(\mathbf{1}, \hat{L}(\lambda_1)\tilde{\otimes} \dots \tilde{\otimes}\hat{L}(\lambda_m))^{\heartsuit}.$$

If $\sum_{i=1}^m \lambda_i \notin \mathbb{N}[I] \subset X$ then it vanishes by the above comparison with its "quantum group" incarnation.

9.6. **Theorem.** *Suppose that $\sum_{i=1}^m \lambda_i = \alpha \in \mathbb{N}[I]$. Then the local system of conformal blocks restricted to $T\mathcal{A}^{om}$ is isomorphic to a canonical subquotient of a "geometric" local system*

$$\xi^* R^{-2m-2}\bar{\eta}^{\alpha}_{m+1*} j_{!*}\mathcal{I}^{\alpha}_{\vec{\lambda}}.$$

Proof. This follows from Theorem 8.11 and the previous discussion. □

9.7. **Corollary.** *The above local system of conformal blocks is semisimple. It is a direct summand of the geometric local system above.*

Proof. The geometric system is semisimple by Decomposition theorem, [BBD], Théorème 6.2.5. □

9.8. Example 5.12 shows that in general a local system of conformal blocks is a *proper* direct summand of the corresponding geometric system.

Part V. MODULAR STRUCTURE ON THE CATEGORY $\mathcal{FS}$

1. Introduction

1.1. Let $C \longrightarrow S$ be a smooth proper morphism of relative dimension 1. Let $\vec{x} = (x_1, \dots, x_m)$ be an m-tuple of disjoint sections $x_k : S \longrightarrow C$. In this part we will show how to localize $\mathfrak{u}$-modules to the sections $\vec{x}$. To this end we will need a version of cohesive local system on the space of (relative) configurations on C.

The main difference from the case $C = \mathbb{A}^1$ is that the local systems are in general no more abelian one-dimensional. In fact, the monodromy in these local systems factors through the finite Heisenberg group. To stress the difference we will call them the *Heisenberg local systems.*

These local systems are constructed in Chapter 1.

1.2. Given a K-tuple of $\mathfrak{u}$-modules (or, equivalently, factorizable sheaves) $\{\mathcal{X}_k\}$ we study the sheaf $\mathfrak{g}(\{\mathcal{X}_k\})$ on C obtained by gluing the sheaves $\{\mathcal{X}_k\}$. Namely, we study its behaviour when the curve C degenerates into a stable curve $\underline{C}$ with nodes.

It appears that the sheaf $\mathfrak{g}(\{\mathcal{X}_k\})$ degenerates into a sheaf $\underline{\mathfrak{g}}(\{\mathcal{X}_k\}, \{\mathcal{R}_j\})$ obtained by gluing the sheaves $\{\mathcal{X}_k\}$ and a few copies of the sheaf $\mathcal{R}$: one for each node of $\underline{C}$ (Theorem 17.3).

The sheaf $\mathcal{R}$ is not an object of $\mathcal{FS}$, but rather of $\mathcal{FS}^{\otimes 2}$ (or, strictly speaking, of $\mathrm{Ind}\mathcal{FS} \otimes \mathrm{Ind}\mathcal{FS}$). It corresponds to the regular $\mathfrak{u}$-bimodule $\mathbf{R}$ under the equivalence Φ.

The Theorem 17.3 is the central result of this part. Its proof occupies Chapters 2–5. We study the degeneration away from the nodes in Chapter 2. We study the degeneration near the nodes in Chapter 4, after we collect the necessary information about the regular bimodule $\mathbf{R}$ in Chapter 3.

As a byproduct of geometric construction of the regular bimodule we derive the hermitian autoduality of $\mathbf{R}$ and the adjoint representation $\mathbf{ad}$.

1.3. In Chapter 5 we investigate the global sections of the sheaf $\mathfrak{g}(\{\mathcal{X}_k\})$. They form a local system on the moduli space of curves with K marked points and nonzero tangent vectors at these points (strictly speaking, the local system lives on the punctured determinant line bundle over this space). The collection of all such local systems equips the category $\mathcal{FS}$ with the *fusion*, or *modular*, structure in the terminology of [BFM] (Theorem 18.2).

Historically, first examples of modular categories appeared in the conformal field theory (WZW models), see e.g. [MS] and [TUY]. Namely, the category $\tilde{\mathcal{O}}_\kappa$ of integrable $\hat{\mathfrak{g}}$-modules of central charge $\kappa - h$ has a natural modular structure.

As far as we know, the category $\mathcal{FS}$ is the first example of nonsemisimple modular category.

1.4. In Chapter 6 we study the connection between modular categories $\tilde{\mathcal{O}}_\kappa$ and $\mathcal{C}_\zeta$ for $\zeta = \exp(\frac{\pi\sqrt{-1}}{d\kappa})$.

It appears that the modular structure on the former category can be reconstructed in terms of the modular structure on the latter one.

As a corollary we get a description of local systems of conformal blocks in WZW models in arbitrary genus as natural subquotients of some semisimple local systems of geometric origin (Theorem 19.8).

The geometric local systems are equipped with natural hermitian nondegenerate fiberwise scalar product (being direct images of perverse sheaves which are Verdier-autodual up to the replacement $\zeta \mapsto \zeta^{-1}$) which gives rise to a hermitian nondegenerate scalar product on conformal blocks in WZW models.

1.5. Our work on this part began 5 years ago as an attempt to understand the remarkable paper [CFW]. In fact, the key ingredients — Heisenberg local system, and adjoint representation — were already present in this paper.

V.Ginzburg has drawn our attention to this paper. D.Kazhdan's interest to our work proved extremely stimulating.

During these years we benefited a lot from discussions with many people. The idea of Chapter 4 is due to P.Deligne. The idea to study the degeneration of Heisenberg local system is due to B.Feigin. We are grateful to R.Hain, J.Harris and T.Pantev who took pain of answering our numerous questions about various line bundles on the moduli spaces. The second author is obliged to V.Ostrik for useful discussions of adjoint representation.

Chapter 1. Heisenberg local system

2. Notations and statement of the main result

2.1. Let $\alpha \in \mathbb{N}[X]$, $\alpha = \sum a_\mu \mu$. We denote by $\operatorname{supp}\alpha$ the subset of X consisting of all μ s.t. $a_\mu \neq 0$. Let $\pi : J \longrightarrow X$ be an unfolding of α, that is $\sharp\pi^{-1}(\mu) = a_\mu$ for any $\mu \in X$. As always, Σ_π denotes the group of automorphisms of J preserving the fibers of π.

2.1.1. The fibered product $C \times_S \ldots \times_S C$ (J times) will be denoted by C^J. The group Σ_π acts naturally on C^J, and the quotient space C^J/Σ_π will be denoted by C^α.

$\overset{\circ}{C}{}^J$ (resp. $\overset{\circ}{C}{}^\alpha$) stands for the complement to diagonals in C^J (resp. in C^α).

2.1.2. TC^J stands for the complement to the zero sections in the relative (over S) tangent bundle. So $TC^J \longrightarrow C^J$ is a $(\mathbb{C}^*)^J$-torsor. We denote by $T\overset{\circ}{C}{}^J$ its restriction to $\overset{\circ}{C}{}^J$. The group Σ_π acts freely on $T\overset{\circ}{C}{}^J$, and we denote the quotient $T\overset{\circ}{C}{}^J/\Sigma_\pi$ by $T\overset{\circ}{C}{}^\alpha$.

The natural projection $T\overset{\circ}{C}{}^J \longrightarrow T\overset{\circ}{C}{}^\alpha$ will be denoted by π, or sometimes π_J.

2.1.3. Given $j \in J$ we consider the relative tangent bundle on C^J along the j-th coordinate T_j. It is a line bundle on C^J. For $\mu \in \operatorname{supp}\alpha$ we consider the line bundle $\bigotimes_{j\in\pi^{-1}(\mu)} T_j$ on C^J. It has a natural Σ_π-equivariant structure and can be descended to the line bundle T_μ on C^α.

2.1.4. Given $\varepsilon > 0$ we denote by D_ε the standard disk of radius ε. If there is no danger of confusion we will omit ε from our notations and will denote D_ε simply by D.

The definitions of $D^J, D^\alpha, \overset{\circ}{D}{}^J, \overset{\circ}{D}{}^\alpha, T\overset{\circ}{D}{}^J, T\overset{\circ}{D}{}^\alpha$ simply copy the above definitions, and we do not reproduce them.

2.1.5. Given a surjection $\tau : J \longrightarrow K$ we consider the map $\pi_K : K \longrightarrow X$, $k \mapsto \sum_{j\in\tau^{-1}(k)} \pi(j)$. We will use the notation α_K for $\sum_{\mu\in X} \sharp\pi_K^{-1}(\mu)\mu \in \mathbb{N}[X]$.

We consider the following (infinite dimensional) manifold

$$T\overset{\circ}{C}{}^\tau = \widetilde{T\overset{\circ}{C}{}^K} \times \prod_{k\in K} T\overset{\circ}{D}{}^{\tau^{-1}(k)}$$

where $\widetilde{T\overset{\circ}{C}{}^K}$ is the space of analytic open embeddings $S_K \times D \hookrightarrow C$ such that the restriction to $S_K \times 0$ is just a K-tuple of sections $S \longrightarrow C$. Here S_K denotes the disjoint union of K copies of S.

We have an evident projection $p_K : \widetilde{T\overset{\circ}{C}{}^K} \longrightarrow T\overset{\circ}{C}{}^K$ taking the first jet. Note that p_K is a homotopy equivalence.

We denote by q_τ the natural substitution map $\overset{\circ}{T}C^\tau \longrightarrow \overset{\circ}{T}C^J$.

2.1.6. Recall (see IV.9.1) that ℓ stands for $\frac{l}{2}$ in case l is even, and for l in case l is odd. Recall (see *loc. cit.*) that

$$Y_\ell := \{\lambda \in X \mid \lambda \cdot \mu \in \ell\mathbb{Z}\ \forall \mu \in X\}$$

We denote by d the cardinality of X/Y, and by d_ℓ the determinant of the form $\frac{1}{\ell}?\cdot?$ restricted to the sublattice $Y_\ell \subset X$ Note that if ℓ is divisible by $d := \max_{i\in I} d_i$ then $\mathrm{d}_\ell = \sharp(X/Y_\ell)$. This will be the case in our applications to conformal blocks.

To handle the general case we need to introduce some new characters. We define

$$X_\ell := \{\mu \in X \otimes \mathbb{Q} \mid \mu \cdot Y_\ell \in \ell\mathbb{Z}\}$$

Evidently, $Y_\ell \subset X \subset X_\ell$, and X_ℓ is generated by X and $\{\frac{\ell}{d_i} i', i \in I\}$. So if $d|\ell$ then $X_\ell = X$ but in general this is not necessarily the case.

Note that $\mathrm{d}_\ell = \sharp(X_\ell/Y_\ell)$.

To study the modular properties of the Heisenberg local system and the category $\mathcal{FS}$ (cf. especially the Theorem 7.6(b)) we will have to modify slightly the definition of the latter one, and, correspondingly, of the category $\mathcal{C}$. We start with the category $\mathcal{C}$.

Consider the subalgebra $\mathfrak{u}' \subset \mathfrak{u}$ (see II.12.3) generated by $\theta_i, \epsilon_i, \tilde{K}_i^{\pm 1}$, $i \in I$ (notations of *loc. cit.*).

We define $\mathcal{C}'$ to be a category of finite dimensional X_ℓ-graded vector spaces $V = \oplus V_\lambda$ equipped with a structure of a left $\mathfrak{u}'$-module compatible with X_ℓ-gradings and such that

$$\tilde{K}_i x = \zeta^{\langle d_i i, \lambda\rangle} x$$

for $x \in V_\lambda$, $i \in I$.

This is well defined since $\langle d_i i, X_\ell\rangle \in \mathbb{Z}$ for any $i \in I$.

We have a natural inclusion of a full subcategory $\mathcal{C} \hookrightarrow \mathcal{C}'$.

We define the category $\mathcal{FS}'$ exactly as in III.5.2, just replacing all occurences of X by X_ℓ.

We have a natural inclusion of a full bbrt subcategory $\mathcal{FS} \hookrightarrow \mathcal{FS}'$, and the equivalence $\Phi: \mathcal{FS} \longrightarrow \mathcal{C}$ extends to the same named equivalence $\Phi: \mathcal{FS}' \longrightarrow \mathcal{C}'$. The proof is the same as in III; one only has to replace X by X_ℓ everywhere.

Recall that in case $d|\ell$ (the case of interest for applications to conformal blocks) we have $X_\ell = X, \mathcal{FS}' = \mathcal{FS}, \mathcal{C}' = \mathcal{C}$.

From now on we will restrict ourselves to the study of the categories $\mathcal{FS}', \mathcal{C}'$. However, in order not to scare the reader away by a bunch of new notations we will denote them by $\mathcal{FS}, \mathcal{C}$.

The interested reader will readily perform substitutions in the text below.

2.1.7. Consider a function $n: X_\ell \longrightarrow \mathbb{Z}[\frac{1}{2\mathrm{d}}]$ such that $n(\mu+\nu) = n(\mu) + n(\nu) + \mu\cdot\nu$. We will choose n of the following form:

$$n(\mu) = \frac{1}{2}\mu\cdot\mu + \mu\cdot\nu_0$$

for some $\nu_0 \in X_\ell$. From now on we fix such a function n and the corresponding ν_0.

Let g be the genus of our relative curve $C \longrightarrow S$.

2.1.8. *Definition.* (a) We will denote the unique homomorphism

$$\mathbb{N}[X_\ell] \longrightarrow X_\ell$$

identical on X_ℓ by $\alpha \mapsto \alpha^\sim$;

(b) $\alpha \in \mathbb{N}[X_\ell]$ is called *g-admissible* if $\alpha^\sim = (2g-2)\nu_0$.

2.1.9. From now on we assume that ζ is a primitive root of unity of degree l. Then $\zeta = \exp(2\pi\sqrt{-1}\frac{k}{l})$ for some integer k prime to l. We fix k, and for a rational number q we define $\zeta^q := \exp(2\pi\sqrt{-1}q\frac{k}{l})$.

2.1.10. Given $\alpha = \sum a_\mu \mu \in \mathbb{N}[Y]$ and its unfolding $\pi: J \longrightarrow Y$, we consider the following one-dimensional local system $\mathcal{I}^J$ on $\overset{\circ}{T}D^J$:

by definition, its monodromies are as follows:

around diagonals: $\zeta^{2\pi(j_1)\cdot\pi(j_2)}$;

around zero sections of tangent bundle: $\zeta^{2n(\pi(j))}$.

We define the one-dimensional local system $\mathcal{I}^\alpha$ on $\overset{\circ}{T}D^\alpha$ as $\mathcal{I}^\alpha := (\pi_{J*}\mathcal{I}^J)^{\Sigma_\pi,-}$ (cf. III(46)).

2.1.11. For a line bundle $\mathcal{L}$ we denote the corresponding $\mathbb{C}^*$-torsor by $\dot{\mathcal{L}}$.

2.2. Statement of the main result.

2.2.1. *Definition.* The *Heisenberg local system* $\mathcal{H}$ is the following collection of data:

1) A local system $\mathcal{H}^\alpha$ on $\overset{\circ}{T}C^\alpha$ for each admissible $\alpha \in \mathbb{N}[X_\ell]$;

2) *Factorization isomorphisms:* for each $\alpha \in \mathbb{N}[X_\ell]$, unfolding $\pi: J \longrightarrow X_\ell$, surjection $\tau: J \longrightarrow K$, the following isomorphisms are given:

$$\phi_\tau: q_\tau^*\pi_J^*\mathcal{H}^\alpha \xrightarrow{\sim} p_K^*\pi_K^*\mathcal{H}^{\alpha_K} \boxtimes \boxed{\times}_{k\in K} \mathcal{I}^{\tau^{-1}(k)}$$

satisfying the usual associativity constraints.

2.2.2. *Theorem.* Let $\delta \longrightarrow S$ denote the *determinant* line bundle of the family $C \longrightarrow S$ (see e.g. [KM]). Then after the base change

$$\begin{array}{ccc} C_\delta & \longrightarrow & C \\ \downarrow & & \downarrow \\ \dot{\delta} & \longrightarrow & S \end{array}$$

there exists a Heisenberg Local System $\mathcal{H}$.

The dimension of $\mathcal{H}$ is equal to d_ℓ^g, and the monodromy around the zero section of δ is equal to $(-1)^{\mathrm{rk} X_\ell}\zeta^{12\nu_0\cdot\nu_0}$.

2.2.3. *Remark.* In the case $g = 1$ the line bundle δ^{12} is known to be trivial. It is easy to see that there exists a one-dimensional local system on $\dot{\delta}$ with any given monodromy around the zero section.

We will construct the Heisenberg local system $\mathcal{H}$ over S (as opposed to $\dot{\delta}$). Lifting it to $\dot{\delta}$ and twisting by the above one-dimensional systems we can obtain a Heisenberg local system with any given scalar monodromy around the zero section.

The construction of the desired Heisenberg System will be given in the rest of this Chapter.

3. The scheme of construction

3.1. First note that it suffices to construct the desired local system $\mathcal{H}$ for $\zeta = \exp(\frac{\pi\sqrt{-1}}{\ell})$. If k is prime to l, and $\zeta' = \exp(2\pi\sqrt{-1}\frac{k}{l})$, then $\mathcal{H}_{\zeta'}$ is obtained from $\mathcal{H}_\zeta$ just by application of a Galois automorphism of our field k. So till the end of construction we will assume that $\zeta = \exp(\frac{\pi\sqrt{-1}}{\ell})$.

3.2. In what follows everything is relative over the base S. To unburden the notations we will pretend though that C is an absolute curve. Thus $H^1(C)$ stands, say, for the local system of $\mathbb{Z}$-modules of rank $2g$ over S.

For $\alpha \in \mathbb{N}[X_\ell]$ we introduce the following divisor $\mathcal{D}^\alpha$ on C^α:

$$\mathcal{D}^\alpha = \frac{\mathrm{d}_\ell^g}{\ell}\Big(\sum_{\mu\neq\nu} \mu\cdot\nu\Delta_{\mu\nu} + \frac{1}{2}\sum_{\mu}\mu\cdot\mu\Delta_{\mu\mu}\Big)$$

where $\Delta_{\mu\nu}$, $\mu, \nu \in \operatorname{supp}\alpha$, stands for the corresponding diagonal in C^α. Note that for $g \geq 2$ all the coefficients of the above sum are integers. To simplify the exposition we will assume that $g \geq 2$. For the case of $g = 1$ see 3.6.

Given an unfolding $\pi : J \longrightarrow X_\ell$ we denote the pullback of $\mathcal{D}^\alpha$ under π_J by $\mathcal{D}^J$; this is a divisor on C^J.

We consider the (relative) Picard scheme $\operatorname{Pic}(C) \otimes X_\ell$. The group of its connected components is naturally isomorphic to X_ℓ. Each component carries a canonical polarization ω which we presently describe. It is a skew-symmetric bilinear form on $H_1(\operatorname{Pic}^0(C) \otimes X_\ell) = H_1(C) \otimes X_\ell$ equal to the tensor product of the canonical cup-product form on $H_1(C)$ and the symmetric bilinear form $\frac{\mathrm{d}_\ell^g}{\ell}(?\cdot?)$ on X_ℓ. Note that $\frac{\mathrm{d}_\ell^g}{\ell}(?\cdot?)$ is positive definite, so ω is (relatively) ample.

We denote by $\mathrm{AJ}_\alpha : \; C^\alpha \longrightarrow \operatorname{Pic}(C) \otimes X_\ell$ the Abel-Jacobi map $\mathrm{AJ}_\alpha(\sum \mu x_\mu) = \sum(x_\mu) \otimes \mu$.

The admissibility condition implies that the Abel-Jacobi map lands into the connected component $(\operatorname{Pic}(C)\otimes X_\ell)_{(2g-2)\nu_0}$ to be denoted by A. Note that the projection $A \longrightarrow S$ has a canonical section $\Omega \otimes \nu_0$, so A is a genuine abelian variety, not just a torsor over one. Here Ω denotes the (relative) canonical line bundle on C.

3.2.1. *Definition.* We define the following line bundle $\mathcal{L}_\alpha$ on C^α:

$$\mathcal{L}_\alpha = \otimes_{\mu\in\operatorname{supp}(\alpha)} T_\mu^{\otimes \frac{\mathrm{d}_\ell^g}{\ell} n(\mu)} \otimes \mathcal{O}(\mathcal{D}^\alpha)$$

The desired construction is an easy consequence of the following Propositions:

3.3. **Proposition.** There is a unique line bundle $\mathcal{L}$ on A such that for any g-admissible α we have $\mathcal{L}_\alpha = \mathrm{AJ}^*_\alpha(\mathcal{L})$. The first Chern class $c_1(\mathcal{L}) = -[\omega]$.

3.4. **Proposition.** There exists a local system $\mathfrak{H}$ on $\dot{\mathcal{L}} \times_S \dot{\delta}$ such that $\dim \mathfrak{H} = \mathrm{d}^g_\ell$ (see 2.1.6); the monodromy around the zero section of $\mathcal{L}$ is equal to $\zeta^{\frac{2\ell}{\mathrm{d}^g_\ell}}$; and the monodromy around the zero section of δ is equal to $(-1)^{\mathrm{rk} X_\ell} \zeta^{12\nu_0 \cdot \nu_0}$.

3.5. In the remainder of this section we derive the desired construction from the above Propositions.

We fix a g-admissible α. We denote by $\dot{\mathrm{AJ}}_\alpha$ the natural map between the total spaces of the corresponding $\mathbb{C}^*$-torsors:

$$\dot{\mathrm{AJ}}_\alpha : (\mathrm{AJ}^*_\alpha(\mathcal{L}))^{\cdot} \longrightarrow \dot{\mathcal{L}}$$

By the Proposition 3.3 we have an isomorphism

$$\mathrm{AJ}^*_\alpha(\mathcal{L}) \xrightarrow{\sim} \mathcal{L}_\alpha \tag{287}$$

It is clear from the definition of $\mathcal{L}_\alpha$ that the pullback of $\mathcal{L}_\alpha$ to TC^α has the canonical meromorphic section s_α. The restriction of this section to $\overset{\circ}{T}C^\alpha$ does not have poles nor zeros, and hence it defines the same named section s_α of the $\mathbb{C}^*$-torsor $(\mathrm{AJ}^*_\alpha(\mathcal{L}))^{\cdot}$.

We change the base to $\dot{\delta}$, and preserve the notations $\mathrm{AJ}_\alpha, s_\alpha$ for the base change of the same named morphism and section. By the Proposition 3.4 we have the local system $\mathfrak{H}$ on $\dot{\mathcal{L}} \times_S \dot{\delta}$.

We define $\mathcal{H}^\alpha$ to be $s^*_\alpha \dot{\mathrm{AJ}}_\alpha{}^* \mathfrak{H}$ twisted by the one-dimensional sign local system.

The proof of the above Propositions and the construction of factorization isomorphisms will be given in the following sections.

3.6. The above construction does not work as stated in the case of elliptic curves: the line bundle $\mathcal{O}(\mathcal{D}^\alpha)$ in the Definition 3.2.1 does not make sense since the coefficients $\frac{\mathrm{d}_\ell}{2\ell}\mu \cdot \mu$ of the divisor $\mathcal{D}^\alpha$ apriori may be halfintegers.

We will indicate how to carry out the construction in this case.

For any $N \in \mathbb{Z}$ such that $N\frac{\mathrm{d}_\ell}{2\ell}\mu \cdot \mu, N\frac{\mathrm{d}_\ell}{\ell}n(\mu) \in \mathbb{Z}\ \forall \mu \in X_\ell$ we define

$$\mathcal{L}^N_\alpha := \bigotimes_{\mu \in \mathrm{supp}\alpha} T^{N\frac{\mathrm{d}_\ell}{\ell}n(\mu)} \otimes \mathcal{O}(N\mathcal{D}^\alpha).$$

We will prove the following versions of the above Propositions.

3.6.1. **Proposition.** There is a unique line bundle $\mathcal{L}^N$ on A such that for any g-admissible $\alpha \in \mathbb{N}[X_\ell]$ we have $\mathcal{L}^N_\alpha = \mathrm{AJ}^*_\alpha \mathcal{L}^N$. The first Chern class $c_1(\mathcal{L}^N) = -N[\omega]$.

3.6.2. **Proposition.** There exists a local system $\mathfrak{H}^N$ on $(\mathcal{L}^N)^{\cdot}$ such that $\dim \mathfrak{H}^N = \mathrm{d}_\ell$. The monodromy of $\mathfrak{H}^N$ around the zero section of $\mathcal{L}^N$ is equal to $\zeta^{\frac{2}{N\mathrm{d}_\ell}}$.

Moreover, for any $N|N'$ we have $\mathfrak{H}^N = [\frac{N'}{N}]^* \mathfrak{H}^{N'}$ where $[\frac{N'}{N}] : \mathcal{L}^N \longrightarrow \mathcal{L}^{N'}$ is the map of raising to the $(\frac{N'}{N})$-the power.

3.6.3. Now $\mathcal{H}^\alpha$ is defined as $s^*_{\alpha,N}\dot{\mathrm{AJ}}^*\mathfrak{H}^N$ for any N as above exactly as in 3.5. In fact, it is enough to take $N = 2$.

4. The universal line bundle

In this section we will give a proof of the Propositions 3.3 and 3.6.1.

4.1. First we formulate a certain generalization. Suppose given a free $\mathbb{Z}$-module Λ of finite rank with an even symmetric bilinear pairing $(,): \Lambda \times \Lambda \longrightarrow \mathbb{Z}$. We fix an element $\nu \in \Lambda$, and a function $\mathrm{b}: \Lambda \longrightarrow \mathbb{Z}$, $\mathrm{b}(\lambda) := \frac{1}{2}(\lambda, \lambda) + (\lambda, \nu)$.

4.1.1. For $\alpha \in \mathbb{N}[\Lambda]$ we introduce the following divisor $\mathcal{D}^\alpha$ on C^α:

$$\mathcal{D}^\alpha = \sum_{\mu \neq \lambda} (\mu, \lambda)\Delta_{\mu\lambda} + \frac{1}{2}\sum_{\mu} (\mu, \mu)\Delta_{\mu\mu}$$

where $\Delta_{\mu\lambda}$, $\mu, \lambda \in \mathrm{supp}\alpha$, stands for the corresponding diagonal in C^α. Note that all the coefficients of the above sum are integers.

Given an unfolding $\pi: J \longrightarrow \Lambda$ we denote the pullback of $\mathcal{D}^\alpha$ under π_J by $\mathcal{D}^J$; this is a divisor on C^J.

We consider the (relative) Picard scheme $\mathrm{Pic}(C) \otimes \Lambda$. The group of its connected components is naturally isomorphic to Λ. Each component carries a canonical polarization ω which we presently describe. It is a skew-symmetric bilinear form on $H_1(\mathrm{Pic}^0(C) \otimes \Lambda) = H_1(C) \otimes \Lambda$ equal to the tensor product of the canonical cup-product form on $H_1(C)$ and the symmetric bilinear form $(,)$ on Λ.

We denote by $\mathrm{AJ}_\alpha: C^\alpha \longrightarrow \mathrm{Pic}(C) \otimes \Lambda$ the Abel-Jacobi map $\mathrm{AJ}_\alpha(\sum \mu x_\mu) = \sum (x_\mu) \otimes \mu$.

The admissibility condition implies that the Abel-Jacobi map lands into the connected component $(\mathrm{Pic}(C) \otimes \Lambda)_{(2g-2)\nu}$ to be denoted by A_Λ.

Note that the projection $A_\Lambda \longrightarrow S$ has a canonical section $\Omega \otimes \nu$, so A_Λ is a genuine abelian variety, not just a torsor over one. Here Ω denotes the (relative) canonical line bundle on C.

4.1.2. *Definition.* We define the following line bundle $\mathcal{L}_\alpha$ on C^α:

$$\mathcal{L}_\alpha = \otimes_{\mu \in \mathrm{supp}(\alpha)} T_\mu^{\otimes \mathrm{b}(\mu)} \otimes \mathcal{O}(\mathcal{D}^\alpha)$$

An element $\alpha = \sum a_\lambda \lambda \in \mathbb{N}[\Lambda]$ (formal sum) is called g-admissible if $\alpha^\sim = (2g - 2)\nu$.

Now we are able to formulate the Proposition generalizing 3.3.

4.2. **Proposition.** There is a unique line bundle $\mathcal{L}(\Lambda, (,), \nu)$ on A_Λ such that for any g-admissible α we have $\mathcal{L}_\alpha = \mathrm{AJ}^*_\alpha(\mathcal{L}(\Lambda, (,), \nu))$. The first Chern class $c_1(\mathcal{L}(\Lambda, (,), \nu)) = -[\omega]$.

4.3. We start the proof of the Proposition 4.2 with the following Lemma. Let $\pi_J : J \longrightarrow \Lambda$ be an unfolding of α. Let $\tau: J \longrightarrow K$ be a surjection, and $\pi_K : K \longrightarrow \Lambda$ be an unfolding of α_K as in 2.1.5. To simplify the notations we will denote α_K by β. Let σ_τ denote the natural ("diagonal") embedding $C^K \hookrightarrow C^J$.

4.3.1. *Lemma.* There is a canonical isomorphism

$$\sigma_\tau^* \pi_J^*(\mathcal{L}_\alpha) = \pi_K^*(\mathcal{L}_\beta)$$

Proof. It is enough to prove the Lemma in the case $|J| = |K| + 1$. So we fix $i, j \in J$ such that $\tau(i) = \tau(j) = k$, and we denote $\pi_J(i)$ by μ_i, and $\pi_J(j)$ by μ_j.

We have $\pi_J^*(\mathcal{D}^\alpha) = \mu_i \cdot \mu_j \Delta_{ij} + \mathcal{D}'$ where $\Delta_{ij} \not\subset \operatorname{supp}(\mathcal{D}')$.

Moreover, it is clear that $\mathcal{D}' \cap \Delta_{ij} = \sigma_\tau(\pi_K^*(\mathcal{D}^\beta))$, and hence

$$\sigma_\tau^* \pi_\tau^*(\mathcal{O}(\mathcal{D}')) = \pi_K^*(\mathcal{O}(\mathcal{D}^\beta))$$

On the other hand, for any smooth divisor $\mathcal{D}$ we have a canonical isomorphism $\mathcal{O}(\mathcal{D})|_\mathcal{D} = \mathcal{N}_\mathcal{D}$ (the normal bundle).

In particular, we have $\sigma_\tau^*(\mathcal{O}(\Delta_{ij})) = \sigma_\tau^*(T_i) = \sigma_\tau^*(T_j) = T_k$.

Thus $\sigma_\tau^*(T_i^{\mathrm{b}(\mu_i)} \otimes T_j^{\mathrm{b}(\mu_j)} \otimes \mathcal{O}((\mu_i, \mu_j)\Delta_{ij})) = T_k^{\mathrm{b}(\mu_i) + \mathrm{b}(\mu_j) + (\mu_i, \mu_j)} = T_k^{\mathrm{b}(\mu_i + \mu_j)} = T_k^{\mathrm{b}(\pi(k))}$.

Finally, if $m \neq i, j$, then evidently $\sigma_\tau^*(T_m) = T_{\tau(m)}$.

Putting all this together we obtain the statement of the Lemma. □

4.4. To prove the Proposition we have to check a necessary condition: that the first Chern class of $\mathcal{L}_\alpha$ is a pullback of some cohomology class on A_Λ under the Abel-Jacobi map AJ_α. This is the subject of the following Lemma.

4.4.1. *Lemma.* If α is admissible, then $c_1(\mathcal{L}_\alpha) = \mathrm{AJ}_\alpha^*(-[\omega])$.

Proof. Let us choose an unfolding $\pi: J \longrightarrow \Lambda$. We denote by π_J the corresponding projection $C^J \longrightarrow C^\alpha$.

It is enough to prove that the pullback of the both sides to C^J under π_J coincide.

We introduce the following family of 2-cycles in C^J. For 1-cycles a, b in C, and $i \neq j \in J$, we denote by $a_i \times b_j$ the following product cycle: the i-th coordinate runs along the cycle a, the j-th coordinate runs along the cycle b, all the rest coordinates are fixed. The homology class of $a_i \times b_j$ depends only on i, j and the classes of a and b.

We denote by f_i the following 2-cycle: the i-th coordinate runs along the fundamental cycle of C, all the rest coordinates are fixed. The homology class of f_i depends on i only.

It suffices to check that the pairings of both sides of 4.4.1 with this family of cycles coincide.

We have:

$$\langle -\omega, a_i \times b_j \rangle = (\pi(i), \pi(j)) a \cap b = (a_i \times b_j) \cap ((\pi(i), \pi(j))\Delta_{ij}) = \langle c_1(\mathcal{L}_\alpha), a_i \times b_j \rangle; \tag{288}$$

$$\langle -\omega, f_i \rangle = -g(\pi(i), \pi(j)); \tag{289}$$

$$\langle c_1(\mathcal{L}_\alpha), f_i \rangle = (2-2g)\mathrm{b}(\pi(i)) + f_i \cap \pi^*(\mathcal{D}^\alpha) = (2-2g)\mathrm{b}(\pi(i)) + (\pi(i), (\sum_{j\in J} \pi(j) - \pi(i))). \tag{290}$$

To assure the equality of (289) and (290) we must have

$$(1-g)(\pi(i), \pi(i)) = (2-2g)\mathrm{b}(\pi(i)) + (\pi(i), \sum_{j\in J} \pi(j)), \tag{291}$$

that is,

$$(1-g)((\pi(i), \pi(i)) - 2\mathrm{b}(\pi(i))) = (\pi(i), \sum_{j\in J} \pi(j)) \tag{292}$$

which is precisely the admissibility condition. □

4.5. Let us choose a basis I of Λ.

Using the Lemma 4.3.1 we see that it suffices to prove the Proposition for α of a particular kind, namely

$$\alpha = \sum_{i\in I} a_i i + \sum_{i\in I} a_{-i}(-i)$$

where all the positive integers a_i, a_{-i} are big enough.

4.5.1. We define $\alpha_+ := \sum_{i\in I} a_i i$, and $\alpha_- := \sum_{i\in I} a_{-i}(-i)$.

We consider the following Abel-Jacobi maps:

$$\mathrm{AJ}_+ : \; C^{\alpha_+} \longrightarrow (\mathrm{Pic}(C) \otimes \Lambda)_{\alpha_+} =: A_+,$$

and

$$\mathrm{AJ}_- : \; C^{\alpha_-} \longrightarrow (\mathrm{Pic}(C) \otimes \Lambda)_{\alpha_-} =: A_-.$$

We have

$$\mathrm{AJ}_\alpha = m \circ (\mathrm{AJ}_+ \times \mathrm{AJ}_-)$$

where $m : \; A_+ \times_S A_- \longrightarrow A_\Lambda$ is the addition map.

If all the $a_{\pm i}$ are bigger than g, then the map

$$(\mathrm{AJ}_+ \times \mathrm{AJ}_-)^* : \; \mathrm{Pic}^0(A_+ \times_S A_-) \longrightarrow \mathrm{Pic}^0(C^{\alpha_+} \times_S C^{\alpha_-})$$

is an isomorphism, and the induced map on the whole Picard groups is an inclusion.

According to the Lemma 4.4.1, $c_1(\mathcal{L}_\alpha) = \mathrm{AJ}_\alpha^*(-[\omega]) = (\mathrm{AJ}_+ \times \mathrm{AJ}_-)^*(m^*(-[\omega]))$. So we deduce that there exists a unique line bundle $\mathcal{L}'$ on $A_+ \times_S A_-$ such that $\mathcal{L}_\alpha = (\mathrm{AJ}_+ \times \mathrm{AJ}_-)^*\mathcal{L}'$.

It remains to show that $\mathcal{L}' = m^*\mathcal{L}$ for some line bundle $\mathcal{L}$ on A_Λ (necessarily uniquely defined). To this end it is enough to verify that the restrictions of $\mathcal{L}'$ to the fibers of m are trivial line bundles.

4.5.2. We choose an unfolding J of α. We choose a surjection $\tau : J \longrightarrow K$ with the following property: $K = K_0 \sqcup K_1$; τ is one-to-one over K_1, and for any $k \in K_0$ we have $\tau^{-1}(k) = \{i, -i\}$ for some $i \in I$.

Moreover, we assume that α is big enough so that for each $i \in I$ both i and $-i$ appear at least g times in τ^{-1} of both K_0 and K_1.

Recall that σ_τ stands for the diagonal embedding $C^K \hookrightarrow C^J$.

It is clear that $C^K = C^{K_0} \times_S C^{K_1}$, and the Abel-Jacobi map $\mathrm{AJ}_J \circ \sigma_\tau : C^K \longrightarrow A_\Lambda$ factors through the projection onto the second factor.

Fix a point $a \in A_\Lambda$. Let us choose $p \in C^{K_1}$ such that $\mathrm{AJ}_J(\sigma_\tau(C^{K_0} \times p)) = a$. Then $(\mathrm{AJ}_+ \times \mathrm{AJ}_-) \circ \sigma_\tau$ maps $C^{K_0} \times p$ to the fiber $m^{-1}(a)$.

It is easy to see that the induced maps on the Picard groups $((\mathrm{AJ}_+ \times \mathrm{AJ}_-) \circ \sigma_\tau)^* : \mathrm{Pic}(m^{-1}(a)) \longrightarrow \mathrm{Pic}(C^{K_0})$ is injective.

Thus we only have to check that

$$((\mathrm{AJ}_+ \times \mathrm{AJ}_-) \circ \sigma_\tau)^* \mathcal{L}'|_{C^{K_0} \times p} = (\sigma_\tau \circ \pi_J)^* \mathcal{L}_\alpha|_{C^{K_0} \times p}$$

is trivial.

According to the Lemma 4.3.1, this line bundle is equal to $\pi_K^*(\mathcal{L}_{\alpha_K})$. It is clear from the definition that $\mathcal{L}_{\alpha_K}$ is lifted from the projection to the factors carrying nonzero weights. In particular, the restriction of $\mathcal{L}_{\alpha_K}$ to a fiber of this projection is trivial.

This completes the proof of the Proposition 4.2.

4.6. To prove the Proposition 3.3 it suffices to apply Proposition 4.2 to the case $\Lambda = X_\ell$; $(?,?) = \frac{\mathrm{d}_\ell^g}{\ell} ?\cdot?$; $\nu = \nu_0$; $\mathrm{b}(\lambda) = \frac{\mathrm{d}_\ell^g}{\ell} n(\lambda)$.

To prove the Proposition 3.6.1 we take $\Lambda = X_\ell$; $(?,?) = N\frac{\mathrm{d}_\ell^g}{\ell} ?\cdot?$; $\nu = \nu_0$; $\mathrm{b}(\lambda) = N\frac{\mathrm{d}_\ell^g}{\ell} n(\lambda)$. □

5. The universal local system

In this section we will give a proof of the Propositions 3.4 and 3.6.2.

5.1. Recall the notations of 4.1. Let us take $\Lambda = X_\ell \oplus Y_\ell$; $((x_1, y_1), (x_2, y_2)) = -\frac{1}{\ell}(y_1 \cdot x_2 + y_2 \cdot x_1 + y_1 \cdot y_2)$; $\nu = (\nu_0, 0)$.

We denote A_Λ by A'. We have an obvious projection $\mathrm{pr}_1 : A' \longrightarrow A$. We denote the line bundle $\mathcal{L}(\Lambda, (,), \nu)$ (notations of 4.2) by $\mathcal{L}'$.

5.2. **Theorem.** (a) $\mathcal{L}'$ is relatively ample with respect to pr_1.

(b) The direct image $\mathcal{E} := \mathrm{pr}_{1*}\mathcal{L}'$ is a locally free sheaf of rank d_ℓ^g.

(c) In the situation of 3.4 (that is, $g > 1$) we have $\det(\mathcal{E}) = \mathcal{L} \otimes (p^*\delta)^{\mathrm{d}_\ell^g(-\frac{1}{2}\mathrm{rk}X_\ell + 6\frac{\nu_0 \cdot \nu_0}{\ell})}$, where p stands for the projection $A \longrightarrow S$.

(d) Assume $g = 1$. Then in the notations of 3.6 we have for any N as in *loc. cit.* $(\det(\mathcal{E}))^{\otimes N} = \mathcal{L}^N \otimes p^*\delta^{\otimes iN}$ for some $i \in \mathbb{Z}$.

5.2.1. Let us construct the desired local system assuming the Theorem 5.2. By the virtue of [BK], Corollary 3.4 and Corollary 4.2, $\mathcal{E}$ is naturaly equipped with the flat projective connection. Hence its lifting to $(\det(\mathcal{E}))^{\cdot}$ carries a flat connection with scalar monodromy around the zero section equal to $\exp(\frac{2\pi\sqrt{-1}}{\mathrm{d}_\ell^g})$. Isomorphism 5.2.(c) yields the map $m : \dot{\mathcal{L}} \times_S \dot{\delta} \longrightarrow (\det(\mathcal{E}))^{\cdot}$, where $m(\lambda, t) = \lambda \otimes t^{\mathrm{d}_\ell^g(-\frac{1}{2}\mathrm{rk} X_\ell + 6\frac{\nu_0 \cdot \nu_0}{\ell})}$. It is clear that $m^*\mathcal{E}$ is a locally free sheaf with flat connection, whose monodromy around the 0-section of $\mathcal{L}$ is equal to $\exp(\frac{2\pi\sqrt{-1}}{\mathrm{d}_\ell^g}) = \zeta^{\frac{2\ell}{\mathrm{d}_\ell^g}}$, and monodromy around the zero section of δ is equal to $(-1)^{\mathrm{rk} X_\ell}\zeta^{12\nu_0 \cdot \nu_0}$. This proves the Proposition 3.4.

The proof of the Proposition 3.6.2 is even simpler. Note that for any M the lifting of $\mathcal{E}$ to $((\det(\mathcal{E}))^{\otimes M})^{\cdot}$ carries a flat connection with scalar monodromy around the zero section equal to $\exp(\frac{2\pi\sqrt{-1}}{M\mathrm{d}_\ell})$.

Now 5.2(d) implies the isomorphism $(\det(\mathcal{E}))^{\otimes 12N} = (\mathcal{L}^N)^{\otimes 12}$ for any N as in 3.6 (recall that $\delta^{\otimes 12}$ is trivial).

So for any N we can define $\mathfrak{H}^N$ to be $m^*\mathcal{E}$ where $m : \mathcal{L}^N \longrightarrow (\det(\mathcal{E}))^{\otimes 12N}$ is the composition of raising to the 12-th power and the above isomorphism. □

5.3. We now proceed with the proof of the Theorem 5.2. The statements (a) and (b) follow immediately from the Proposition 4.2 and the Riemann-Roch formula for abelian varieties (see e.g. [M1]).

To prove (c) we need two auxilliary Propositions.

As usually, it is more convenient to work in greater generality. Thus suppose given Λ, $(,)$, ν as in 4.1.

We will denote the canonical section $\Omega \otimes \nu$ of $A_\Lambda = (\mathrm{Pic}(C) \otimes \Lambda)_{(2g-2)\nu}$ by s. This section identifies A_Λ with $\mathrm{Pic}^0(C) \otimes \Lambda$, and we will use this identification in what follows.

5.4. **Proposition.** $s^*(\mathcal{L}(\Lambda, (,), \nu) = \delta^{-6(\nu,\nu)}$ (notations of 4.2).

5.5. **Corollary.** $\mathcal{L}(\Lambda, (,), \nu) = \mathcal{L}(\Lambda, (,), 0) \otimes p^*\delta^{-6(\nu,\nu)}$.

Proof. Let first $g = 1$. Then note that the condition of g-admissibility reads $\alpha^\sim = 0$ independently of ν. To stress the dependence of $\mathcal{L}^\alpha$ on ν we will include ν as the subindex for a moment.

We have $T_\lambda = p^*(\delta^{-1})$ for any $\lambda \in \Lambda$.

For any 1-admissible α, and any ν we have

$$\mathcal{L}_\nu^\alpha = \mathcal{L}_0^\alpha \otimes \bigotimes_{\lambda \in \mathrm{supp}\alpha} T_\lambda^{(\lambda,\nu)} = \mathcal{L}_0^\alpha \otimes p^*(\delta^{-(\alpha^\sim,\nu)}) = \mathcal{L}_0^\alpha$$

It follows that $\mathcal{L}(\Lambda, (,), \nu) = \mathcal{L}(\Lambda, (,), 0)$ for any ν. On the other hand $6(\nu, \nu) \in 12\mathbb{Z}$, and $\delta^{\otimes 12}$ is known to be trivial.

This takes care of the case $g = 1$.

Assume now $g > 1$.

In view of the Proposition 5.4 it is enough to show that $\mathcal{L}(\Lambda, (,), \nu) = \mathcal{L}(\Lambda, (,), 0) \otimes p^*\Xi$ for some line bundle Ξ on S. So we can assume that S is a point.

Let us choose a nonzero holomorphic form σ with zero divisor (σ) on C. Then, for any $\alpha \in \mathbb{N}[\Lambda]$, we have an embedding

$$\iota_\sigma : C^\alpha \hookrightarrow C^{\alpha+(2g-2)\nu};\ x \mapsto x + (\sigma) \otimes \nu.$$

Recall the Definition 4.1.2 of g-admissibility. Since g is fixed, while ν varies, till the end of the proof we will replace this term by *ν-admissibility.*

Note that if α is 0-admissible, then $\alpha + (2g-2)\nu$ is ν-admissible.

It is clear that for 0-admissible α we have

$$\iota_\sigma^* \mathcal{L}^{\alpha+(2g-2)\nu} = \bigotimes_{\mu\in\mathrm{supp}\alpha} T_\mu^{\mathrm{b}_\nu(\mu)} \otimes \mathcal{O}(\mathcal{D}^\alpha) \otimes \bigotimes_{\mu\in\mathrm{supp}\alpha} T_\mu^{-(\mu,\nu)} =$$

$$= \bigotimes_{\mu\in\mathrm{supp}\alpha} T_\mu^{\frac{1}{2}(\mu,\mu)} \otimes \mathcal{O}(\mathcal{D}^\alpha) = \bigotimes_{\mu\in\mathrm{supp}\alpha} T_\mu^{\mathrm{b}_0(\mu)} \otimes \mathcal{O}(\mathcal{D}^\alpha) = \mathcal{L}^\alpha$$

(notations of 4.1).

Now the Corollary follows from the definition (and uniqueness) of $\mathcal{L}(\Lambda, (,), \nu)$ (Proposition 4.2). □

5.6. **Proposition.** Assume $g > 1$. Then $[Rp_{\Lambda*}(\mathcal{L}(\Lambda, (,), 0))]^1 = \delta^{-\frac{1}{2}d^g \mathrm{rk}\Lambda}$ where d stands for $\det(-(,)\ )$, while p_Λ denotes the projection $A_\Lambda \longrightarrow S$, and $[Rp_{\Lambda*}(?)]^1$ denotes the determinant of the complex $Rp_{\Lambda*}(?)$ (see e.g. [KM]).

Note that d equals the Euler characteristic of $Rp_{\Lambda*}(\mathcal{L}(\Lambda, (,), 0))$.

5.7. Let us derive the Theorem 5.2(c) assuming the above Propositions.

Note that the scalar product $((\nu_0, 0), (\nu_0, 0)) = 0$, and hence the Corollary 5.5 implies that $\mathcal{L}'$ does not depend on ν_0. It follows from *loc. cit.* that the RHS of 5.2(c) does not depend on ν_0 either.

We will assume that $\nu_0 = 0$ until the end of the proof.

For $M \in \mathbb{Z}$ and an abelian group G let $[M] : G \longrightarrow G$ denote the multiplication by M. We choose M such that $MX_\ell \subset Y_\ell$.

5.7.1. *Lemma.* $[M]^* \det(\mathcal{E}) = [M]^*(\mathcal{L} \otimes (p^*\delta)^{d_\ell^g(-\frac{1}{2}\mathrm{rk}X_\ell + 6\frac{\nu_0\cdot\nu_0}{\ell})})$.

Proof. Let us define $+_M : X_\ell \oplus Y_\ell \longrightarrow Y_\ell$ as $+_M(x, y) = Mx + y$.

We have an equality of quadratic forms

$$([M] \times \mathrm{id})^*(,) = (+_M)^*(-\frac{1}{\ell}?\cdot?) + \mathrm{pr}_1^*(\frac{M^2}{\ell}?\cdot?)$$

(note that all the quadratic forms involved are integer valued and even).

It implies the equality of line bundles:

$$([M] \times \mathrm{id})^*\mathcal{L}' = (+_M \otimes \mathrm{id})^*(\mathcal{L}(Y_\ell, -\frac{1}{\ell}?\cdot?, 0) \otimes \mathrm{pr}_1^*\mathcal{L}(X_\ell, \frac{M^2}{\ell}?\cdot?, 0))$$

Note that $(\mathrm{pr}_1, +_M) : X_\ell \oplus Y_\ell \longrightarrow X_\ell \oplus Y_\ell$ is an automorphism. Hence

$$[M]^*\det(\mathcal{E}) = \det(\mathrm{pr}_{1*}(\mathrm{pr}_2^*\mathcal{L}(Y_\ell, -\frac{1}{\ell}?\cdot?,0) \otimes \mathrm{pr}_1^*\mathcal{L}(X_\ell, \frac{M^2}{\ell}?\cdot?,0)) =$$
$$= \det(\mathrm{pr}_{1*}\mathrm{pr}_2^*\mathcal{L}(Y_\ell, -\frac{1}{\ell}?\cdot?,0)) \otimes \mathcal{L}(X_\ell, \frac{M^2}{\ell}?\cdot?,0)^{\mathrm{d}_\ell^g} \quad (293)$$

by the projection formula (note that d_ℓ^g is the Euler characteristic of $\mathrm{pr}_{1*}\mathrm{pr}_2^*\mathcal{L}(Y_\ell, -\frac{1}{\ell}?\cdot?,0)$).

The first tensor multiple is equal to

$$p_A^* \det((p_{\mathrm{Pic}^0(C)\otimes Y_\ell})_*\mathcal{L}(Y_\ell, -\frac{1}{\ell}?\cdot?,0)) = p_A^*\delta^{-\frac{1}{2}\mathrm{d}_\ell^g \mathrm{rk} Y_\ell}$$

by the Proposition 5.6.

The second tensor multiple is identified with

$$\mathcal{L}(X_\ell, \frac{\mathrm{d}_\ell^g M^2}{\ell}?\cdot?,0) = [M]^*\mathcal{L}(X_\ell, \frac{\mathrm{d}_\ell^g}{\ell}?\cdot?,0)$$

This completes the proof of the Lemma. □

5.7.2. Now the difference of the LHS and RHS of 5.2(c) is a line bundle Ξ on $\mathrm{Pic}^0(C)\otimes X_\ell$ which is defined functorially with respect to S. Moreover, we have seen that $[M]^*\Xi$ is trivial. This implies that Ξ is trivial itself.

In effect, it defines a section of $\mathrm{Pic}^0(C)_M \otimes X_\ell$ (where the subscript $_M$ stands for M-torsion).

But $\mathrm{Pic}^0(C)_M \otimes X_\ell = R^1 p_{C*}(\mathbb{Z}/M\mathbb{Z}) \otimes X_\ell$, and it is well known that $R^1 p_{C*}(\mathbb{Z}/M\mathbb{Z})$ does not have nonzero functorial sections. Hence the above section vanishes, so the difference of the LHS and RHS is a line bundle lifted from the base S.

But we know that the restrictions of the LHS and RHS to the zero section of $\mathrm{Pic}^0(C)\otimes X_\ell$ coincide.

This completes the proof of 5.2(c).

5.7.3. Now we will prove 5.2(d).

According to the proof of Corollary 5.5 both sides are independent of ν, so we put $\nu = 0$.

The formula (293) and the argument just above it applies to the case of elliptic curve as well provided $N|M$. The first tensor multiple of (293) is obviously lifted from the base. Since the Picard group of the moduli space $\mathcal{M}_1$ is generated by δ, *loc. cit.* implies that

$$[M]^*\det(\mathcal{E}) = p^*(\delta^i) \otimes \mathcal{L}(X_\ell, \frac{\mathrm{d}_\ell M^2}{\ell}?\cdot?,0)$$

for some i. Hence

$$[M]^*(\det(\mathcal{E}))^{\otimes N} = p^*(\delta^{iN}) \otimes \mathcal{L}(X_\ell, \frac{N\mathrm{d}_\ell M^2}{\ell}?\cdot?,0) = [M]^*(p^*(\delta^{iN}) \otimes \mathcal{L}^N)$$

Exactly as in 5.7.2 this implies that

$$(\det(\mathcal{E}))^{\otimes N} = p^*(\delta^{iN}) \otimes \mathcal{L}^N$$

This completes the proof of 5.2(d). □

5.8. We start with the proof of Proposition 5.4.

If $g = 1$ then both sides are trivial. Indeed, the RHS is trivial since $6(\nu, \nu) \in 12\mathbb{Z}$, and $\delta^{\otimes 12}$ is trivial.

To see that the LHS is trivial it suffices to consider the case $\alpha = \{0\}$ (neutral element of Λ with multiplicity one).

¿From now on we assume that $g > 1$.

Put $P := \mathbb{P}((p_C)_*(\Omega_{C/S}))$ where p_C denotes the projection $C \longrightarrow S$. We have the natural inclusion $i : P \hookrightarrow C^{(2g-2)}$. Let $\alpha \in \mathbb{N}[\Lambda]$ be the multiset consisting of $(2g-2)$ copies of ν. We will identify $C^{(2g-2)}$ with C^α. Of course $\mathrm{AJ}_\alpha \circ i$ maps P to the image of s. So it suffices to prove that

$$(\mathrm{AJ}_\alpha \circ i)^*(\mathcal{L}(\Lambda,\ (,),\ \nu)) = p_P^*(\delta^{-6(\nu,\nu)}) \tag{294}$$

where p_P denotes the projection $P \longrightarrow S$.

Using the fact that $\mathfrak{b}(\nu) = \frac{3}{2}(\nu, \nu)$ we deduce

$$(\mathrm{AJ}_\alpha \circ i)^*(\mathcal{L}(\Lambda,\ (,),\ \nu)) = i^*(\mathcal{L}_\alpha) = i^*(S^{2g-2}(\Omega^{-\frac{3}{2}(\nu,\nu)})(\frac{1}{2}(\nu,\nu)\Delta)). \tag{295}$$

Here $\Delta \subset C^{(n)}$ is the diagonal divisor, and for an invertible sheaf $\mathcal{F}$ on C we denote by $S^n(\mathcal{F})$ the invertible sheaf $\pi_*(\mathcal{F}^{\boxtimes n})^{\Sigma_n}$ on $C^{(n)}$ (where $\pi : C^n \to C^{(n)}$ is the projection).

Let us introduce some more notation. Let $I \subset C \times C^{(n)}$ be the incidence divisor. We will denote the projection of $C \times C^{(n)}$ to the i-th factor by pr_i till the further notice. This will not cause any confusion with our previous use of the notation pr_1. For an invertible sheaf $\mathcal{F}$ on C we define a rank n locally free sheaf $\mathcal{F}^{(n)}$ on $C^{(n)}$ by: $\mathcal{F}^{(n)} := \mathrm{pr}_{2*}(\mathcal{O}_I \otimes \mathrm{pr}_1^*(\mathcal{F}))$.

5.8.1. *Lemma.* (a) The map $\mathcal{F} \mapsto S^n(\mathcal{F})$ defines a homomorphism from $\mathrm{Pic}(C)$ to $\mathrm{Pic}(C^{(n)})$;

(b) For any line bundle $\mathcal{F}$ on C we have $((\pi_*(\mathcal{F}^{\boxtimes n}))^{\Sigma_n,-})^{\otimes 2} = S^n(\mathcal{F}^{\otimes 2})(-\Delta)$;

(c) $\det(\mathcal{F}^{(n)}) = (\pi_*(\mathcal{F}^{\boxtimes n}))^{\Sigma_n,-}$.

Proof. (a) is clear.

(b) We obviously have a morphism from the LHS to $S^n(\mathcal{F}^2)$. We have to check that the cokernel is supported on Δ, and its stalk at the generic point of Δ is 1-dimensional. For this we can assume that $C = \mathbb{A}^1$, and $\mathcal{F} = \mathcal{O}$, and check the statement directly.

(c) Let us first construct a morphism from the LHS to the RHS. Let $\tilde{I} \subset C^{n+1}$ be the union of diagonals $x_1 = x_i$; let $\tilde{\mathrm{pr}}_2$ (resp. $\tilde{\mathrm{pr}}_1$) be the projection of C^{n+1} onto the last n (resp. first) coordinates. Then

$$\pi^*(\det(\mathcal{F}^{(n)})) = \det(\tilde{\mathrm{pr}}_{2*}(\tilde{\mathrm{pr}}_1^*\mathcal{F} \otimes \mathcal{O}_{\tilde{I}})) \tag{296}$$

because the square is Cartesian.

We also have the arrow: $\mathcal{O}_{\tilde{I}} \to \oplus_{i=2,..,n+1}\mathcal{O}_{\{x_1=x_i\}}$ which yields the morphism

$$\det(\tilde{\mathrm{pr}}_{2*}(\tilde{\mathrm{pr}}_1^*\mathcal{F}\otimes\mathcal{O}_{\tilde{I}}) \longrightarrow \det(\tilde{\mathrm{pr}}_{2*}(\tilde{\mathrm{pr}}_1^*\mathcal{F}\otimes(\oplus\mathcal{O}_{\{x_1=x_i\}}))) = \mathcal{F}^{\boxtimes n} \tag{297}$$

It is clear that the morphism (297) anticommutes with the action of Σ_n, hence, by (296), it defines the desired arrow.

It is easy to see that the image of the monomorphism (297) is $\mathcal{F}^{\boxtimes n}(-D) \subset \mathcal{F}^{\boxtimes n}$, where D is the union of all diagonals. But the latter inclusion induces isomorphism

$$(\pi_*(\mathcal{F}^{\boxtimes n}(-D)))^{\Sigma_n,-} = (\pi_*(\mathcal{F}^{\boxtimes n}))^{\Sigma_n,-}$$

This completes the proof of the Lemma. □

5.8.2. We return to the proof of Proposition 5.4. It suffices to prove it assuming that the line bundle $\Omega^{1/2}$ exists over S. Indeed, let $\mathcal{M}_g$ denote the moduli space of curves of genus g. Then it is known that $\mathrm{Pic}(\mathcal{M}_g)$ is torsion free (see e.g. [M2], Lemma 5.14). Hence it injects into the Picard group of the moduli space of curves with θ-characteristics.

We have

$$i^*(\mathcal{L}_\alpha) = i^*(S^{2g-2}(\Omega^{-\frac{3}{2}(\nu,\nu)})(\frac{1}{2}(\nu,\nu)\Delta)) = i^*(S^{2g-2}(\Omega^3)(-\Delta))^{-\frac{1}{2}(\nu,\nu)} =$$

$$i^*(\det{}^{\otimes 2}((\Omega^{\frac{3}{2}})^{(2g-2)}))^{-\frac{1}{2}(\nu,\nu)} = i^*(\det((\Omega^{\frac{3}{2}})^{(2g-2)}))^{-(\nu,\nu)}$$

So the Proposition follows from

5.8.3. *Lemma.* $\det(i^*(\Omega^{\frac{3}{2}})^{(2g-2)}) = p_P^*\delta^6$.

Proof. Consider the exact sequence of sheaves on $C\times C^{(n)}$:

$$0 \longrightarrow \mathcal{F} \longrightarrow \mathrm{pr}_1^*\Omega^{\frac{3}{2}} \longrightarrow \mathrm{pr}_1^*\Omega^{\frac{3}{2}}\otimes\mathcal{O}_I \longrightarrow 0$$

where $\mathcal{F}$ denotes the kernel. Let now (until the end of the proof) pr_i denote the projections of $C\times P$ to the respective factors. We see that

$$(id\times i)^*(\mathcal{F}) = \mathrm{pr}_1^*\Omega^{\frac{1}{2}}\otimes\mathrm{pr}_2^*(?)$$

We apply the functor $R\mathrm{pr}_{2*}\circ(id\times i)^*$ to this exact sequence (note that $[R\mathrm{pr}_{2*}(\Omega^{\frac{1}{2}}\otimes\mathrm{pr}_2^*(?))]^1$ does not depend on ? since the Euler characteristic of $\Omega^{\frac{1}{2}}$ is 0).

Using Mumford's formula (see [M2], Theorem 5.10) we get

$$\det(i^*(\Omega^{\frac{3}{2}})^{(2g-2)}) = [R\mathrm{pr}_{2*}(\mathrm{pr}_1^*\Omega^{\frac{3}{2}}\otimes i^*(\mathcal{O}_I))]^1 =$$

$$= [R\mathrm{pr}_{2*}(\mathrm{pr}_1^*\Omega^{\frac{3}{2}})]^1 - [R\mathrm{pr}_{2*}(\mathrm{pr}_1^*\Omega^{\frac{1}{2}})]^1 =$$

$$= [6((\frac{3}{2})^2 - \frac{3}{2}) + 1 - (6((\frac{1}{2})^2 - \frac{1}{2}) + 1)]p_P^*\delta = 6p_P^*\delta$$

This completes the proof of the Lemma along with the Proposition 5.4.

5.9. We start the proof of the Proposition 5.6.

5.9.1. *Lemma.* Let $\mathfrak{A} \to S$ be a family of abelian varieties over a smooth base S. Let $\mathfrak{a} : \mathfrak{A}' \longrightarrow \mathfrak{A}$ be an isogeny. Let $\mathcal{L}$ be a line bundle over $\mathfrak{A}$. Then $\deg(\mathfrak{a})[Rp_{\mathfrak{A}*}\mathcal{L}]^1 - [Rp_{\mathfrak{A}'*}\mathfrak{a}^*\mathcal{L}]^1$ lies in the torsion of $\mathrm{Pic}(S)$.

Proof. By the relative Riemann-Roch theorem (see e.g. [Fu] Theorem 15.2) we have the equalities in the Chow ring $A^S_{\mathbb{Q}}$

$$ch[R(p_{\mathfrak{A}'*}\mathfrak{a}^*\mathcal{L}] = p_{\mathfrak{A}'*}(ch(\mathfrak{a}^*\mathcal{L})td_{\mathfrak{A}'/S}) = p_{\mathfrak{A}*}\mathfrak{a}_*\mathfrak{a}^*[ch(\mathcal{L})td_{\mathfrak{A}/S}] =$$

$$= \deg(\mathfrak{a})p_{\mathfrak{A}*}[ch(\mathcal{L})td_{\mathfrak{A}/S}] = \deg(\mathfrak{a})ch[Rp_{\mathfrak{A}*}(\mathcal{L})]$$

Taking the components of degree 1 we get the Lemma. □

5.9.2. *Remark.* We will apply the Lemma to the situation $\mathfrak{A} = \mathrm{Pic}^0(C) \otimes \Lambda$. In this case the hypothesis of smoothness of S is redundant. Indeed, it is enough to take for S the moduli space of genus g curves $\mathcal{M}_g$. It is well known that there exists a smooth covering $\pi : \widetilde{\mathcal{M}}_g \longrightarrow \mathcal{M}_g$ inducing injection on Picard groups. One can take for $\widetilde{\mathcal{M}}_g$ the moduli space of curves with a basis in $H^1(C, \mathbb{Z}/3\mathbb{Z})$.

We continue the proof of the Proposition. We define the integer function $m(\Lambda, \ (,)\ )$ by the requirement

$$[Rp_{\Lambda*}(\mathcal{L}(\Lambda, \ (,), \ 0))]^1 = \delta^{m(\Lambda, \ (,)\ )}$$

We will need one more Lemma.

5.9.3. *Lemma.* The function m satisfies the following properties:
(a) $m(\Lambda_1 \oplus \Lambda_2, \ (,)_1 + (,)_2) = m(\Lambda_1, \ (,)_1)d_2^g + m(\Lambda_2, \ (,)_2)d_1^g$ (for the notation d see Proposition 5.6);
(b) Let $\iota : \ \Lambda' \hookrightarrow \Lambda$ be an embedding with finite cokernel. Then $m(\Lambda', \iota^*(,)\ ) = \#(\Lambda/\Lambda')^g m(\Lambda, \sigma)$;
(c) If $\Lambda = \mathbb{Z}$ and $(1,1) = -16$ then $m(\Lambda, \ (,)\ ) = -\frac{16^g}{2}$.

5.9.4. We will prove the Lemma below, and now we derive the Proposition from the Lemma.

We call 2 lattices isogenic if they contain isomorphic sublattices of maximal rank. From (a) it follows that the Proposition holds for the sum of 2 lattices if it holds for each of them. From (b) it follows that the Proposition holds for a lattice if it holds for an isogenic one. Thus it is enough to prove it for $(\Lambda, \ (,)\ ) = (\mathbb{Z}, n)$.

Note that the statement for the lattice $(\mathbb{Z}, n)$ is equivalent to the statement for the lattice $(\mathbb{Z} \oplus \mathbb{Z}, n \oplus (-n))$ since $m(\mathbb{Z} \oplus \mathbb{Z}, n \oplus (-n)) = 2(-n)^g \cdot m(\mathbb{Z}, n)$. But the lattices $(\mathbb{Z} \oplus \mathbb{Z}, n \oplus (-n))$ are isogenic for various n, and (c) says that the Proposition is true for $n = -16$.

The Proposition is proved. This completes the proof of the Theorem 5.2(c). □

5.10. It remains to prove the Lemma 5.9.3.

(a) is easy.

(b) follows from the Lemma 5.9.1.

To prove (c) we will need the following Lemma.

Let $C \longrightarrow S$ be a family such that a square root $\Omega^{\frac{1}{2}}$ exists globally over C. It gives the section $\varrho : S \longrightarrow \mathrm{Pic}^{g-1}(C)$. On the other hand, $\mathrm{Pic}^{g-1}(C)$ posesses the canonical theta line bundle $\mathcal{O}(\theta)$.

5.10.1. *Lemma.* $\varrho^*(\mathcal{O}(\theta)) = \delta^{\frac{1}{2}}$

Proof. (J.Harris and T.Pantev) Consider a relative curve C over base B, and a line bundle L over C of relative degree $g-1$. This gives the section from B to $\mathrm{Pic}^{g-1}(C)$, and we can restrict $\mathcal{O}(\theta)$ to B. Evidently, the class of $\mathcal{O}(\theta)|_B$ equals the class of divisor $D \subset B$: $b \in D$ iff $h^0(L_b) > 0$.

On the other hand, one checks readily that $\mathcal{O}(-D) = [Rp_{C*}L]^1$.

We apply this equality to the case $B = S$, $L = \Omega^{\frac{1}{2}}$. Then $[Rp_{C*}L]^1$ is computed by Mumford formula (see [M2], Theorem 5.10). Namely, we get $[Rp_{C*}\Omega^{\frac{1}{2}}]^1 = \delta^k$ where $k = 6(\frac{1}{2})^2 - 6(\frac{1}{2}) + 1 = -\frac{1}{2}$.

We conclude that $\varrho^*(\mathcal{O}(\theta)) = \delta^{\frac{1}{2}}$. □

5.10.2. Now we are able to prove the Lemma 5.9.3(c).

We can assume that $\Omega^{\frac{1}{2}}$ exists globally over C. The corresponding section $\varrho : S \longrightarrow \mathrm{Pic}^{g-1}(C)$ identifies $\mathrm{Pic}^{g-1}(C)$ with $\mathrm{Pic}^0(C)$. We will use this identification, and we will denote the projection $\mathrm{Pic}^0(C) \longrightarrow S$ by p.

We have $Rp_*(\mathcal{O}(\theta)) = \mathcal{O}$ and $\varrho^*(\mathcal{O}(\theta)) = \delta^{\frac{1}{2}}$ by the Lemma 5.10.1. We define $\mathcal{L}_1 := \mathcal{O}(\theta) \otimes p^*\delta^{-\frac{1}{2}}$.

Then $(4)^*\mathcal{L}_1 = \mathcal{L}(\mathbb{Z}, -16)$. Indeed, the θ-divisor, and hence $\mathcal{L}_1$ is (-1)-invariant. Thus $(2)^*\mathcal{L}_1$ and $\mathcal{L}(\mathbb{Z}, -4)$ are the line bundles with the same Chern class, and both are (-1)-invariant. So fiberwise they can differ only by a 2-torsion element. We conclude that $\mathcal{L}(\mathbb{Z}, -16) = (2)^*\mathcal{L}(\mathbb{Z}, -4) = (4)^*\mathcal{L}_1$ fiberwise. But the restriction of both these line bundles to the 0-section is trivial. Hence they are isomorphic.

Now (c) follows from the Lemma 5.9.1 since obviously $p_*\mathcal{L}_1 = \delta^{-\frac{1}{2}}$. □

5.11. We finish this section with a proof of one simple property of the Heisenberg local system $\mathcal{H}^\alpha, \alpha \in \mathbb{N}[X_\ell]$. It states, roughly speaking, that $\mathcal{H}^\alpha$ depends only on the class of α modulo $\mathbb{N}[Y_\ell]$. More precisely, the following Lemma holds.

5.11.1. *Lemma.* Let $\alpha = \sum_{k\in K} a_k\mu_k, \alpha' = \sum_{k\in K} a_k\lambda_k \in \mathbb{N}[X_\ell]$. Suppose $\lambda_k - \mu_k \in Y_\ell$ for every $k \in K$. Then the natural identification $C^\alpha = C^{\alpha'}$ lifts to a canonical isomorphism $\mathcal{H}^\alpha = \mathcal{H}^{\alpha'}$.

Proof. Let $\gamma \in \mathbb{N}[Y_\ell]$, $\gamma^\sim = 0$. We define $\beta \in \mathbb{N}[X_\ell]$ as $\beta = \alpha + \gamma$. We have an obvious projection $\mathrm{pr} : C^\beta \longrightarrow C^\alpha$. It is enough to construct a canonical isomorphism $\mathcal{H}^\beta \xrightarrow{\sim} \mathcal{H}^\alpha$.

5.11.2. Consider the addition homomorphism of abelian varieties

$$a:\ (\mathrm{Pic}(C)\otimes X_\ell)_{(2g-2)\nu_0}\times(\mathrm{Pic}(C)\otimes Y_\ell)_0\longrightarrow(\mathrm{Pic}(C)\otimes X_\ell)_{(2g-2)\nu_0}$$

We will compute the inverse image $a^*\mathcal{E}$.

To this end we introduce the following bilinear form σ on $X_\ell\oplus Y_\ell\oplus Y_\ell$:

$$\sigma((x,y_1,y_2),(x',y_1',y_2')) := -\frac{1}{\ell}(x\cdot y_2'+x'\cdot y_2+y_1\cdot y_2'+y_1'\cdot y_2+y_2\cdot y_2').$$

The automorphism $(x,y_1,y_2)\mapsto(x,y_1,y_1+y_2)$ of $X_\ell\oplus Y_\ell\oplus Y_\ell$ carries this form into σ', where

$$\sigma'((x,y_1,y_2),(x',y_1',y_2')) := -\frac{1}{\ell}(x\cdot y_2'+x'\cdot y_2+y_2\cdot y_2'-y_1'\cdot y_1-x\cdot y_1'-x'\cdot y_1).$$

Note that $\sigma'=\sigma_1+\sigma_2$ where

$$\sigma_1((x,y_1,y_2),(x',y_1',y_2')) := -\frac{1}{\ell}(x\cdot y_2'+x'\cdot y_2+y_2'\cdot y_2)$$

and

$$\sigma_2((x,y_1,y_2),(x',y_1',y_2')) := -\frac{1}{\ell}(-y_1'\cdot y_1-x\cdot y_1'-x'\cdot y_1)$$

5.11.3. We will denote by pr_{12} the projection of $(\mathrm{Pic}(C)\otimes X_\ell)_{(2g-2)\nu_0}\times(\mathrm{Pic}(C)\otimes Y_\ell)_0\times(\mathrm{Pic}(C)\otimes Y_\ell)_0$ to $(\mathrm{Pic}(C)\otimes X_\ell)_{(2g-2)\nu_0}\times(\mathrm{Pic}(C)\otimes Y_\ell)_0$ (the product of first two factors).

We also denote by pr_1 the projection of $(\mathrm{Pic}(C)\otimes X_\ell)_{(2g-2)\nu_0}\times(\mathrm{Pic}(C)\otimes Y_\ell)_0$ to $(\mathrm{Pic}(C)\otimes X_\ell)_{(2g-2)\nu_0}$.

Then we have

$$a^*\mathcal{E}=\mathrm{pr}_{12*}\mathcal{L}(X_\ell\oplus Y_\ell\oplus Y_\ell,\sigma,(\nu_0,0,0))=\mathrm{pr}_{12*}\mathcal{L}(X_\ell\oplus Y_\ell\oplus Y_\ell,\sigma',(\nu_0,0,0))=$$
$$=\mathrm{pr}_{12*}(\mathcal{L}(X_\ell\oplus Y_\ell\oplus Y_\ell,\sigma_1,(\nu_0,0,0))\otimes\mathcal{L}(X_\ell\oplus Y_\ell\oplus Y_\ell,\sigma_2,(\nu_0,0,0)))=$$
$$=\mathrm{pr}_1^*\mathcal{E}\otimes\mathcal{L}(X_\ell\oplus Y_\ell,\sigma_2,(\nu_0,0))$$

In the last line we view σ_2 as a bilinear form on $X_\ell\oplus Y_\ell$ since anyway it factors through the projection of $X_\ell\oplus Y_\ell\oplus Y_\ell$ to the sum of the first two summands. The last equality follows from the projection formula.

5.11.4. Let us denote the line bundle $\mathcal{L}(X_\ell\oplus Y_\ell,\sigma_2,(\nu_0,0))$ by $\mathfrak{L}$. Then we have

$$\det(a^*\mathcal{E})=\det(\mathrm{pr}_1^*\mathcal{E})\otimes\mathfrak{L}^{d_\ell^g}$$

Hence we can define a map

$$m:\ (\det(\mathrm{pr}_1^*\mathcal{E}))^{\cdot}\times\dot{\mathfrak{L}}\longrightarrow(\det(a^*\mathcal{E}))^{\cdot},\ (t,u)\mapsto t\otimes u^{d_\ell^g}.$$

Recall that $\mathfrak{H}$ is the universal local system on $(\det\mathcal{E})^{\cdot}$. We obviously have the equality

$$m^*a^*\mathfrak{H}=\mathrm{pr}_1^*\mathfrak{H}$$

(To be more precise, one could put the exterior tensor product of $\mathrm{pr}_1^*\mathfrak{H}$ with the constant sheaf on $\dot{\mathfrak{L}}$ in the RHS).

5.11.5. We are ready to finish the proof of the Lemma.

Recall that $\mathcal{H}^\beta$ is defined as $s_\beta^*\mathrm{AJ}_\beta^*\mathfrak{H}$ (notations of 3.5).

We consider also the Abel-Jacobi map $\mathrm{AJ}_{\alpha,\gamma}: C^\beta \longrightarrow (\mathrm{Pic}(C)\otimes X_\ell)_{(2g-2)\nu_0}\times(\mathrm{Pic}(C)\otimes Y_\ell)_0$. As in *loc. cit.* we have the canonical section ς of $\mathrm{AJ}_{\alpha,\gamma}^*(\det(\mathrm{pr}_1^*\mathcal{E}))$ and the canonical section s of $\mathrm{AJ}_{\alpha,\gamma}^*(\det(a^*\mathcal{E}))$.

We have

$$\mathcal{H}^\beta = s^*\mathrm{AJ}_{\alpha,\gamma}^* a^*\mathfrak{H}$$

and

$$\mathrm{pr}^*\mathcal{H}^\alpha = \varsigma^*\mathrm{AJ}_{\alpha,\gamma}^*\mathrm{pr}_1^*\mathfrak{H}$$

It is easy to see that there exists a section $\mathfrak{s}$ of $\mathrm{AJ}_{\alpha,\gamma}^*\mathfrak{L}$ such that

$$m(\varsigma,\mathfrak{s}) = s.$$

Hence $\mathcal{H}^\beta = s^*\mathrm{AJ}_{\alpha,\gamma}^* a^*\mathfrak{H} = (\varsigma,\mathfrak{s})^*\mathrm{AJ}_{\alpha,\gamma}^*\mathrm{pr}_1^*\mathfrak{H} = \varsigma^*\mathrm{AJ}_{\alpha,\gamma}^*\mathrm{pr}_1^*\mathfrak{H} = \mathrm{pr}^*\mathcal{H}^\alpha$.

This completes the proof of the Lemma. □

6. Factorization isomorphisms

6.1. We need to introduce some more notation. Recall the setup of 2.1.

First, we will need the space TC^τ containing $\overset{\circ}{T}C^\tau$ as an open subset. It is defined in the same way as $\overset{\circ}{T}C^\tau$, only we do not throw away nor the diagonals, neither the zero sections.

The space TC^τ decomposes into the direct product

$$TC^\tau = \widetilde{TC^K} \times \prod_{k\in K} TD^{\tau^{-1}(k)}$$

The projection to the first (resp. second) factor will be denoted by pr_1 (resp. pr_2). We have the natural substitution map $TC^\tau \longrightarrow TC^J$. We will denote it by q_τ when there is no risk of confusing it with the same named map $\overset{\circ}{T}C^\tau \longrightarrow \overset{\circ}{T}C^J$.

The evident projection map $TC^J \longrightarrow C^J$ will be denoted by pr. We will use the same notation for the projections $\overset{\circ}{T}C^J \longrightarrow \overset{\circ}{C}^J$, and $\overset{\circ}{T}C^\alpha \longrightarrow \overset{\circ}{C}^\alpha$ when there is no risk of confusion.

The open embedding $\overset{\circ}{C}^J \hookrightarrow C^J$ will be denoted by j. We will keep the same notation for the open embedding $\overset{\circ}{C}^\alpha \hookrightarrow C^\alpha$.

6.2. Recall that we made a choice in the definition of $\mathcal{H}^\alpha$: the isomorphism (287) was defined up to a scalar multiple.

Let us explain how to make a consistent family of choices.

Given a surjection $\tau: J \longrightarrow K$ we denote α_K by β for short.

6.2.1. *Definition.* The isomorphisms $\theta_\alpha:\ \mathcal{L}_\alpha \xrightarrow{\sim} \mathrm{AJ}^*_\alpha(\mathcal{L})$, and $\theta_\beta:\ \mathcal{L}_\beta \xrightarrow{\sim} \mathrm{AJ}^*_\beta(\mathcal{L})$ are called *compatible* if the following diagram commutes:

$$\begin{array}{ccc}
(\pi_J\circ\sigma_\tau)^*\mathcal{L}_\alpha & \overset{\pi_J^*(\theta_\alpha)}{\underset{\sim}{\longrightarrow}} & (\mathrm{AJ}_\alpha\circ\sigma_\tau)^*\mathcal{L} \\
\| & & \| \\
\pi_K^*\mathcal{L}_\beta & \overset{\pi_K^*(\theta_\beta)}{\underset{\sim}{\longrightarrow}} & \mathrm{AJ}^*_\beta\mathcal{L}
\end{array}$$

Here the left vertical equality is just the Lemma 4.3.1, and the right vertical equality is tautological since $\mathrm{AJ}_\beta = \mathrm{AJ}_\alpha\circ\sigma_\tau$.

Clearly, we can choose the isomorphisms (287) for all J in a compatible way. We will assume such a choice made once and for all.

This gives rise to the local system $\tilde{\mathcal{H}}^\alpha$ on $\dot{\mathcal{L}}_\alpha$.

6.3. Now we are ready for the construction of factorization isomorphisms.

6.3.1. Let us start with the case of disk.

Given an unfolding $\pi:\ J \longrightarrow X_\ell$ we define the meromorphic function F_J on TD^J as follows. We fix a total order $<$ on J. The standard coordinates on TD^J are denoted by $(x_j,\xi_j;\ j\in J)$. We define

$$F_J(x_j,\xi_j) = \prod_{\pi(i)\neq\pi(j),i<j}(x_i-x_j)^{2\pi(i)\cdot\pi(j)}\times\prod_{\pi(i)=\pi(j)}(x_i-x_j)^{\pi(i)\cdot\pi(j)}\times\prod_{j\in J}\xi_j^{2n(\pi(j))} \tag{298}$$

The function does not depend on the order on J.

Recall the setup of 2.1.5 for the case $C=D$. We have the following two functions on TD^τ:

$$F_J(q_\tau),\ \text{and}\ F_K(p_K)\times\prod_{k\in K}F_{\tau^{-1}(k)} \tag{299}$$

It is easy to see that the quotient of these two functions $TD^\tau \longrightarrow \mathbb{C}$ actually lands to $\mathbb{C}^*$; moreover the quotient function $TD^\tau \longrightarrow \mathbb{C}^*$ is "canonically" homotopic to the constant map $TD^\tau \longrightarrow 1$.

Let $\mathcal{Q}$ denote the one dimensional local system on $\mathbb{C}^*$ with monodromy ζ around the origin and with fiber at 1 trivialized.

We define $\mathcal{I}^J := F_J^*(\mathcal{Q})$.

Let $m:\ \mathbb{C}^*\times\mathbb{C}^* \longrightarrow \mathbb{C}^*$ denote the multiplication map. Then $m^*(\mathcal{Q})\cong\mathcal{Q}\boxtimes\mathcal{Q}$ canonically.

Hence the homotopy between the functions 299 yields the factorization isomorphisms for the local systems $\mathcal{I}^J$.

The associativity of these factorization isomorphisms follows from associativity of the above mentioned homotopies between functions 299.

6.3.2. We return to the case of an arbitrary curve. We denote the "zero section" $C^K \longrightarrow TC^\tau$ by z.

Consider the following line bundle on TC^τ:

$$\mathcal{L}_\tau = (q_\tau\circ pr)^*\mathcal{L}_\alpha$$

$\mathcal{L}_\tau$ carries the local system $\mathcal{H}^\tau := (q_\tau \circ pr)^*\tilde{\mathcal{H}}^\alpha$.

Since $z \circ q_\tau \circ pr = \sigma_\tau$, we have $z^*\mathcal{L}_\tau = \mathcal{L}_\beta$.

The latter line bundle over C^K has a meromorphic section $\pi_K^* s_\beta$ constructed in 3.5. We will extend this section to the whole TC^τ.

We do have a meromorphic section $s'_\tau : TC^\tau \longrightarrow \mathcal{L}_\tau$, $s'_\tau = q_\tau^*\pi_J^*(s_\alpha)$, where $s_\alpha : C^\alpha \longrightarrow \mathcal{L}_\alpha$ was defined in 3.5.

Recall that in 6.3.1 we have defined the meromorphic function F_M on TD^M for any unfolding $M \longrightarrow X_\ell$.

We have

$$\operatorname{div}(s'_\tau) = pr_1^{-1}(\operatorname{div}(p_K^*\pi_K^* s_\beta)) + pr_2^{-1}(\operatorname{div}(\prod_{k\in K} F_{\tau^{-1}(k)}))$$

We define the meromorphic section

$$s_\tau = s'_\tau \prod_{k\in K} F^{-1}_{\tau^{-1}(k)} \tag{300}$$

Then $\operatorname{div}(s_\tau) = pr_1^{-1}(\operatorname{div}(p_K^*\pi_K^* s_\beta))$, and it is easy to see that $z^*s_\tau = \pi_K^* s_\beta$.

Since pr_1 is a projection with contractible fibers it follows that restricting to $\overset{\circ}{T}C^\tau$ we have

$$s_\tau^*(\mathcal{H}^\tau) \simeq pr_1^* p_K^* \pi_K^* s_\beta^* \tilde{\mathcal{H}}^\beta = pr_1^* p_K^* \pi_K^* \mathcal{H}^\beta \tag{301}$$

6.3.3. Let $m : \mathbb{C}^* \times \mathcal{L}_\tau \longrightarrow \mathcal{L}_\tau$ denote the multiplication map. Then $m^*\mathcal{H}^\tau = \mathcal{Q} \boxtimes \mathcal{H}^\tau$.

So putting all the above together, we get

$$q_\tau^*\pi_J^*\mathcal{H}^\alpha = (s'_\tau)^*\mathcal{H}^\tau = (s_\tau \prod_{k\in K} F_{\tau^{-1}(k)})^*\mathcal{H}^\tau =$$

$$= (\prod_{k\in K} F_{\tau^{-1}(k)}; s_\tau)^* m^*\mathcal{H}^\tau = (\prod_{k\in K} F_{\tau^{-1}(k)}; s_\tau)^*(\mathcal{Q} \boxtimes \mathcal{H}^\tau) =$$

$$= (\prod_{k\in K} F_{\tau^{-1}(k)})^*\mathcal{Q} \boxtimes s_\tau^*\mathcal{H}^\tau = (\prod_{k\in K} F_{\tau^{-1}(k)})^*\mathcal{Q} \boxtimes s_\beta^*\mathcal{H}^\beta =$$

$$= p_K^*\pi_K^*\mathcal{H}^{\alpha_K} \boxtimes \boxed{\times}_{k\in K} \mathcal{I}^{\tau^{-1}(k)}$$

which completes the construction of the factorization isomorphisms.

6.4. The compatibility of the constructed factorization isomorphism for $\mathcal{H}$ with the ones for $\mathcal{I}$ is immediate.

Chapter 2. The modular property of the Heisenberg local system

7. Degeneration of curves: recollections and notations

7.1. Let $\mathcal{M}_g$ be the moduli space of curves of genus g. Let $\mathfrak{D}$ be a smooth locus of an irreducible component of the divisor at infinity in the compactification $\overline{\mathcal{M}_g}$. It has one of the following types.

(a) Either we have a decomposition $g = g_1 + g_2,\ g_1, g_2 > 0$. Then $\mathfrak{D}$ is the moduli space of the following objects: curves C_1, C_2 of genera g_1, g_2 respectively; points $x_1 \in C_1, x_2 \in C_2$. Thus, C_g degenerates into $C_1 \sqcup C_2/(x_1 = x_2)$.

(b) Or we put $g_0 = g - 1$, and then $\mathfrak{D}$ is the moduli space of the following objects: curve C_0 of genus g_0 with two distinct points $x_1 \neq x_2 \in C_0$. Thus, C_g degenerates into $C_0/(x_1 = x_2)$.

We denote by $C_{\mathfrak{D}}$ the universal family over $\mathfrak{D}$.

7.2. Let $\tilde{\mathfrak{D}} \longrightarrow \mathfrak{D}$ denote the moduli space of objects as above plus nonzero tangent vectors v_1 at x_1, and v_2 at x_2. Let $\dot{T}_{\mathfrak{D}}\mathcal{M}_g$ denote the normal bundle to $\mathfrak{D}$ in $\overline{\mathcal{M}_g}$ with the zero section removed. There is the canonical map $\wp : \tilde{\mathfrak{D}} \to \dot{T}_{\mathfrak{D}}\overline{\mathcal{M}_g}$ (see e.g. [BFM], §4).

7.3. Let $\overline{C}$ be the universal curve over $\overline{\mathcal{M}_g}$. Let $\alpha \in \mathbb{N}[X_\ell]$.

Consider $\overset{\circ}{T}\overline{C}^\alpha \to \overline{\mathcal{M}_g}$ (see 2.1.2). Note that $\overline{C}$ has singularities (nodes) over $\overline{\mathcal{M}_g} - \mathcal{M}_g$. To simplify the exposition we just throw these nodal points away. That is, $\overset{\circ}{T}\overline{C}^\alpha$ is formed by configurations of nonsingular points with nonzero tangent vectors.

Let us describe the preimage of $\mathfrak{D}$ in $\overset{\circ}{T}\overline{C}^\alpha$, to be denoted by $\mathcal{T}C^\alpha_{\mathfrak{D}}$.

In case 7.1(a) $\mathcal{T}C^\alpha_{\mathfrak{D}}$ is a union of connected components numbered by the decompositions $\alpha = \alpha_1 + \alpha_2$. The connected component $\mathcal{T}C^{\alpha_1\alpha_2}_{\mathfrak{D}}$ is the product $\tilde{C}^{\alpha_1}_1 \times \tilde{C}^{\alpha_2}_2$. Here $\tilde{C}^{\alpha_r}_r$ denotes the configurations of α_r distinct points with nonzero tangent vectors on $C_r - x_r,\ r = 1, 2$.

In case 7.1(b) $\mathcal{T}C^\alpha_{\mathfrak{D}}$ is the space of configurations of α distinct points with nonzero tangent vectors on $C_0 - \{x_1, x_2\}$.

7.4. Thus we obtain the map $\wp_\alpha : \tilde{\mathfrak{D}} \times_{\mathfrak{D}} \mathcal{T}C^\alpha_{\mathfrak{D}} \longrightarrow \dot{T}_{\mathcal{T}C^\alpha_{\mathfrak{D}}}\overset{\circ}{T}\overline{C}^\alpha$ — the normal bundle to $\mathcal{T}C^\alpha_{\mathfrak{D}}$ in $\overset{\circ}{T}\overline{C}^\alpha$ with the zero section removed.

Note that if we prescribe a weight μ_1 to the point x_1, and a weight μ_2 to the point x_2 then

in case 7.1(a) we can identify $\tilde{\mathfrak{D}} \times_{\mathfrak{D}} \mathcal{T}C^{\alpha_1\alpha_2}_{\mathfrak{D}}$ with $\overset{\circ}{T}C_1^{\alpha_1+\mu_1} \times \overset{\circ}{T}C_2^{\alpha_2+\mu_2}$ (notations of 2.1.2);

in case 7.1(b) we can identify $\tilde{\mathfrak{D}} \times_{\mathfrak{D}} \mathcal{T}C^\alpha_{\mathfrak{D}}$ with $\overset{\circ}{T}C_0^{\alpha+\mu_1+\mu_2}$

(the exponents are formal sums, and we do not perform symmetrization over μ_1 and μ_2 if they happen to be equal or contained in supp α).

7.5. It is known that the determinant line bundle δ_g over $\mathcal{M}_g$ extends to the line bundle $\overline{\delta_g}$ over $\overline{\mathcal{M}_g}$.

In case 7.1(a) let $\varpi_i, i = 1, 2$, denote the natural projection from $\mathfrak{D}$ to $\mathcal{M}_{g_i}$.

In case 7.1(b) let ϖ_0 denote the natural projection from $\mathfrak{D}$ to $\mathcal{M}_{g_0}$.

Then $\overline{\delta_g}|_{\mathfrak{D}} = \varpi_1^*\delta_{g_1} \otimes \varpi_2^*\delta_{g_2}$ (resp. $\overline{\delta_g}|_{\mathfrak{D}} = \varpi_0^*\delta_{g_0}$) (see [BFM]).

Summing up we obtain the map also denoted by $\wp_\alpha$:

in case 7.1(a):

$$(\overset{\circ}{T}C_1^{\alpha_1+\mu_1} \times_{\mathcal{M}_{g_1}} \delta_{g_1}) \times (\overset{\circ}{T}C_2^{\alpha_2+\mu_2} \times_{\mathcal{M}_{g_2}} \delta_{g_2}) \longrightarrow \dot{T}_{\mathcal{T}C^\alpha_{\mathfrak{D}} \times_{\overline{\mathcal{M}_g}} \overline{\delta_g}}(\overset{\circ}{T}\overline{C}^\alpha \times_{\overline{\mathcal{M}_g}} \overline{\delta_g}); \tag{302}$$

in case 7.1(b):

$$\overset{\circ}{T}C_0^{\alpha+\mu_1+\mu_2} \times_{\mathcal{M}_{g_0}} \delta_{g_0} \longrightarrow \dot{T}_{\mathcal{T}C^\alpha_{\mathfrak{D}} \times_{\overline{\mathcal{M}_g}} \overline{\delta_g}}(\overset{\circ}{T}\overline{C}^\alpha \times_{\overline{\mathcal{M}_g}} \overline{\delta_g}). \tag{303}$$

7.6. Consider the specialization $\mathbf{Sp}\mathcal{H}_g^\alpha$ of the Heisenberg local system along the boundary component $\mathcal{T}C^\alpha_{\mathfrak{D}} \times_{\overline{\mathcal{M}_g}} \overline{\delta_g}$. It is a sheaf on $\dot{T}_{\mathcal{T}C^\alpha_{\mathfrak{D}} \times_{\overline{\mathcal{M}_g}} \overline{\delta_g}}(\overset{\circ}{T}\overline{C}^\alpha \times_{\overline{\mathcal{M}_g}} \overline{\delta_g})$. We will describe its inverse image under the map $\wp_\alpha$.

Theorem. (a) In case 7.1(a) we have

$$\wp_\alpha^*\mathbf{Sp}\mathcal{H}_g^\alpha|_{(\overset{\circ}{T}C_1^{\alpha_1+\mu_1} \times_{\mathcal{M}_{g_1}} \delta_{g_1}) \times (\overset{\circ}{T}C_2^{\alpha_2+\mu_2} \times_{\mathcal{M}_{g_2}} \delta_{g_2})} = \mathcal{H}_{g_1}^{\alpha_1+\mu_1} \boxtimes \mathcal{H}_{g_2}^{\alpha_2+\mu_2}$$

for $\mu_1 = (2g_1 - 2)\nu_0 - \alpha_1^\sim$, and $\mu_2 = (2g_2 - 2)\nu_0 - \alpha_2^\sim$ (notations of 2.1.8);

(b) In case 7.1(b) we have

$$\wp_\alpha^*\mathbf{Sp}\mathcal{H}_g^\alpha = \oplus_{\mu_1+\mu_2=-2\nu_0} \mathcal{H}_{g_0}^{\alpha+\mu_1+\mu_2}$$

Here the sum is taken over the set X_ℓ/Y_ℓ (see IV.9.1.2 and 2.1.6). The statement makes sense by the virtue of the Lemma 5.11.1. Thus there are exactly d_ℓ summands.

The proof occupies the rest of this Chapter.

8. Proof of Theorem 7.6(a)

We are in the situation of 7.1(a). Let us denote $\mathcal{M}_g \cup \mathfrak{D}$ by $\overline{\mathcal{M}}^{\circ} \subset \overline{\mathcal{M}}_g$. In this section we will extend the constructions of sections 4 and 5 to the universal family of curves over $\overline{\mathcal{M}}^{\circ}$.

8.1. It is well known that $\operatorname{Pic}(C/\mathcal{M}_g)$ extends to the family of algebraic groups over $\overline{\mathcal{M}}^{\circ}$ which will be denoted by $\overline{\operatorname{Pic}}(\overline{C}/\overline{\mathcal{M}}^{\circ})$.

Restricting to $\mathfrak{D}$ we get

$$\overline{\operatorname{Pic}}(\overline{C}/\overline{\mathcal{M}}^{\circ})|_{\mathfrak{D}} = \operatorname{Pic}(C_1/\mathfrak{D}) \times_{\mathfrak{D}} \operatorname{Pic}(C_2/\mathfrak{D})/\langle (x_1) - (x_2) \rangle, \tag{304}$$

and $\operatorname{Pic}(\overline{C}/\mathfrak{D}) = \operatorname{Pic}(C_1/\mathfrak{D}) \times_{\mathfrak{D}} \operatorname{Pic}(C_2/\mathfrak{D})$.

In particular, $\overline{\operatorname{Pic}}(\overline{C}/\mathfrak{D})$ is an extension of $\mathbb{Z}$ by an abelian scheme.

8.1.1. Recall the notations of 4.1. Given Λ, $(,)$, ν as in 4.1 one constructs the linear bundle $\mathcal{L}(\Lambda, (,), \nu)$ over A_Λ (see 4.2). To unburden the notations we will denote this line bundle simply by $\mathcal{L}$ when it does not cause confusion.

$\overline{A}_\Lambda$ will denote $(\overline{\operatorname{Pic}}(\overline{C}/\overline{\mathcal{M}}^{\circ}) \otimes \Lambda)_{(2g-2)\nu}$, and p will denote the projection $\overline{A}_\Lambda \longrightarrow \overline{\mathcal{M}}^{\circ}$.

We have an isomorphism

$$\mho : A_\Lambda(C_1/\mathfrak{D}) \times_{\mathfrak{D}} A_\Lambda(C_2/\mathfrak{D}) \xrightarrow{\sim} \overline{A}_\Lambda|_{\mathfrak{D}}$$

where

$$\mho(a,b) := (a+x_1\otimes\nu, b+x_2\otimes\nu) \in (\operatorname{Pic}(C_1/\mathfrak{D})\otimes\Lambda)_{(2g_1-1)\nu} \times_{\mathfrak{D}} (\operatorname{Pic}(C_2/\mathfrak{D})\otimes\Lambda)_{(2g_2-1)\nu} =$$
$$= \overline{\operatorname{Pic}}(\overline{C}/\mathfrak{D}) \otimes \Lambda)_{(2g-2)\nu}$$

(the latter isomorphism is obtained from (304)).

8.1.2. *Lemma.* There exists a line bundle $\overline{\mathcal{L}}$ on $\overline{A}_\Lambda$ such that $\overline{\mathcal{L}}|_{p^{-1}(\mathcal{M}_g)} = \mathcal{L}$, and

$$\mho^*(\overline{\mathcal{L}}|_{\mathfrak{D}}) = \mathcal{L}_{C_1}^{\mathrm{d}_\ell^{g_2}} \boxtimes \mathcal{L}_{C_2}^{\mathrm{d}_\ell^{g_1}}.$$

Proof. Note that the line bundle $\mathcal{L}_\alpha$ (see 4.1.2) defined over C^α extends to $C^\alpha \cup (C_1 - x_1)^{\alpha_1} \times (C_2 - x_2)^{\alpha_2} \subset \overline{C}^\alpha$ by the same formula 4.1.2. Moreover, $\mathcal{L}_\alpha$ automatically extends to $C^\alpha \cup C_1^{\alpha_1} \times C_2^{\alpha_2}$ since the complement to $C^\alpha \cup (C_1 - x_1)^{\alpha_1} \times (C_2 - x_2)^{\alpha_2} \subset \overline{C}^\alpha$ in $C^\alpha \cup C_1^{\alpha_1} \times C_2^{\alpha_2}$ has codimension two.

We define the following line bundle on $C^\alpha \cup C_1^{\alpha_1} \times C_2^{\alpha_2}$:

$$\overline{\mathcal{L}}_\alpha := \mathcal{L}_\alpha(\mathrm{b}(\mu_1)(C_1^{\alpha_1} \times C_2^{\alpha_2})) \tag{305}$$

Here $C_1^{\alpha_1} \times C_2^{\alpha_2}$ is viewed as a divisor on $C^\alpha \cup C_1^{\alpha_1} \times C_2^{\alpha_2}$. Note also that $\mathrm{b}(\mu_1) = \mathrm{b}(\mu_2)$ since $\mu_1 + \mu_2 = -2\nu$.

The line bundle $\mathcal{L}$ on A_Λ extends uniquely to a line bundle $\overline{\mathcal{L}}$ on $\overline{A}_\Lambda$ in such a way that

$$\mathrm{AJ}^*\overline{\mathcal{L}} = \overline{\mathcal{L}}_\alpha. \tag{306}$$

In effect, any two extensions of $\mathcal{L}$ to $\overline{A}_\Lambda$ differ by the twist by some power of $\mathcal{O}(p^{-1}\mathfrak{D})$ — the line bundle lifted from $\overline{\mathcal{M}}^\circ$. But evidently, $\mathrm{AJ}^*\mathcal{O}(p^{-1}\mathfrak{D}) = \mathcal{O}(C_1^{\alpha_1} \times C_2^{\alpha_2})$, so we can choose our twist uniquely to satisfy (306).

8.1.3. Let us choose a basis I of Λ. Let us fix $\alpha_1, \alpha_2 \in \mathbb{N}[\Lambda]$; $\alpha_1 = \sum_{i\in I} b_i i + \sum_{i\in I} b_{-i}(-i), \alpha_2 = \sum_{i\in I} c_i i + \sum_{i\in I} c_{-i}(-i)$ such that $b_i, b_{-i} > g_1$; $c_i, c_{-i} > g_2$.

We have the closed embeddings $\sigma_1 : C_1^{\alpha_1} \hookrightarrow C_1^{\alpha_1+\mu_1}$ (resp. $\sigma_2 : C_2^{\alpha_2} \hookrightarrow C_2^{\alpha_2+\mu_2}$) adding to a configuration on C_1 (resp. C_2) μ_1 copies of x_1 (resp. μ_2 copies of x_2).

It is clear that $\mathrm{AJ}_1 \circ \sigma_1$ (resp. $\mathrm{AJ}_2 \circ \sigma_2$) induces injection from $\mathrm{Pic}(A_{1\Lambda}/\mathfrak{D})$ to $\mathrm{Pic}(C_1^{\alpha_1}/\mathfrak{D})$ (resp. from $\mathrm{Pic}(A_{2\Lambda}/\mathfrak{D})$ to $\mathrm{Pic}(C_2^{\alpha_2}/\mathfrak{D})$).

Hence it is enough to check that

$$(\mho \circ (\mathrm{AJ}_1 \times \mathrm{AJ}_2) . \circ (\sigma_1 \times \sigma_2))^*\overline{\mathcal{L}}|_{p^{-1}\mathfrak{D}} = \sigma_1^*\mathcal{L}_{\alpha_1+\mu_1}^{\mathrm{d}_\ell^{g_2}} \boxtimes \sigma_2^*\mathcal{L}_{\alpha_2+\mu_2}^{\mathrm{d}_\ell^{g_1}} \tag{307}$$

Consider unfoldings $\pi_1 : J_1 \longrightarrow \Lambda$ of α_1, and $\pi_2 : J_2 \longrightarrow \Lambda$ of α_2. It is enough to check that

$$(\pi_1 \times \pi_2)^*(\mho \circ (\mathrm{AJ}_1 \times \mathrm{AJ}_2) \circ (\sigma_1 \times \sigma_2))^*\overline{\mathcal{L}}|_{p^{-1}\mathfrak{D}} = (\pi_1 \times \pi_2)^*(\sigma_1^*\mathcal{L}_{\alpha_1+\mu_1}^{\mathrm{d}_\ell^{g_2}} \boxtimes \sigma_2^*\mathcal{L}_{\alpha_2+\mu_2}^{\mathrm{d}_\ell^{g_1}}) \tag{308}$$

One checks readily that the isomorphism (308) holds after restriction to $(C_1 - x_1)^{J_1} \times (C_2 - x_2)^{J_2}$.

The necessary condition to extend the isomorphism across the complement divisor is the equality of (relative) first Chern classes:

$$c_1((\mho \circ (\mathrm{AJ}_1 \times \mathrm{AJ}_2) \circ (\sigma_1 \times \sigma_2))^*\overline{\mathcal{L}}|_{p^{-1}\mathfrak{D}}) = c_1(\sigma_1^*\mathcal{L}_{\alpha_1+\mu_1}^{\mathrm{d}_\ell^{g_2}} \boxtimes \sigma_2^*\mathcal{L}_{\alpha_2+\mu_2}^{\mathrm{d}_\ell^{g_1}})$$

This is immediate since $c_1(\mho^*(\overline{\mathcal{L}}_{\mathfrak{D}})) = c_1(\mathcal{L}_{C_1}^{\mathrm{d}_\ell^{g_2}}) \boxtimes c_1(\mathcal{L}_{C_2}^{\mathrm{d}_\ell^{g_1}})$. If the complement divisor were irreducible, this condition would be sufficient as well. If the divisor is reducible, one has to check that the fundamental classes of different irreducible components are linearly independent. All the irreducible components of our divisor are diagonals $z_j = x_r$, $r = 1, 2; j \in J_r$. Their fundamental classes are linearly independent in $H^2(C_1^{J_1} \times C_2^{J_2})$.

This completes the proof of the Lemma. □

8.2. We proceed with the proof of the Theorem 7.6(a).

Let us consider the line bundle $\overline{\mathcal{L}} := \overline{\mathcal{L}}(X_\ell, \frac{\mathrm{d}_\ell^g}{\ell}?\cdot?, \nu_0)$ on $\overline{A} := \overline{A}_{X_\ell}$. On the other hand, let us consider the triple Λ, $(,)$, ν defined in 5.1, and the corresponding line bundle $\overline{\mathcal{L}}' := \overline{\mathcal{L}}(\Lambda, (,), \nu)$ on $\overline{A}' := \overline{A}_{X_\ell \oplus Y_\ell}$. We denote by $\mathfrak{p}$ (resp. $\mathfrak{p}_r$, $r = 1, 2$) the obvious projection $\overline{A}' \longrightarrow \overline{A}$ (resp. $A'(C_r) \longrightarrow A(C_r)$).

We define $\overline{\mathcal{E}} := \mathfrak{p}_*\overline{\mathcal{L}}'$, $\mathcal{E}_{C_r} := \mathfrak{p}_{r*}\mathcal{L}'_{C_r}$. It is obvious that

$$\overline{\mathcal{E}}|_{p^{-1}(\mathfrak{D})} = (\mho^{-1})^*(\mathcal{E}_{C_1} \boxtimes \mathcal{E}_{C_2}) \tag{309}$$

and hence

$$\det(\overline{\mathcal{E}})|_{p^{-1}(\mathfrak{D})} = (\mho^{-1})^*((\det(\mathcal{E}_{C_1}))^{\mathrm{rk}(\mathcal{E}_{C_2})} \boxtimes (\det(\mathcal{E}_{C_2}))^{\mathrm{rk}(\mathcal{E}_{C_1})}) = \\ = (\mho^{-1})^*((\det(\mathcal{E}_{C_1}))^{\mathrm{d}_\ell^{g_2}} \boxtimes (\det(\mathcal{E}_{C_2}))^{\mathrm{d}_\ell^{g_1}}) \tag{310}$$

Recall that the line bundle δ_g on $\mathcal{M}_g$ extends to the line bundle $\overline{\delta_g}$ on $\overline{\mathcal{M}_g}$, and $\overline{\delta_g}|_{\mathfrak{D}} = \delta_{g_1} \boxtimes \delta_{g_2}$.

Proposition. $\det(\overline{\mathcal{E}}) = \overline{\mathcal{L}} \otimes (p^*\overline{\delta_g})^{\mathrm{d}_\ell^g(-\frac{1}{2}\mathrm{rk}X_\ell + 6k\frac{\nu_0\cdot\nu_0}{\ell})}$ where p denotes the projection $\overline{A} \longrightarrow \overline{\mathcal{M}}^\circ$.

Proof. If we restrict both sides to $p^{-1}\mathcal{M}_g \subset p^{-1}\overline{\mathcal{M}}^\circ$ then we just have the Theorem 5.2.

On the other hand, it is easy to see that the normal line bundle to $\mathfrak{D}$ in $\overline{\mathcal{M}}^\circ$ is nontrivial. This means that two extensions of a line bundle from $p^{-1}\mathcal{M}_g$ to $p^{-1}\overline{\mathcal{M}}^\circ$ coincide iff their restrictions to $p^{-1}\mathfrak{D}$ coincide.

So it remains to compare the LHS and RHS restricted to $p^{-1}\mathfrak{D}$.

By the virtue of (310) above we have

$$\det(\overline{\mathcal{E}})|_{p^{-1}(\mathfrak{D})} = (\mho^{-1})^*((\det(\mathcal{E}_{C_1}))^{\mathrm{d}_\ell^{g_2}} \boxtimes (\det(\mathcal{E}_{C_2}))^{\mathrm{d}_\ell^{g_1}}) = \\ = (\mho^{-1})^*((\mathcal{L}_{C_1}^{\mathrm{d}_\ell^{g_2}} \boxtimes \mathcal{L}_{C_2}^{\mathrm{d}_\ell^{g_1}}) \otimes (\delta_{g_1}^{\mathrm{d}_\ell^g(-\frac{1}{2}\mathrm{rk}X_\ell + 6\frac{\nu_0\cdot\nu_0}{\ell})} \boxtimes \delta_{g_2}^{\mathrm{d}_\ell^g(-\frac{1}{2}\mathrm{rk}X_\ell + 6\frac{\nu_0\cdot\nu_0}{\ell})})) = \\ = \overline{\mathcal{L}}|_{p^{-1}(\mathfrak{D})} \otimes p^*\overline{\delta}^{\mathrm{d}_\ell^g(-\frac{1}{2}\mathrm{rk}X_\ell + 6\frac{\nu_0\cdot\nu_0}{\ell})}$$

Here the second equality follows from the Theorem 5.2, and the third one follows from the Lemma 8.1.2.

This completes the proof of the Proposition. □

8.3. We are ready to finish the proof of the Theorem 7.6(a).

8.3.1. Let $\mathcal{N}$ be the standard deformation of $\mathring{T}\overline{C}^\alpha \times_{\overline{\mathcal{M}_g}} \overline{\delta}_g$ to the normal cone of $\mathcal{T}C^\alpha_{\mathfrak{D}} \times_{\overline{\mathcal{M}_g}} \overline{\delta}_g$ (see e.g. [Fu]).

So we have

$$T_{\mathcal{T}C^\alpha_{\mathfrak{D}} \times_{\overline{\mathcal{M}_g}} \overline{\delta}_g}(\mathring{T}\overline{C}^\alpha \times_{\overline{\mathcal{M}_g}} \overline{\delta}_g) \hookrightarrow \mathcal{N} \hookleftarrow \mathring{T}\overline{C}^\alpha \times_{\overline{\mathcal{M}_g}} \overline{\delta}_g \times \mathbb{C}^*$$

Let $t: \mathcal{N} \longrightarrow \mathbb{C}$; $\mathrm{pr}: \mathcal{N} \longrightarrow \mathring{T}\overline{C}^\alpha \times_{\overline{\mathcal{M}_g}} \overline{\delta}_g$ denote the canonical projections. We define an open subset $U \subset \mathcal{N}$ as

$$U := (\mathring{T}C^\alpha \times_{\mathcal{M}_g} \dot{\delta}) \times \mathbb{C}^* \sqcup \mathring{T}_{\mathcal{T}C^\alpha_{\mathfrak{D}} \times_{\overline{\mathcal{M}_g}} \overline{\delta}_g}(\mathring{T}\overline{C}^\alpha \times_{\overline{\mathcal{M}_g}} (\overline{\delta_g}))$$

To finish the proof it is enough to construct a local system $\mathcal{H}_\mathcal{N}$ on U such that

$$\mathcal{H}_\mathcal{N}|_{(\mathring{T}C^\alpha \times_{\mathcal{M}_g} \dot{\delta}_g) \times \mathbb{R}^{>0}} = \mathcal{H}^\alpha_g; \tag{311}$$

$$\wp_\alpha^*(\mathcal{H}_N|_{(\overset{\circ}{T}C_1^{\alpha_1+\mu_1}\times_{\mathfrak{D}}\dot{\delta}_{g_1})\times(\overset{\circ}{T}C_2^{\alpha_2+\mu_2}\times_{\mathfrak{D}}\dot{\delta}_{g_2})}) = \mathcal{H}_{g_1}^{\alpha_1+\mu_1}\boxtimes\mathcal{H}_{g_2}^{\alpha_2+\mu_2}. \tag{312}$$

Recall the notations of 3.5. The canonical meromorphic section s_α of $\mathrm{AJ}_\alpha^*(\det(\mathcal{E}))$ extends to the same named meromorphic section of $\mathrm{AJ}_\alpha^*(\det(\overline{\mathcal{E}}))$ over $T\overline{C}^\alpha\times_{\overline{\mathcal{M}_g}}(\overline{\delta_g})^{\cdot}$.

8.3.2. *Lemma.* The section $s'_\alpha := t^{-\frac{\mathrm{d}_\ell^g}{\ell}n(\mu_1)}\mathrm{pr}^*s_\alpha$ is regular on U, and we have

$$\wp_\alpha^*(s'_\alpha|_{p^{-1}(\mathfrak{D})}) = s_{\alpha_1}^{\mathrm{d}_\ell^{g_2}}\otimes s_{\alpha_2}^{\mathrm{d}_\ell^{g_1}}$$

(see (310))

Proof. Immediate from the Proposition 8.2 and the construction of $\overline{\mathcal{L}}$ in 8.1.2. Note also that $n(\mu_1)=n(\mu_2)$. □

8.3.3. *Remark.* If one of g_1, g_2 is equal to 1, then the following remark is in order. Suppose, say, $g_1 = 1$. Then $N = \mathrm{d}_\ell^{g_2}$ satisfies the assumptions of 3.6, and $s_{\alpha_1}^{\mathrm{d}_\ell^{g_2}}$ should be understood as the canonical section of $\mathrm{AJ}_{\alpha_1}^*(\det(\mathcal{E}_{C_1}))^{\otimes\mathrm{d}_\ell^{g_2}}$ arising from the Theorem 5.2(d).

8.3.4. We construct the local system $\overline{\mathfrak{H}}$ on $(\det(\overline{\mathcal{E}}))^{\cdot}$ as in 5.2.1. Now we define

$$\mathcal{H}_N := (s'_\alpha)^*\overline{\mathfrak{H}}$$

The property (311) is clear. Let us prove the property (312).

It is clear that the isomorphism (309) is the isomorphism of bundles with flat projective connections.

Suppose $g_1, g_2 > 1$. Let us define a map

$$\widetilde{\mathfrak{V}}:\ (\det(\mathcal{E}_{C_1}))^{\cdot}\times_{\mathfrak{D}}(\det(\mathcal{E}_{C_1}))^{\cdot}\longrightarrow(\det(\overline{\mathcal{E}}|_{\mathfrak{D}}))^{\cdot}$$

as a composition of

$$(\det(\mathcal{E}_{C_1}))^{\cdot}\times_{\mathfrak{D}}(\det(\mathcal{E}_{C_1}))^{\cdot}\longrightarrow(\det(\mathcal{E}_{C_1})^{\otimes\mathrm{d}_\ell^{g_2}})^{\cdot}\otimes(\det(\mathcal{E}_{C_1})^{\otimes\mathrm{d}_\ell^{g_1}})^{\cdot},\ (\lambda_1,\lambda_2)\mapsto\lambda_1^{\mathrm{d}_\ell^{g_2}}\otimes\lambda_2^{\mathrm{d}_\ell^{g_1}},$$

the isomorphism (310), and the obvious map $(\mathfrak{V}^*(\det(\overline{\mathcal{E}}|_{\mathfrak{D}})))^{\cdot}\longrightarrow(\det(\overline{\mathcal{E}}|_{\mathfrak{D}}))^{\cdot}$.

Then we see that

$$\widetilde{\mathfrak{V}}^*(\overline{\mathfrak{H}}|_{\mathfrak{D}}) = \mathfrak{H}_{C_1}\boxtimes\mathfrak{H}_{C_2}$$

The property (312) follows.

This completes the proof of 7.6 in case $g_1, g_2 > 1$.

The minor changes in case one of g_1, g_2 equals 1 are safely left to the interested reader. □

9. Proof of Theorem 7.6(b)

We start the proof with the following weaker statement.

9.1. **Proposition.** $\wp_\alpha^*\mathbf{Sp}\mathcal{H}_g^\alpha = \oplus_{\mu_1+\mu_2=-2\nu_0}\mathcal{H}_{g_0}^{\alpha+\mu_1+\mu_2}\otimes\mathcal{P}_{\mu_1}$ for some one-dimensional local systems $\mathcal{P}_{\mu_1}$.

9.2. To prove the proposition, let us introduce some notations. We denote $\mathcal{M}_g \cup \mathfrak{D}$ by $\overline{\mathcal{M}}^\circ \subset \overline{\mathcal{M}_g}$. It is well known that $\mathrm{Pic}(C/\mathcal{M}_g)$ extends to the family of algebraic groups $\mathrm{Pic}(\overline{C}/\overline{\mathcal{M}}^\circ)$ over $\overline{\mathcal{M}}^\circ$.

Restricting to $\mathfrak{D}$ we get the projection

$$\pi : \mathrm{Pic}(\overline{C}/\overline{\mathcal{M}}^\circ)|_{\mathfrak{D}} \longrightarrow \mathrm{Pic}(C_0/\mathfrak{D})$$

which identifies $\mathrm{Pic}(\overline{C}/\overline{\mathcal{M}}^\circ)|_{\mathfrak{D}}$ with a $\mathbb{C}^*$-torsor over $\mathrm{Pic}(C_0/\mathfrak{D})$.

The class of this torsor is equal to $(x_1) - (x_2) \in \mathrm{Pic}^0(C_0/\mathfrak{D}) = \mathrm{Pic}^0(\mathrm{Pic}^n(C_0/\mathfrak{D}))$ for any $n \in \mathbb{Z}$.

For any $\mu_1, \mu_2 \in X_\ell$ such that $\mu_1 + \mu_2 = -2\nu_0$ we have a subtraction map

$$r_{\mu_1,\mu_2} : (\mathrm{Pic}(C_0/\mathfrak{D}) \otimes X_\ell)_{(2g-2)\nu_0} \longrightarrow (\mathrm{Pic}(C_0/\mathfrak{D}) \otimes X_\ell)_{(2g_0-2)\nu_0},$$
$$x \mapsto x - \mu_1 \otimes x_1 - \mu_2 \otimes x_2 \quad (313)$$

Recall that in 5.2 we have defined for any curve C of genus g a vector bundle with flat projective connection $\mathcal{E}$ on $(\mathrm{Pic}(C) \otimes X_\ell)_{(2g-2)\nu_0}$. For the curve C_0 we will denote this bundle by $\mathcal{E}_{C_0}$.

We define the vector bundle $\mathcal{E}_{\mu_1} := r^*_{\mu_1,\mu_2}\mathcal{E}_{C_0}$ on $(\mathrm{Pic}(C_0/\mathfrak{D}) \otimes X_\ell)_{(2g-2)\nu_0}$. It carries a flat projective connection.

If $\mu_1' - \mu_1 \in Y_\ell$ then the vector bundles $\mathcal{E}_{\mu_1}$ and $\mathcal{E}_{\mu_1'}$ differ by a twist by a line bundle, and the corresponding flat projective connections are identified by the Lemma 5.11.1. In particular, the locally constant sheaf of algebras $\underline{\mathrm{End}}(\mathcal{E}_{\mu_1})$ canonically depends only on the class of μ_1 modulo Y_ℓ. We will use the notation $\underline{\mathrm{End}}(\mathcal{E}_\mu)$ for $\mu \in X_\ell/Y_\ell$ from now on.

9.3. The plan of the proof is as follows.

Consider the sheaf of algebras $\underline{\mathrm{End}}(\mathcal{E})$ on $(\mathrm{Pic}(C) \otimes X_\ell)_{(2g-2)\nu_0}$. The data of a flat projective connection on $\mathcal{E}$ is equivalent to the data of a flat genuine connection on $\underline{\mathrm{End}}(\mathcal{E})$ preserving the algebra structure.

Let $\mathcal{U}$ be a small (punctured) neighbourhood (in classical topology) of $p^{-1}(\mathfrak{D})$ in $(\mathrm{Pic}(C) \otimes X_\ell)_{(2g-2)\nu_0}$. The monodromy around $p^{-1}(\mathfrak{D})$ acts on $\mathfrak{H}$, or, equivalently, on $\mathcal{E}$, semisimply. It acts on the sheaf of algebras $\underline{\mathrm{End}}(\mathcal{E})|_{\mathcal{U}}$ by the algebra automorphism

$$\mathfrak{m} : \underline{\mathrm{End}}(\mathcal{E}) \longrightarrow \underline{\mathrm{End}}(\mathcal{E})$$

We will construct a locally constant sheaf of algebras $\mathfrak{M} \subset \underline{\mathrm{End}}(\mathcal{E})$ with the following properties.

9.3.1. (i) $\mathfrak{M}$ can be extended to a locally constant sheaf of algebras $\overline{\mathfrak{M}}$ on $(\mathrm{Pic}(\overline{C}/\overline{\mathcal{M}}^\circ) \otimes X_\ell)_{(2g-2)\nu_0}$.

(ii) The sheaf $\mathcal{E}|_{\mathcal{U}}$ decomposes into direct sum of d_ℓ eigensheaves of $\mathfrak{m}$: $\mathcal{E}|_{\mathcal{U}} = \oplus_\lambda \mathcal{E}_\lambda$; and $\mathfrak{M} = \oplus_\lambda \underline{\mathrm{End}}(\mathcal{E}_\lambda)$.

(iii) $\overline{\mathfrak{M}}|_{p^{-1}(\mathfrak{D})} = \oplus_{\mu \in X_\ell/Y_\ell} \underline{\mathrm{End}}(\mathcal{E}_\mu)$.

9.3.2. The construction and the proof of the above properties will occupy the rest of this section. Let us derive the Proposition 9.1 from these properties.

The decomposition (ii) of $\mathcal{E}|_{\mathcal{U}}$ induces a decomposition: $\mathbf{Sp}\mathfrak{H}_g = \oplus_\mu \mathfrak{H}_\mu$. It follows from (ii),(iii) that there is a bijection between the set of summands and X_ℓ/Y_ℓ such that $\overline{\mathfrak{M}}|_{p^{-1}(\mathfrak{D})} = \oplus_{\mu\in X_\ell/Y_\ell}\underline{\mathrm{End}}(\mathfrak{H}_\mu)$. Moreover, $\underline{\mathrm{End}}(\mathfrak{H}_\mu) = \underline{\mathrm{End}}(\mathcal{E}_\mu)$.

Now for any g-admissible $\alpha \in X_\ell$ we have an isomorphism of local systems

$$\bigoplus_{\mu_1+\mu_2=-2\nu_0}^{\mu_1,\mu_2\in X_\ell/Y_\ell} \underline{\mathrm{End}}(\mathcal{H}_{g_0}^{\alpha+\mu_1+\mu_2}) = \bigoplus_{\mu_1+\mu_2=-2\nu_0}^{\mu_1,\mu_2\in X_\ell/Y_\ell} s^*_{\alpha+\mu_1+\mu_2}\mathrm{AJ}^*_{\alpha+\mu_1+\mu_2}\underline{\mathrm{End}}(\mathcal{E}_{\mu_1}) =$$

$$= \bigoplus_{\mu_1+\mu_2=-2\nu_0}^{\mu_1,\mu_2\in X_\ell/Y_\ell} s^*_{\alpha+\mu_1+\mu_2}\mathrm{AJ}^*_{\alpha+\mu_1+\mu_2}\underline{\mathrm{End}}(\mathfrak{H}_{\mu_1})$$

Hence $s^*_{\alpha+\mu_1+\mu_2}\mathrm{AJ}^*_{\alpha+\mu_1+\mu_2}\mathfrak{H}_{\mu_1} = \mathcal{H}_{g_0}^{\alpha+\mu_1+\mu_2}\otimes\mathcal{P}_{\mu_1}$ for some one-dimensional local system $\mathcal{P}_{\mu_1}$.

On the other hand, obviously,

$$\bigoplus_{\mu_1+\mu_2=-2\nu_0}^{\mu_1,\mu_2\in X_\ell/Y_\ell} s^*_{\alpha+\mu_1+\mu_2}\mathrm{AJ}^*_{\alpha+\mu_1+\mu_2}\mathfrak{H}_{\mu_1} = s^*_\alpha\mathrm{AJ}^*_\alpha\mathbf{Sp}\mathfrak{H}_g = \mathbf{Sp}\mathcal{H}_g^\alpha$$

This completes the proof of the Proposition 9.1. □

9.4. Before we prove the properties 9.3.1(i)–(iii) we need to introduce the following general construction.

Let $p_{\mathfrak{A}} : \mathfrak{A} \longrightarrow S$ be an abelian scheme over a base S. Let L be a relatively ample line bundle on $\mathfrak{A}$. It defines a morphism ϕ_L from $\mathfrak{A}$ to the dual abelian variety $\check{\mathfrak{A}} :\ a \mapsto T_aL\otimes L^{-1}$ where T_a denotes the translation by a. Let $i_G :\ G \hookrightarrow \mathfrak{A}$ be the kernel of this homomorphism. Then $p_G :\ G \longrightarrow S$ is an etale cover.

Let $L_\heartsuit$ denote the line bundle $L\otimes p_{\mathfrak{A}}^*0^*L^{-1}$ where 0 stands for the zero section $S \longrightarrow \mathfrak{A}$. Recall that $\dot{L}$ denotes the $\mathbb{C}^*$-torsor corresponding to the line bundle L.

9.4.1. The sheaf $p_{G*}i_G^*\dot{L}_\heartsuit$ has a natural structure of a group scheme acting on the sheaf $p_{\mathfrak{A}*}L$. There is an exact sequence

$$0 \longrightarrow \mathcal{O}^* \longrightarrow p_{G*}i_G^*\dot{L}_\heartsuit \xrightarrow{\mathfrak{p}} G \longrightarrow 0$$

The subsheaf $\mathcal{O}^*$ acts on $p_{\mathfrak{A}*}L$ by multiplications with functions. The action of an element $h \in p_{G*}i_G^*\dot{L}_\heartsuit$ covers the translation by $\mathfrak{p}(h) \in G \subset \mathfrak{A}$.

The group scheme $p_{G*}i_G^*\dot{L}_\heartsuit$ carries a unique flat connection compatible with multiplication. In effect, the elements of finite order form a Zariski dense union of etale group subschemes.

9.4.2. The sheaf $p_{G*}i_G^*L_\heartsuit$ has a natural structure of algebra acting on the sheaf $p_{\mathfrak{A}*}L$. The natural inclusion $p_{G*}i_G^*\dot{L}_\heartsuit \hookrightarrow p_{G*}i_G^*L_\heartsuit$ is multiplicative and compatible with the action on $p_{\mathfrak{A}*}L$. The sheaf of algebras $p_{G*}i_G^*L_\heartsuit$ carries a unique flat connection compatible with the one on $p_{G*}i_G^*\dot{L}_\heartsuit$.

The map

$$p_{G*}i_G^* L_\heartsuit \xrightarrow{\sim} \underline{\mathrm{End}}(p_{\mathfrak{A}*}L) \tag{314}$$

is an isomorphism.

9.4.3. *Remark.* Note that the above construction gives an alternative way to define the flat projective connection on $p_{\mathfrak{A}*}L$ — without referring to [BK].

9.5. We apply the above construction to the following situation. We take $S = (\mathrm{Pic}(C/\mathcal{U}) \otimes X_\ell)_{(2g-2)\nu_0}$; $\mathfrak{A} = (\mathrm{Pic}(C/\mathcal{U}) \otimes (X_\ell \oplus Y_\ell))_{(2g-2)\nu_0,0}$; $\pi = \mathrm{pr}_1$; $L = \mathcal{L}'$ as in 5.1.

Then $G = p^*_{(\mathrm{Pic}(C/\mathcal{U})\otimes X_\ell)_{(2g-2)\nu_0}} R^1 p_{C*}(\underline{X_\ell/Y_\ell})$.

Recall that by the Picard-Lefschetz theory the monodromy $\mathfrak{m}$ acts on $R^1p_{C*}(\underline{\mathbb{Z}})$ by the formula

$$y \mapsto y \pm \langle y, x\rangle x$$

where x is the vanishing cycle. In particular, the invariants of $\mathfrak{m}$ in $R^1p_{C*}(\underline{\mathbb{Z}})$ form a sublattice of codimension 1, namely, the orthogonal complement of the vanishing cycle.

We define $\mathfrak{M} \subset \underline{\mathrm{End}}(\mathcal{E}) = \mathfrak{p}_{\mathfrak{G}*}\mathfrak{i}^*_{\mathfrak{G}}\mathcal{L}'_\heartsuit$ as follows:

$$\mathfrak{M} := \mathfrak{p}_{\mathfrak{G}^{\mathfrak{m}}*}\mathfrak{i}^*_{\mathfrak{G}^{\mathfrak{m}}}\mathcal{L}'_\heartsuit$$

where $G^{\mathfrak{m}}$ stands for the invariants of $\mathfrak{m}$ on G.

9.6. Let us check the properties 9.3.1.

The sheaf $G^{\mathfrak{m}} = (X_\ell/Y_\ell) \otimes (R^1p_{C*}(\underline{\mathbb{Z}}))^{\mathfrak{m}}$ extends to a sheaf $\overline{G}$ on $\overline{\mathcal{M}}^\circ$. Moreover, we have an embedding

$$i_{\overline{G}} : \overline{G} \hookrightarrow (\mathrm{Pic}(\overline{C}/\overline{\mathcal{M}}^\circ) \otimes Y_\ell)_0$$

Let us choose some extension $\overline{\mathcal{L}'}$ of $\mathcal{L}'$ to $(\mathrm{Pic}(\overline{C}/\overline{\mathcal{M}}^\circ) \otimes (X_\ell \oplus Y_\ell))_{(2g-2)\nu_0,0}$. Recall that $\overline{\mathcal{L}'}_\heartsuit$ denotes $\overline{\mathcal{L}'} \otimes p^*0^*\overline{\mathcal{L}'}^{-1}$. We define

$$\overline{\mathfrak{M}} := p_{\overline{G}*}i^*_{\overline{G}}\overline{\mathcal{L}'}_\heartsuit$$

Evidently, $\overline{\mathfrak{M}}|_{\mathcal{U}} = \mathfrak{M}$.

It is easy to see that the algebra structure on $\mathfrak{M}$ extends uniquely to $\overline{\mathfrak{M}}$. This proves (i).

Property (i) implies that $\mathfrak{M} \subset (\underline{\mathrm{End}}(\mathcal{E}))^{\mathfrak{m}}$. On the other hand one can check that there exists a unique decomposition of $\mathcal{E}$ in the sum of d_ℓ summands of equal dimension: $\mathcal{E} = \oplus\mathcal{E}_\lambda$ such that $\mathfrak{M} = \oplus\underline{\mathrm{End}}(\mathcal{E}_\lambda) \subset \underline{\mathrm{End}}(\mathcal{E})$. Since $\mathfrak{m}$ commutes with $\mathfrak{M}$ it follows that each $\mathcal{E}_\lambda$ is an eigenspace of $\mathfrak{m}$. (ii) is proved.

9.7. It remains to check 9.3.1(iii). Recall that for $\mu_1, \mu_2 \in X_\ell$ such that $\mu_1 + \mu_2 = -2\nu_0$ we have introduced the substraction map r_{μ_1,μ_2} in (313).

Let us denote the line bundle introduced in 5.1 for the curve C_0 by $\mathcal{L}'_{C_0}$.

9.7.1. *Lemma.* For any $\mu_1, \mu_2 \in X_\ell$ such that $\mu_1 + \mu_2 = -2\nu_0$ we have a canonical isomorphism

$$\overline{\mathcal{L}'}_\heartsuit|_{p^{-1}(\mathfrak{D})} = (r_{\mu_1,\mu_2} \times \mathrm{id})^*(\mathcal{L}'_{C_0})_\heartsuit$$

Proof. Just as in the proof of the Lemma 8.1.2 we define the extension of the line bundle $\mathcal{L}_\alpha$ on C^α to the line bundle $\overline{\mathcal{L}}_\alpha$ on $C^\alpha \cup (C_0 - x_1 - x_2)^\alpha$ for any $\alpha \in \mathbb{N}[X_\ell \oplus Y_\ell]$. The extension is given by the same formula (305).

We necessarily have

$$\mathrm{AJ}_\alpha^* \overline{\mathcal{L}'} = \overline{\mathcal{L}}_\alpha(n(C_0 - x_1 - x_2)^\alpha)$$

for some $n \in \mathbb{Z}$ where $(C_0 - x_1 - x_2)^\alpha$ is viewed as a divisor in $\overline{C}^\alpha$.

On the other hand,

$$\overline{\mathcal{L}}_\alpha|_{p^{-1}(\mathfrak{D})} = \mathrm{AJ}_\alpha^* \pi^* r_{\mu_1,\mu_2}^*(\mathcal{L}'_{C_0}) \otimes \mathcal{N}^m$$

for some $m \in \mathbb{Z}$ where $\mathcal{N}$ denotes the normal line bundle to $(C_0 - x_1 - x_2)^\alpha$ in $C^\alpha \cup (C_0 - x_1 - x_2)^\alpha$.

The line bundle $\mathcal{N}$ is lifted from the base $\mathfrak{D}$. Hence $(L \otimes \mathcal{N})_\heartsuit = L_\heartsuit$ for any line bundle L on $(\mathrm{Pic}(C_0/\mathfrak{D}) \otimes (X_\ell \oplus Y_\ell))_{(2g-2)\nu_0,0}$.

Finally, two line bundles on $(\mathrm{Pic}(C_0/\mathfrak{D}) \otimes (X_\ell \oplus Y_\ell))_{(2g-2)\nu_0,0}$ are isomorphic iff their inverse images with respect to AJ_α are isomorphic for any α.

This completes the proof of the Lemma. □

9.8. Now we can use the above Lemma together with (314) to define for any $\mu \in X_\ell/Y_\ell$ the morphism of algebras

$$\varphi_\mu : \overline{\mathfrak{M}}|_{p^{-1}(\mathfrak{D})} \longrightarrow \underline{\mathrm{End}}(\mathcal{E}_\mu)$$

It remains to show that the morphism

$$\varphi := \sum_{\mu \in X_\ell/Y_\ell} \varphi_\mu : \overline{\mathfrak{M}}|_{p^{-1}(\mathfrak{D})} \longrightarrow \bigoplus_{\mu \in X_\ell/Y_\ell} \underline{\mathrm{End}}(\mathcal{E}_\mu)$$

is an isomorphism. It is enough to prove that φ is surjective.

Since each morphism φ_μ is obviously surjective, it suffices to check that all the idempotents of $\bigoplus_{\mu \in X_\ell/Y_\ell} \underline{\mathrm{End}}(\mathcal{E}_\mu)$ lie in the image of φ.

Consider the subgroup $Z \subset \overline{G}|_{p^{-1}(\mathfrak{D})}$,

$$Z := \mathrm{Ker}\,\pi \cap \overline{G}|_{p^{-1}(\mathfrak{D})}$$

We have $Z \simeq X_\ell/Y_\ell$.

Clearly, the morphism φ_μ maps $p_{Z*} i_Z^* \overline{\mathcal{L}'}_\heartsuit \subset p_{\overline{G}*} i_{\overline{G}}^* \overline{\mathcal{L}'}_\heartsuit$ to the one-dimensional subspace of $\underline{\mathrm{End}}(\mathcal{E}_\mu)$ generated by id. Let us denote the corresponding linear functional on $p_{Z*} i_Z^* \overline{\mathcal{L}'}_\heartsuit$ by $\aleph_\mu$.

It is enough to prove that the set of functionals $\{\aleph_\mu, \mu \in X_\ell/Y_\ell\}$ is linearly independent.

Note that the quadratic form ?·? on X_ℓ induces a perfect pairing $(X_\ell/Y_\ell) \times (X_\ell/Y_\ell) \longrightarrow \mathbb{Q}/l\mathbb{Z}$, and thus identifies the finite abelian group X_ℓ/Y_ℓ with its dual $(X_\ell/Y_\ell)^\vee$.

Since $p_{Z*} i_Z^* \overline{\mathcal{L}'}_\heartsuit$ is X_ℓ/Y_ℓ-graded, the group $(X_\ell/Y_\ell)^\vee$ acts on it by multiplication, and the corresponding module is isomorphic to the dual regular representation $B[X_\ell/Y_\ell]$.

Consider $\mu = 0 \in X_\ell/Y_\ell$. The functional $\aleph_0$ does not vanish on any graded component of $p_{Z*}i_Z^*\overline{\mathcal{L}'}_\heartsuit$. Hence $\aleph_0$ generates $(p_{Z*}i_Z^*\overline{\mathcal{L}'}_\heartsuit)^\vee$ as a $(X_\ell/Y_\ell)^\vee$-module. On the other hand, it is easy to see that for any $\lambda \in (X_\ell/Y_\ell)^\vee$ we have $\lambda(\aleph_0) = \aleph_\lambda$.

This completes the proof of the property 9.3.1(iii) along with the statement the Proposition 9.1. □

9.9. Now we will derive the Theorem 7.6(b) from the Proposition 9.1.

Given $\mu \in X_\ell/Y_\ell$ we will denote $-2\nu_0 - \mu$ by μ'.

We have proved that $\mathbf{Sp}\mathcal{E} = \oplus_{\mu \in X_\ell/Y_\ell}\mathcal{E}_\mu$, and the flat projective connections on $\mathcal{E}, \mathcal{E}_\mu$ agree (note that the sum of bundles with flat projective connections does not have a natural projective connection).

Recall that in order to define the local system $\mathcal{H}^{\alpha+\mu+\mu'}$ we have used the canonical section $s_{\alpha+\mu+\mu'}$ of the line bundle $\mathrm{AJ}^*_{\alpha+\mu+\mu'}\det(\mathcal{E}_\mu)$.

The Theorem follows immediately from the next two statements.

9.10. **Lemma.** The canonical section s_α of $\mathrm{AJ}^*_\alpha\det(\mathcal{E})$ can be extended to a section $\overline{s}_\alpha$ over $\overline{\mathcal{M}}^\circ$ in such way that

$$\overline{s}_\alpha|_{p^{-1}(\mathfrak{D})} = \prod_{\mu \in X_\ell/Y_\ell} s_{\alpha+\mu+\mu'}$$

The proof is completely parallel to the one of 8.3.2. □

9.11. We consider $\oplus_{\mu \in X_\ell/Y_\ell}\mathcal{E}_\mu$ as a projectively flat bundle via the isomorphism with $\mathbf{Sp}\mathcal{E}$. Then $\oplus_{\mu \in X_\ell/Y_\ell}\det(\mathcal{E}_\mu)$ is a direct summand of an exterior power of $\oplus_{\mu \in X_\ell/Y_\ell}\mathcal{E}_\mu$ and thus inherits a flat projective connection.

Proposition. The collection $(s_{\alpha+\mu+\mu'},\ \mu \in X_\ell/Y_\ell)$ forms a projectively flat section of $\oplus_{\mu \in X_\ell/Y_\ell}\det(\mathcal{E}_\mu)$.

9.12. **Proof.** The flat projective connection on $\oplus_{\mu \in X_\ell/Y_\ell}\det(\mathcal{E}_\mu)$ induces a flat genuine connection on $\det(\mathcal{E}_\mu) \otimes \det(\mathcal{E}_\lambda)^{-1}$ for any $\mu, \lambda \in X_\ell/Y_\ell$. We have to prove that $s_\mu s_\lambda^{-1}$ is a flat section of $\det(\mathcal{E}_\mu) \otimes \det(\mathcal{E}_\lambda)^{-1}$.

It suffices to prove that $s_\mu^N s_\lambda^{-N}$ is a flat section of $(\det(\mathcal{E}_\mu) \otimes \det(\mathcal{E}_\lambda)^{-1})^N$ for some positive integer N.

This follows immediately from the next two Lemmas.

9.12.1. *Lemma.* There exists $N > 0$ such that the one-dimensional local system $(\det(\mathcal{E}_\mu) \otimes \det(\mathcal{E}_\lambda)^{-1})^N$ is trivial for any $\mu, \lambda \in X_\ell/Y_\ell$.

9.12.2. *Lemma.* Any invertible function on $\mathring{T}C_0^{\alpha+\mu+\lambda} \times_{\mathcal{M}_{g_0}} \dot{\delta}$ is constant if $g_0 > 1$.

9.12.3. We will prove the Lemmas in the following subsections. Let us derive the Proposition now. First of all, if $g_0 = 0$ there is nothing to prove. The case $g_0 = 1$ is left to the reader. Otherwise, we conclude that $s_\mu^N s_\lambda^{-N}$ being invertible over $\mathring{T}C_0^{\alpha+\mu+\lambda} \times_{\mathcal{M}_{g_0}} \dot{\delta}$ is a constant function, and thus a flat section of a trivial local system $(\det(\mathcal{E}_\mu) \otimes \det(\mathcal{E}_\lambda)^{-1})^N$. □

9.13. To prove the Lemma 9.12.1 it is enough to show that the image of the monodromy representation of the fundamental group $\pi_1(\overset{\bullet}{T}_{TC^{\alpha}_{\mathfrak{D}} \times_{\overline{\mathcal{M}_g}} \overline{\delta_g}}(\overset{\circ}{T}\overline{C}^{\alpha} \times_{\overline{\mathcal{M}_g}} \overline{\delta_g}))$ (see (303)) in $PGL(\oplus_{\mu \in X_\ell / Y_\ell} \mathcal{E}_\mu)$ is finite.

To this end it suffices to show that the image of $\pi_1(\overset{\circ}{T}C^{\alpha})$ in $PGL(\mathcal{E})$ is finite.

Recall the notations of 9.4, 9.5. Let us choose a finite subgroup $\tilde{G} \subset p_{G*} i_G^* \dot{\mathcal{L}}'_{\heartsuit}$ projecting surjectively onto G and such that the kernel of this projection contains at least three elements.

Then it follows from the discussion in 9.4 that the action of $\pi_1(\overset{\circ}{T}C^{\alpha})$ on $\mathcal{E}$ factors through the group $\mathrm{Aut}(\tilde{G})$.

This proves the Lemma 9.12.1. □

9.14. Let us prove the Lemma 9.12.2. To unburden the notation we will omit the subindex $_0$. Assume there is a nonconstant invertible function. We can lift it to the nonsymmetrized Cartesian power and obtain a nonconstant invertible function on $(\overset{\bullet}{T}C)^J \times_{\mathcal{M}_g} \dot{\delta}$ for some set J of positive cardinality n.

Let us choose a subset $K = J - j$ of cardinality $n-1$ and consider the projection

$$\mathrm{pr}: (\overset{\bullet}{T}C)^J \times_{\mathcal{M}_g} \dot{\delta} \longrightarrow (\overset{\bullet}{T}C)^K \times_{\mathcal{M}_g} \dot{\delta}$$

9.14.1. *Claim.* $\mathrm{pr}_* \mathcal{O}^* = \mathcal{O}^*$.

Proof. Consider a projection

$$\mathrm{pr}': (\overset{\bullet}{T}C)^K \times_{\mathcal{M}_g} \dot{\delta} \times_{\mathcal{M}_g} C_j \longrightarrow (\overset{\bullet}{T}C)^K \times_{\mathcal{M}_{g0}} \dot{\delta}$$

Then we have

$$\mathrm{pr}_* \mathcal{O}^* = \bigoplus_{i \in \mathbb{Z}} \mathrm{pr}'_* (\dot{\Omega}^i (C_j / \mathcal{M}_g))$$

But for generic $(C, z_k \in C,\ k \in K)$ the subgroup of $\mathrm{Pic}(C)$ generated by $\{\Omega, (z_k),\ k \in K\}$ is free. That is, for any $i \in \mathbb{Z}$ there are no meromorphic sections of Ω^i invertible on $C - \{z_k,\ k \in K\}$.

This completes the proof of the Claim. □

9.14.2. Using the Claim inductively we conclude that

$$\Gamma((\overset{\bullet}{T}C)^J \times_{\mathcal{M}_g} \dot{\delta}, \mathcal{O}^*) = \Gamma(\dot{\delta}, \mathcal{O}^*) = \bigoplus_{i \in \mathbb{Z}} \Gamma(\mathcal{M}_g, \dot{\delta}^i)$$

Since $g > 1$, we know that for $i \neq 0$ the line bundle δ^i is nontrivial, and hence does not have any invertible sections.

For $i = 0$ we know that that the only invertible functions on $\mathcal{M}_g$ are constants (see e.g. [BM] 2.3).

This completes the proof of Lemma 9.12.2 along with the Theorem 7.6. □

Chapter 3. Regular Representation

10. A characterization of the Regular bimodule

This section belongs really to the Chapter 2 of IV.

10.1. Consider the algebra $\tilde{\mathfrak{u}}$ defined in [AJS], Remark 1.4. It is an infinite dimensional k-algebra containing $\mathfrak{u}$ as a subalgebra. We have $\tilde{\mathfrak{u}} = \mathfrak{u}^- \otimes \tilde{\mathfrak{u}}^0 \otimes \mathfrak{u}^+$ where $\tilde{\mathfrak{u}}^0$ is formed by some divided power expressions in K_i.

We have $\mathcal{C} = \tilde{\mathfrak{u}} -$ mod, and $\tilde{\mathfrak{u}}$ itself viewed as a left regular $\mathfrak{u}$-module is isomorphic to $\oplus_{\lambda\in Y}\ \mathfrak{u} \otimes L(l\lambda)$ where $\mathfrak{u}$ is considered as a left regular $\mathfrak{u}$-module.

Thus $\tilde{\mathfrak{u}}$ does not belong to $\mathcal{C}$ being infinite dimensional, but belongs to Ind$\mathcal{C}$, see [D4], §4. If we take into account the right regular action of $\mathfrak{u}$ an $\tilde{\mathfrak{u}}$ as well, then we can regard $\tilde{\mathfrak{u}}$ as an object of Ind$\mathcal{C} \otimes$ Ind$\mathcal{C}_r$, that is, an infinite dimensional $\mathfrak{u}$-bimodule. Finally, composing the right action with the antipode, we may view $\tilde{\mathfrak{u}}$ as an object of Ind$\mathcal{C} \otimes$ Ind$\mathcal{C}$. We will denote this object by $\mathbf{R}$.

10.2. Given $V, W \in \mathcal{C}$ we define a functor $F_{V\otimes W} : \mathcal{C} \longrightarrow \mathcal{C}$,

$$F_{V\otimes W}(M) = (sM \otimes_{\mathcal{C}} V) \otimes W.$$

The functor $\mathcal{C} \times \mathcal{C} \longrightarrow \mathrm{F}(\mathcal{C},\mathcal{C}),\ (V,W) \mapsto F_{V\otimes W}$, is exact in W and right exact in V. Therefore, by the universal property it extends to a functor

$$\mathcal{C} \otimes \mathcal{C} \longrightarrow \mathrm{F}(\mathcal{C},\mathcal{C}),\ \mathcal{O} \mapsto F_{\mathcal{O}}.$$

This in turn extends to a functor

$$\mathrm{Ind}\mathcal{C} \otimes \mathrm{Ind}\mathcal{C} \longrightarrow \mathrm{F}(\mathrm{Ind}\mathcal{C}, \mathrm{Ind}\mathcal{C}),\ \mathcal{O} \mapsto F_{\mathcal{O}}.$$

We have $F_{\mathbf{R}} = \mathrm{Id}$.

In concrete terms, $\mathcal{C}\otimes\mathcal{C}$ is equivalent to $X_\ell \times X_\ell$-graded $\mathfrak{u}$-bimodules. This equivalence sends $V \otimes W$ to $V \otimes sW$ with an evident bimodule structure (the latter tensor product is understood as a product of vector spaces). For a bimodule $\mathcal{O}$ and $M \in \mathcal{C}$ we have

$$F_{\mathcal{O}}(M) = s(sM \otimes_{\mathcal{C}} \mathcal{O}).$$

Here $sM \otimes_{\mathcal{C}} \mathcal{O}$ has a right $\mathfrak{u}$-module structure and X_ℓ-grading inherited from $\mathcal{O}$. So, for the bimodule $\mathbf{R}$ we have

$$s(sM \otimes_{\mathcal{C}} \mathbf{R}) = M$$

for all $M \in \mathcal{C}$.

10.3. Let $\mathcal{C}^{\mathbb{Z}}$ denote the category of $\mathbb{Z}$-graded objects of $\mathcal{C}$. For an $X_\ell \times X_\ell$-graded $\mathfrak{u}$-bimodule $\mathcal{O}$ we define a functor $F_{\mathcal{O};\frac{\infty}{2}+\bullet} : \mathcal{C} \longrightarrow \mathcal{C}^{\mathbb{Z}}$,

$$F_{\mathcal{O};\frac{\infty}{2}+\bullet}(M) := s\mathrm{Tor}^{\mathcal{C}}_{\frac{\infty}{2}+\bullet}(sM, \mathcal{O}).$$

Here $\mathrm{Tor}^{\mathcal{C}}_{\frac{\infty}{2}+\bullet}(sM, \mathcal{O})$ has a right $\mathfrak{u}$-module structure and X_ℓ-grading inherited from $\mathcal{O}$.

10.4. **Lemma.** (i) *Suppose $\mathcal{O}$ is $\mathfrak{u}^+$-induced as a left $\mathfrak{u}$-module. Then* $F_{\mathcal{O};\frac{\infty}{2}+n} = 0$ *for* $n < 0$, *and* $F_{\mathcal{O};\frac{\infty}{2}+0} = F_{\mathcal{O}}$.

(ii) *Suppose $\mathcal{O}$ is projective as a left $\mathfrak{u}$-module. Then* $F_{\mathcal{O};\frac{\infty}{2}+n} = 0$ *for* $n \neq 0$, *and* $F_{\mathcal{O};\frac{\infty}{2}+0} = F_{\mathcal{O}}$.

Proof follows immediately from the next

10.4.1. **Lemma.** *Let $V \in \mathcal{C}_r$, and assume that $N \in \mathcal{C}$ is $\mathfrak{u}^+$-induced. Then*

$$\mathrm{Tor}^{\mathcal{C}}_{\frac{\infty}{2}+\bullet}(V, N) = \mathrm{Tor}^{\mathcal{C}}_{\bullet}(V, N).$$

In particular, if N is projective then

$$\mathrm{Tor}^{\mathcal{C}}_{\frac{\infty}{2}+\bullet}(V, N) = \begin{cases} 0 & \textit{if } n \neq 0 \\ V \otimes_{\mathcal{C}} N & \textit{if } n = 0. \end{cases}$$

Proof. According to Theorem IV.4.9 we may choose any $\mathfrak{u}^+$-induced right convex resolution of N for the computation of $\mathrm{Tor}^{\mathcal{C}}_{\frac{\infty}{2}+\bullet}(V, N)$. So, let us choose N itself. Then we have

$$H^\bullet(R^\bullet_{\swarrow}(V) \otimes_{\mathcal{C}} N) = H^\bullet((R^\bullet_{\swarrow}(V) \otimes sN) \otimes_{\mathcal{C}} \mathsf{k}).$$

But $R^\bullet_{\swarrow}(V) \otimes sN$ is a left $\mathfrak{u}$-projective resolution of $V \otimes sN$, so

$$H^{-\bullet}((R^\bullet_{\swarrow}(V) \otimes sN) \otimes_{\mathcal{C}} \mathsf{k}) = \mathrm{Tor}^{\mathcal{C}}_{\bullet}((V \otimes sN), \mathsf{k}) = \mathrm{Tor}^{\mathcal{C}}_{\bullet}(V, N). \ \square$$

10.5. **Corollary.**

$$F_{\mathbf{R};\frac{\infty}{2}+n} = \begin{cases} 0 & \textit{if } n \neq 0 \\ \mathrm{Id} & \textit{if } n = 0. \end{cases}$$

Proof. The bimodule $\mathbf{R}$ is projective as a left $\mathfrak{u}$-module. $\square$

10.6. **Lemma.** *Let $M_1, \dots, M_n, N \in \mathcal{C}$. There is a canonical perfect pairing between* $\mathrm{Ext}_{\mathcal{C}}^{\frac{\infty}{2}+k}(M_1 \otimes \ldots \otimes M_n, N)$ *and* $\mathrm{Tor}^{\mathcal{C}}_{\frac{\infty}{2}+k}(\mathsf{k}, M_1 \otimes \ldots \otimes M_n \otimes N^*)$ *for every* $k \in \mathbb{Z}$.

Proof. This is just IV.4.8. $\square$

10.7. **Converse Theorem.** Suppose $Q \in \mathrm{Ind}\,\mathcal{C} \otimes \mathrm{Ind}\,\mathcal{C}$, and an isomorphism of functors $\phi : \ F_{\mathbf{R},\frac{\infty}{2}+\bullet} \xrightarrow{\sim} F_{Q,\frac{\infty}{2}+\bullet}$ is given. Then ϕ is induced by the isomorphism $\bar{\phi} : \mathbf{R} \xrightarrow{\sim} Q$.

Proof. We start with the following Lemma:

10.7.1. *Lemma.* Let $M \in \mathrm{Ind}\,\mathcal{C}$. Suppose $\mathrm{Ext}_{\mathcal{C}}^{\frac{\infty}{2}+n}(M, N) = 0$ for $n \neq 0$ and arbitrary $N \in \mathcal{C}$. Suppose further that $\mathrm{Ext}_{\mathcal{C}}^{\frac{\infty}{2}+0}(M, N)$ is exact in N. Then M is projective, and $\mathrm{Ext}_{\mathcal{C}}^{\frac{\infty}{2}+0}(M, N) = \mathrm{Hom}_{\mathcal{C}}(M, N)$.

Proof. By the Theorem IV.4.9, $\mathrm{Ext}_{\mathcal{C}}^{\frac{\infty}{2}+\bullet}(M, N) = H^\bullet(\mathrm{Hom}_{\mathcal{C}}C(P^\bullet(M), R^\bullet(N)))$ where $P^\bullet(M)$ is some $\mathfrak{u}^+$-induced convex right resolution of M, and $R^\bullet(N)$ is some $\mathfrak{u}^-$-induced concave right resolution of N.

Since $\mathrm{Ext}_{\mathcal{C}}^{\frac{\infty}{2}+n}(M,?)$ vanishes for $n \neq 0$, and is exact for $n = 0$, we deduce that $\mathrm{Ext}_{\mathcal{C}}^{\frac{\infty}{2}+\bullet}(M,N) = H^\bullet(\mathrm{Hom}_{\mathcal{C}}(P^\bullet(M),N))$. In particular,

$$\mathrm{Ext}_{\mathcal{C}}^{\frac{\infty}{2}+0}(M,N) = \mathrm{Coker}(\mathrm{Hom}_{\mathcal{C}}(P^1(M),N) \xrightarrow{d_0^*} \mathrm{Hom}_{\mathcal{C}}(P^0(M),N))$$

Let us denote this latter functor in N by $h(N)$ for brevity. We denote by $\xi^* : h(N) \longrightarrow \mathrm{Hom}_{\mathcal{C}}(M,N)$ the evident morphism of functors. By the assumption, $h(N)$ is exact and hence representable by some projective $P^0 \in \mathrm{Ind}\,\mathcal{C}$. We denote by $\xi : M \longrightarrow P^0$ the map inducing ξ^* on the functors they represent.

We denote by $\theta : P^0 \longrightarrow P^0(M)$ the map inducing the evident projection $\mathrm{Hom}_{\mathcal{C}}(P^0(M),N) \longrightarrow h(N)$ on the functors they represent. Then $\theta \circ \xi = \varepsilon : M \longrightarrow P^0(M)$ where ε stands for the augmentation.

Now ε is injective, hence $\xi : M \longrightarrow P^0$ is injective. We may extend $\xi : M \longrightarrow P^0$ to a $\mathfrak{u}^+$-induced convex right resolution $P^0 \longrightarrow P^1 \longrightarrow \ldots$ of M, and use this resolution instead of $P^\bullet(M)$ for the calculation of $\mathrm{Ext}_{\mathcal{C}}^{\frac{\infty}{2}+\bullet}(M,N)$. Repeating the above argument for $P^\bullet$ we conclude that

$$\mathrm{Coker}(\mathrm{Hom}_{\mathcal{C}}(P^1,N) \xrightarrow{d_0^*} \mathrm{Hom}_{\mathcal{C}}(P^0,N)) = \mathrm{Hom}_{\mathcal{C}}(P^0,N),$$

i.e. $d_0^* = 0$, whence $0 = d_0 : P^0 \longrightarrow P^1$, whence $M = P^0$. □

10.7.2. *Proof of the Theorem.* Applying the above Lemma together with the Lemma IV.4.8, we see that Q is projective as a left $\mathfrak{u}$-module, and $s\mathrm{Tor}^{\mathcal{C}}_{\frac{\infty}{2}+n}(sN,Q) = s\mathrm{Tor}^{\mathcal{C}}_n(sN,Q) = s(sN \otimes_{\mathcal{C}} Q)$ if $n = 0$, and zero otherwise for any $N \in \mathcal{C}$.

So ϕ boils down to the isomorphism of functors in N:

$$s(sN \otimes_{\mathcal{C}} \mathbf{R}) \xrightarrow{\sim} s(sN \otimes_{\mathcal{C}} Q)$$

Applying the Lemma IV.4.8 again, we obtain the isomorphism of functors

$$\mathcal{C} \longrightarrow \mathcal{C} : \ s(\mathrm{Hom}_{\mathcal{C}}(\mathbf{R},?)) \xrightarrow{\sim} s(\mathrm{Hom}_{\mathcal{C}}(Q,?)),$$

whence the isomorphism of representing objects. □

11. The adjoint representation

11.1. For $\mu \in Y_\ell \subset X_\ell$ (see IV.9.1.2) we denote by T_μ an autoequivalence $\mathcal{C} \longrightarrow \mathcal{C}$ given by twisting by $L(\mu)$: $T_\mu(N) := N \otimes L(\mu)$.

For $\mu, \nu \in Y_\ell \subset X_\ell$ we will consider an autoequivalence $T_\mu \otimes T_\nu$ of $\mathrm{Ind}\mathcal{C} \otimes \mathrm{Ind}\mathcal{C}$.

The objects $\mathbf{R}$ and $T_\mu \otimes T_{-\mu}\mathbf{R}$ give rise to the same functor $F_{\mathbf{R}} = F_{T_\mu \otimes T_{-\mu}\mathbf{R}} = \mathrm{Id} : \mathrm{Ind}\mathcal{C} \longrightarrow \mathrm{Ind}\mathcal{C}$ (notations of 10.2). The equality of functors is induced by the isomorphism of objects $t_\mu : \mathbf{R} \xrightarrow{\sim} T_\mu \otimes T_{-\mu}\mathbf{R}$.

Consider the tensor product functor $\otimes : \mathrm{Ind}\mathcal{C} \otimes \mathrm{Ind}\mathcal{C} \longrightarrow \mathrm{Ind}\mathcal{C}$.

Let us denote by $\hat{\mathbf{ad}}$ the object $\otimes(\mathbf{R}) \in \mathrm{Ind}\mathcal{C}$. For $\mu \in Y_\ell \subset X_\ell$ we have $\otimes(T_\mu \otimes T_{-\mu}\mathbf{R}) = \hat{\mathbf{ad}}$. Thus the isomorphism t_μ induces the same named *automorphism* of $\hat{\mathbf{ad}}$. This way we obtain the action of the group Y_ℓ on $\hat{\mathbf{ad}}$.

We define the *adjoint representation* $\mathbf{ad}$ as the quotient $\hat{\mathbf{ad}}/Y_\ell$.

It is easy to see that $\mathbf{ad} \in \mathcal{C} \subset \mathrm{Ind}\mathcal{C}$, and in fact $\hat{\mathbf{ad}} = \mathbf{ad} \otimes \mathsf{k}[Y_\ell]$ with the trivial action of $\mathfrak{u}$ on the second multiple, and the trivial action of Y_ℓ on the first multiple.

In concrete terms, **ad** coincides with $\mathfrak{u}$ considered as a $\mathfrak{u}$-module via the adjoint action. The X_ℓ-grading on **ad** is compatible with multiplication on $\mathfrak{u}$, and the generators have the following weights:

$$\theta_i \in (\mathbf{ad})_{-i'};\ \epsilon_i \in (\mathbf{ad})_{i'};\ K_i \in (\mathbf{ad})_0.$$

Note that the highest weight of **ad** is equal to $2\rho_\ell$ (notations of IV.9.1.2).

11.2. Recall the notations of IV.9.1, 9.2. Recall that $\tilde{\mathcal{O}}_\zeta$ is a semisimple subcategory of $\mathcal{C}$ consisting of all the direct sums of simples $L(\lambda), \lambda \in \Delta = \Delta_\ell$.

The natural embedding $\tilde{\mathcal{O}}_\zeta \hookrightarrow \mathcal{C}$ admits the right adjoint functor $N \mapsto N^\Delta$ (the maximal $\tilde{\mathcal{O}}_\zeta$-subobject), as well as the left adjoint functor $N \mapsto N_\Delta$ (the maximal $\tilde{\mathcal{O}}_\zeta$-quotient). Finally,

$$N \mapsto \langle N \rangle_\Delta$$

denotes the image functor of the canonical morphism from the right adjoint to the left adjoint.

One checks easily that if $M \subset N$ is a direct summand belonging to $\tilde{\mathcal{O}}_\zeta$ and not properly contained in any of the kind (i.e. maximal), then we have a canonical isomorphism $\langle N \rangle_\Delta \xrightarrow{\sim} M$.

For this reason we call $\langle N \rangle_\Delta$ *the maximal* $\tilde{\mathcal{O}}_\zeta$*-direct summand of* N.

H.Andersen and J.Paradowski introduced the tensor structure $\tilde{\otimes}$ on the category $\tilde{\mathcal{O}}_\zeta$ (see [AP]). Arguing like in the proof of Lemma IV.9.3, one can show easily that for $M, N \in \tilde{\mathcal{O}}_\zeta \subset \mathcal{C}$ we have

$$M \tilde{\otimes} N = \langle M \otimes N \rangle_\Delta$$

11.2.1. *Definition. The semisimple adjoint representation* is the following object of $\tilde{\mathcal{O}}_\zeta$:

$$\tilde{\mathbf{ad}} := \bigoplus_{\lambda \subset \Delta} L(\lambda) \tilde{\otimes} L(\lambda)^*$$

Here $*$ is the rigidity (see e.g. IV.4.6.2).

11.3. **Theorem.** $\tilde{\mathbf{ad}} = \langle \mathbf{ad} \rangle_\Delta$.

Proof. According to [LM], $\mathbf{ad} \in \mathcal{C}$ represents the functor $V \mapsto \mathrm{Nat}(\mathrm{Id}, \mathrm{Id} \otimes V)$ sending $V \in \mathcal{C}$ to the vector space of natural transformations between endofunctors $\mathcal{C} \longrightarrow \mathcal{C}$.

Hence $\langle \mathbf{ad} \rangle_\Delta$ represents the functor sending $V \in \tilde{\mathcal{O}}_\zeta$ to the vector space of natural transformations between endofunctors Id and $\langle \mathrm{Id} \otimes V \rangle_\Delta : \tilde{\mathcal{O}}_\zeta \longrightarrow \tilde{\mathcal{O}}_\zeta$.

Similarly, $\tilde{\mathbf{ad}} \in \tilde{\mathcal{O}}_\zeta$ represents the functor $W \mapsto \mathrm{Nat}(\mathrm{Id}, \mathrm{Id} \tilde{\otimes} W)$.

It remains to recall that for $W \in \tilde{\mathcal{O}}_\zeta \subset \mathcal{C}$ we have $\mathrm{Id} \tilde{\otimes} W = \langle \mathrm{Id} \otimes W \rangle_\Delta$. $\square$

Chapter 4. Quadratic degeneration in genus zero

This chapter belongs really to the part IV. Here we construct a sheaf $\mathcal{R} \in \mathrm{Ind}\mathcal{FS} \otimes \mathrm{Ind}\mathcal{FS}$ corresponding to the regular bimodule $\mathbf{R} \in \mathrm{Ind}\mathcal{C} \otimes \mathrm{Ind}\mathcal{C}$ under the equivalence Φ.

12. I-SHEAVES

Since $\mathcal{R}$ is an object of $\mathrm{Ind}\mathcal{FS} \otimes \mathrm{Ind}\mathcal{FS}$ we start with developing a little machinery to describe certain Ind-sheaves.

12.1. We consider the root datum $Y^\diamond := Y \times Y$;
$X^\diamond = X \times X$;
$I^\diamond := I \sqcup I$;
$\langle (y_1, y_2), (x_1, x_2) \rangle := \langle y_1, x_1 \rangle + \langle y_2, x_2 \rangle$;
$I^\diamond \ni (i_1, i_2) \mapsto (i_1, i_2) \in Y^\diamond$;
$I^\diamond \ni (i_1, i_2) \mapsto (i'_1, i'_2) \in X^\diamond$.

The corresponding category $\mathcal{FS}^\diamond$ is equivalent to $\mathcal{FS} \otimes \mathcal{FS}$. Recall that an object $\mathcal{X}$ of $\mathcal{FS}^\diamond$ is the following collection of data (see III.4.2):

(a) a weight $\lambda(\mathcal{X}) = (\lambda_1, \lambda_2) \in X_\ell \times X_\ell$;

(b) for each $(\alpha_1, \alpha_2) \in \mathbb{N}[I] \times \mathbb{N}[I] = \mathbb{N}[I^\diamond]$, a sheaf $\mathcal{X}^{\alpha_1,\alpha_2} \in \mathcal{M}(\mathcal{A}^{\alpha_1,\alpha_2}_{\lambda_1,\lambda_2}; \mathcal{S})$;

Note that $(\mathcal{A}^{\alpha_1,\alpha_2}_{\lambda_1,\lambda_2}, \mathcal{S}) = (\mathcal{A}^{\alpha_1}_{\lambda_1}, \mathcal{S}_1) \times (\mathcal{A}^{\alpha_2}_{\lambda_2}, \mathcal{S}_2)$;

we will denote by $\mathcal{X}^{\vec{\alpha}_1,\vec{\alpha}_2}(\vec{d})$ perverse sheaves over $\mathcal{A}^{\vec{\alpha},\vec{\alpha}_2}_{\lambda_1,\lambda_2}(\vec{d})$ obtained by taking the restrictions with respect to the embeddings $\mathcal{A}^{\vec{\alpha}_1,\vec{\alpha}_2}_{\lambda_1,\lambda_2}(\vec{d}) \hookrightarrow \mathcal{A}^{\alpha_1,\alpha_2}_{\lambda_1,\lambda_2}$;

(c) for each $(\alpha_1, \alpha_2), (\beta_1, \beta_2) \in \mathbb{N}[I] \times \mathbb{N}[I]$, $d > 0$, a *factorization isomorphism*

$$\psi^{\alpha_1,\alpha_2;\beta_1,\beta_2}(d) : \mathcal{X}^{(\alpha_1,\alpha_2;\beta_1,\beta_2)}(d) \xrightarrow{\sim} \overset{\bullet}{\mathcal{I}}{}^{(\alpha_1,\alpha_2;0,0)}_{\lambda_1-\beta'_1,\lambda_2-\beta'_2}(d) \boxtimes \mathcal{X}^{(0,0;\beta_1,\beta_2)}(d) \tag{315}$$

satisfying the usual associativity conditions.

Note that $\mathcal{A}^{\alpha_1,\alpha_2;0,0}(d) = \mathcal{A}^{\alpha_1;0}(d) \times \mathcal{A}^{\alpha_2;0}(d)$, and

$\overset{\bullet}{\mathcal{I}}{}^{(\alpha_1,\alpha_2;0,0)}_{\lambda_1-\beta'_1,\lambda_2-\beta'_2}(d) = \overset{\bullet}{\mathcal{I}}{}^{(\alpha_1;0)}_{\lambda_1-\beta'_1}(d) \boxtimes \overset{\bullet}{\mathcal{I}}{}^{(\alpha_2;0)}_{\lambda_2-\beta'_2}(d)$.

12.2. **Definition.** An *I-sheaf* is the following collection of data:

(a) a weight $\chi \in X_\ell$, to be denoted by $\chi(X_\ell)$;

(b) for each $(\lambda_1, \lambda_2) \in X_\ell \times X_\ell$ such that $\lambda_1 + \lambda_2 - \chi = \gamma \in \mathbb{N}[I]$, and $\alpha_1, \alpha_2 \in \mathbb{N}[I]$, such that $\alpha_1, \alpha_2 \leq \gamma$, a sheaf $\mathcal{X}^{\alpha_1,\alpha_2}_{\lambda_1,\lambda_2} \in \mathcal{M}(\mathcal{A}^{\alpha_1,\alpha_2}_{\lambda_1,\lambda_2}; \mathcal{S})$;

we will denote by $\mathcal{X}^{\vec{\alpha}_1,\vec{\alpha}_2}(\vec{d})$ perverse sheaves over $\mathcal{A}^{\vec{\alpha},\vec{\alpha}_2}_{\lambda_1,\lambda_2}(\vec{d})$ obtained by taking the restrictions with respect to the embeddings $\mathcal{A}^{\vec{\alpha}_1,\vec{\alpha}_2}_{\lambda_1,\lambda_2}(\vec{d}) \hookrightarrow \mathcal{A}^{\alpha_1,\alpha_2}_{\lambda_1,\lambda_2}$;

(c) for each $\lambda_1, \lambda_2 \in X_\ell$ with $\lambda_1 + \lambda_2 - \chi = \gamma \in \mathbb{N}[I]$, and $\alpha_1, \alpha_2, \beta_1, \beta_2 \in \mathbb{N}[I]$ such that $\alpha_1 + \beta_1, \alpha_2 + \beta_2 \leq \gamma$, $d > 0$, a *factorization isomorphism*

$$\psi^{\alpha_1,\alpha_2;\beta_1,\beta_2}(d) : \mathcal{X}^{(\alpha_1,\alpha_2;\beta_1,\beta_2)}(d) \xrightarrow{\sim} \overset{\bullet}{\mathcal{I}}{}^{(\alpha_1,\alpha_2;0,0)}_{\lambda_1-\beta_1',\lambda_2-\beta_2'}(d) \boxtimes \mathcal{X}^{(0,0;\beta_1,\beta_2)}(d) \tag{316}$$

satisfying the usual associativity conditions;

(d) for each $(\lambda_1, \lambda_2) \in X_\ell \times X_\ell$ such that $\lambda_1 + \lambda_2 - \chi = \gamma \in \mathbb{N}[I]$, and $\alpha_1, \alpha_2 \in \mathbb{N}[I]$ with $\alpha_1, \alpha_2 \leq \gamma$, and $\beta_1, \beta_2 \in \mathbb{N}[I]$, an isomorphism

$\mathcal{X}^{\alpha_1+\beta_1,\alpha_2+\beta_2}_{\lambda_1+\beta_1',\lambda_2+\beta_2'} \xrightarrow{\sim} \sigma_* \mathcal{X}^{\alpha_1,\alpha_2}_{\lambda_1,\lambda_2}$

satisfying the usual compatibilities.

Here σ stands for the closed embedding $\mathcal{A}^{\alpha_1,\alpha_2}_{\lambda_1,\lambda_2} \hookrightarrow \mathcal{A}^{\alpha_1+\beta_1,\alpha_2+\beta_2}_{\lambda_1+\beta_1',\lambda_2+\beta_2'}$.

12.2.1. *Remark.* Note that an I-sheaf $\mathcal{X}$ can be uniquely reconstructed from the partial set of data (b),(c),(d), namely the data (b),(c),(d) given only for $\alpha_1 = \alpha_2, \beta_1 = \beta_2$. In what follows we will describe I-sheaves by these partial data.

13. Degeneration of quadrics

13.1. We recall the construction of [KL]II, 15.2, 15.3. Consider $\mathbb{P}^1 \times \mathbb{P}^1 \times \mathbb{A}^1$ with coordinates p, q, t. Consider the subvariety $Q \subset \mathbb{P}^1 \times \mathbb{P}^1 \times \mathbb{A}^1$ given by the equation $pq = t$. The fiber Q_t of Q over $t \in \mathbb{A}^1$ is a projective line if $t \neq 0$, and the union of two projective lines if $t = 0$.

Each fiber Q_t has two marked points: $(p = 0, q = \infty)$, and $(p = \infty, q = 0)$. For $t = 0$ each irreducible component of Q_0 has exactly one marked point, and the marked points lie away from the singularity of Q_0. The irreducible components of Q_0 are denoted by Q_{0v} (for vertical), and Q_{0h} (for horizontal).

There are two maps f_1, f_2 from the standard projective line $\mathbb{P}^1_{st}$ to Q_t, taking $0 \in \mathbb{P}^1_{st}$ to the first (resp. second) marked point on Q_t, and mapping $\mathbb{P}^1_{st}$ isomorphically onto the irreducible component of Q_t containing this marked point. If z is the standard coordinate on $\mathbb{P}^1_{st}$, then

$$f_1(z) = \left(\frac{tz}{z-1}, \frac{z-1}{z}\right); \; f_2(z) = \left(\frac{z-1}{z}, \frac{tz}{z-1}\right).$$

Thus, restricting to the open subset $'Q \subset Q$ given by the inequality $t \neq 0$, we obtain a map $\mathbb{C}^* \longrightarrow \tilde{\mathcal{P}}^2$.

The limit for $t \longrightarrow 0$ is a boundary point of $\tilde{\mathcal{P}}^2$ (notations of IV.2.3). Composing with the "1-jet at 0" projection $\tilde{\mathcal{P}}^2 \longrightarrow T\mathcal{P}^{\circ 2}$ (notations of IV.2.7) we obtain a map $\mathbb{C}^* \longrightarrow T\mathcal{P}^{\circ 2}$; and the limit for $t \longrightarrow 0$ is a boundary point of $T\mathcal{P}^{\circ 2}$.

13.2. For $\alpha \in \mathbb{N}[I]$ let $Q^\alpha \xrightarrow{\pi^\alpha} \mathbb{A}^1$ denote the space of relative configurations: $(\pi^\alpha)^{-1}(t) = Q^\alpha_t$.

13.2.1. Q^α contains the open subset ${}'Q^\alpha := (\pi^\alpha)^{-1}(\mathbb{C}^*)$.

$Q^{\bullet\alpha} \subset Q^\alpha$ is the open subset of configurations where none of the points equals the marked ones.

$Q^{\circ\alpha} \subset Q^{\bullet\alpha}$ is the open subset of configurations where all the points are distinct.

${}'Q^{\bullet\alpha} := {}'Q^\alpha \cap Q^{\bullet\alpha}; \quad {}'Q^{\circ\alpha} := {}'Q^\alpha \cap Q^{\circ\alpha}$.

The open inclusion ${}'Q^{\circ\alpha} \hookrightarrow {}'Q^{\bullet\alpha}$ is denoted by j.

The projection $Q^{\bullet\alpha} \longrightarrow \mathbb{A}^1$ is denoted by $\pi^{\bullet\alpha}$.

13.2.2. Let $\mu_1, \mu_2 \in X_\ell$ be a pair of weights such that the triple (μ_1, μ_2, α) is admissible (see IV.3.1).

The above construction defines the map $\theta^\alpha : {}'Q^{\circ\alpha} \longrightarrow \mathcal{P}^{\circ\alpha}_{\mu_1\mu_2}$ (notations of IV.3.2), and we define the local system $\mathcal{J}^\alpha_{\mu_1\mu_2}$ on ${}'Q^{\circ\alpha}$ as $\theta^{\alpha*}\mathcal{I}^\alpha_{\mu_1\mu_2}$ (notations of IV.3.4).

We denote by $\mathcal{J}^{\bullet\alpha}_{\mu_1\mu_2}$ the perverse sheaf $j_{!*}\mathcal{J}^\alpha_{\mu_1\mu_2}[\dim Q^\alpha]$ on ${}'Q^{\bullet\alpha}$.

13.3. The aim of this Chapter is to compute the nearby cycles $\Psi_{\pi^{\bullet\alpha}}\mathcal{J}^{\bullet\alpha}_{\mu_1\mu_2}$ as a perverse sheaf on $Q_0^{\bullet\alpha}$. The problem is not quite trivial, but it appears that one can combine the sheaves $\Psi_{\pi^{\bullet\alpha}}\mathcal{J}^{\bullet\alpha}_{\mu_1\mu_2}$ for different α, μ_1, μ_2 into a single I-sheaf; and this I-sheaf is already easy to compute.

14. The I-sheaf $\mathcal{R}$

14.1. We denote by $Q_0^\bullet$ the open subset of the special fiber $Q_0^\bullet := Q_0 - \{(0,\infty), (\infty, 0)\}$ obtained by throwing away the marked points. Evidently, $Q_0^\bullet$ is the union of two irreducible components $\mathbb{A}^1_v$ and $\mathbb{A}^1_h$ given by the equations $p = 0$ (vertical component), and $q = 0$ (horizontal component) respectively. They intersect at the point $(0,0) \in Q_0^\bullet$.

The irreducible components of $Q_0^{\bullet\alpha}, \alpha \in \mathbb{N}[I]$, are numbered by decompositions $\alpha = \alpha_1 + \alpha_2;\ \alpha_1, \alpha_2 \in \mathbb{N}[I]$. Namely,

$$Q_0^{\bullet\alpha} = \bigcup_{\alpha_1+\alpha_2=\alpha} \mathcal{A}_v^{\alpha_1} \times \mathcal{A}_h^{\alpha_2}$$

Each irreducible component embeds into $\mathcal{A}^\alpha_v \times \mathcal{A}^\alpha_h$.

Namely, $(q_1, \ldots, q_{|\alpha_1|}; p_1, \ldots, p_{|\alpha_2|})$ goes to $(q_1, \ldots, q_{|\alpha_1|}, \overbrace{0, \ldots, 0}^{|\alpha_2|}; p_1, \ldots, p_{|\alpha_2|}, \overbrace{0, \ldots, 0}^{|\alpha_1|})$.

We see readily that these embeddings agree on the intersections of irreducible components, whence they combine together to the closed embedding

$$\varsigma^\alpha : Q_0^{\bullet\alpha} \hookrightarrow \mathcal{A}^\alpha_v \times \mathcal{A}^\alpha_h \tag{317}$$

14.2. We define an I-sheaf $\mathcal{R}$ as follows.

(a) We set $\chi(\mathcal{R}) = (l-1)2\rho$;

(b) For $\mu_1, \mu_2 \in X_\ell$ such that $\mu_1 + \mu_2 - (l-1)2\rho = \alpha \in \mathbb{N}[I]$, we define

$$\mathcal{R}^{\alpha,\alpha}_{\mu_1\mu_2} := \varsigma^\alpha_* \Psi_{\pi^{\bullet\alpha}} \mathcal{J}^{\bullet\alpha}_{\mu_2\mu_1} \tag{318}$$

(Note the reverse order of μ_1, μ_2!)

$\mathcal{R}^{\alpha,\alpha}_{\mu_1\mu_2}$ is a perverse sheaf on $\mathcal{A}^\alpha_{\mu_1} \times \mathcal{A}^\alpha_{\mu_2} = \mathcal{A}^{\alpha,\alpha}_{\mu_1\mu_2}$.

For $\beta \in \mathbb{N}[I]$ we define $\mathcal{R}^{\alpha+\beta,\alpha+\beta}_{\mu_1+\beta,\mu_2+\beta} := \sigma_* \mathcal{R}^{\alpha,\alpha}_{\mu_1,\mu_2}$ where σ stands for the closed embedding $\mathcal{A}^{\alpha,\alpha}_{\mu_1\mu_2} \hookrightarrow \mathcal{A}^{\alpha+\beta,\alpha+\beta}_{\mu_1+\beta,\mu_2+\beta}$.

Thus the isomorphisms 12.2(d) are simultaneously defined.

(c) We will construct the factorization isomorphisms 12.2(c) for the case $\alpha_1 = \alpha_2 = \alpha; \beta_1 = \beta_2 = \beta$ (cf. Remark 12.2.1).

14.2.1. For $d > 0$ we introduce the following analytic open subsets $Q_{<d,<d}, Q_{>d,>d}, Q_{<d,>d}, Q_{>d,<d}$:

$Q_{<d,<d} := \{(p,q,t) \big| \, |p| < d > |q|\}$; $Q_{>d,>d} := \{(p,q,t) \big| \, |p| > d < |q|\}$;

$Q_{<d,>d} := \{(p,q,t) \big| \, |p| < d < |q|\}$; $Q_{>d,<d} := \{(p,q,t) \big| \, |p| > d > |q|\}$.

For $\alpha, \beta, \gamma \in \mathbb{N}[I]$ we introduce the following open subset $Q^{\alpha,\beta,\gamma}(d) \subset Q^{\alpha+\beta+\gamma}$: it is formed by the configurations such that exactly α points of configuration lie in $Q_{<d,<d}$; exactly β points lie in $Q_{<d,>d}$, and exactly γ points lie in $Q_{<d,>d}$.

We have evident decomposition

$$Q^{\alpha,\beta,\gamma}(d) = Q^{0,\beta,0}(d) \times Q^{0,0,\gamma}(d) \times Q^{\alpha,0,0}(d).$$

Intersecting the above subsets with $'Q^{\alpha+\beta+\gamma}$, $'Q^{\bullet\alpha+\beta+\gamma}$, $'Q^{\circ\alpha+\beta+\gamma}$, $Q^{\bullet\alpha+\beta+\gamma}$, etc. we obtain the opens $'Q^{\alpha,\beta,\gamma}(d)$, etc. with similar decompositions.

We denote by $\mathcal{J}^{\bullet\alpha,\beta,\gamma}_{\mu_1\mu_2}(d)$ the restriction of $\mathcal{J}^{\bullet\alpha+\beta+\gamma}_{\mu_1\mu_2}$ to $'Q^{\bullet\alpha,\beta,\gamma}(d)$.

14.2.2. We have the canonical isomorphisms

$$\mathcal{J}^{\bullet\alpha,\beta,\gamma}_{\mu_1\mu_2}(d) \xrightarrow{\sim} \dot{\mathcal{I}}^{(0,\beta)}_{\mu_1}(d) \boxtimes \dot{\mathcal{I}}^{(0,\gamma)}_{\mu_2} \boxtimes \mathcal{J}^{\bullet\alpha,0,0}_{\mu_1-\beta',\mu_2-\gamma'}(d) \tag{319}$$

satisfying the standard associativity constraints. Here $\dot{\mathcal{I}}^{(0,\beta)}_{\mu_1}(d), \dot{\mathcal{I}}^{(0,\gamma)}_{\mu_2}$ have the following meaning.

The set $Q_{<d,>d}$ (resp. $Q_{>d,<d}$) projects to the vertical component Q_{0v} (resp. horizontal component Q_{0h}) of the special fiber $Q_0 : \ (p,q,t) \mapsto (0,q,0)$ (resp. $(p,q,t) \mapsto (p,0,0)$).

This induces the projections

$$v^\beta : \ Q^{0,\beta,0}(d) \longrightarrow Q^\beta_{0v};$$
$$h^\gamma : \ Q^{0,0,\gamma}(d) \longrightarrow Q^\gamma_{0h}.$$

Recall that Q_{0v} (resp. Q_{0h}) has the marked point $(0,\infty)$ (resp. $(\infty,0)$) lying on it, and $v^\beta(Q^{0,\beta,0}(d))$ (resp. $h^\gamma(Q^{0,0,\gamma}(d))$) is a standard open subset of Q^β_{0v} (resp. Q^γ_{0h}).

Restricting our maps to $'Q^\bullet$ we get

$$v^\beta : \ 'Q^{\bullet 0,\beta,0}(d) \longrightarrow Q^{\bullet\beta}_{0v};$$

$$h^\gamma : {}'Q^{\bullet 0,0,\gamma}(d) \longrightarrow Q^{\bullet\gamma}_{0h}.$$

Note that $v^\beta({}'Q^{\bullet 0,\beta,0}(d)) = Q^{\bullet 0,\beta}_{0v}(d)$ – the standard open formed by the configurations with all β points running in the annular neighbourhood of the marked point on Q_{0v}.

Similarly, $h^\gamma({}'Q^{\bullet 0,0,\gamma}(d)) = Q^{\bullet 0,\gamma}_{0h}(d)$ — the standard open formed by the configurations with all γ points running in the annular neighbourhood of the marked point on Q_{0h}.

The sheaf $\overset{\bullet}{\mathcal{I}}{}^{0,\beta}_{\mu_1}(d)$ (resp. $\overset{\bullet}{\mathcal{I}}{}^{0,\gamma}_{\mu_2}(d)$) is defined in III.3.5. We keep the same notation for its inverse image to ${}'Q^{\bullet 0,\beta,0}(d)$ (resp. ${}'Q^{\bullet 0,0,\gamma}(d)$) with respect to v^β (resp. h^γ).

14.2.3. One final remark is in order. Note that $Q^\bullet_{0v} = \mathbb{A}^1_v$; $Q^\bullet_{0h} = \mathbb{A}^1_h$ (notations of 14.1). Hence $Q^\bullet\beta_{0v} = \mathcal{A}^\beta_v$; $Q^\bullet\gamma_{0h} = \mathcal{A}^\gamma_h$.

Under this identification the open $Q^{\bullet 0,\beta}_{0v}(d)$ corresponds to the standard open $\mathcal{A}^{\beta,0}_v(d)$, and the open $Q^{\bullet 0,\gamma}_{0h}(d)$ corresponds to the standard open $\mathcal{A}^{\gamma,0}_h(d)$ (notations of III.2.2).

Warning. Note the opposite roles played by the marked points on Q_{0v} and $\mathbb{A}^1_v$: the open $Q^{\bullet 0,\beta}_{0v}(d) = \mathcal{A}^{\beta,0}_v(d)$ is formed by configurations where all points run near *the marked point of Q_{0v}, or equivalently, where all points run* far from *the marked point $0 \in \mathbb{A}^1$.*

The sheaf $\overset{\bullet}{\mathcal{I}}{}^{0,\beta}_{\mu_1}(d)$ on $Q^{\bullet 0,\beta}_{0v}(d)$ corresponds to the sheaf $\overset{\bullet}{\mathcal{I}}{}^{\beta,0}_{(l-1)2\rho+\beta-\mu_1}(d)$ on $\mathcal{A}^{\beta,0}_v(d)$ (notations of III.3.5).

The sheaf $\overset{\bullet}{\mathcal{I}}{}^{0,\gamma}_{\mu_2}(d)$ on $Q^{\bullet 0,\gamma}_{0h}(d)$ corresponds to the sheaf $\overset{\bullet}{\mathcal{I}}{}^{\gamma,0}_{(l-1)2\rho+\gamma-\mu_2}(d)$ on $\mathcal{A}^{\gamma,0}_h(d)$.

14.2.4. Putting all the above together, the desired factorization isomorphisms 12.2(c) for the sheaf $\mathcal{R}$ are induced by the factorization isomorphisms (319) for the sheaf $\mathcal{J}^\bullet$.

15. Convolution

15.1. Let us throw away the line $p = \infty, q = 0$ from Q. What remains is the degenerating family Q_a of affine lines with the marked point $(0, \infty)$. The notations ${}'Q_a, Q^\bullet_a$, etc. speak for themselves.

Given a finite factorizable sheaf $\mathcal{X} \in \mathcal{FS}$ we can glue it into the marked point $(0, \infty)$ and obtain the factorizable sheaf $\tilde{\mathcal{X}}$ over the family ${}'Q_a$ of affine lines. For each $t \in \mathbb{C}^*$ the restriction of $\tilde{\mathcal{X}}$ to the fiber $(\mathcal{Q}_a)_t = \mathcal{A}$ is the FFC $\mathcal{X}$ we started with.

More precisely, let $\lambda = \lambda(\mathcal{X})$, $\alpha \in \mathbb{N}[I]$. We have the sheaf $\tilde{\mathcal{X}}^\alpha_\lambda$ over ${}'Q^\alpha_a$ such that for any $t \in \mathbb{C}^*$ its restriction to each fiber $({}'Q^\alpha_a)_t = \mathcal{A}^\alpha$ coincides with $\mathcal{X}^\alpha_\lambda$.

We have the projection

$$\mathrm{pr} : Q_a \longrightarrow \mathbb{A}^1_h \times \mathbb{C}, \ \mathrm{pr}(p, q, t) = (p, t).$$

The restriction of pr to ${}'Q_a$ is one-to-one onto $\mathbb{A}^1_h \times \mathbb{C}^*$, so $\mathrm{pr}_*\tilde{\mathcal{X}}$ is the constant family of FFC on $\mathcal{A}_h \times \mathbb{C}^*$.

Recall that π denotes the projection to the t-coordinate. The square

$$\begin{array}{ccc} Q_a & \xrightarrow{\pi} & \mathbb{C} \\ {\scriptstyle\mathrm{pr}}\downarrow & & \uparrow{\scriptstyle\mathrm{id}} \\ \mathbb{A}^1\times\mathbb{C} & \xrightarrow{\pi} & \mathbb{C} \end{array}$$

commutes, and the map pr is proper.

By the proper base change for nearby cycles we have

$$\mathrm{pr}^\alpha_*\Psi_{\pi^\alpha}\tilde{\mathcal{X}}^\alpha_\lambda = \Psi_{\pi^\alpha}(\mathrm{pr}^\alpha_*\tilde{\mathcal{X}}^\alpha_\lambda) = \mathcal{X}^\alpha_\lambda$$

Here $\pi^\alpha, \mathrm{pr}^\alpha$ form the evident commutative diagram

$$\begin{array}{ccc} \mathcal{Q}^\alpha_a & \xrightarrow{\pi^\alpha} & \mathbb{C} \\ {\scriptstyle\mathrm{pr}^\alpha}\downarrow & & \uparrow{\scriptstyle\mathrm{id}} \\ \mathcal{A}^\alpha_h\times\mathbb{C} & \xrightarrow{\pi^\alpha} & \mathbb{C} \end{array}$$

15.2. Let us compute $\mathrm{pr}^\alpha_*\Psi_{\pi^\alpha}\tilde{\mathcal{X}}^\alpha_\lambda$ in another way. First we describe $\Psi_{\pi^\alpha}\tilde{\mathcal{X}}^\alpha_\lambda$. It is a sheaf on $(\mathcal{Q}^\alpha_a)_0$ which will be denoted by $\mathcal{Q}^\alpha_{0a}$ for short.

Note that $Q_{0a} = \mathbb{P}^1_v \cup \mathbb{A}^1_h$ is the union of two irreducible components intersecting at the point $(0,0)$. Hence $\mathcal{Q}^\alpha_{0a}$ is the union of irreducible components numbered by the decompositions $\alpha = \alpha_1 + \alpha_2,\ \alpha_1, \alpha_2 \in \mathbb{N}[I]$. Namely,

$$\mathcal{Q}^\alpha_{0a} = \bigcup_{\alpha_1+\alpha_2=\alpha} \mathcal{P}^{\alpha_1}_v \times \mathcal{A}^{\alpha_2}_h.$$

Here $\mathcal{P}^{\alpha_1}_v$ stands for the space of configurations on $\mathbb{P}^1_v$.

Each irreducible component embeds into $\mathcal{P}^\alpha_v \times \mathcal{A}^\alpha_h$ as in 14.1, and one checks readily that these embeddings agree on intersections, whence they combine to the closed embedding

$$\varsigma^\alpha:\ \mathcal{Q}^\alpha_{0a} \hookrightarrow \mathcal{P}^\alpha_v \times \mathcal{A}^\alpha_h.$$

We will describe $\varsigma^\alpha_*\Psi_{\pi^\alpha}\tilde{\mathcal{X}}^\alpha_\lambda$.

15.2.1. Note that $\mathbb{P}^1_v$ is equipped with two marked points with tangent vectors. The marked points are $\infty, 0$. The tangent vector at 0 is ∂_q, and the tangent vector at ∞ was defined in 13.1.

Recall that for $\mu \in X_\ell$ such that the triple (λ, μ, α) is admissible, the sheaf $\varsigma^\alpha_*\Psi_{\pi^{\bullet\alpha}}\mathcal{J}^{\bullet\alpha}_{\lambda\mu}$ on $\mathcal{A}^\alpha_v \times \mathcal{A}^\alpha_h$ was introduced in 14.2(b).

According to the Theorem IV.3.6 we can glue the sheaves $\varsigma^\alpha_*\Psi_{\pi^{\bullet\alpha}}\mathcal{J}^{\bullet\alpha}_{\lambda\mu}$ and $\mathcal{X}^\alpha_\lambda$ to obtain the sheaf $\bar{\mathcal{X}}^\alpha_{\lambda\mu}$ on the space $\mathcal{P}^\alpha_v \times \mathcal{A}^\alpha_h$.

Here $\mathcal{X}^\alpha_\lambda$ is viewed as a sheaf on $Q^{0,\alpha}_{0v}(d) \times \mathcal{A}^\alpha_h$ (notations of 13.1 and 14.2.2) constant along $\mathcal{A}^\alpha_h$. So it is glued into $\infty \times \mathcal{A}^\alpha_h$, while $\varsigma^\alpha_*\Psi_{\pi^{\bullet\alpha}}\mathcal{J}^{\bullet\alpha}_{\lambda\mu}$ is glued into $0 \times \mathcal{A}^\alpha_h$.

15.2.2. *Claim.*

$$\varsigma^\alpha_*\Psi_{\pi^{\bullet\alpha}}\tilde{\mathcal{X}}^\alpha_\lambda = \bar{\mathcal{X}}^\alpha_{\lambda\mu}$$

Proof. Evident. □

15.3. We proceed with the computation of $\mathrm{pr}^{\alpha}_{*}\Psi_{\pi^{\alpha}}\tilde{\mathcal{X}}^{\alpha}_{\lambda}$.

Recall the sheaf $\mathcal{R}$ introduced in 14.2.

15.3.1. We define the *convolution* $\mathcal{X} \star \mathcal{R}$ on $\mathcal{P}_v \times \mathcal{A}_h$ as the following collection of data:

For $\mu \in X_\ell$ such that $\lambda + \mu - (l-1)2\rho = \alpha \in \mathbb{N}[I]$, the sheaf

$$(\mathcal{X} \star \mathcal{R})^{\alpha}_{\lambda\mu\lambda} := \bar{\mathcal{X}}^{\alpha}_{\lambda\mu}$$

on the space $\mathcal{P}^{\alpha}_{\lambda\mu} \times \mathcal{A}^{\alpha}_{\lambda}$.

The notation $\mathcal{P}^{\alpha}_{\lambda\mu}$ suggests that the monodromy around ∞ is λ, and the monodromy around 0 is μ.

We leave the formulation of factorization isomorphisms to the interested reader.

15.3.2. Denote by pr^{α} the projection $\mathcal{P}^{\alpha} \times \mathcal{A}^{\alpha} \longrightarrow \mathcal{A}^{\alpha}$. Then, evidently,

$$\mathrm{pr}^{\alpha}_{*}(\mathcal{X} \star \mathcal{R})^{\alpha}_{\lambda\mu\lambda} = \mathrm{pr}^{\alpha}_{*}\bar{\mathcal{X}}^{\alpha}_{\lambda\mu} = \mathrm{pr}^{\alpha}_{*}\varsigma^{\alpha}_{*}\Psi_{\pi^{\alpha}}\tilde{\mathcal{X}}^{\alpha}_{\lambda} = \mathcal{X}^{\alpha}_{\lambda}$$

Here the second equality holds by the Claim 15.2.2, and the last equality was explained in 15.1.

15.4. Thus, for each $\mathcal{X} \in \mathcal{FS}$, we have the natural isomorphism $\mathrm{pr}_*(\mathcal{X} \star \mathcal{R}) = \mathcal{X}$. That is, for $\lambda = \lambda(\mathcal{X}), \alpha \in \mathbb{N}[I], \mu = \alpha + (l-1)2\rho - \lambda$, we have $\mathrm{pr}_*(\mathcal{X} \star \mathcal{R})^{\alpha}_{\lambda\mu\lambda} = \mathcal{X}^{\alpha}_{\lambda}$.

15.4.1. *Lemma.* There is a natural isomorphism of functors in $\mathcal{X}$:

$$\Phi\mathrm{pr}_*(\mathcal{X} \star \mathcal{R}) \xrightarrow{\sim} F_{\Phi(\mathcal{R});\frac{\infty}{2}+\bullet}(\Phi(\mathcal{X}))$$

(notations of 10.3).

Proof. This is just the Theorem IV.8.4. □

15.5. **Theorem.** There is a natural isomorphism in $\mathrm{Ind}\mathcal{C} \otimes \mathrm{Ind}\mathcal{C}$:

$$\Phi(\mathcal{R}) \xrightarrow{\sim} \mathbf{R}.$$

Proof. This is just the above Lemma combined with the Converse Theorem 10.7. □

Chapter 5. Modular functor

16. Gluing over C

This section is quite parallel to the Chapter 1 of IV.

We will change our notations slightly to make them closer to those of *loc. cit.*

16.1. Let again $C \longrightarrow S$ be a smooth proper relative curve of genus g.

Let us assume that $g > 1$ for a moment. It is well known that the isomorphism classes of complex structures on a surface of genus g correspond bijectively to the conformal equivalence classes of Riemann metrics on this surface.

Each conformal equivalence class contains a unique metric of constant curvature -1. Thus we may assume that each fiber C_s is equipped with such a metric.

Given $0 < \varepsilon \leq 1$ and a finite set K, let $\tilde{C}_\varepsilon^K$ denote the space of K-tuples $(u_k)_{k\in K}$ of analytic morphisms over S : $S \times D_\varepsilon \longrightarrow C$ inducing isometry on each fiber, such that the images $u_k(S \times D_\varepsilon)$ do not intersect. Here we equip the disk D_ε with the Poincaré metric of constant curvature -1.

Given a K-tuple $\vec{\alpha} = (\alpha_k) \in \mathbb{N}[X_\ell]^K$ such that $\alpha = \sum_k \alpha_k$, define a space

$$C_\varepsilon^{\vec{\alpha}} := \tilde{C}_\varepsilon^K \times \prod_{k\in K} \dot{T} D_\varepsilon^{\alpha_k}$$

and an open subspace

$$C_\varepsilon^{o\vec{\alpha}} := \tilde{C}_\varepsilon^K \times \prod_{k\in K} T D_\varepsilon^{o\alpha_k}.$$

We have an evident "substitution" map

$$q_{\vec{\alpha}} : C_\varepsilon^{\vec{\alpha}} \longrightarrow \dot{T} C^\alpha$$

which restricts to $q_{\vec{\alpha}} : C_\varepsilon^{o\vec{\alpha}} \longrightarrow TC^{o\alpha}$.

16.2. Let us define an element $\rho_\ell \in X_\ell$ by the condition $\langle i, \rho_\ell\rangle = \ell_i - 1$ for all $i \in I$ (see IV.9.1.2). ¿From now on we choose a balance function n in the form

$$n(\mu) = \frac{1}{2}\mu \cdot \mu - \mu \cdot \rho_\ell.$$

In other words, we set

$$\nu_0 = -\rho_\ell.$$

We pick a corresponding Heisenberg system $\mathcal{H}$.

16.2.1. *Remark.* The monodromy of $\mathcal{H}$ around the zero section of determinant line bundle is equal to $(-1)^{\mathrm{rk} X_\ell}\zeta^{12\rho_\ell\cdot\rho_\ell}$. According to the strange formula of Freudenthal–deVries (see e.g. [FZ]) we have $12\rho\cdot\rho = dh \dim \mathfrak{g}$. Here $\mathfrak{g}$ is the Lie algebra with the root datum $X, Y, \ldots$, and $\check{h}$ is the dual Coxeter number of $\mathfrak{g}$, while $d := \max_{i\in I} d_i$.

Let $\zeta = \exp(\frac{\pi\sqrt{-1}}{d\kappa})$ for some positive integer κ. Then the above monodromy equals $\exp(\pi\sqrt{-1}(\mathrm{rk}\mathfrak{g} + \frac{h\dim\mathfrak{g}}{\kappa}))$ which coincides with the multiplicative central charge of the conformal field theory associated with the affine Lie algebra $\hat{\mathfrak{g}}$ at level κ (see e.g. [BFM]).

We will offer an explanation of this coincidence in Chapter 6.

16.2.2. Given $\alpha = \sum a_i i \in \mathbb{N}[I]$ and $\vec{\mu} = (\mu_k) \in X_\ell{}^K$, we define an element

$$\alpha_{\vec{\mu}} = \sum a_i\cdot(-i') + \sum_k \mu_k \in \mathbb{N}[X_\ell]$$

where the sum in the right hand side is a formal one. We say that a pair $(\vec{\mu}, \alpha)$ is *g-admissible* if

$$\sum_k \mu_k - \alpha \equiv (2-2g)\rho_\ell \bmod Y_\ell.$$

Note that given $\vec{\mu}$, there exists $\alpha \in \mathbb{N}[I]$ such that $(\vec{\mu}, \alpha)$ is admissible if and only if $\sum_k \mu_k \in Y$; if this holds true, such elements α form an obvious countable set.

We will denote by

$$e : \mathbb{N}[I] \longrightarrow \mathbb{N}[X_\ell]$$

a unique homomorphism sending $i \in I$ to $-i' \in X_\ell$.

16.3. Let us consider the space $\dot{T}C^K \times \dot{T}C^{e(\alpha)}$; its points are quadruples $((z_k), (\tau_k), (x_j), (\omega_j))$ where $(z_k) \in C^K$, τ_k — a non-zero (relative) tangent vector to C at z_k, $(x_j) \in C^{e(\alpha)}$, ω_j — a non-zero tangent vector at x_j. To a point z_k is assigned a weight μ_k, and to x_j — a weight $-\pi(j)'$. Here $\pi : J \longrightarrow I$ is an unfolding of α (implicit in the notation $(x_j) = (x_j)_{j\in J}$).

We will be interested in some open subspaces:

$$\dot{T}C^\alpha_{\vec{\mu}} := TC^{oK} \times \dot{T}C^{e(\alpha)} \subset \dot{T}C^{\alpha_{\vec{\mu}}}$$

and

$$TC^{o\alpha}_{\vec{\mu}} \subset \dot{T}C^\alpha_{\vec{\mu}}$$

whose points are quadruples $((z_k), (\tau_k), (x_j), (\omega_j)) \in \dot{T}C^\alpha_{\vec{\mu}}$ with all $z_k \neq x_j$. We have an obvious symmetrization projection

$$p^\alpha_{\vec{\mu}} : TC^{o\alpha}_{\vec{\mu}} \longrightarrow TC^{o\alpha_{\vec{\mu}}}.$$

Define a space

$$C^\alpha_{\vec{\mu}} = TC^{oK} \times C^{e(\alpha)};$$

its points are triples $((z_k), (\tau_k), (x_j))$ where (z_k), (τ_k) and (x_j) are as above; and to z_k and x_j the weights as above are assigned. We have the canonical projection

$$\dot{T}C^\alpha_{\vec{\mu}} \longrightarrow C^\alpha_{\vec{\mu}}.$$

We define the open subspaces

$$C^{o\alpha}_{\vec{\mu}} \subset C^{\bullet\alpha}_{\vec{\mu}} \subset C^\alpha_{\vec{\mu}}.$$

Here the $\bullet$-subspace (resp., o-subspace) consists of all $((z_k),(\tau_k),(x_j))$ with $z_k \neq x_j$ for all k, j (resp., with all z_k and x_j distinct).

We define the *principal stratification* $\mathcal{S}$ of $C^{\alpha}_{\vec{\mu}}$ as the stratification generated by subspaces $z_k = x_j$ and $x_j = x_{j'}$ with $\pi(j) \neq \pi(j')$. Thus, $C^{o\alpha}_{\vec{\mu}}$ is the open stratum of $\mathcal{S}$. As usually, we will denote by the same letter the induced stratifications on subspaces.

The above projection restricts to

$$TC^{o\alpha}_{\vec{\mu}} \longrightarrow C^{o\alpha}_{\vec{\mu}}.$$

16.4. Factorization structure.

16.4.1. Suppose we are given $\vec{\alpha} \in \mathbb{N}[I]^K$, $\beta \in \mathbb{N}[I]$; set $\alpha := \sum_k \alpha_k$. Define a space

$$C^{\vec{\alpha},\beta}_{\vec{\mu},\varepsilon} \subset \tilde{C}^K_{\varepsilon} \times \prod_k D^{\alpha_k}_{\varepsilon} \times C^{e(\beta)}$$

consisting of all collections $((u_k),((x^{(k)}_j)_k),(y_j))$ where $(u_k) \in \tilde{C}^K_{\varepsilon}$, $(x^{(k)}_j)_k \in D^{\alpha_k}$, $(y_j) \in C^{e(\beta)}$, such that

$$y_j \in C - \bigcup_{k\in K} \overline{u_k(S \times D_{\varepsilon})}$$

for all j (the bar means closure).

We have canonical maps

$$q_{\vec{\alpha},\beta} : C^{\vec{\alpha},\beta}_{\vec{\mu},\varepsilon} \longrightarrow C^{\alpha+\beta}_{\vec{\mu}},$$

assigning to $((u_k),((x^{(k)}_j)_k),(y_j))$ a configuration $(u_k(0)),(\overset{\bullet}{u}_k(\tau)),(u_k(x^{(k)}_j)),(y_j))$, where τ is the unit tangent vector to D_{ε} at 0, and

$$p_{\vec{\alpha},\beta} : C^{\vec{\alpha},\beta}_{\vec{\mu},\varepsilon} \longrightarrow \prod_k D^{\alpha_k}_{\varepsilon} \times C^{\bullet\beta}_{\vec{\mu}-\vec{\alpha}}$$

sending $((u_k),((x^{(k)}_j)_k),(y_j))$ to $((u_k(0)),(\overset{\bullet}{u}_k(\tau)),(y_j))$.

16.5. *From now on we change the base to $\dot{\delta} \longrightarrow S$ in all the above spaces and morphisms. We preserve the old notations.*

Given a g-admissible pair $(\vec{\mu},\alpha)$ we can find another K-tuple of weights $\vec{\mu}' \in {X_\ell}^K$ such that $\vec{\mu}'$ is congruent to $\vec{\mu}$ modulo Y_ℓ (see IV.9.1.2), and $\alpha_{\vec{\mu}'}$ is g-admissible in the sense of 2.1.8.

Let us consider a local system $p^{\alpha *}_{\vec{\mu}'}\mathcal{H}^{\alpha_{\vec{\mu}'}}$ over $TC^{o\alpha}_{\vec{\mu}'}$.

Note that all the spaces $TC^{o\alpha}_{\vec{\mu}'}$ (for different choices of $\vec{\mu}'$) are identified with $TC^{o\alpha}_{\vec{\mu}}$, and the local system $p^{\alpha *}_{\vec{\mu}'}\mathcal{H}^{\alpha_{\vec{\mu}'}}$ does not depend on the choice of $\vec{\mu}'$ by the virtue of the Lemma 5.11.1. Thus we will identify all these local systems and call the result $p^{\alpha *}_{\vec{\mu}}\mathcal{H}^{\alpha_{\vec{\mu}}}$.

By our choice of the balance function n, its monodromies with respect to the rotating of tangent vectors ω_j at points x_j corresponding to negative simple roots, are trivial.

Therefore it descends to a unique local system over $C^{o\alpha}_{\vec{\mu}}$, to be denoted by $\mathcal{H}^{\alpha}_{\mu}$.

We define a perverse sheaf

$$\mathcal{H}^{\bullet\alpha}_{\vec{\mu}} := j_{!*}\mathcal{H}^{\alpha}_{\vec{\mu}}[\dim C^{\alpha}_{\vec{\mu}}] \in \mathcal{M}(C^{\bullet\alpha}_{\vec{\mu}};\mathcal{S}).$$

16.6. Factorizable sheaves over C. Suppose we are given a K-tuple of FFS's $\{\mathcal{X}_k\}$, $\mathcal{X}_k \in \mathcal{FS}_{c_k}$, $k \in K$, $c_k \in X_\ell/Y$, where $\sum_k c_k = 0$. Let us pick $\vec{\mu} = (\vec{\mu}_k) \geq (\lambda(\mathcal{X}_k))$.

Let us call a *factorizable sheaf over C obtained by gluing the sheaves $\mathcal{X}_k$* the following collection of data which we will denote by $\mathrm{g}(\{\mathcal{X}_k\})$.

(i) For each $\alpha \in \mathbb{N}[I]$ such that $(\vec{\mu}, \alpha)$ is g-admissible, a sheaf $\mathcal{X}^\alpha_{\vec{\mu}} \in \mathcal{M}(C^\alpha_{\vec{\mu}}; \mathcal{S})$.

(ii) For each $0 < \varepsilon \leq 1, \vec{\alpha} = (\alpha_k) \in \mathbb{N}[I]^K$, $\beta \in \mathbb{N}[I]$ such that $(\vec{\mu}, \alpha+\beta)$ is g-admissible (where $\alpha = \sum \alpha_k$), a *factorization isomorphism*

$$\phi_{\vec{\alpha},\beta} : q^*_{\vec{\alpha},\beta}\mathcal{X}^{\alpha+\beta}_{\vec{\mu}} \xrightarrow{\sim} p^*_{\vec{\alpha},\beta}((\boxtimes_{k\in K} \mathcal{X}^{\alpha_k}_{\mu_k}) \boxtimes \mathcal{H}^{\bullet\beta}_{\vec{\mu}-\vec{\alpha}}).$$

These isomorphisms should satisfy the standard associativity property.

16.7. **Theorem.** *There exists a unique up to a canonical isomorphism factorizable sheaf over C obtained by gluing the sheaves $\{\mathcal{X}_k\}$.*

Proof is similar to III.10.3. □

16.8. We will need the following version of the above Theorem.

Suppose we are given a K-tuple $\{\mathcal{X}_k\}$ of objects of $\mathcal{FS}$, and a J-tuple $\{\mathcal{Y}_j\}$ of I-sheaves (see 12.2).

Let us pick $\vec{\lambda} = (\vec{\lambda}_k) \geq (\lambda(\mathcal{X}_k))$, and $(\vec{\mu}, \vec{\nu}) = (\vec{\mu}_j, \vec{\nu}_j)$ such that $(\vec{\mu}_j + \vec{\nu}_j) \geq (\chi(\mathcal{Y}_j))$. We will denote the concatenation of $\vec{\lambda}$, $\vec{\mu}$, and $\vec{\nu}$ by $\vec{\lambda}\vec{\mu}\vec{\nu}$.

Let us call a *factorizable sheaf over C obtained by gluing the sheaves $\mathcal{X}_k$ and I-sheaves $\mathcal{Y}_j$* the following collection of data which we will denote by $\mathrm{g}(\{\mathcal{X}_k\}, \{\mathcal{Y}_j\})$.

(i) For each $\alpha \in \mathbb{N}[I]$ such that $(\vec{\lambda}\vec{\mu}\vec{\nu}, \alpha)$ is g-admissible, a sheaf $\mathcal{X}^\alpha_{\vec{\lambda}\vec{\mu}\vec{\nu}} = \mathrm{g}(\{\mathcal{X}_k\}, \{\mathcal{Y}_j\})^\alpha_{\vec{\lambda}\vec{\mu}\vec{\nu}} \in \mathcal{M}(C^\alpha_{\vec{\lambda}\vec{\mu}\vec{\nu}}; \mathcal{S})$.

(ii) For each $0 < \varepsilon \leq 1, \vec{\alpha} = (\alpha_k) \in \mathbb{N}[I]^K$, $(\vec{\beta}, \vec{\gamma}) = (\beta_j, \gamma_j) \in \mathbb{N}[I]^{J\sqcup J}$, $\xi \in \mathbb{N}[I]$ such that $\beta_j, \gamma_j \leq \vec{\mu}_j + \vec{\nu}_j - \chi(\mathcal{Y}_j)$ for any $j \in J$, and $(\vec{\lambda}\vec{\mu}\vec{\nu}, \alpha + \beta + \gamma + \xi)$ is g-admissible (where $\alpha = \sum \alpha_k$, $\beta = \sum \beta_j$, $\gamma = \sum \gamma_j$), a *factorization isomorphism*

$$\phi_{\vec{\alpha}\vec{\beta}\vec{\gamma},\xi} : q^*_{\vec{\alpha}\vec{\beta}\vec{\gamma},\xi}\mathcal{X}^{\alpha+\beta+\gamma+\xi}_{\vec{\lambda}\vec{\mu}\vec{\nu}} \xrightarrow{\sim} p^*_{\vec{\alpha}\vec{\beta}\vec{\gamma},\xi}((\boxtimes_{k\in K} \mathcal{X}^{\alpha_k}_{\lambda_k}) \boxtimes (\boxtimes_{j\in J} \mathcal{Y}^{\beta_j,\gamma_j}_{\mu_j,\nu_j}) \boxtimes \mathcal{H}^{\bullet\xi}_{\vec{\lambda}\vec{\mu}\vec{\nu}-\vec{\alpha}\vec{\beta}\vec{\gamma}}).$$

These isomorphisms should satisfy the standard associativity property.

Theorem. *There exists a unique up to a canonical isomorphism factorizable sheaf over C obtained by gluing the sheaves $\{\mathcal{X}_k\}$ and I-sheaves $\{\mathcal{Y}_j\}$.*

Proof is similar to III.10.3. □

16.9. We will apply the above Theorem exclusively to the case $\mathcal{Y}_j = \mathcal{R}$ (see 14.2).

16.9.1. Recall (see 16.3) that $C^\alpha_{\vec{\lambda}\vec{\mu}\vec{\nu}} = TC^{0K\sqcup J\sqcup J} \times C^{e(\alpha)}$ is the space of 7-tuples

$$((z_k), (\tau_k); (x_j), (\omega_j); (y_j), (\eta_j); (t_m)).$$

Here $(t_m) \in C^{e(\alpha)}$; $(z_k) \in C^K$; $(x_j), (y_j) \in C^J$. All the points z_k, x_j, y_j are distinct, and τ_k is a nonzero tangent vector at z_k; ω_j is a nonzero tangent vector at x_j; η_j is a nonzero tangent vector at y_j.

We define

$$\underline{C}^{\alpha}_{\vec{\lambda}\vec{\mu}\vec{\nu}} := TC^{0K\sqcup J\sqcup J} \times (C/[(x_j) = (y_j)])^{e(\alpha)}$$

Note that the natural projection $\mathfrak{p} : C^{\alpha}_{\vec{\lambda}\vec{\mu}\vec{\nu}} \longrightarrow \underline{C}^{\alpha}_{\vec{\lambda}\vec{\mu}\vec{\nu}}$ is the normalization map.

16.9.2. Suppose we are in the situation of 16.8, and $\{\mathcal{Y}_j\}$ is just a set of J copies of the I-sheaf $\mathcal{R}$ (see 14.2). Thus, $\{\mathcal{Y}_j\} = \{\mathcal{R}_j\}$; $\mathcal{R}_j = \mathcal{R}$ for any $j \in J$.

It is immediate from the definition of $\mathcal{R}$ that the sheaf $\mathrm{g}(\{\mathcal{X}_k\}, \{\mathcal{R}_j\})^{\alpha}_{\vec{\lambda}\vec{\mu}\vec{\nu}}$ on $C^{\alpha}_{\vec{\lambda}\vec{\mu}\vec{\nu}}$ descends to a sheaf on $\underline{C}^{\alpha}_{\vec{\lambda}\vec{\mu}\vec{\nu}}$ to be denoted by $\underline{\mathrm{g}}(\{\mathcal{X}_k\}, \{\mathcal{R}_j\})^{\alpha}_{\vec{\lambda}\vec{\mu}\vec{\nu}}$.

16.9.3. *Remark.* All the above constructions generalize immediately to the case where C is a union of smooth connected components C_n of genera g_n, $\sum g_n = g$.

Then C^{α} is a union of connected components numbered by the partitions $\alpha = \sum \alpha_n$. Each connected component is the product $\prod C_n^{\alpha_n}$. The Heisenberg local system is just the product of $\mathcal{H}_{g_n}$. A factorizable sheaf $\mathrm{g}(\{\mathcal{X}_k\})$ on each connected component also decomposes into external product of the corresponding sheaves on factors. Note that each connected component gives rise to its own admissibility condition.

On the other hand, a sheaf $\underline{\mathrm{g}}(\{\mathcal{X}_k\}, \{\mathcal{R}_j\})$ does not necessarily decompose into external product. In effect, if for some j the sections x_j and y_j lie on the different connected components of C, then some connected components of $C^{\alpha}_{\vec{\lambda}\vec{\mu}\vec{\nu}}$ will be glued together in $\underline{C}^{\alpha}_{\vec{\lambda}\vec{\mu}\vec{\nu}}$.

17. Degeneration of factorizable sheaves

17.1. **Definition.** An m-tuple of weights $\vec{\mu} \in X_\ell{}^m$ is called *g-positive* if

$$\sum_{j=1}^{m} \mu_j + 2(g-1)\rho_\ell \in \mathbb{N}[I] \subset X_\ell$$

(notations of IV.9.1.2). If this is so, we will denote

$$\alpha_g(\vec{\mu}) := \sum_{j=1}^{m} \mu_j + 2(g-1)\rho_\ell$$

17.2. We will use freely all the notations of section 7. In particular, C is the universal curve of genus g.

Given a K-tuple of FFS's $\{\mathcal{X}_k\}$ and a g-positive K-tuple of weights $\vec{\mu} = (\vec{\mu}_k) \geq (\lambda(\mathcal{X}_k))$ we define $\alpha := \alpha_g(\vec{\mu})$.

We consider a sheaf $\mathcal{X}^{\alpha}_{\vec{\mu}} = \mathrm{g}^{\alpha}_{\vec{\mu}}(\{\mathcal{X}_k\})$ on $C^{\alpha}_{\vec{\mu}}$ obtained by gluing the sheaves $\mathcal{X}_k$ (see 16.6). We will study its specialization along the boundary component $\mathfrak{D}$.

Let C denote the following object:

in case 7.1(a) C is the universal curve over $\mathcal{M}_{g_1} \times \mathcal{M}_{g_2}$. It consists of two connected components C_1 and C_2.

in case 7.1(b) C is the universal curve C_0 over $\mathcal{M}_{g_0}$.

The maps (302) and (303) give rise to the following morphisms also denoted by $\wp_\alpha$:

$$(C_1)^{\alpha_1}_{\vec{\mu}_1,\mu_1} \times (C_2)^{\alpha_2}_{\vec{\mu}_2,\mu_2} \longrightarrow \dot{T}_{(\overline{C}|_{\mathfrak{D}})^\alpha_{\vec{\mu}}}\overline{C}^\alpha_{\vec{\mu}}; \tag{320}$$

$$(C_0)^\alpha_{\vec{\mu},\mu_1,\mu_2} \longrightarrow \dot{T}_{(\overline{C}|_{\mathfrak{D}})^\alpha_{\vec{\mu}}}\overline{C}^\alpha_{\vec{\mu}} \tag{321}$$

(recall that by default all the bases are changed to the punctured determinant line bundles). In (320) the sets $\vec{\mu}_1$ and $\vec{\mu}_2$ form an arbitrary partition of the set $\vec{\mu}$. Finally, we can choose μ_1, μ_2 arbitrarily: the spaces we consider do not depend on a choice of μ_r.

Recall that $C^\alpha_{\vec{\mu},\mu_1,\mu_2}$ is the set of tuples $((z_k),(\tau_k);x,\omega;y,\eta;(t_m))$. Here $(t_m) \in C^{e(\alpha)}$; $(z_k) \in C^K$; $(x_j),(y_j) \in C^J$. All the points z_k, x, y are distinct, and τ_k is a nonzero tangent vector at z_k; ω is a nonzero tangent vector at x; η is a nonzero tangent vector at y (see 16.9.1).

We defined the nodal curve $\underline{C}$ in *loc. cit.*. One can see that the morphisms (320) for different connected components (resp. the morphism (321)) can be glued together into (resp. factor through) the same named morphism $\wp_\alpha$:

$$\underline{C}^\alpha_{\vec{\mu},\mu_1,\mu_2} \longrightarrow \dot{T}_{(\overline{C}|_{\mathfrak{D}})^\alpha_{\vec{\mu}}}\overline{C}^\alpha_{\vec{\mu}} \tag{322}$$

17.3. **Theorem.** There is a canonical isomorphism

$$\wp_\alpha^*\mathbf{Spg}^\alpha_{\vec{\mu}}(\{\mathcal{X}_k\}) = \mathrm{g}^\alpha_{\vec{\mu},\mu_1,\mu_2}(\{\mathcal{X}_k\},\mathcal{R})$$

(notations of 16.9.2 and 16.9.3).

We have to specify μ_1, μ_2 in the formulation of the Theorem.

17.3.1. In case (320) the choice of μ_1, μ_2 depends on the connected component (see 16.9.3). Namely, we set $\mu_r = \alpha_r - \alpha_{g_r}(\vec{\mu}_r)$, $r = 1, 2$.

In case (321) we sum up over all the choices of $(\mu_1, \mu_2) \in (X_\ell \times X_\ell)/Y_\ell$ (antidiagonal action) such that $\mu_1 + \mu_2 = 2\rho_\ell$. Thus, if we identify all the spaces $\underline{C}^\alpha_{\vec{\mu},\mu_1,\mu_2}$ with one abstract space $\underline{C}^\alpha_{\vec{\mu},?,?}$, then $\wp_\alpha^*\mathbf{Spg}^\alpha_{\vec{\mu}}(\{\mathcal{X}_k\})$ is a direct sum of d_ℓ summands. Or else we can view $\wp_\alpha^*\mathbf{Spg}^\alpha_{\vec{\mu}}(\{\mathcal{X}_k\})$ as a collection of sheaves on different spaces $\underline{C}^\alpha_{\vec{\mu},\mu_1,\mu_2}$.

17.3.2. *Proof.* By the definition of gluing (16.8) it suffices to construct the desired isomorphism "near the points x, y" and, separately, "away from the points x, y".

The situation "near the points x, y" is formally equivalent to (a direct sum of a few copies of) the situation of 14.2. Thus we obtain a copy of the I-sheaf $\mathcal{R}$ glued into the node $x = y$.

The situation "away from the points x, y" was the subject of the Theorem 7.6.

We leave it to the reader to work out the values of μ_1, μ_2. We just note that these are exactly the values making the tuple $(\alpha, \vec{\mu}, \mu_1, \mu_2)$ g-admissible. □

18. Global sections over C

18.1. Given a K-tuple of weights $\vec{\mu}$ and $\alpha = \alpha_g(\vec{\mu})$, consider the projection

$$\eta:\ C^{\alpha}_{\vec{\mu}} \longrightarrow TC^{\circ K} = \overset{\circ}{T}C^{\vec{\mu}}$$

(see 16.3).

Given a K-tuple of FFS's $\{\mathcal{X}_k\}$ with $(\vec{\mu}_k) \geq (\lambda(\mathcal{X}_k))$ we consider the sheaf $\mathrm{g}(\{\mathcal{X}_k\})^{\alpha}_{\vec{\mu}}$ on $C^{\alpha}_{\vec{\mu}}$ obtained by gluing the sheaves $\mathcal{X}_k$.

In this section we will study the complex of sheaves with smooth cohomology $R\eta_*\mathrm{g}(\{\mathcal{X}_k\})^{\alpha}_{\vec{\mu}}$ on $\overset{\circ}{T}C^{\vec{\mu}}$. Note that it does not depend on a choice of $(\vec{\mu}_k) \geq (\lambda(\mathcal{X}_k))$ (if we identify the spaces $\overset{\circ}{T}C^{\vec{\mu}}$ for different choices of $\vec{\mu}$).

Namely, we want to compute its specialization along the boundary component $\mathfrak{D}$. Recall the notations of 7.

In case 7.1(a) for any partition $\vec{\mu} = \vec{\mu}_1 \sqcup \vec{\mu}_2$ we have the map (cf. (302)) $\wp_{\vec{\mu}}$:

$$\overset{\circ}{T}\mathrm{C}_1^{\vec{\mu}_1,\mu_1} \times \overset{\circ}{T}\mathrm{C}_2^{\vec{\mu}_2,\mu_2} \longrightarrow \overset{\bullet}{T}_{\overset{\circ}{T}\overline{C}^{\vec{\mu}}|_{\mathfrak{D}}} \overset{\circ}{T}\overline{C}^{\vec{\mu}}; \tag{323}$$

in case 7.1(b) we have the map (cf. (303)) $\wp_{\vec{\mu}}$:

$$\overset{\circ}{T}\mathrm{C}_0^{\vec{\mu},\mu_1,\mu_2} \longrightarrow \overset{\bullet}{T}_{\overset{\circ}{T}\overline{C}^{\vec{\mu}}|_{\mathfrak{D}}} \overset{\circ}{T}\overline{C}^{\vec{\mu}}. \tag{324}$$

On the other hand, consider the space $\underline{\mathrm{C}}^{\alpha}_{\vec{\mu},\mu_1,\mu_2}$ (see 17.2). In the case 7.1(b) it projects to $\overset{\circ}{T}\mathrm{C}_0^{\vec{\mu},\mu_1,\mu_2}$. We denote the projection by $\underline{\eta}$.

In the case 7.1(a) $\underline{\mathrm{C}}^{\alpha}_{\vec{\mu},\mu_1,\mu_2}$ is disconnected. Its connected components are numbered by the partitions $\vec{\mu} = \vec{\mu}_1 \sqcup \vec{\mu}_2$. Each component projects to the corresponding space $\overset{\circ}{T}\mathrm{C}_1^{\vec{\mu}_1,\mu_1} \times \overset{\circ}{T}\mathrm{C}_2^{\vec{\mu}_2,\mu_2}$. Each one of those projections will be denoted by $\underline{\eta}$, as well as their totality.

18.2. **Theorem.** In the notations of 17.3 and 17.3.1 we have a canonical isomorphism

$$\wp^*_{\vec{\mu}}\mathbf{Sp}R\eta_*\mathrm{g}(\{\mathcal{X}_k\})^{\alpha}_{\vec{\mu}} = R\underline{\eta}_*\underline{\mathrm{g}}^{\alpha}_{\vec{\mu},\mu_1,\mu_2}(\{\mathcal{X}_k\},\mathcal{R})$$

Proof. This is just the Theorem 17.3 plus the proper base change for nearby cycles. □

One can apply Theorems 17.3 and 18.2 iteratively — degenerating the curve C more and more, and inserting more and more copies of $\mathcal{R}$ into new nodes.

In the terminology of [BFM] 4.5 (see especially 4.5.6) the Theorem 18.2 means that the category $\mathcal{FS}$ is equipped with the structure of *fusion category*. The correspondence associating to any K-tuple of FFS's $\{\mathcal{X}_k\}$ the cohomology local systems $R\eta_*\mathrm{g}(\{\mathcal{X}_k\})^{\alpha}_{\vec{\mu}}$ on the space $TC^{\circ K}$, satisfying the compatibility conditions 18.2, is called *fusion functor*, or *modular functor*.

Chapter 6. Integral representations of conformal blocks

19. Conformal blocks in arbitrary genus

19.1. Recall the notations of IV.9. We fix a positive integer κ and consider the category $\tilde{\mathcal{O}}_\kappa$ of integrable $\hat{\mathfrak{g}}$-modules of central charge $\kappa - h$. We set $\zeta = \exp(\frac{\pi\sqrt{-1}}{d\kappa})$, so that $l = 2d\kappa$, and $\ell = d\kappa$.

We have an equivalence of bbrt categories $\phi : \tilde{\mathcal{O}}_\kappa \xrightarrow{\sim} \tilde{\mathcal{O}}_\zeta$ where $\tilde{\mathcal{O}}_\zeta$ is a semisimple subcategory of $\mathcal{C}$ defined in *loc. cit.* It is well known that $\tilde{\mathcal{O}}_\kappa$ has a structure of *fusion category* (see e.g. [BFM] or [TUY]). The category $\tilde{\mathcal{O}}_\zeta$ is also equipped with the structure of fusion category via the equivalence ϕ. From now on we will not distinguish between the categories $\tilde{\mathcal{O}}_\kappa$ and $\tilde{\mathcal{O}}_\zeta$.

19.2. Here is a more elementary way to think of fusion functors. Recall that a fusion functor is a correspondence associating to any K-tuple of objects $\{L_k\}$ of our category the local systems (or, more generally, the complexes of sheaves with smooth cohomology) $\langle\{L_k\}\rangle_C^K$ on the spaces $TC^{\circ K}$, satisfying a certain compatibility conditions.

We can view those local systems as representations of Modular Teichmüller groups $T_{g,K}$ in the stalks at some base points.

We will choose the base points "at infinity" (see [D4]). Namely, we degenerate the curve C into the projective line $\mathbb{P}^1$ with g nodal points. More precisely, we choose the real numbers $x_1 < y_1 < x_2 < \ldots < x_g < y_g$ and degenerate C into $\mathbb{P}^1/(x_j = y_j,\ 1 \leq j \leq g)$. The marked points degenerate into the real numbers $z_1 < \ldots < z_m$ (m is the cardinality of K) such that $z_m < x_1$. The tangent vectors at all the points are real and directed to the right. We will denote this homotopy base point at infinity by $\infty_{g,K}$.

19.3. For $L_1, \ldots, L_m \in \tilde{\mathcal{O}}_\zeta$ the stalk $(\langle\{L_k\}\rangle_C^K)_{\infty_{g,K}}$ is canonically isomorphic to $\langle L_1\tilde{\otimes}\ldots\tilde{\otimes}L_m\tilde{\otimes}\tilde{\mathbf{ad}}^{\tilde{\otimes}g}\rangle$ (see [TUY]). Here we use the notation $\langle ? \rangle$ introduced in IV.5.9 and IV.9 (maximal trivial summand), and the notation $\tilde{\mathbf{ad}}$ introduced in 11.2.1 (semisimple adjoint representation).

The action of $T_{g,K}$ on $\langle L_1\tilde{\otimes}\ldots\tilde{\otimes}L_m\tilde{\otimes}\tilde{\mathbf{ad}}^{\tilde{\otimes}g}\rangle$ is induced by its action on $L_1\tilde{\otimes}\ldots\tilde{\otimes}L_m\tilde{\otimes}\tilde{\mathbf{ad}}^{\tilde{\otimes}g}$: generators act by braiding, balance, and Fourier transforms (see [LM]) of the adjoint representation.

19.4. The Theorem 18.2 equips the category $\mathcal{FS}$ (and, via the equivalence Φ, the category $\mathcal{C}$) with the fusion structure. To unburden notations and to distinguish from the case of $\tilde{\mathcal{O}}_\zeta$ we will denote the corresponding local systems (or, rather, complexes of sheaves with smooth cohomology) by $[L_1, \ldots, L_m]_C^K$.

Theorem. The stalk $([L_1, \ldots, L_m]_C^K)_{\infty_{g,K}}$ is canonically isomorphic to

$$\mathrm{Tor}^{\mathcal{C}}_{\frac{\infty}{2}+\bullet}(\mathsf{k}, L_1 \otimes \ldots \otimes L_m \otimes \mathbf{ad}^{\otimes g})$$

(notations of IV.4.7 and 10.1).

Proof. In effect, the stalk in question is the global sections of some sheaf on the space C^α where $\mathrm{C} = \mathbb{P}^1/(x_j = y_j,\ 1 \le j \le g)$. The global sections will not change if we merge the points y_j into x_j for any j along the real line. The specialization of our sheaf will live on the space $(\mathbb{P}^1)^\alpha$ (no nodal points anymore).

It will be the sheaf obtained by gluing $\{\mathcal{L}_k\}, \mathcal{T}_1, \dots, \mathcal{T}_g$ into the marked points $z_1, \dots, z_m, x_1, \dots, x_g$. Here $\mathcal{L}_k = \Phi^{-1} L_k$, and the sheaves $\mathcal{T}_1, \dots, \mathcal{T}_g$ are just g copies of the same sheaf $\mathcal{T} = \Phi^{-1}\mathbf{ad}$. This follows immediately from the Theorem 17.3, the Definition of **ad** in 11.1, and the Theorem 15.5.

It remains to apply the Theorem IV.8.4. □

19.5. The action of $T_{g,K}$ on $\mathrm{Tor}^{\mathcal{C}}_{\frac{\infty}{2}+\bullet}(\mathsf{k}, L_1 \otimes \dots \otimes L_m \otimes \mathbf{ad}^{\otimes g})$ is induced by its action on $L_1 \otimes \dots \otimes L_m \otimes \mathbf{ad}^{\otimes g}$: generators act by braiding, balance, and Fourier transforms of the adjoint representation.

Now suppose $L_k \in \tilde{\mathcal{O}}_\zeta$ for all $k \in K$. Then $L_1 \tilde{\otimes} \dots \tilde{\otimes} L_m \tilde{\otimes} \tilde{\mathbf{ad}}^{\tilde{\otimes} g} = \langle L_1 \otimes \dots \otimes L_m \otimes \mathbf{ad}^{\otimes g}\rangle_\Delta$ (notations of 11.2) by the Theorem 11.3. The RHS is a natural subquotient of $L_1 \otimes \dots \otimes L_m \otimes \mathbf{ad}^{\otimes g}$, hence the action of $T_{g,K}$ on the latter module induces some action of $T_{g,K}$ on the RHS. This action coincides with the one of 19.3 by the uniqueness result of [MS] which claims that any two extensions of bbrt structure on a category to a fusion structure coincide.

19.6. By the virtue of Corollary IV.5.11, $\langle L_1 \tilde{\otimes} \dots \tilde{\otimes} L_m \tilde{\otimes} \tilde{\mathbf{ad}}^{\tilde{\otimes} g}\rangle$ is canonically a subquotient of $\mathrm{Tor}^{\mathcal{C}}_{\frac{\infty}{2}+\bullet}(\mathsf{k}, (L_1 \otimes \dots \otimes L_m \otimes \mathbf{ad}^{\otimes g}) \otimes L(2\rho_\ell))$.

The above discussion implies that the action 19.3 of $T_{g,K}$ on the former space is induced by its action 19.5 on the latter space. More precisely, $T_{g,K}$ acts on $\mathrm{Tor}^{\mathcal{C}}_{\frac{\infty}{2}+\bullet}(\mathsf{k}, (L_1 \otimes \dots \otimes L_m \otimes \mathbf{ad}^{\otimes g}) \otimes L(2\rho_\ell))$ in the evident way, leaving $L(2\rho_\ell)$ alone.

19.7. Let us translate the above discussion back into geometric language. Let us consider the set $J := K \sqcup j$. We denote by ξ the natural projection $TC^{\circ J} \longrightarrow TC^{\circ K}$.

Summarizing the above discussion, we obtain

Theorem. Let $L_k \in \tilde{\mathcal{O}}_\zeta,\ k \in K$. Then the local system $\xi^*\langle L_1, \dots, L_m\rangle^K_C$ on $TC^{\circ J}$ is a natural subquotient of the local system $[L_1, \dots, L_m, L(2\rho_\ell)]^J_C$.

19.8. Let us reformulate the above Theorem in more concrete terms.

Recall that $\nabla_\kappa = \Delta_\ell$ is the first alcove parametrizing the irreducibles in $\tilde{\mathcal{O}}_\kappa$. Let $\lambda_1, \dots, \lambda_m \in \nabla_\kappa$ be a K-tuple of weights. We consider the corresponding irreducibles $L(\lambda_1), \dots, L(\lambda_m) \in \tilde{\mathcal{O}}_\kappa$, and the corresponding local system of conformal blocks $\langle L(\lambda_1), \dots, L(\lambda_m)\rangle^K_C$ on the space $TC^{\circ K}$.

On the other hand, consider $\alpha := \lambda_1 + \dots + \lambda_m + 2g\rho_\ell$. We define $J := K \sqcup j$, and $\lambda_j := 2\rho_\ell$. Let $\vec{\lambda}$ denote $\lambda_1, \dots, \lambda_m, \lambda_j$.

Denote by $\mathcal{X}^\alpha_{\vec{\lambda}}$ the sheaf on $C^\alpha_{\vec{\lambda}}$ obtained by gluing $\mathcal{L}(\lambda_1), \dots, \mathcal{L}(\lambda_m), \mathcal{L}(\lambda_j)$.

Note that $\mathcal{X}^\alpha_{\vec{\lambda}} = u_{!*}\mathcal{H}^\alpha_{\vec{\lambda}}$ where u stands for the open embedding $C^{\circ\alpha}_{\vec{\lambda}} \hookrightarrow C^\alpha_{\vec{\lambda}}$.

Recall that η stands for the projection $C^\alpha_{\vec{\lambda}} \longrightarrow TC^{\circ J}$.

Theorem. The local system of conformal blocks $\xi^*\langle L(\lambda_1), \ldots, L(\lambda_m)\rangle_C^K$ is isomorphic to a canonical subquotient of a "geometric" local system $R\eta_* u_{!*}\mathcal{H}_{\bar\lambda}^{\alpha}$. □

19.9. **Corollary.** *The above local system of conformal blocks is semisimple. It is a direct summand of the geometric local system above.*

Proof. The geometric system is semisimple by Decomposition theorem, [BBD], Théorème 6.2.5. □

REFERENCES

[A] H.Andersen, Representations of quantum groups, invariants of 3-manifolds and semisimple tensor categories, *Isr. Math. Conf. Proc.*, **7**(1993), 1-12.

[AJS] H.Andersen, J.Jantzen, W.Soergel, Representations of quantum groups at p-th root of unity and of semisimple groups in characteristic p: independence of p, *Astérisque* **220**(1994).

[AP] H.Andersen, J.Paradowski, Fusion categories arising from semisimple Lie algebras, *Commun. Math. Phys.*, **169**(1995), 563-588.

[APW1] H.Andersen, P.Polo, Wen K., Representations of quantum algebras, *Invent. math.* **104** (1991), 1-59.

[APW2] H.Andersen, P.Polo, Wen K., Injective modules for quantum algebras, *Amer. J. Math.* **114**(1992), 571-604.

[AW] H.Andersen, Wen K., Representations of quantum algebras. The mixed case, *J. Reine Angew. Math.* **427** (1992), 35-50.

[Ark] S.Arkhipov, Semiinfinite cohomology of quantum groups, *Commun. Math. Phys.*, **188** (1997), 379-405.

[B1] A.Beilinson, On the derived category of perverse sheaves, in: K-theory, Arithmetic and Geometry, Yu.I.Manin (Ed.), *Lect. Notes in Math.*, **1289**(1987), 27-41.

[B2] A.Beilinson, Lectures on R. MacPherson's work, Independent Moscow University, Spring 1994.

[BBD] A.Beilinson, J.Bernstein, P.Deligne, Faisceaux pervers, *Astérisque* **100**(1982).

[BD] A.Beilinson, V.Drinfeld, Chiral algebras, Preprint (1995).

[BP] A.Beilinson, A.Polishchuk, Heisenberg extensions and abelian algebraic field theories, in preparation.

[BFM] A.Beilinson, B.Feigin, B.Mazur, Introduction to algebraic field theory on curves, Preprint (1993).

[BK] A.Beilinson, D.Kazhdan, Flat projective connections, Preprint.

[BM] A.Beilinson, Yu.Manin, The Mumford form and the Polyakov Measure in String Theory, *CMP*, **107** (1986), 359-376.

[Br] E.Brieskorn, Sur les groupes des tresses (d'apres V.Arnold), Sém. Bourbaki, no. 401, *Lect. Notes Math.* **317**(1973), Springer-Verlag, Berlin et al., 21-44.

[C] D.C.Cohen, Cohomology and intersection cohomology of complex hyperplane arrangements, *Adv. Math.*, **97**(1993), 231-266.

[CFW] M.Crivelli, G.Felder, C.Wieczerkowski, Generalized hypergeometric functions on the torus and the adjoint representation of $\mathbf{U}_q(sl_2)$, *Comm. Math. Phys.*, **154** (1993), 1-23.

[D1] P.Deligne, Une description de catégorie tressée (inspiré par Drinfeld), Letter to V.Drinfeld (1990). Lectures at IAS (1990), unpublished.

[D2] P.Deligne, Catégories Tannakiennes, The Grothendieck Festschrift II, Progr. Math. **87**, Birkhäuser, Boston (1990), 111-195.

[D3] P.Deligne, Le formalisme des cycles évanescents, in: P. Deligne, A. Grothendieck, N. Katz, Groupes de monodromie en Géometrie Algébrique (SGA 7), Part II, Lect. in Math., **340**, Exp. XIII, 82-115.

[D4] P.Deligne, Le groupe fondamental de la droite projective moins trois points, in: Y.Ihara, K.Ribet, J.-P. Serre (ed.), Galois Groups over $\mathbb{Q}$, *MSRI Publ.*, **16** (1989), 79-298.

[DM] P.Deligne, D.Mumford, The irreducibility of the space of curves of a given genus, *Publ. Math. IHES*, **36** (1969), 75-109.

[Dr] V.Drinfeld, Quasihopf algebras, *Algebra and Analysis*, **1**, no. 6(1989), 114-149(russian).

[F1] M.Finkelberg, Thesis, Harvard, 1993.

[F2] M.Finkelberg, An equivalence of fusion categories, *J. Geom. and Funct. Anal.*, **6**, No. 2 (1996), 249-267.

[FS] M.Finkelberg, V.Schechtman, Localization of u-modules. I. Intersection cohomology of real arrangements, Preprint hep-th/9411050 (1994), 1-23; II. Configuration spaces and quantum groups, Preprint q-alg/9412017 (1994), 1-59; III. Tensor categories arising from configuration spaces, Preprint q-alg/9503013 (1995), 1-59; IV. Localization on $\mathbb{P}^1$, Preprint q-alg/9506011 (1995), 1-29.

[FZ] B.Feigin, A.Zelevinsky, Representations of contragredient Lie algebras and Macdonald identities, in: Representations of Lie groups and Lie algebras, A.A.Kirillov ed. *Ak. Kiado*, Budapest (1985), 25-77.

[FW] G.Felder, C.Wieczerkowski, Topological representations of the quantum group $\mathbf{U}_q(sl_2)$, *Comm. Math. Phys.*, **138** (1991), 583-605.

[Fu] W.Fulton, Intersection Theory, *Ergebnisse der Mathematik* **2**, Springer-Verlag, 1984.

[H] J.L.Harer, The cohomology of the Moduli Spaces of Curves, unpublished lectures, 1985.

[K] A.Kirillov, Jr. On inner product in modular tensor categories, *J. Amer. Math. Soc.*, **9**, No. 4 (1996), 1135-1169.

[KL] D.Kazhdan, G.Lusztig, Tensor structures arising from affine Lie algebras. I-IV, *J. Amer. Math. Soc.*, **6**(1993), 905-947; **6**(1993), 949-1011; **7**(1994), 335-381; **7**(1994), 383-453.

[KM] F.Knudsen and D.Mumford, The projectivity of the moduli space of stable curves I: Preliminaries on "det" and "Div", *Mathematica Scandinavica* **39** (1976), 19-35.

[Kn] F.Knudsen, The projectivity of the moduli space of stable curves, II: The stacks $M_{g,n}$, *Math. Scand.*, **52** (1983), 161-199; III: The line bundles on $M_{g,n}$, and a proof of the projectivity of $\bar{M}_{g,n}$ in characteristic 0, *ibid.*, **52** (1983), 200-212.

[KS] M.Kashiwara, P.Schapira, Sheaves on Manifolds, *Grund. math. Wiss.* **292**, Springer-Verlag, Berlin et al., 1994.

[L1] G.Lusztig, Introduction to quantum groups, Boston, Birkhäuser, 1993.

[L2] G.Lusztig, Quantum groups at roots of 1, *Geom. Dedicata* **35**(1990), 89-114.

[L3] G.Lusztig, Monodromic systems on affine flag varieties, *Proc. Royal Soc. Lond.* A **445** (1994), 231-246; Errata, **450** (1995), 731-732.

[LM] V.Lyubashenko, S.Majid, Braided groups and quantum Fourier transform, preprint DAMTP/91-26.

[M] S.MacLane, Homology, Springer-Verlag, Berlin et al., 1963.

[M1] D.Mumford, Abelian varieties, Oxford University Press, Oxford, 1970.

[M2] D.Mumford, Stability of projective varieties, *L'Enseignement Mathematique* **23** (1977), 39-110.

[MS] G.Moore, N.Seiberg, Classical and conformal field theory, *CMP*, **123** (1989), 117-254.

[PW] B.Parshall, J.-P.Wang, Quantum linear groups, *Mem. Amer. Math. Soc.* **439** (1991).

[R] M.Rosso, Certaines formes bilinéaires sur les groupes quantiques et une conjecture de Schechtman et Varchenko, *C. R. Acad. Sci. Paris*, **314**, Sér. I(1992), 5-8.

[Sa] M. Salvetti, Topology of the complement of real hyperplanes in $\mathbb{C}^N$, *Inv. Math.*, **88**(1987), 603-618.

[S] V.Schechtman, Vanishing cycles and quantum groups II, *Int. Math. Res. Notes*, **10**(1992), 207-215.

[SV1] V.Schechtman, A.Varchenko, Arrangements of hyperplanes and Lie algebra homology, *Inv. Math.*, **106**(1991), 134-194.

[SV2] V.Schechtman, A.Varchenko, Quantum groups and homology of local systems, *ICM-90 Satellite Conference Proceedings Algebraic Geometry and Analytic Geometry (Tokyo 1990)*, Springer(1991), 182-191.

[TUY] A.Tsuchia, K.Ueno, Y.Yamada, Conformal field theory on the universal family of stable curves with gauge symmetries, *Adv. Stud. Pure Math.* **19** (1989), 459-565.

[V] A.Varchenko, Multidimensional hypergeometric functions and representation theory of Lie algebras and quantum groups, *Adv. Ser. Math. Phys.* **21**, World Scientific Publishers (1995).

Index of Notation

$p(\nu_1,\ldots,\nu_n):\ D^{\nu_1,\ldots,\nu_n}\longrightarrow D^{\nu_1}\times\overset{\bullet}{D}{}^{\nu_2}\times\ldots\times\overset{\bullet}{D}{}^{\nu_n}$, 0.3.3

$p(J_1,\ldots,J_n):\ D^{J_1,\ldots,J_n}\longrightarrow D^{J_1}\times\overset{\bullet}{D}{}^{J_2}\times\ldots\times\overset{\bullet}{D}{}^{J_n}$, 0.3.3

$p^{\beta}_{\lambda}:\ \mathcal{FS}_{c;\le\lambda}\longrightarrow\mathcal{M}(\mathcal{A}^{\beta}_{\lambda};\mathcal{S})$, III.5.4

$p:\ \mathcal{J}ac_{\Lambda}\longrightarrow\mathcal{J}ac$, 0.14.8

${}^{n}p:\ {}^{n}\mathbb{A}\longrightarrow\mathbb{A}^{(n)}$, II.9.1

$p_K:\ {}^{K}\mathcal{FS}\longrightarrow\mathcal{FS}^{\otimes K}$, III.10.2

$p_K:\ \widetilde{\overset{\circ}{T}C^K}\longrightarrow\overset{\circ}{T}C^K$, V.2.1

$p_{\vec{\alpha},\beta}:\ C^{\vec{\alpha},\beta}_{\vec{\mu},\varepsilon}\longrightarrow\prod_k D^{\alpha_k}_{\varepsilon}\times C^{\bullet\beta}_{\vec{\mu}-\vec{\alpha}}$, V.16.4

$p^{\alpha}_{\vec{\mu}}:\ TC^{o\alpha}_{\vec{\mu}}\longrightarrow TC^{o\alpha_{\vec{\mu}}}$, V.16.3

$\mathcal{Q}$, V.6.3

$\mathcal{Q}_2(J)$, II.9.2

Q_v, 0.13.1

Q_{0v}, V.13.1

Q_{0h}, V.13.1

Q_u, 0.13.1

Q_t, 0.13.1

Q, 0.13.1

$Q_{>d,<d}$, V.14.2

$Q_{<d,>d}$, V.14.2

$Q_{>d,>d}$, V.14.2

$Q_{<d,<d}$, V.14.2

$Q^{\alpha,\beta,\gamma}(d)$, V.14.2

$'Q$, 0.13.1

$'Q^{\nu\bullet}$, 0.13.2

$'Q^{\bullet\alpha}$, V.13.2

$'Q^{o\alpha}$, V.13.2

$Q:\ \tilde{\mathcal{C}}\longrightarrow\mathcal{C}$, II.11.7, II.12.5

$\mathbf{q}:\ \mathcal{H}\longrightarrow\mathbb{C}^*$, I.4.2

$\mathbf{q}(C,C')$, I.4.11

q_{ij}, II.2.25

q_{λ}, III.14.1

$q^{*}_{\lambda\le\mu}$, III.14.8

$q^{!}_{\lambda\le\mu}$, III.14.8

$^{\pi}\phi^{(\eta)}_{\Delta,*}: \Phi_\Delta(^\pi\mathcal{I}_*) \longrightarrow {}^{\pi\,\prime}\mathfrak{f}^*_{\chi_J}$, II.6.10

$^{\pi}\phi^{(\eta)}_{\Delta,!*}: \Phi_\Delta(^\pi\mathcal{I}_{!*}) \xrightarrow{\sim} {}^\pi\mathfrak{f}_{\chi_J}$, II.6.11

$^\pi\phi^{(\eta)}_{\varrho,!}: \Phi^+_{F_\varrho}(\mathcal{I}(^\pi\Lambda)_!) \xrightarrow{\sim} {}_\varrho C^{-r}_{\pi\ \prime\mathfrak{f}}(V(^\pi\Lambda))$, II.8.9

$^\pi\phi^{(\eta)}_!: {}^+C^\bullet(^\pi\mathbb{A};\mathcal{I}(^\pi\Lambda)_!) \xrightarrow{\sim} {}_{\chi_J}C^\bullet_{\pi\ \prime\mathfrak{f}}(V(^\pi\Lambda))$, II.8.10

$^\pi\phi^{(\eta)}_{0,*}: \Phi^+_0(\mathcal{I}(^\pi\Lambda)_*) \xrightarrow{\sim} V(^\pi\Lambda)^*_{\chi_J}$, II.8.14

$\phi^{(\eta)}_{\nu,\Lambda,!*}: \Phi_\nu(\mathcal{I}_\nu(\Lambda)_{!*}) \xrightarrow{\sim} L(\Lambda)_\nu$, II.8.18

$^\pi\phi^{(\eta)}_{\varrho,*}: \Phi^+_{F_\varrho}(\mathcal{I}(^\pi\Lambda)_*) \xrightarrow{\sim} {}_\varrho C^{-r}_{\pi\ \prime\mathfrak{f}^*}(V(^\pi\Lambda)^*)$, II.8.20

$^\pi\phi^{(\eta)}_*: {}^+C^\bullet(^\pi\mathbb{A};\mathcal{I}(^\pi\Lambda)_*) \xrightarrow{\sim} {}_{\chi_J}C^\bullet_{\pi\ \prime\mathfrak{f}^*}(V(^\pi\Lambda)^*)$, II.8.21

${}_\Lambda\phi^{(\eta)}_{\nu,!*,0}: \mathcal{I}_\nu(\Lambda)_{!*,0} \xrightarrow{\sim} {}_\nu C^\bullet_{\mathfrak{f}}(L(\Lambda))$, II.8.23

$\phi^{(\eta)}_{!*}: \Phi_\nu(^\psi\mathcal{I}_\nu(\Lambda_0,\ldots,\Lambda_{-n})_{!*}) \xrightarrow{\sim} (L(\Lambda_0)\otimes\ldots\otimes L(\Lambda_{-n}))_\nu$, II.10.6, II.12.7

$\phi^{(\eta)}_{!*,0}: {}^\psi\mathcal{I}_\nu(\Lambda_0,\ldots,\Lambda_{-n})_{!*0} \xrightarrow{\sim} {}_\nu C^\bullet_{\mathfrak{f}}(L(\Lambda_0)\otimes\ldots\otimes L(\Lambda_{-n}))$, II.10.8, II.12.8

$\phi_i = \phi^{\vec\alpha}_{\mu;i}(\vec d): \mathcal{I}^{\vec\alpha}_\mu(\vec d) \xrightarrow{\sim} \mathcal{I}^{\vec\alpha_{\le i}}_{\mu_{\le i}}(\vec d_{\le i}) \boxtimes \mathcal{I}^{\vec\alpha_{\ge i}}_{\mu_{\ge i}}(\vec d_{\ge i})$, III.3.2

$\phi_i = \phi^{\vec\alpha}_{\mu;i}(\vec d): \overset{\bullet}{\mathcal{I}}{}^{\vec\alpha}_\mu(\vec d) \xrightarrow{\sim} \overset{\bullet}{\mathcal{I}}{}^{\vec\alpha_{\le i}}_{\mu_{\le i}}(\vec d_{\le i}) \boxtimes \overset{\bullet}{\mathcal{I}}{}^{\vec\alpha_{\ge i}}_{\mu_{\ge i}}(\vec d_{\ge i})$, III.3.5

$\phi_i = \phi^{\vec\alpha}_{i;\vec\mu}(\tau): \mathcal{I}^{\vec\alpha}_{\vec\mu}(\tau) \cong \mathcal{I}^{\vec\alpha_{\le i}}_{\vec\mu_{\le i}}(\tau_{\le i}) \boxtimes \boxed{\times}_{k\in K_i} \mathcal{I}^{\vec\alpha_{\ge k}}_{\vec\mu_{\ge k}}(\tau_{\ge k})$, III.8.2

$\phi_i = \phi^{\vec\alpha}_{i;\vec\mu}(\tau): \overset{\bullet}{\mathcal{I}}{}^{\vec\alpha}_{\vec\mu}(\tau) \cong \overset{\bullet}{\mathcal{I}}{}^{\vec\alpha_{\le i}}_{\vec\mu_{\le i}}(\tau_{\le i}) \boxtimes \boxed{\times}_{k\in K_i} \overset{\bullet}{\mathcal{I}}{}^{\vec\alpha_{\ge k}}_{\vec\mu_{\ge k}}(\tau_{\ge k})$, III.8.5

$\phi^{\vec\alpha_1,\vec\alpha_2}_{d_1,d_2}: \mathcal{X}^{\vec\alpha_1}(\tau_{d_1})|_{\mathcal{A}^{\vec\alpha_1}_{\vec\lambda}(\tau_{d_1})\cap\mathcal{A}^{\vec\alpha_2}_{\vec\lambda}(\tau_{d_2})} \xrightarrow{\sim} \mathcal{X}^{\vec\alpha_2}(\tau_{d_2})|_{\mathcal{A}^{\vec\alpha_1}_{\vec\lambda}(\tau_{d_1})\cap\mathcal{A}^{\vec\alpha_2}_{\vec\lambda}(\tau_{d_2})}$, III.10.3

$\phi_\pi: \otimes_K X_k \xrightarrow{\sim} \otimes_L (\otimes_{K_l} X_k)$, III.11.3

$\phi_{K\longrightarrow L}: r_\pi\circ\Psi_K \xrightarrow{\sim} \Psi_L\circ\Psi_{K\longrightarrow L}$, III.11.10

$\phi_{\vec\alpha}: q^*_{\vec\alpha}\mathcal{I}^\alpha \xrightarrow{\sim} p^*_K\pi^*_K\mathcal{I}^{\alpha_K} \boxtimes \boxed{\times}_k \mathcal{I}^{\widetilde{\alpha_k}}_D$, IV.2.8

$\phi_{\vec\alpha,\beta}: q^*_{\vec\alpha,\beta}\mathcal{X}^{\alpha+\beta}_{\vec\mu} \xrightarrow{\sim} p^*_{\vec\alpha,\beta}((\boxed{\times}_{k\in K} \mathcal{X}^{\alpha_k}_{\mu_k}) \boxtimes \mathcal{I}^{\bullet\beta}_{\vec\mu-\vec\alpha})$, IV.3.5

$\phi: \tilde{\mathcal{O}}_\kappa \xrightarrow{\sim} \tilde{\mathcal{O}}_\zeta$, IV.9.2

$\phi_\tau: q^*_\tau\pi^*_J\mathcal{H}^\alpha \xrightarrow{\sim} p^*_K\pi^*_K\mathcal{H}^{\alpha_K} \boxtimes \boxed{\times}_{k\in K} \mathcal{I}^{\tau^{-1}(k)}$, V.2.2

$\phi_{\vec\alpha,\beta}: q^*_{\vec\alpha,\beta}\mathcal{X}^{\alpha+\beta}_{\vec\mu} \xrightarrow{\sim} p^*_{\vec\alpha,\beta}((\boxed{\times}_{k\in K} \mathcal{X}^{\alpha_k}_{\mu_k}) \boxtimes \mathcal{H}^{\bullet\beta}_{\vec\mu-\vec\alpha})$, V.16.6

$\phi_{\vec\alpha\vec\beta\vec\gamma,\xi}: q^*_{\vec\alpha\vec\beta\vec\gamma,\xi}\mathcal{X}^{\alpha+\beta+\gamma+\xi}_{\vec\lambda\vec\mu\vec\nu} \xrightarrow{\sim} p^*_{\vec\alpha\vec\beta\vec\gamma,\xi}(((\boxed{\times}_{k\in K} \mathcal{X}^{\alpha_k}_{\lambda_k}) \boxtimes (\boxed{\times}_{j\in J} \mathcal{Y}^{\beta_j;\gamma_j}_{\mu_j;\nu_j}) \boxtimes \mathcal{H}^{\bullet\xi}_{\vec\lambda\vec\mu\vec\nu-\vec\alpha\vec\beta\vec\gamma})$, V.16.8

χ_K, II.3.3, II.4.1

Ψ_f, I.2.11

Ψ_K, III.11.8

$\Psi_{\pi;d}: \mathcal{M}(\mathcal{A}^\alpha(\tau_{\pi;d})) \longrightarrow \mathcal{M}(\mathcal{A}^\alpha(L)_d \times \prod_{l\in L}\mathcal{O}(K_l))$, III.11.9

$\Psi_{K\longrightarrow L}: \mathcal{M}(\mathcal{A}^\alpha(K)) \longrightarrow \mathcal{M}(\mathcal{A}^\alpha(L)\times\prod_l \mathcal{O}(K_l))$, III.11.9

$\Psi_K: {}^K\mathcal{FS} \longrightarrow \mathcal{L}ocsys(\mathcal{O}(K);\mathcal{FS})$, III.11.11

Index of Terminology

Druck: Strauss Offsetdruck, Mörlenbach
Verarbeitung: Schäffer, Grünstadt